# Frequently Used Formulas



## Chapter 2

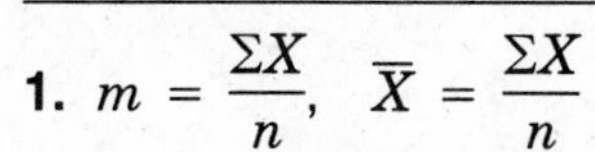

1. $m = \frac{\Sigma X}{n}, \quad \overline{X} = \frac{\Sigma X}{n}$

2. $s = \sqrt{\frac{\Sigma(X - m)^2}{n - 1}}$

3. $s = \sqrt{\frac{\Sigma X^2 - \frac{(\Sigma X)^2}{n}}{n - 1}}$

4. $z_X = \frac{X - \mu}{\sigma}, \quad z_X = \frac{X - \mu}{s}, \quad z_X = \frac{X - m}{s}$

5. $X = \mu + z\sigma, \quad X = \mu + zs, \quad X = m + zs$

6. $PR_X = \frac{B + \frac{1}{2}E}{n}$ (100)

## Chapter 4

1. $p = P(\text{event}) = \frac{F}{T}$

2. $q = 1 - p$

## Chapter 7

1. $\mu_S = np$
2. $\sigma_S = \sqrt{npq}$
3. $S_? = \mu_S + z\sigma_S$

## Chapter 8

1. $\mu_S = np$
2. $\sigma_S = \sqrt{npq}$
3. $S_c = \mu_S + z_c\sigma_S$
4. Power $= 1 - \beta$

## Chapter 9

1. $\mu_{d\hat{p}} = p_1 - p_2$
2. $\hat{p} = \frac{S_1 + S_2}{n_1 + n_2}$
3. $\hat{\sigma}_{d\hat{p}} = \sqrt{\hat{p}\hat{q}\left(\frac{1}{n_1} + \frac{1}{n_2}\right)}$
4. Sample difference $d\hat{p} = \hat{p}_1 - \hat{p}_2$
5. Critical difference $d\hat{p}_c = \mu_{d\hat{p}} + z_c\hat{\sigma}_{d\hat{p}}$

## Chapter 10

*One-Sample Tests*

1. $\mu_m = \mu_{\text{pop}}$
2. $s = \sqrt{\frac{\Sigma X^2 - \frac{(\Sigma X)^2}{n}}{n - 1}}$
3. $s_m = \frac{s}{\sqrt{n}}$
4. $m_c = \mu_m + z_c s_m$
5. Experimental outcome, $m = \frac{\Sigma X}{n}$

*Two-Sample Tests*

6. $\mu_{\text{dm}} - \mu_1 - \mu_2$
   (if $H_0$ states that $\mu_1 = \mu_2$, then $\mu_1 - \mu_2 = 0$)
7. $s_{\text{dm}} = \sqrt{\frac{s_1^2}{n_1} + \frac{s_2^2}{n_2}}$
8. $\text{dm}_c = \mu_{\text{dm}} + z_c s_{\text{dm}}$
9. Experimental outcome, $\text{dm} - m_1 - m_2$

*(Continued on back cover)*

# Understanding Statistics

# Understanding Statistics

Arnold Naiman
Late Professor Nassau Community College

Robert Rosenfeld
Nassau Community College

Gene Zirkel
Nassau Community College

**The McGraw-Hill Companies, Inc.**
New York St. Louis San Francisco Auckland Bogotá Caracas
Lisbon London Madrid Mexico City Milan Montreal
New Delhi San Juan Singapore Sydney Tokyo Toronto

*McGraw-Hill*

*A Division of The* ***McGraw-Hill*** *Companies*

Understanding Statistics



This book is printed on acid-free paper.

7 8 9 10 DOC/DOC 0 9 8 7 6 5 4 3 2

ISBN 0-07-045915-0

This book was set in New Aster by GTS Graphics.
The editors were Jack Shira, Maggie Lanzillo, and Scott Amerman;
the text was designed by Paula Goldstein;
the cover was designed by BC Graphics.
Cover photo by John P. Kelly for the Image Bank.
The production supervisor was Louise Karam.
New drawings were done by FineLine.
"Stat Kat" cartoons drawn by Jody Jobe.
R. R. Donnelley & Sons Company was printer and binder.

*About the Cover*

The inset on the cover shows a sample of all the runners in the larger picture, which, in turn, shows a sample of all the runners in this marathon.

Library of Congress Cataloging-in-Publication Data

Naiman, Arnold.
Understanding statistics / Arnold Naiman, Robert Rosenfeld, Gene Zirkel. — 4th ed.
p. cm.
Includes index.
ISBN 0-07-045915-0 (hardcover)
1. Statistics. I. Rosenfeld, Robert. II. Zirkel, Gene. III. Title.
QA276.12.N34 1996
519.5—dc20 95-39801

INTERNATIONAL EDITION



When ordering this title, use ISBN 0-07-114534-6.

# About the Authors

**Robert Rosenfeld** is Professor of Mathematics and Statistics at Nassau Community College. He received an M.S. and M.Phil. in biostatistics from Columbia University, and an M.A.T. in mathematics from Harvard University. He has received NSF and NIH awards to study statistics and to improve statistical education. Professor Rosenfeld has worked as a statistical consultant in medicine, and has taught mathematics and statistics at Nassau Community College since 1966. He is the author of the McGraw-Hill 36 Hour Course in Business Statistics, as well as the co-author of several algebra textbooks.

**Gene Zirkel** teaches in the Math/Statistics/Computer Processing Department of Nassau Community College. He holds degrees in math and computer science from St. John's University and the New York Institute of Technology, respectively. The author of four textbooks and dozens of articles, and former columnist for *Math and Computer Education,* he has more than 40 years of classroom experience. He is former president of the Dozenal Society of America and still serves on their Board of Directors. He is the founder of the New Horizon Learning Center and leads personal growth and development seminars. Although a layman, he is the Catholic Chaplain at Nassau Community College.

For over 900 combined years of encouragement and support thanks to:

Rose and Max, Bess and Marty, Syl and Jake, Mollie and Bernie, Marilyn and Is, Betty and Mare, and in memory of Izzy.

And of course to Leda.

–R. R. R.

To Pat, George, and
Gladys, Buddy, Don, Re, Glo, Jerry, Ev, Ray, Pat, Vic, Kathy, Peg, and Mike and Jean. Whose love and concern helped make my life better.

–G. Z.

# Contents

## 4 Probability 101



## 5 The Binomial Distribution 125



## 6 The Normal Distribution 148

## 7 Approximation of the Binomial Distribution by Use of the Normal Distribution 173



## 8 Hypothesis Testing: One-Sample Test of Percentages in Binomial Experiments 194

# 11 Hypothesis Testing with Sample Means: Small Samples 283



# 12 Confidence Intervals 308

## 13 Chi Square 329



## 14 Correlation and Prediction 363



## 15 Tests Involving Variance 394

# Preface

This textbook offers a thorough but lighthearted introduction to the basic concepts of descriptive and inferential statistics. When the first edition of this book appeared in 1972 we were pleased that so many instructors found it useful in their introductory courses, especially courses for students with weaker math backgrounds. We have kept that same target audience in mind through three revisions and have been told by many users that the major strengths of this book are its readability, its often humorous approach, and the variety of its exercise sets.

Our book definitely has a lighter tone than many other introductory texts. There are many puns and silly names scattered through the text that are included for good pedagogical reasons. First of all, this style makes the material friendlier and not as menacing to many new students who expect the material to be difficult. Secondly, it makes certain points and examples easy to refer to—you can just refer to the characters in the problem by name. And third, it creates a bond between the teacher and the students who can complain together about the awful humor of the authors. If this approach is foreign to you, the book may take some getting used to. We realize no single approach is right for all instructors, but we have been pleased that so many instructors have told us the text "worked" for them. We have heard more than once from instructors who on first glance thought that the book looked too simple for their course but who were then gratified by how well the students came to understand and remember the material.

As in the earlier editions, our overall objective is to show readers how statistics connects to their lives, and how the simpler statistical methods should be used properly. It is not our intention to train the readers to be statisticians. Our major emphasis is on understanding the types of statistics that appear frequently in the media, and on learning how to interpret them.

In addition to using "real" examples, we believe it is beneficial to teach the basics with "made up" problems which are LIKE real ones. We do not consider this a course in number crunching, and we use real problems for their inherent interest, not because they contain messy numbers that demand machine calculation. We expect students to use calculators routinely, and encourage them to look for any special features that can help in statistical calculations, but this is not a major point of the text. For those instructors who want their students to use computer software or graphing calculators, special supplements are available.

In concert with our notion of the book as a liberal arts text, we have included many text boxes with historical vignettes, odd tidbits about statistics, and other remarks which help portray statistics as a human endeavor. For instructors and students who are interested in further material of this type we strongly recommend these books:

1. *The History of Statistics,* Stephen M. Stigler (Harvard University Press, 1986).
2. *Games, Gods and Gambling,* F. N. David (Charles Griffin & Co., 1962).
3. *A History of Mathematics,* V. J. Katz (HarperCollins, 1993).

## New to This Edition

In this fourth edition we have tried to maintain those strengths that users of the text have praised while improving the book in the following ways:

- **Coverage of hypothesis testing** is more clearly presented in a **step-by-step** fashion.
- **Classroom Experiments:** The book contains suggestions in many chapters for classroom activities that yield statistical data, often based on characteristics of the students in your classroom. Most students find statistics more interesting when the numbers refer to themselves.
- Increased clarity in **describing the population** in each problem as well as in distinguishing between the sample and the population.
- Improved presentation of **rounding-off calculations.**
- **Redesigned,** two-color format includes improved use of graphics, marginal notes and definitions, highlight boxes for key concepts, and a new cartoon character, StatKat.
- **Chapter goals** are listed at the beginning of each chapter to familiarize students with upcoming topics.
- **Chapter summaries** at the end of each chapter provide students with a quick study reference and reinforce important topics.
- **Historical notes** and anecdotes are included throughout the text.
- A **new appendix** on the use of a **random number** table has been included.

## Major Pedagogical Features

- **Writing Style:** We maintain a light and conversational tone throughout the text.
- **Exercise Sets:** A mix of serious and silly problems are provided. Answers to odd-numbered exercises are included at the back of the textbook.
- **Survey Data:** A class of students can complete the survey given in Chapter 1 to create a viable data set. There are survey exercises in almost every chapter which refer to this original collection of data, allowing new concepts to be illustrated with real data from the class. Alternatively, we have also included the data from a fictitious survey of "typical" students that can also be used for this purpose.
- **Field Projects:** For instructors who like to assign data-collecting experiments, there are suggestions in most chapters for field projects outside the class. These projects provide an excellent opportunity for students to work in groups, if the instructor prefers.

## Content of the Book

In general, concepts are introduced through concrete examples. We use the criterion of "reasonableness" based on specific illustrations as opposed to mathematical rigor. The first three chapters deal mainly with the tools of descriptive statistics, stressing tables and graphs, and a few summary statistics. Two notable sections in this part of the book are one that deals with reasonable use of approximate numbers and one that deals with vital statistics and other rates. Chapter 1 includes a discussion of sampling bias, with an optional appendix on using a random number table to pick simple random samples.

Chapter 4 deals with the basic concepts of probability and contains just enough material to guide the student over the next two chapters from the use of the binomial distribution to the normal distribution. Beginning in Chapter 8, we include various topics in statistical inference: Chapters 8 through 11 cover standard topics in hypothesis tests for percentages and means, and Chapter 12 includes the corresponding material on confidence intervals. The remaining four chapters cover chi-square tests, correlation and regression, tests based on the variance including ANOVA, and a few nonparametric tests.

In this edition, we have tried to maintain a very consistent presentation of the logic and structure of hypothesis tests. As part of this approach, for instance, we always expect hypotheses and conclusions to be stated in words and in both mathematical symbols and English sentences.

We have limited our text to a few basic topics. Many introductory books, though they are excellent reference books, are too long for the courses where they are actually used. We instead have opted for a book which is manageable and does not appear overwhelming to a beginning student. For a typical semester or quarter course we suggest that all instructors use Chapters 1 to 8 as a base, covering descriptive statistics, probability, and the simplest one-sample hypothesis tests. Remaining topics of interest can then be included from the rest of the book.

Three optional appendixes are included in the text.

1. *Arithmetic Review:* This material may be used as a refresher for working with percents and decimals.
2. *How to Use a Random Number Table:* Useful for classes where exercises in picking simple random samples are used.
3. *Probability:* Further development than is used in the main part of the text for probability is included for courses requiring a more thorough coverage of this topic. This appendix covers independent vs. dependent events, issues of sampling with and without replacement, and mutually exclusive events.

## Supplements

### Student Solutions Manual

The *Student Solutions Manual* contains detailed solutions to the odd-numbered end-of-section and end-of-chapter exercises.

### Instructor's Resource Manual

The *Instructor's Resource Manual* contains detailed solutions to the even-numbered exercises in the textbook and teaching suggestions from the authors.

### Print Test Bank

A test bank of approximately 1,000 questions is available to instructors.

### Computerized Test Bank

A computerized version of the testbank is available in both IBM and Macintosh formats.

### McGraw-Hill's Statistics Discovery Series: A Guide to Learning Statistics

This supplement is intended to help students enhance their understanding of introductory statistics. Each section of this study guide contains study objectives, an overview of the topics covered, key terms and definitions, worked-out examples, helpful hints to the student, and new exercises and their solutions.

### McGraw-Hill's Statistics Discovery Series: A Guide to Minitab

This supplement helps the student gain a better understanding of statistics through the use of the statistics software Minitab. Worked-out examples and new exercises for use with Minitab are presented, along with a data disk containing data sets ready for use with Minitab. The supplement contains command information for DOS, Windows, and Macintosh platforms, and is packaged with either an IBM or a Macintosh disk.

### McGraw-Hill's Statistics Discovery Series: A Guide to TI Graphing Calculators for Statistics

This supplement contains instructions for the student using a graphing calculator in an introductory statistics course through worked-out examples and new exercises. Appendixes for Texas Instruments 81, 82, and 85 graphing calculators are also included.

### *Against All Odds* Videotapes

Videotapes depicting the use of statistics in our world, produced by the Annenberg/CPB Project, are available.

### Mystat

This student version of the statistical software program Systat is available through McGraw-Hill.

### Acknowledgments

We have received much encouragement from our colleagues in the Department of Mathematics at Nassau Community College, especially Dennis Christy. They have been generous with their praise and not shy with their criticisms. We hope they find merit in this fourth edition. As usual, each author claims credit for what our colleagues like and blames the other for what they dislike.

The original concept of this book was a joint effort of the two current authors and our late friend and colleague Arnold Naiman. We hope that this edition maintains the spirit and the clarity that were so important to him, and we hope that his family still finds the book worthy.

In the preparation of this manuscript we have benefited particularly from editorial support by Maggie Lanzillo and editing assistance coordinated by Scott Amerman at McGraw-Hill. For critical reviews of the manuscript we wish to thank Philip C. Almes, Wayland Baptist University; Pat C. Cook, Weatherford College; Margaret S. Davis, Floyd College; Nirmal Devi, Embry-Riddle Aeronautical University; Ernest L. East, Northwestern Michigan College; John R. Formsma, Los Angeles City College; Raymond P. Guzman, Pasadena City College; Clarence G. Hanley, North Country Community College; Barbara Hatfield, University of Rio Grande; Rebecca M. Howard, Roane State Community College; Ronald G. Kendis, Los Angeles City College; J. Patrick Lang, Virginia Wesleyan College; Denise M. Pocta, Virginia Wesleyan College; T. Michael Rhodes, University of Rio Grande; Robert L. Sartain, Wayland Baptist University; Jack C. Sharp, Floyd College; Billy L. Smith, Roane State Community College; Eileen P. Stitt, University of Rio Grande; Ara B. Sullenberger,

Tarrant County Junior College; John Vangor, Gary L. Wright, Roane State Community College; and especially William J. Rickert, Ocean County College. For expert problem checking and preparation of supplements, we are grateful to Robert K. Smidt and Michael R. Lasarev, both of California Polytechnic State University—San Luis Obispo. For skillful copyediting we thank Za Za Ziemba.

Robert Rosenfeld
Gene Zirkel

# Understanding Statistics

# 1

# Introduction

The average family in this country has 2.7 children.

**CONTENTS**



**GOALS**

At the end of this chapter you will have learned:

- The meaning of the word "statistics," and some ways it is important in your life.
- Several basic concepts vital to the beginning student of statistics. In particular, you will understand the difference between a population, a sample, and a distribution; between inferential statistics and descriptive statistics; and between random samples and biased samples.
- How to interpret approximate numbers, and how to round off such numbers.

## 1.1 Statistics in Everyday Life

When you applied to college, you probably filled out a form designed to establish your "financial need." Maybe you took the SAT or ACT exam. Then some stranger used these numbers to help decide what college you got into and how much money you had to pay. To a large degree, you were being treated as a collection of numerical values, a collection of statistics.

You should understand how statistics are used because decisions that affect you personally are based on statistics. "Your grade point average is only 1.47; sorry, kid, we'll have to put you on probation. I know that you just had a bad time with your parents, but let's face it, everybody's got problems." That hurts.

There may be times you have mixed feelings about being treated as a source of statistics.

♥AMICABLE DATING SERVICE♥

14 Valentine Lane
Lonelyville, OR 12345

FEBRUARY 14

DEAR FRIEND:

OUR RELIABLE COMPUTERIZED MATCHING SYSTEM HAS PROCESSED YOUR VITAL STATISTICS AND WE ARE PLEASED TO ENCLOSE HEREWITH THE NAMES, ADDRESSES (WITH ZIP CODES) AND PHONE NUMBERS (WITH AREA CODES) OF SIX IDEAL MATCHES.

WE ARE HOPEFUL THAT YOU CAN ESTABLISH A LASTING RELATIONSHIP WITH AT LEAST ONE OF THEM.

WE REMIND YOU THAT YOUR FEE IS NOT REFUNDABLE.

SINCERELY,
IMA FRIEND, COORDINATOR

### A Newspaper Item

"The government's handling of the deportation came under criticism in Israel, although a poll for the newspaper *Yediot Aharanot* found 91 percent *of the country's Jews* support the mass expulsion itself." [Emphasis added]

—*Newsday,* December 19, 1992, p. 10

Some questions that you might wish to ask the reporter:

1. How was this poll taken?
2. How many people were questioned?
3. Does 91% refer to all "the country's Jews," or only to the sample polled?
4. What was the exact question each person was asked?
5. Was the sample representative?
6. What is the probable margin of error between the sample results and the population?
7. How long after the expulsion was the poll taken?

The information gathered from Gallup polls provides statistics which are used to determine which products will or will not be manufactured and made available to you. Nielsen surveys help TV networks decide which shows you will or will not be able to watch.

Products are sold to you all the time with numbers thrown at you: (1) "I used Grit toothpaste, and now I have 20% fewer cavities." (Fewer than what?), or (2) "Hey, kids! Start your day with Daystart Cereal! It has twice as much iron as a delicious slice of toast and more vitamin C than two slices of bacon!" (So who said that bread was a good source of iron in the first place, or that bacon has a lot of vitamin C?)

Doctors prescribe medicine and treatment for you, basing their judgment on statistical information: (1) "Use of this pill will cause deleterious side effects in 1.4% of its users." (Is the risk worth taking?) or (2) "There is a 40% chance that an adult suffering from a herniated spinal disk will recover spontaneously." (Should we go ahead with the operation?)

If you have ever donated blood, you know that from a few drops of your blood, a test for iron is performed, and inferences are made about the iron quality of ALL the blood in your body. Some prospective donors are then accepted, while others are turned away.

How can the medical profession judge the fitness of all your blood when they have taken only a few drops in a pipette? Can TV executives accurately predict the popularity of a particular TV show when they have asked the opinion of only some of their viewers? Can manufacturers be assured of how well a product will sell if they sample only some of their possible customers?

The answer is that they often cannot know for sure. However, they can be *very certain* that the information gathered from part of your blood, part of their customers, or part of their viewers is *probably* representative of the information in which they are interested. Their assurance is based on the assumption that the sample gathered was *representative* of the information needed.

The science of statistics affects your life in many ways. You can ignore it, but you cannot avoid it. Others are using statistics, and their decisions influence your life. This text provides you with information about the pervasive effects the use (and the misuse) of statistics has upon you and those you love. In this instance, ignorance is certainly NOT bliss!

## 1.2 Some Statistical Terms

Words are important. In statistics some words have technical meanings which may differ from everyday usage. Here are a few which may be new to you.

statistics

> The **science of statistics** is the study of collections of numbers in order to (1) describe them accurately and concisely and (2) draw valid inferences from them.

population

The word **population** is used in everyday speech to refer to a group of people or things in which we are interested. To the statistician the technical meaning of the word refers to a group of **numbers** or **categories** derived from these people. Both uses are correct, and the meaning will be clear from the context.

For example, we could be interested in the *population of children* in Los Angeles in the fourth grade. If we administer a reading comprehension test to each child, we might talk about which children scored in the top 10%. These test scores form a *population of numbers.*

> A **population** is all the values, people, or things in which we are interested for a particular study.

A population of interest can be large or small. The crucial point is that the population contains ALL the subjects or values that we are interested in.

A population is determined by the topic in which you are interested. You can't define the population for a study until you know the purpose of your collecting data.

### Illustrations of Populations

❑ EXAMPLE 1.1 First-Year Dropouts

Suppose that you wish to find out the percentage of people who start college and then drop out before the first year is over. What is the population under consideration?

SOLUTION

First ask yourself some questions. Clarify what you mean by "college." Are you including 2-year schools and 4-year schools? Are you thinking of schools all over the world, just in the United States, or only in your state? When? What time frame do you wish to consider? Your answer might well be THE POPU-

LATION OF ALL STUDENTS WHO STARTED AT EITHER A 2-YEAR OR A 4-YEAR COLLEGE IN THE UNITED STATES LAST FALL, and the category of interest would be WHETHER OR NOT THEY DROPPED OUT OF COLLEGE BEFORE THE YEAR WAS OVER. ❑

### ❑ EXAMPLE 1.2 Trucks on the N.J. Turnpike

If you were interested in discovering the average weight of trucks that use the New Jersey Turnpike, what would you use as your population?

**SOLUTION**

Ask pertinent questions to clarify your purpose. What is a truck? What is the time period? Your answer might be THE WEIGHTS OF ALL SEMITRAILERS WHICH WILL USE THE NEW JERSEY TURNPIKE BETWEEN JANUARY 1 AND MARCH 31 OF NEXT YEAR. ❑

**population vs. sample**

We have seen that the word **population** is used to refer to *all* the persons, objects, scores, or measurements under consideration. The word **sample** refers to any portion of the population.

**sample**

A **sample** is any part of a population.

### Illustrations of samples

1. The group of students who started at *your* college last fall is a sample of the population described in Example 1.1.
2. All semitrailers that enter the New Jersey Turnpike at *Exit 10 next February 2nd* are a sample of the population described in Example 1.2 above.
3. The *bankers* in Ohio are a sample of all the residents of Ohio.

### Why Do Statisticians Use Samples?

Since the purpose of a statistical study is to learn about a population, why do statisticians use samples? Why don't they use the information from the whole population?

In fact, some studies do use entire populations. When an attempt is made to include ALL the members of a population in a study, the study is called a **census.** You are probably familiar with the United States Census, which is taken every 10 years. But there are good reasons that this approach cannot always be used, when sampling is the only way to proceed.

Sometimes the population is too large to be studied completely within the time and budget constraints of the study. This would be the case in many political opinion surveys, where a sense of the community is needed quickly.

### *Statistics and the Moon*

Since ancient times, astronomy has fascinated human beings, and astronomical tables have long been a source of statistical data. By noting specific craters that appeared and disappeared, Galileo Galilei (1564–1642) determined that, in spite of the popular belief that the moon always presents the same face to the earth, we actually view about 60% of its surface.

In 1714 a commission was formed in England to award grants and prizes for new methods of determining the position of a ship at sea, information which was crucial for commerce and military purposes. One such method depended upon using tables of the position of the moon, and in 1765 a £3000 prize was eventually awarded to the widow of Johann Tobias Mayer (1723–1762) for his work in 1750 wherein he accurately described the path of the moon. In his work, Mayer used averages derived from 27 sets of observations as key values in his equations, attaching what he reasoned were justified *margins of error.* This is one of the earliest successful attempts to describe the accuracy of a statistic.

Sometimes nothing important is gained by including every piece of data and, in fact, if you try to study too much data in too little time, you are more likely to commit data-processing errors. In bank or corporate audits, for example, an accountant often decides to study carefully a wisely chosen sample of the accounts instead of all the accounts. This audit will probably be just as informative as a complete enumeration of accounts, and the smaller amount of data can be prepared, analyzed, and interpreted more effectively and with less chance of error.

Other research demands sampling because the study itself alters or destroys the very things you are studying. This can easily happen in industrial quality control studies. For instance, auto manufacturers crash cars as part of safety testing. They couldn't possibly crash-test EVERY car they made; there would be nothing left to sell. Therefore they must assume that the test cars are a representative sample of the production run.

Some populations don't even physically exist at the time of the study. For instance, suppose a scientist is trying to determine the average weight of all 1-year-old male white rabbits which are or will be raised in laboratories using a certain diet. It is impossible to weigh every rabbit in the population, because the population never exists completely at any one time. More rabbits will always be born who are potential members of this population. If 50 of these rabbits are selected and their average weight is determined, these 50 would be thought of as a sample from the population.

### *Randomness*

During World War II, putting heavy armor on airplanes was a problem. Abraham Wald drew a diagram of an airplane and marked each spot where a bullet hole occurred on a plane returning from a bombing mission. Some parts of the plane appeared very black from many bullet holes. Other parts were practically white, indicating very few bullet holes. Assuming that the bullet holes are random, where would you place the armor? (See Abraham Wald's solution on page 8.)

Note, the same group may be considered a population for one purpose and a sample for another purpose. For instance, suppose that the average age in this class is 19.7.

Considering this class as a **population** of interest, 19.7 is the precise answer to the question: *"What is the average age of the members of this class?"*

Considering this class as a **sample** of students from the entire college population, we may wish to infer that the average age of all students at this college is *probably about* 19.7.

**random process**

No word is more important to the theory of statistics than the word **random**. An item is chosen "at random" from a population if the *process* of selection does not favor any particular item either intentionally or inadvertently. Entire textbooks have been written describing procedures for selecting random samples, and the process can become quite technical. In this text it will be sufficient to think of a random sample as one that has been picked

## *Nonrepresentative Samples?*

"More than three-quarters of *Americans* [emphasis added] believe that a retiree whose annual income exceeds $100,000 should not get Social Security benefits according to the responses of *Parade* readers. . . .

In all, 99,744 of our readers responded to an invitation in the Feb. 21 issue to *call a 900 number* [emphasis added]. . . . (In addition, some voted by mail after they had difficulty calling.)"

—*Parade* Magazine
April 4, 1993, page 21

"The *Journal* of the AMA published a study by Walter Willet ["Dietary Fat and Fiber in Relation to Risk of Breast Cancer," October 21] suggesting that diet has no bearing upon one's risk for breast cancer. . . . In the study the median percent of calories from fat was in the 30s. Among the women who had the lowest percentage of calories from fat in their diet, the median percentage was 27. The reason Willet found little difference in the cancer rates was that the diets were still too high in fat. In effect, Willet studied a high-fat diet versus a very high-fat diet. The Physicians Committee for Responsible Medicine has always held that the percent of calories from fat in one's diet should not be higher than 10. . . ."

—Neal D. Barnard, M.D.
[From a letter to the editor of *Newsday,* December 19, 1992]

"According to a recent study (1991) done by Simon LeVay, a neuroscientist at the Salk Institute in San Diego, and recorded in the August issue of *Science* magazine, it was found that the anterior hypothalamus (a region of the brain that governs sexual behavior) in homosexual men has the same anatomical form found in women rather than the form found in heterosexual men. . . . LeVay was able to perform this study *by using the brains of homosexual men who had died of AIDS.* [Emphasis added.] He found that in a majority of the 19 gay men studied, one of the regions called INAH-3 was smaller than that of a heterosexual man, but was the same size as a woman's (*Science,* August 1991). . . . Could it be that the AIDS virus has affected the size of the INAH-3 instead of the smaller size being the result of being homosexual?"

—*Vignette,* Nassau Community College student newspaper

### Wald's Answer

Place the armor where the white areas are. Planes hit in the black areas were able to return to base. Presumably, those hit by random bullets in the white areas were severely damaged or destroyed.

The use of a graph—a picture—to help with the interpretation of data is a common statistical device.

Wald was an important member of the Columbia University group of statisticians who did secret war work. Ironically, he was killed in 1950 in a plane crash.

"fairly"—without prejudicing the chances of any member of the population to be chosen.*

For example, if we want to pick a random sample of 20 people from some population, then every possible grouping of 20 people should have an equal chance of being selected as the sample. The practice of putting paper slips into a large drum, mixing them well, and then picking one without looking is a simple model of random selection.

**random sample**

A **random sample** is a sample in which each member or item in the population had the *same probability* or *chance* of being selected for the sample.

Assuming that there are more than two in your family, then if you put the names of all the members of your family in a hat, mixed them well, and drew out two names, you would have a (very small) random sample of that population.

## Bias

Statistical testing is frequently based upon the assumption that the sample was picked randomly. If it turns out that the sample was not random, the results may not be useful. Hidden, unsuspected bias can completely destroy the usefulness of statistical information and statistical inferences made from such information. A sampling *method* is **biased** when it systematically under- or overrepresents some portion of the population. Consequently, we also say that the resulting sample is a **biased sample**. When you use a biased sampling procedure, the conclusions you draw about the population are likely to be wrong. No honest person uses a biased sampling method on purpose, but carelessness or lack of insight can unintentionally lead to a bias.

**sample**

A survey sample can be biased for many reasons, but perhaps the most common reason is convenience. If you choose people for your survey who happen to be easy to contact (and the worst case is to use volunteers), then it is quite likely that they are not representative of the population under discussion. Well-known examples are call-in surveys on TV shows, and volunteer response-mail surveys that come to your house. In both cases, people with

*See Appendix B for one approach to the selection of random samples.

strong feelings are much more likely to respond, so the survey will tell you how *they* feel (which may be very interesting and important), **but** you will not know much about how the *population in general* feels. Such convenience samples are almost sure to be biased because significant segments of the population will not be heard from.

Sometimes a sampling procedure is biased because the people who actually have a chance to be chosen (they are called the **sampling frame**) are not representative of the population as a whole. For example, suppose a pollster for a political candidate decides to interview people selected at random from a phone book to see if they intend to vote for the candidate. Voters who have no phone have no chance to be picked. If those people differ in some significant way from the people who have phones, then the results of the survey will not reflect the opinions of the entire population.

If random phone calls were made at 1 P.M. to sample the population of all voters, many people with full-time day jobs would be missed.

Often, the bias is not as obvious as the phone illustrations above. For example, it took a while before pollsters realized that surveys taken on Friday nights are more likely to find people at home who will vote Democrat. People who vote Republican are more likely to be out for the evening, thus biasing Friday night polls in favor of Democratic candidates. (Another subtle bias occurs in phone surveys where the caller speaks only to the person who answers the phone; commonly in such surveys men are underrepresented. In many homes, it seems, men just don't answer the phone if someone else is home.)

***Note*** The word "random" describes the *process* by which the sample was chosen. This does *not guarantee* that the sample will turn out to be representative, but, as we shall see, it allows us to determine the probability that the sample is representative. This is usually given in the newspaper as the *margin of error*. However, if the sample was inadvertently biased in some way, it is possible that the sample results are far from the truth.

### *Margin of Error*

The following disclaimer appeared at the end of a newspaper article reporting on a Gallup poll:

For findings based on samples of this size, one can say with 95 percent confidence that the error attributable to sampling and other random effects could be two percentage points in either direction. In addition to sampling error, the reader should bear in mind that question wording and practical difficulties in conducting surveys can introduce error or bias in the findings.

### Two Uses of Statistics: Descriptive and Inferential

**descriptive vs. inferential**

The applications of statistics fall into two broad categories: **descriptive statistics** and **inferential**, or **sampling**, **statistics**.

Consider the following example. (Maybe you have participated in this kind of survey if you live near a big shopping mall.)

## Descriptive Statistics

An extraordinarily good use of descriptive statistics is to be found at the Ellis Island Immigration Museum in New York Harbor. Here one finds a "statistical portrait of the pageant of immigration" in the Peopling of America Exhibit.

A number of colorful displays depict immigration data in various ways. These graphic exhibits range from a frequency polygon contrasting the rise and fall of the number of immigrants over the years (see photo), to a 3-D word tree presenting the ethnic roots of many words in American English, to a signpost indicating the number of people entering and leaving our country each decade.

There is a 3-D pictograph plotting numbers of immigrants against sex and time in 15-year periods (see photo on opposite page) and another with numbers of immigrants vs. geographic origins and time in 20-year intervals.

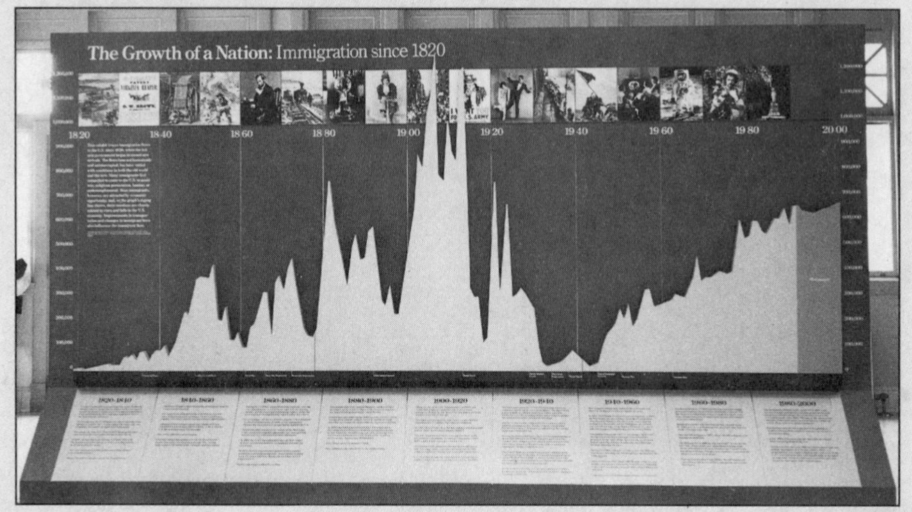


*(Photo by Patricia Zirkel)*

*A Frequency Polygon Entitled "The Growth of a Nation: Immigration Since 1820."*

A market research organization interviewed a random sample of adults in a large shopping plaza in the city of Niles. Of those questioned, 15% used NoCav brand toothpaste.

Subsequently, a concentrated advertising campaign was undertaken to sell NoCav to the public. Then, another survey was taken in the same plaza 3 weeks after this campaign. This second poll showed that 19% of those questioned used NoCav toothpaste.

Are we correct in assuming that the rise from 15% in the first sample to 19% in the second sample is due to advertising? If we have doubts that the advertising caused a substantial increase in the use of NoCav toothpaste, what questions should we ask concerning these findings? What about any data that were omitted from the presentation?

## Descriptive Statistics (continued)

There is a 2-D electronic map showing, by state, where the diverse ethnic groups have settled and another illustrating the original Native American population prior to 1492. An electrified 3-D globe portrays worldwide immigration trends in 25-year periods.

An interesting map of the slave trade from Africa reveals that less than 6% of the slaves came to North America, while over 31% went to Brazil, and more than 53% went to the Caribbean. A pictograph using 20-year intervals from 1620 to 1780 shows that more than three-fourths of all immigrants in those years were either slaves or indentured servants.

(Photo by Patricia Zirkel)

*A 3-D Pictograph Entitled "A Changing Pattern: Male/Female Immigration Trends."*

In this example we see numbers used in two different ways. The number 15% is used to describe the fraction of people in the first sample who used NoCav. As such, it summarizes with conciseness and clarity the unreported fact that of 140 persons interviewed, 21 used NoCav. This is an example of descriptive statistics.

> **Descriptive statistics** is the use of numbers to summarize information which is known about some situation.

The information could be summarized by a pictorial method such as a bar graph or in numerical format as a percentage or an average.

In contrast to this use of the data, if we use the above sample to imply that approximately 15% of *all* the adults who patronize the shopping plaza used NoCav, then we are using the number to infer something about a larger population for which we do not have complete information. This use of the number 15% is an example of statistical inference.

**Statistical inference** is the use of numbers to give numerical information about larger groups than those from which the original raw data were taken.

In characterizing a large amount of data by a few descriptive statistics, we gain clarity and compactness, but we lose detail. We *describe* the data relating to some population concisely. For example, in a university with 20,000+ students, a list of all the grade point averages (GPAs) could be less helpful than a few descriptive statistics such as the average GPA, along with the lowest and the highest GPAs. We often display our descriptive statistics in a table such as the following one, which gives the percent of students in various categories.

| GPA | % of Students |
|---|---|
| 3.50–4.00 | 18 |
| 3.00–3.49 | 22 |
| 2.50–2.99 | 27 |
| 2.00–2.49 | 11 |
| 1.50–1.99 | 8 |
| 0.00–1.49 | 14 |

The following descriptive statistics, by summarizing information, describe, in some way, the populations from which they were taken.

1. The average SAT score for Nostrum College is 503.2.
2. The marks on the last exam in this class ranged from 51 to 98.

The following are examples of statistical inference. We might infer from appropriate samples that:

1. Between 20% and 25% of American college students are married.
2. High cholesterol levels are associated with increased risk of heart disease.

The same numbers may be used for either describing a smaller distribution or making inferences about a larger distribution:

3a. Nielsen reports that 24.7% of those *who were interviewed* watched the President's news conference last Sunday night.

3b. Probably about 24.7% of *all* television viewers watched the President's news conference last Sunday.

4a. The average age of students enrolled in this class is 19.7 years.
4b. The average age of students enrolled at this college is probably about 19.7 years.

Here is a more detailed example of statistical inference. Suppose there were a disease in which three-fourths of the patients recovered without treatment within 1 month of contracting the disease. Suppose also that a doctor claims to have discovered a new drug to cure this disease. We shall administer the drug to 100 patients. Even if the drug were useless, we would still expect about 75 (three-fourths) of these people to recover within 1 month. Due to chance variations, more or less than 75 may recover.

One of the problems of statistical inference in this example is to decide how many must recover before we are willing to accept the drug as a cure. Certainly, if all 100 recovered, we would be enthusiastic about the drug's potential. But how about 95 or 90 or 80 recoveries? Where should we draw the line?

The job of deciding where to draw the line is an important one for the statistician. It is one of the main skills we hope you will develop from this book. Can we confidently say that the new drug saves lives, or is it likely that this result occurred by chance? Even if all 100 recovered, it is possible

### *How Many Fish Are in the Sea? Estimating the Size of a Population*

How can we estimate the size of the population of tuna fish in a specific part of the Atlantic? Suppose that we knew that 200 of those tuna had red tails. Then if we captured a random sample of tuna, the proportion of red tails in our sample could be used to estimate the size of the entire school of tuna.

For example, suppose we captured a random sample of 1000 tuna, and found that 50 of them had red tails. Since 50 is 25% of 200, we have captured 25% of the red-tailed tuna. We might therefore conclude that our entire sample makes up about 25% of the entire school. Hence, there are probably about 4000 tuna in our population of interest. The only thing left for us to do is to first capture 200 tuna, dye their tails red, and release them into the sea. Then, after allowing them to mix with the other tuna, we would catch a sample of 1000 fish and calculate the necessary proportions.

(though very unlikely) that they would have recovered anyway. Perhaps, just by luck, this particular group of 100 patients was unusually resistant to the disease.

It is important for the statistician to pick the sample in an impartial way. If by chance we happened to test the drug on only mild cases, our results would be misleading. We would hope that the sample is truly a mirror of the population we want to learn about (in this case, *all* victims of this disease).

Sample surveys, polls, and statistical tests have become a part of our way of life. Every day, people present figures to prove or disprove some claim: Does a certain food additive cause cancer? Does smoking lead to heart disease? In this book we will study some of the tests that statisticians use when making claims. We hope to show you how such tests should be properly done and how to interpret the "proof" of such claims.

### More Statistical Terms

**raw score**

If we are listing the ages of students in a certain school, then each age is called a **raw score**.

> A **raw score** is a basic observation on an individual in a study.

**distribution**

A collection of scores is often called a **distribution**. The term "distribution" may refer to *either* a population or a sample. Thus we might be interested in the distribution of the ages of all the freshmen in the United States. We could learn something about this population by considering the distribution of the ages in a sample from this population.

> A **distribution** is a collection of values which measure something of interest.

In a statistical experiment, we often collect raw scores about several different things. We may, for instance, collect information on heights, weights, and ages of the people who belong to a particular organization. The entire collection of information in a study is called the **data** for the study.

> **Data** are all the information we have—that is, one or more distributions of raw scores.

#### *Data*

Incidentally, "data" is a plural noun. For example, you should say, "The data show that $X$ is more popular than $Y$." It is incorrect to say, "The data *shows* that . . ." The singular of "data" is "datum."

# 1.3 Approximate Numbers

Suppose 7 people organize a party as follows:

John bought chips $ 7.49
Jean bought dip 12.37
Tony bought bread 13.71
Toni bought cold cuts 35.09
Sam bought paper goods 9.11
Sammi will buy cake for about $10.00.

Pat, who laid out no money, adds these figures and divides by 7, obtaining an average cost of $\$87.77 \div 7 = \$12.5385714\ldots$, and claims that the party will cost each person $12.54.

Of course, you can see the foolishness of Pat's claim to know the cost of the party *correct to the nearest penny* when the cost of the cake is not precisely known. Computer scientists use the acronym "GIGO" to express the fact that one's results can never be better than the data one starts with. "GIGO" stands for Garbage In, Garbage Out.

### The Meaning of "About"

If "about" $10 means that the cost of the cake is between $9.50 and $10.50, the average would be between $12.46 and $12.61, and the best that Pat could claim would be an average cost of about $12.50, give or take a few pennies.

However, if "about" $10.00 means that the cost of the cake is somewhere between $5.00 and $15.00, the average would be between $11.82 and $13.25, and the best that Pat could claim would be an average cost of about $12.50, give or take almost one dollar.

The meaning of "about" often depends upon how precisely the original data are known. In general, the context of a specific problem will suggest reasonable interpretation for "about."

### Rounding

During this course you will have to do a lot of numerical exercises, and most likely you will use a calculator for some of the arithmetic. One of the characteristics of calculators is that they routinely display a lot of digits, often more than make sense in a particular problem. (And sometimes they display fewer.) This means that you will want to round off your results to some convenient *approximate* value.

If we round 1204.69079 to hundredths, we obtain 1204.69. If we round to units, we obtain 1205. The rule is to look at the digit to the right of the place that we wish to round off to. If it is 5 or greater, we round up by adding 1 to the digit to its left. If it is 4 or less, we leave the digit to the left alone. Then we drop the remaining digits, replacing any to the left of the decimal point by zeros.

Thus, if we round 6.493 to tenths, we obtain 6.5. Rounding to units yields 6. Notice that the result is 6 and not 7. It is not correct to round 6.493 (which is less than 6.5) to 6.49, then to 6.5, and finally to 7. We round only once.

If we round 500127 to hundreds, we write 500100 (and not just 5001), keeping the zeros to maintain the *magnitude* of the number.

When a number ends in trailing zeros, we may use a bar to show where we rounded off. Thus if we rounded 500127 to thousands, we would write $50\bar{0}000$. The round-off error is at most half of the place marked. Thus $50\bar{0}000$ means that the thousands place is correct and the error is at most one-half thousand, that is, ±500.

---

### ❑ EXAMPLE 1.3 Rounding to Two Decimal Places

Round these numbers to hundredths: 16.837, 8.00319, 9.105, 5.999, and 10.1349.

**SOLUTION**

| Original Data | Next Digit | Rule | Result |
|---|---|---|---|
| 16.837 | $7 \geq 5$ | Add 1 to 3 | 16.84 |
| 8.00319 | $3 < 5$ | Drop 3 | 8.00 |
| 9.105 | $5 \geq 5$ | Add 1 to 0 | 9.11 |
| 5.999 | $9 \geq 5$ | Add 1 to 9 | 6.00 |
| 10.1349 | $4 < 5$ | Drop 4 | 10.13 |

❑

---

### ❑ EXAMPLE 1.4 Rounding to Hundreds

Round off these numbers to the nearest hundred: 5826, 9084, 1163.7, 4841, and 5032.

**SOLUTION**

| Original Data | Next Digit | Rule | Result |
|---|---|---|---|
| 5826 | 2 | Drop 2 | 5800 |
| 9084 | 8 | Add 1 to 0 | 9100 |
| 1163.7 | 6 | Add 1 to 1 | 1200 |
| 4841 | 4 | Drop 4 | 4800 |
| 5032 | 3 | Drop and add bar | $5\bar{0}00$ |

❑

### *Stupid and Confusing*

The following quotation from *Statistical Method in Biological Assay* by D. J. Finney (London: C. Griffin, 1964) underscores our point.

> Bioassays are seldom sufficiently precise to warrant quotation of results to more than 4 significant digits: a statement that a test preparation is estimated to have a potency of 35.71685 units per milligram is both stupid and confusing.

An illustration of rounding 1024.69079 follows:

| | |
|---|---|
| to ten thousandths | 1024.6908 |
| to thousandths | 1024.691 |
| to hundredths | 1024.69 |
| to tenths | 1024.7 |
| to units | 1025 |
| to tens | 1020 |
| to hundreds | $1\bar{0}00$ |
| to thousands | 1000 |

As we stated previously, different calculators may yield different numbers of digits in their answers, and if you use pencil and paper, you will probably use fewer digits than you would using a calculator. This can sometimes lead to a slightly different final answer.

Which answer is wrong? Neither one. Remember we are dealing with approximate numbers. Do not be concerned if your answer differs a little in the least-accurate digit from the answer obtained by your neighbor, your instructor, or from an answer key.

### *Helpful Hint*

Don't round off results in your calculator until the final result is obtained. Make use of the memory to store intermediate results.

### *Helpful Hint*

Include final zeros in your data, so that all numbers appear with the same number of decimal places.

Throughout this book we will often assume that all the data in any one problem are given to the same degree of accuracy or precision. This is in keeping with ordinary usage. In a given study, statisticians do not usually measure some people to the nearest half inch and others to the nearest inch. Thus, if some data are given as 17, 21, 19, and 30, we presume that 30 is also correct to the nearest unit. Similarly, 18.02, 191, 19.61, and 10 imply that the 10 and the 191 are correct to two decimal places. *It is better* to write them as 10.00 and 191.00 to indicate this.

A helpful rule of thumb is to round off *after* the final result is obtained. Thus, if we desire the quotient of 4854.0678 divided by 13.747 rounded off to hundredths, we divide BEFORE we round off, and not vice versa. Always carry *at least one more digit* than desired until the final answer is obtained.

### *Statistics Packages and Statistics Calculators*

Many—but not all—students use computer statistical software packages (such as Minitab). Others use inexpensive calculators specifically made for statistics. These tools relieve a lot of the drudgery of arithmetic calculations—and often produce the correct answer! Some problems in this text will be considerably easier if you have one, and we encourage their use.

The calculator keyboard will contain symbols such as $n$, $\bar{X}$, $\Sigma X$, $\Sigma X^2$, $\sigma_x$, and $s_x$. The time spent learning to use such devices will be more than repaid by the time saved in performing calculations.

However, since not all students have these wonderful instruments, our text is independent of them. They are *very, very* helpful, but not required for this book.

## STUDY AIDS

### Chapter Summary

*In this chapter you have learned:*

- Why a knowledge of statistics is essential to an educated person in today's world
- Some important words and phrases indispensable to a student of statistics
- The distinction between inferential and descriptive statistics
- How to round approximate numbers, to always keep more significant digits during a calculation than you will require for the final result, and to question the accuracy of results displayed by a calculating device

### Vocabulary

You should be able to explain the meaning of each of these terms:

1. The science of statistics
2. Population
3. Sample
4. Random
5. Random sample, random sampling procedure
6. Biased sample, biased sampling procedure
7. Descriptive statistics
8. Statistical inference
9. Inferential statistics
10. Sampling statistics
11. Raw score
12. Distribution
13. Data

### Helpful Hints

List any *Helpful Hints* found in this chapter. Perhaps you want to start a special section in your notes for this purpose.

## EXERCISES

1-1 Give an illustration in each of these fields where statistics may be used.

(a) Medicine (b) Politics (c) Education (d) Sports

(e) Manufacturing (f) Sales (g) Government (h) Science

1-2 Name some aspect of your own life where decisions are made about you based on statistics. Describe how this is done. Is this OK with you?

1-3 Why do people both admire and fear statistics? What are some of the advantages to the use of statistics? Are there any disadvantages?

### *Lies, Lies, Lies*

"There are three kinds of lies: lies, damned lies, and statistics."
—attributed by Mark Twain to Benjamin Disraeli, prime minister of England (1804–1881)

1-4 For each of the following statements, describe the *population* or *populations* that should have been sampled to get this information. If necessary, clarify the question until it is clear what population is meant.

(a) 40 percent of American adults support the President's new plan.

(b) English girls learn to speak before English boys do.

(c) A dosage of 50cc of this pesticide killed half of the rats in the experiment.

(d) At Hudson University, psychology majors have higher grade point averages than do chemistry majors.

(e) Too much cholesterol is bad for your heart.

1-5 Suppose the leaders of your town or city wanted to know how the citizens felt about building a new sports arena. If they surveyed 100 people in *your neighborhood,* would that give them a good idea of how the citizens feel? Using the words "population" and "sample," explain your answer.

1-6 Are the farmers in California a sample of the residents of California? Are they a random sample of the residents of California? Explain.

1-7 Michelangelo tosses a fair coin 4 times and obtains 4 heads. Is this sample of the results of a coin-tossing experiment a *random* sample? Explain.

1-8 (a) Read the box on page 7 concerning the poll taken by *Parade* magazine. Why is this *not* a random sample of the opinion of *Americans,* as implied in the article? In what way is the sample likely to be biased? Who would respond to this survey?

(b) Which one of the three samples in this box is an example of convenience? Which one is an example of volunteers? Which one is an example of sampling the wrong population?

1-9 (a) A Beverly Hills survey reported that restaurants lost 30% of their normal business during a smoking ban in 1987. (*Newsday,* August 12, 1994, page A39.) The 30% figure surfaced again in a survey of restaurants in the Los Angeles suburb of Bellflower, which banned smoking from March 1991 to June 1992.

Is the fact that the same 30% figure appeared twice important? What questions would you want to ask with regard to these surveys?

(b) The Beverly Hills survey was taken by the Beverly Hills Restaurant Association. They asked restaurants how much business they *thought* they lost during the ban.

> A study of taxable sales of Beverly Hills Restaurants conducted by the accounting firm of Laventhol & Horwath showed a more modest drop of only 6.7%, and this was only temporary.

The Bellflower survey was taken by Restaurants for a Sensible Policy, a group supported by the Tobacco Institute.

A study of sales receipts commissioned by the city of Bellflower showed that receipts rose by 2.4% during the smoking ban.

What is your opinion on the impartiality of these four surveys?

1-10 An advertisement states that three-fourths of doctors interviewed recommended brand X. What is your reaction?

1-11 In the weeks before a gun control bill is to be voted on, Senator Target, a state legislator, receives 400 letters from constituents on this issue. Of these, 300 letters urge the senator to vote for the bill and 100 are opposed to it. Should the senator decide that probably about 75% of all the constituents favor the bill? Why or why not? In this situation, what is the population, and what is the sample?

1-12 One Monday morning the advertising agency for EyeHi eyeliner sets up a table in an outdoor flea market in Pretty City and interviews every woman who stops at the table. Each one who stops is asked whether she has ever tried EyeHi. By noon, 200 people have been interviewed, of whom 50 said yes, they had tried EyeHi. Should the advertiser conclude that about 25% of the women in Pretty City have tried EyeHi? In this situation, what is the population? What is the sample?

1-13 Find a newspaper or magazine report of a survey and find out how the sample subjects were chosen. Clearly distinguish between the population and the sample for this study.

1-14 (a) Find some uses of statistics (sample, average, percentile, etc.) in texts that you use in other courses. Classify them as either descriptive or inferential.

(b) Find some uses of statistics in current magazines and newspapers. Classify them as descriptive or inferential.

1-15 Classify each of the following as either statistical inference or descriptive statistics:

(a) Dan Drather predicts the results of a statewide election after looking at the votes in 15 of the 97 districts.

(b) Dr. Bea Kareful, an ecologist, says that the flesh of fish in Lake Beyond Woe contains an average of 400 units of mercury.

(c) At Webelo Normal High School last year, the average SAT score was 528.

(d) The safety councils of Pessam and Mystic counties predict 600 automobile accidents for the next July 4th weekend.

(e) Last year 72% of the workers in Scrooge and Marley's accounting firm missed at least 1 day of work.

1-16 Write a brief report giving your reaction to one of the following. Cite the source of your information. (Photocopy the data, if possible.) Focus on the question of representative samples.

(a) Find some reference to the 1936 survey by *Literary Digest* which predicted that Alf Landon would easily win the United States presidential election (For example, see Moore, *The Superpollsters,* 1992).

(b) Find some reference to the polls on the June 18, 1970, British election. (Check newspapers of that week.)

(c) Find some references which discuss the randomness of the December 1, 1969, draft lottery. [Check newspapers of that week, or see the book *A Sampler on Sampling* by Bill Williams (New York: Wiley, 1978)].

(d) Find some reference to the "Hite Report," a study of female sexuality published in 1976 by Shere Hite (e.g., Moore, *The Superpollsters*).

(e) One study on the drug ribavirin for HIV treatment was set up so that patients with mild cases got the drug and those with advanced cases got the placebo. (See the *New York Times,* June 3, 1987, or *F-D-C Reports,* June 1, 1987.)

1-17 Here are two references to the history of probability and statistics. You may enjoy reading more about them.

(a) The following remarks were written in Latin about 2000 years ago:

> Nothing is so unpredictable as a throw of the dice, and yet every man who plays often will at some time or other make a Venus cast: now and then indeed he will make it twice and even thrice in succession. Are we going to be so feeble-minded then as to aver that such a thing happened by the personal intervention of Venus rather than by pure luck?
>
> —from *De Divinatione* by Cicero, as quoted on page 25 of *Games, Gods, and Gambling* by Florence Nightingale David.

Marcus Tullius Cicero (106–43 B.C.) was a Roman statesman and writer. In his culture many people explained improbable events by saying that one of the gods or goddesses must have stepped in to cause it. In the common game of dice (which were actually heel bones of a domestic animal), the player tossed 4 dice; if none of them showed the same face, that was called the "Venus cast." The Venus cast was a rare outcome. Judging from the quotation above, what was Cicero's explanation for someone's getting 3 Venus casts in a row? To learn more about how we know that there were board games that used these gambling bones as far back as 3500 B.C., see the first chapter of David's *Games, Gods, and Gambling.*

(b) For some insight into the early history of data collection and interpretation, read an encyclopedia article or some other reference on the *Doomsday Book,* the report of a general census in England in the years 1085 to 1086, or the *Bills of Mortality,* which list the causes of death in England in the sixteenth and seventeenth centuries.

### *F. N. David*

Statistician Florence Nightingale (F. N.) David (1909–1993) was the author of more than 100 papers and 9 books, including *Games, Gods, and Gambling.* This latter work was allowed into the United States only after customs officials removed the dustcover with its sketch of a scantily clad Tyche, the Greek goddess of chance.

Born in England, David was a fellow of the American Statistical Association. She worked with many famous statisticians, such as Jerzy Neyman and Maurice G. Kendall, and was the last research assistant to Karl Pearson—famous for Pearson's "coefficient of correlation." She earned her doctorate at University College in London, and taught both there and at the University of California's Berkeley and Riverside campuses.

1-18 We repeat the party data from the example given in this chapter:

Suppose 7 people organize a party as follows:

| | |
|---|---|
| John bought chips | \$ 7.49 |
| Jean bought dip | 12.37 |
| Tony bought bread | 13.71 |
| Toni bought cold cuts | 35.09 |
| Sam bought paper goods | 9.11 |
| Sammi will buy cake for about | \$10.00 |

If Pat spent nothing in advance, what is the average cost per person for the party if the cake is estimated to cost between \$9.75 and \$10.25?

1-19 Round off as indicated.

(a) 16.43 (to tenths)

(b) 50,631 (to hundreds)

(c) 40,538 (to tens)

(d) 18.062 (to tenths)

(e) 40,100 (to thousands)

(f) 19.8963 (to hundredths)

(g) 1060.4 (to units)

(h) 1060.4 (to tens)

(i) 1060.4 (to hundreds)

(j) 1060.4 (to thousands)

1-20 (a) Calculate $z = \dfrac{90.3 - 46.12}{20.3}$ to two decimal places.

(b) Calculate $z = \dfrac{1031 - 982.8}{2.41}$ to two decimal places.

(c) Calculate $X_f = 10.4 + 1.96(83.12)$ correct to the nearest fathom, where $X_f$ represents a distance measured in fathoms.

(d) Calculate $X_w = 0.136 - 2.58(0.0617)$ correct to the nearest hundredth of a watt, where $X_w$ represents wattage.

1-21 (a) Multiply $0.18422 \times 1.9$, and round to the nearest tenth.

(b) Round to 0.1842, multiply by 1.9, and round to the nearest tenth.

(c) Comment on your answers to parts (a) and (b).

1-22 Find examples of numbers which have been rounded off in a newspaper, magazine, or textbook.

1-23 We know that $(123{,}456{,}789)^2 + 5 - (123{,}456{,}789)^2 = 5$. Evaluate $(123{,}456{,}789)^2 + 5 - (123{,}456{,}789)^2$ on your calculator. Did you obtain 5? If not, can you explain what happened? (What good is an expensive machine that gives wrong answers?)

1-24 Describe a situation where a random sample of one person would be useful.

1-25 What is the point of the NoCav example in this chapter? What questions would you want to ask to help you decide if the advertising was effective? What are some other possible explanations for the increase in users of NoCav?

1-26 Of the people who use heroin, 95% started out by using marijuana regularly. Therefore, using marijuana regularly leads to using heroin. Comment.

1-27 Of the people who use marijuana, 98% first drank milk on a regular basis. Therefore, drinking milk on a regular basis leads to using marijuana. Comment.

## CLASS SURVEY

Learning the basic ideas of statistics is more interesting when you can work with information that you collected yourself, and the simplest such data to obtain are your class's own personal data. The following brief survey will provide enough raw data from your own class to let you consider some interesting questions. Perhaps you will want to add one or two questions of your own. Save this data! At the end of most chapters we will include a question or two using this data gathered from your own class.

Write X if you don't know some answer.

1. Your sex
2. Your age today
3. Your height in inches
4. Your father's height in inches
5. Your mother's height in inches
6. Fifth (or middle) digit of your 9-digit Social Security number
7. Last digit of your Social Security number
8. Your hair color
9. Your eye color
10. Have you ever broken a bone in your body?
11. Are you left-handed, right-handed, or ambidextrous?
12. Do you smoke cigarettes regularly?
13. Add one or more questions of interest to you here.

Here is a convenient way to collect data so that each student gets a full copy. First, have each student record his or her own responses on a sheet of paper. Second, pass around a master sheet onto which students copy their responses. You should end up with one piece of paper which looks like the one below. Copies of this piece of paper can be given to each student.

**Survey results**

**Results of a Class Survey**

| No. | 1 Sex | 2 Age | 3 Ht | 4 Dad's ht | 5 Mom's ht | 6 5th | 7 Last | 8 Hair | 9 Eyes | 10 Bone | 11 Hand | 12 Smoke |
|---|---|---|---|---|---|---|---|---|---|---|---|---|
| 1 | F | 19 | 66 | 69 | 67 | 8 | 9 | Brown | Blue | N | Left | N |
| 2 | F | 17 | 67 | 68 | 66 | 4 | 2 | Brown | Gray | N | Right | N |
| 3 | M | 27 | 65 | 65 | 65 | 0 | 2 | Brown | Blue | N | Right | N |
| 4 | F | 20 | 69 | 64 | 67 | 4 | 2 | Brown | Brown | N | Right | N |
| 5 | F | 22 | 68 | 70 | 70 | 6 | 6 | Brown | Black | Y | Right | N |

**Results of a Class Survey (continued)**

| No. | 1 Sex | 2 Age | 3 Ht | 4 Dad's ht | 5 Mom's ht | 6 5th | 7 Last | 8 Hair | 9 Eyes | 10 Bone | 11 Hand | 12 Smoke |
|---|---|---|---|---|---|---|---|---|---|---|---|---|
| 6 | M | 19 | 64 | 71 | 66 | 8 | 7 | Black | Blue | N | Right | Y |
| 7 | M | 33 | 66 | 67 | 67 | 8 | 8 | Black | Brown | N | Right | N |
| 8 | M | 21 | 67 | 66 | 71 | 2 | 7 | Brown | Blue | N | Right | N |
| 9 | F | 45 | 63 | 67 | 63 | 2 | 3 | Red | Blue | N | Right | Y |
| 10 | F | 20 | 65 | 70 | 68 | 4 | 5 | Brown | Blue | N | Right | N |
| 11 | F | 20 | 68 | 66 | 69 | 2 | 5 | Blonde | Blue | N | Left | N |
| 12 | F | 18 | 66 | 67 | 68 | 2 | 1 | Brown | Blue | N | Right | N |
| 13 | M | 22 | 70 | 66 | 69 | 8 | 8 | Brown | Blue | Y | Right | N |
| 14 | F | 19 | 65 | 67 | 68 | 6 | 6 | Brown | Black | N | Right | N |
| 15 | F | 17 | 66 | 71 | 69 | 6 | 5 | Red | Blue | N | Right | N |
| 16 | F | 19 | 64 | 66 | 66 | 0 | 2 | Black | Black | N | Right | N |
| 17 | F | 19 | 62 | 67 | 68 | 4 | 6 | Black | Blue | N | Right | N |
| 18 | F | 22 | 67 | X | 72 | 4 | 5 | Black | Blue | Y | Ambi. | N |
| 19 | M | 20 | 68 | 69 | 67 | 2 | 6 | Brown | Brown | N | Right | N |
| 20 | M | 28 | 61 | 68 | 65 | 4 | 9 | Brown | Blue | N | Right | N |
| 21 | F | 18 | 67 | 67 | 69 | 0 | 8 | Brown | Brown | Y | Right | N |
| 22 | M | 17 | 63 | 70 | 66 | 4 | 1 | Brown | Blue | N | Right | N |
| 23 | M | 19 | 68 | 66 | 67 | 2 | 2 | Brown | Brown | N | Right | N |
| 24 | F | 31 | 65 | 69 | 66 | 2 | 4 | Blonde | Gray | N | Right | N |
| 25 | F | 21 | 66 | 67 | 68 | 6 | 2 | Brown | Black | N | Right | N |
| 26 | M | 23 | 65 | 68 | 70 | 2 | 1 | Black | Brown | N | Right | N |
| 27 | M | 27 | 66 | 69 | 62 | 2 | 3 | Blonde | Brown | N | Right | N |
| 28 | F | 19 | 65 | 68 | 64 | 4 | 6 | Brown | Gray | N | Right | N |
| 29 | F | 18 | 67 | 69 | X | 4 | 2 | Brown | Brown | N | Left | N |
| 30 | M | 35 | 68 | 66 | 71 | 4 | 8 | Blonde | Gray | N | Right | N |
| 31 | M | 19 | 69 | 65 | 65 | 2 | 0 | Black | Black | N | Right | Y |

**Results of a Class Survey (concluded)**

| No. | 1 Sex | 2 Age | 3 Ht | 4 Dad's ht | 5 Mom's ht | 6 5th | 7 Last | 8 Hair | 9 Eyes | 10 Bone | 11 Hand | 12 Smoke |
|---|---|---|---|---|---|---|---|---|---|---|---|---|
| 32 | M | 21 | 66 | 68 | 69 | 2 | 9 | Blonde | Blue | N | Right | Y |
| 33 | F | 29 | 65 | 69 | 70 | 0 | 1 | Black | Blue | N | Right | N |
| 34 | F | 28 | 67 | 67 | 67 | 4 | 2 | Red | Brown | N | Right | N |
| 35 | M | 22 | 68 | 68 | 66 | 2 | 7 | Brown | Brown | N | Right | N |
| 36 | F | 30 | 67 | 71 | 68 | 2 | 1 | Blonde | Gray | N | Ambi. | N |
| 37 | F | 18 | 66 | 69 | 69 | 8 | 4 | Brown | Blue | N | Ambi | N |
| 38 | F | 20 | 64 | 64 | 68 | 2 | 3 | Blonde | Black | N | Right | N |
| 39 | F | 18 | 68 | 68 | 67 | 8 | 1 | Brown | Brown | N | Right | N |
| 40 | M | 19 | 67 | 69 | 64 | 0 | 6 | Brown | Blue | N | Right | N |
| 41 | F | 18 | 63 | 66 | 69 | 8 | 8 | Brown | Black | N | Right | N |
| 42 | M | 18 | 65 | 68 | 66 | 2 | 9 | Brown | Blue | N | Right | Y |
| 43 | M | 20 | 64 | 67 | 63 | 8 | 3 | Brown | Black | N | Right | N |
| 44 | F | 21 | 66 | X | X | 0 | 8 | Brown | Blue | N | Right | N |
| 45 | F | 20 | 68 | 63 | 68 | 0 | 2 | Blonde | Green | N | Left | Y |
| 46 | F | 69 | 63 | 66 | 65 | 3 | 6 | Blonde | Gray | N | Right | N |
| 47 | M | 19 | 67 | X | 69 | 8 | 8 | Brown | Brown | N | Right | N |
| 48 | M | 22 | 70 | 65 | 68 | 6 | 4 | Black | Brown | N | Left | N |
| 49 | M | 20 | 68 | 68 | 67 | 6 | 5 | Brown | Blue | N | Right | N |
| 50 | F | 18 | 69 | 69 | 66 | 0 | 0 | Blonde | Black | N | Right | N |
| 51 | F | 25 | 67 | 67 | 71 | 4 | 0 | Brown | Blue | N | Left | N |
| 52 | F | 23 | 68 | 68 | 70 | 8 | 2 | Brown | Blue | N | Right | N |
| 53 | M | 17 | 67 | 71 | 69 | 2 | 0 | Red | Brown | N | Right | N |
| 54 | M | 17 | 65 | 72 | 71 | 8 | 1 | Blonde | Blue | N | Right | N |
| 55 | M | 22 | 68 | 66 | 70 | 0 | 1 | Brown | Brown | N | Left | N |
| 56 | M | 19 | 67 | 67 | 68 | 8 | 5 | Brown | Blue | N | Right | N |
| 57 | M | 21 | 68 | 70 | 71 | 0 | 4 | Blonde | Blue | N | Right | N |

## CLASS SURVEY QUESTIONS

Using your own class data (or the survey results above), answer the following questions.

1. Find the number of males, the number of females, and the total number of people in the survey.
2. Find the percentage of males and the percentage of females in the survey. What is the sum of these two percentages? Why?
3. How many of the females smoke cigarettes regularly? What percentage of the females is that? Do you think there is any relationship between sex and cigarette smoking? Explain.
4. Look over all the data. Report one observation that seems interesting to you. Present your finding in a clear fashion.
5. Comment upon the validity of the data. Do people lie about their age? Their height? Their smoking? etc. Do people know their height, or do they guess? Do they know the heights of their parents?

## FIELD PROJECTS

*In many chapters Field Projects are included. They are usually given in two parts: (1) the design of the project and (2) the execution of the project. These foster better understanding of the material that is covered.*

Suppose you had a random sample of the full-time day students at this school. If the sample were representative by age and by sex, then we would be confident that the average age of your sample would be close to the average age of the entire population and that the proportion of males and females in your sample would be near the proportion of males and females in the population. Your assignment, if you decide to accept it, is to devise a method of obtaining a random sample of 150 full-time students so that the age and sex will probably be representative.

1. Outline this method in a clear, detailed, and specific paragraph. Include:
   (a) The population being sampled
   (b) The sample size
   (c) Where you will take your poll
   (d) How you will select the respondents
   (e) The *exact* questions you will ask
   (f) Comments on some of the strengths and some of the weaknesses of your method
2. After your method has been approved by your instructor, gather these data. Include in your report:
   (a) Part 1 above.
   (b) The data.
   (c) The average age of your sample, correct to two decimal places.
   (d) The number of males and the number of females in your sample.

(e) The percentage of males and the percentage of the females in your sample.

(f) The oldest and the youngest ages you encountered.

(g) Comment on anything that occurred that was not expected.

(h) Do you think that the average age of all the students is close to the average you computed for your sample?

(i) Do you think that the proportion of males on campus is close to the proportion in your sample?

***Note:*** Save your data and results for use in projects in future chapters.

## CLASS EXPERIMENTS

1-1 Have someone in the class secretly mix in a large bag any amount of dried yellow split peas with any amount of dried green split peas. Without counting or even seeing *all* the peas, discuss any methods that could be used to estimate what fraction of the peas in the bag are green. Test your methods. Were they successful? In this experiment, what is the population? What is the sample?

1-2 (a) Get a large supply of some item that can be easily marked and easily mixed, such as toothpicks or poker chips. (Note, paper does not always mix easily.) Have each member of the class put *some* of the items into a bag. Now, to estimate how many items are in the bag, draw a sample of size $n_1$ of some of them and mark them with a marking pen. Return the marked items to the bag, mix well, and draw a second sample of size $n_2$. Count how many are marked, and denote this by $M$. The total number $T$ of items (in this case, toothpicks) in the population can now be estimated by

$$T \approx n_2 \div \frac{M}{n_1}$$

In the tuna fish example in the text (page 13), we had

$$T \approx 1000 \div \frac{50}{200} = 1000 \div 0.25 = 4000$$

(b) If you repeat this experiment using a marker of a different color, will you obtain the same value of $T$? Comment.

(c) Describe a specific population whose size is unknown. Detail exactly how you could use this two-sampling method to estimate its size.

# 2 Common Statistical Measures

Which Average?
In 1993 the average salary of major league baseball players was both $1,160,000 and also $490,000! The mean was over $1 million, but the median was less than one-half million dollars.

**CONTENTS**



**GOALS**

In this chapter you will discover how to calculate and interpret:

- Measures of central tendency, or averages
- Measures of variability, or spread
- Measures of individual scores
- The box-and-whisker display
- Rates used in vital statistics.

# 2.1 Measures of Central Tendency

Tommy Tufluque just got his first D. He complained to the head of the mathematics department that Professor Noays grades too low. The grades on the first test were as follows:

100 100 100 63 62 60 12 12 6 2 0

Tommy indicated that the class average was 47, which he felt was rather low. Professor Noays stated that, nevertheless, there were more 100s than any other grade. The department head said that the middle grade was 60, which was not unusual.

**average**

**measure of central tendency**

Each of these three people was looking for one number to represent the general trend of these test grades. Such a number is called an **average**, or a **measure of central tendency**.

> *Definition:* An **average**, or a **measure of central tendency**, is a number used to represent the typical value of a distribution of numbers.

**mean**

**mode**

**median**

Mathematicians use many kinds of averages. Mr. Tufluque used the **mean**, or arithmetic average, which is obtained by adding the grades and dividing by the number of grades. Professor Noays used the **mode**, which is the most frequent value. The department head used the **median**, which is the middle number when the group of numbers is written in numerical order.

The mean is the "center of balance."

**arithmetic mean**

> *Definition:* The **(arithmetic) mean** of a distribution of values is the quotient of the sum of the values divided by the number of values.
>
> $$\text{Mean} = \frac{\text{sum of values}}{\text{number of values}}$$

> *Definition:* The **mode** of a distribution is a value which appears most often.

> *Definition:* The **median** of a distribution is the middle value when the data are arranged in order by size.

### *Weighted Averages*

Roger Cotes (1682–1716) was one of the first statisticians to recognize that not all data had the same importance. He discussed the weighted mean in his work published posthumously in 1722. Cotes worked mainly in the fields of astronomy and navigation.

These are three commonly used averages. Which of them is the best? That depends on the particular situation. Consider these nine numbers: 21, 21, 21, 21, 23, 24, 24, 25, and 45. If they represent style numbers of dresses sold today in the Chic Dress Boutique, you can see that the style number 21 was the most popular. It is the mode. (Why is pie with ice cream called pie à la mode?) The mode would be important in reordering stock. If the nine numbers represent the grades from a psychology quiz, then perhaps you would want the mean, 25, for use in certain statistical testing. If they represent the annual salary, in thousands of dollars, for the employees of Smith's Emporium, then you might take the median, $23,000, as the average salary. Note that the mean salary of $25,000 is larger than seven of the nine salaries.

Each average has certain properties. Depending on the context, these properties may or may not be useful. For example, the median is less affected by extremely large or extremely small values, while the mean is greatly affected by extreme scores. In this book we will have more occasion to use the mean, because it lends itself to much statistical testing.

### *The Median*

The median is intuitively that value which splits a distribution into two equal-sized portions—the upper half and the lower half.

The mean and the median always exist, but there may be no mode—such as for the distribution 5, 7, 1, and 9. (Alternatively, but of less interest, every value could be considered a mode.) However, the mode can be found for non-numerical data such as dress colors.

In the previous illustration we found the median of a distribution with an *odd* number of raw scores. Finding the number in the middle was no problem. For example, the median of 3, 7, 5, 6, 8 is ____? We hope that you did not say 5, since the definition of "median" indicates the middle term *after the numbers are arranged according to size.* Thus, the median of 3, 5, 6, 7, 8 is 6. If the distribution includes an *even* number of raw scores, then there are two scores in the middle, and the median is halfway between them. *Example:* The median of 3, 3, 5, 6, 8, 13 is found by adding the two middle numbers, 5 and 6, and dividing their sum by 2. Thus, the median is 5.5. Note that half the scores are less than 5.5 and the other half are greater.

## 2.2 Symbols and Formulas

We will use $n$ to indicate the *number* of numbers or raw scores in a given sample. For the sample data 3, 1, 8, and 9, $n = 4$.

For the *mean* of a sample distribution, we use the letter $m$. For the mean of a population, we use the Greek letter for $m$, which is $\mu$ (read: mu).

$\mu$

Mu looks like a script $u$ with a long tail in front. Practice writing it a few times.

Recall that, in many studies, samples are used to represent larger populations. For instance, we may want to determine the average income for families in a certain city, and for practical reasons we may decide to estimate this average income based on data collected from a random sample of families from that city. In such a study it is important to distinguish between the mean of the population (which we don't know exactly) and the mean of the sample (which we do know exactly). We use the words "parameter" and "statistic" to make this distinction. Measures that take into account every member of the population are called **parameters;** $\mu$ is a parameter. Measures based on sample data—that is, measures which take into account only *some* members of the population—are called **statistics;** $m$ is a statistic. A basic skill for a statistician is to know what statistic will serve as a good estimate for any particular parameter. For instance, you will not be surprised to learn that the mean of a random sample provides a good estimate for the population mean. Thus, if we know that $m = 30$, we often conclude that $\mu$ is probably close to 30.

**parameters**

**statistics**

### *S and P*

The initial letters of the words are an easy way to recall that Statistics come from Samples while Parameters come from Populations.

*Definition:* A **parameter** is a measure calculated from ALL the data in a population.

*Definition:* A **statistic** is a measure calculated from only SOME of the data in a population.

## Variables

In most applications data are collected and represented in tabular form. For instance, the table below shows the data collected from a sample of five children in a nursery school.

| ID | Name | Age, Years | Height, Inches |
|---|---|---|---|
| 1 | Nora | 4 | 36 |
| 2 | Maya | 4 | 34 |
| 3 | Denise | 3 | 37 |
| 4 | Paulo | 4 | 36 |
| 5 | Becky | 3 | 38 |

Statisticians call each column a "variable" for the study. Thus, the four variables in this table are ID, Name, Age, and Height. Often a single letter is used to name a variable, such as $A$ for age or $H$ for height. In many illustrations it is common to use $X$ and $Y$ as "generic" variables, and we will often use them in this book. Someone who wants to know what kind of information you collected in a study may ask you, "What are the variables in your study?"

When we say, "Consider the variable $X$: 1, 5, 3, 2," this means we are picturing a column labeled $X$ which consists of these four values, as shown below.

| $X$ |
|---|
| 1 |
| 5 |
| 3 |
| 2 |

(At times, we use $X$ to name just one of the numbers in the column, such as $X = 3$. Some texts use $X_i$ for this purpose. It will be clear from the context which way $X$ is being used.)

Σ

We will use Σ, the Greek letter for capital S, to stand for the command "Sum." This Greek letter is uppercase sigma; it looks like a sideways M. Practice writing several at this time.

**It's All Greek to Me**

Σ means "sum."

Using this notation for variable $X$ above, $\Sigma X$ represents the sum of the $X$'s, so $\Sigma X = 1 + 5 + 3 + 2 = 11$.

**subscript**

To represent the mean of a column of sample data called $X$, we use the letter $m$ with a **subscript** $X$. Thus, $m_X = 11/4 = 2.75$.

Whenever we wish to differentiate between two variables in the same example—for instance, we may use $X$ as the label for height and $Y$ as the label for weight—we can use subscripts to distinguish between them. For sample means, we would get

$$m_X = \frac{\Sigma X}{n_X} \quad \text{and} \quad m_Y = \frac{\Sigma Y}{n_Y}$$

In similar fashion, we can attach subscripts to other symbols. In problems where there is only one variable, the subscript is not needed. You could then just write $m = 2.75$. Another popular symbol for the mean of sample variable $X$ is $\bar{X}$ (read: $X$ bar).

A formula for the mean of sample variable $X$ is

$$m = \frac{\Sigma X}{n} \quad \text{or} \quad \bar{X} = \frac{\Sigma X}{n}$$

In the above example, we have

$$m = \frac{\Sigma X}{n} = \frac{1 + 5 + 3 + 2}{4} = \frac{11}{4} = 2.75$$

### The Square of a Sum vs. the Sum of Some Squares

If in a column labeled $X$ we have the numbers 1, 5, 3, and 2, then $X^2$ will indicate a column of numbers obtained by squaring each number in column $X$. The column $X^2$ will consist of 1, 25, 9, and 4, and $\Sigma(X^2)$ or $\Sigma X^2 = 39$. However, the symbol $(\Sigma X)^2$ represents $(11)^2$, or 121. Notice how $(\Sigma X)^2$ is different from $\Sigma X^2$.

**Helpful Hint**

Many calculators have special keys for easily computing statistics. Check your calculator to see if it has special keys for computing sums and means automatically.

| $X$ | $X^2$ |
|---|---|
| 1 | 1 |
| 5 | 25 |
| 3 | 9 |
| 2 | 4 |
| $\Sigma X = 11$<br>$(\Sigma X)^2 = 121$ | $\Sigma X^2 = 39$ |

In the same way, $(X - 1)$ will be the heading for the column obtained by subtracting 1 from each number in column $X$. The column will consist of 0, 4, 2, and 1. Therefore, $\Sigma(X - 1) = 7$.

| $X$ | $X - 1$ |
|---|---|
| 1 | 0 |
| 5 | 4 |
| 3 | 2 |
| 2 | 1 |
| $\Sigma X = 11$ | $\Sigma(X - 1) = 7$ |

Note that $\Sigma(X - 1) = 7$, while $\Sigma X - 1 = 11 - 1 = 10$.

**Greek vs. English symbols**

In many cases statisticians use Greek letters for parameters and English letters for statistics. For instance, $\mu$ is used for the population mean, and $m$ is used for the sample mean. It is important for you to learn how to write a few Greek letters and to know their names.

### The Average Person

Adolfe J. Quetelet (1796–1874), a Belgian astronomer, made the first sustained attempt to apply statistics to the social sciences. Two of his contributions were the invention of the concept of the "average person"—one whose actions and ideas would correspond to the average results obtained for society—and the notion that distributions of naturally occurring data often follow a normal curve.

Using subscripts, we would write $\mu_X$ and $\mu_Y$ to represent the means of populations $X$ and $Y$, respectively.

Suppose, for example, that the president of the local Planned Parenthood Association has 4 boys, ages 18, 11, 15, and 9, and 3 girls, ages 18, 2, and 10. If we let $X$ represent the distribution of the boys' ages, and $Y$ the distribution of the girls' ages, we have

| $X$ | $Y$ |
|---|---|
| 18 | 18 |
| 11 | 2 |
| 15 | 10 |
| 9 | |
| $\Sigma X = 53$ | $\Sigma Y = 30$ |
| $n_X = 4$ | $n_Y = 3$ |

$$\mu_X = \frac{\Sigma X}{n_X} = \frac{53}{4} = 13.25$$

$$\mu_Y = \frac{\Sigma Y}{n_Y} = \frac{30}{3} = 10.00$$

## Exercises

2-1 Salaries in a mathematics department were as follows: 4 people at $25,000, 6 at $26,000, 2 at $41,000, and 1 at $48,000. Compute the mean, median, and mode salaries. Which seems most meaningful for this distribution?

2-2 A family had kept track of the age at death of its members over several generations. The ages are 72, 68, 0, 67, 45, 7, 70, 68, 72, 66, and 70. Compute the mean, median, and mode ages, and decide which you think is most representative for this population.

2-3 Find the mean, median, and mode of the following grade point averages: 2.9, 3.1, 3.4, and 3.8. Give your answers correct to tenths.

2-4 Illustrate by an example using a distribution of five numbers the sentence from this chapter which states, ". . . the median is less affected by extremely large or extremely small values, while the mean is greatly affected by extreme scores."

2-5 Write three different distributions where each distribution contains five numbers and has a mean equal to 70. How many such distributions of numbers is it possible to find?

2-6 Two students trying the previous exercise took for the first four numbers in one distribution: 0, 1, 2, 3. Can they still complete the distribution so that it will have a mean of 70? Could they have started off with *any* four numbers?

2-7 Find the mean of the following arithmetic test grades: 70, 75, 80, 81, 82, 83, 85, 85, 86, 86, 86, 89, 90, 90, 91, 92, 94, and 95.

2-8 Here are the daily high temperatures (°F) for one January week in Vermont. Find the mean and median temperatures.

| | |
|---|---|
| Tuesday | 0° |
| Wednesday | 5° |
| Thursday | −10° |
| Friday | −8° |
| Saturday | 1° |
| Sunday | −6° |
| Monday | −1° |

2-9 A recent newspaper article revealed that over recent years the mean hourly wage of U.S. workers had risen but the median hourly wage of U.S. workers had fallen. Explain how that could happen.

Here are the current wages of some workers.

| ID | Hourly Wage, Dollars |
|---|---|
| 1 | 6 |
| 2 | 8 |
| 3 | 9 |
| 4 | 12 |
| 5 | 15 |

Invent new wages for this group for next year in such a way that the mean increases and the median decreases.

2-10 **Grouped data.** A sample distribution $X$ consists of 25 threes and 6 fives, as shown in this table.

| $X$ | Frequency |
|---|---|
| 3 | 25 |
| 5 | 6 |

(a) What is $n$? (b) What is $\Sigma X$? (c) What is $\Sigma X^2$?

(d) If the 31 values are arranged in numerical order, what is the value of the median?

2-11 A sample distribution $X$ consists of 1000 fives, 500 sixes, and 500 eights. Find the total number of values sampled and the mean, median, and mode.

2-12 A distribution of grades on a statistics test was as follows:

| Grade | Number of Students |
|---|---|
| 100 | 4 |
| 97 | 7 |
| 93 | 12 |
| 90 | 11 |
| 89 | 1 |
| 85 | 5 |
| 83 | 10 |
| 82 | 6 |
| 80 | 3 |
| 77 | 1 |
| 71 | 2 |
| 69 | 1 |
| 3 | 1 |

Find the total number of grades and the mean, median, and mode grades.

2-13 The U.S. government reports each year the median age at first marriage of brides and grooms. Here are the results compiled from several editions of *Statistical Abstract of the United States:*

| Year | Median Age of Bride | Median Age of Groom |
|---|---|---|
| 1970 | 20.6 | 22.5 |
| 1971 | 20.5 | 22.5 |
| 1972 | 20.5 | 22.4 |
| 1973 | 20.6 | 22.5 |
| 1974 | 20.6 | 22.5 |
| 1975 | 20.8 | 22.7 |
| 1976 | 21.0 | 22.9 |
| 1977 | 21.1 | 23.0 |
| 1978 | 21.4 | 23.2 |
| 1979 | 21.6 | 23.4 |
| 1980 | 21.8 | 23.6 |
| 1981 | 22.0 | 23.9 |
| 1982 | 22.3 | 24.1 |
| 1983 | 22.5 | 24.4 |
| 1984 | 22.8 | 24.6 |
| 1985 | 23.0 | 24.8 |
| 1986 | 23.3 | 25.1 |
| 1987 | 23.6 | 25.3 |
| 1988 | 23.7 | 25.5 |

(a) What are the three variables represented in this table?
(b) Do you see any trends?

(c) The results would be different if they reported mean age instead of median. Do you think that they would have been higher or lower? Why?

2-14 The mode was defined to be "the most frequent number appearing in a distribution." Some distributions may have more than one mode. Find the modes in each of the following:
(a) 5, 3, 7, 3, 8, 5, 7, 1, 3, 6, 2, 8, 7
(b) 2, 0, 3, 3, 0, 5, 2, 6, 0, 7, −1, 2, 3
(c) 1, 5, 9, 7

2-15 A billboard ad for *Smithsonian* magazine in 1981 stated that the average income of its subscribers was $42,500. Which average would you guess they used? How do you suppose they got this information?

2-16 **Some properties of the mean.** Knowing some properties of the mean can be useful in calculation. *If the same value is added to every number in a distribution, then the mean goes up by that value.* For example, if your professor adds 2 points to each test score, the average score is increased by 2 points also.

*Example*

| $X$ | $Y = X + 6$ |
|---|---|
| 2 | 8 |
| 3 | 9 |
| 4 | 10 |
| 7 | 13 |
| $\bar{X} = 4$ | $\bar{Y} = \bar{X} + 6 = 10$ |

The same property holds for subtraction, multiplication, and division.

*Use the above idea to answer the following questions:*

(a) A mistake was made in grading the arithmetic papers in Exercise 2-7, and each student is entitled to 5 more points than given. Correct the value for the mean found in the answer to Exercise 2-7.
(b) If the mean of 15, 18, 23, 24 equals 20, what is the mean of 1.5, 1.8, 2.3, 2.4? Why?

2-17 Jorge gets 75, 78, and 82 on his first three exams in underwater basket weaving, then decides he really needs an A for the course. His professor says that an average for all four tests must be at least 89.5 to earn an A. Can Jorge do it?

2-18 Contrast these two statements. Informally they are used interchangeably, but which one is correct mathematically? Explain the distinction between them.
(a) The height of the average American woman is 5 feet, 4 inches.
(b) The average height of American women is 5 feet, 4 inches.

2-19 The average family has 2.3 children. "This cannot be; who ever heard of $\frac{3}{10}$ of a child?" Comment on this often-heard criticism of averages.

2-20 Let $X$ be the label for a variable. What reason is there for denoting its mean by the symbol $m$? By the symbol $\bar{X}$?

2-21 Given $Y$: 2, 3, 4, 5, 6, 7, 8, calculate each of the following quantities:

(a) $\Sigma Y$ (b) $\Sigma Y^2$ (c) $(\Sigma Y)^2$

(d) $\dfrac{\Sigma Y}{n}$ (e) $\bar{Y}$ (f) $\Sigma(Y - 2)$

(g) $\Sigma(Y - \bar{Y})$ (h) $\Sigma Y - \bar{Y}$ (i) $\dfrac{\Sigma(Y - \bar{Y})^2}{n - 1}$

(j) $\dfrac{\Sigma Y^2 - \dfrac{(\Sigma Y)^2}{n}}{n - 1}$

(k) Which is bigger, $(\Sigma Y)^2$ or $\Sigma Y^2$?

2-22 Repeat the previous exercise with $Y$: 3, 4, 5, 6, 7, 8, 9.

2-23 Given a sample distribution of six values, $X$: 4, 4, 3, 0, −1, 2, calculate each of the following quantities:

(a) $\Sigma X$ (b) $m$ (c) $\bar{X}$

(d) $\Sigma(X - m)$ (e) $(\Sigma X)^2$ (f) $\Sigma X^2$

(g) $\dfrac{\Sigma X^2 - \dfrac{(\Sigma X)^2}{n}}{n - 1}$

2-24 Repeat the previous exercise with $X$: 3, 3, 2, −1, −2, 1.

2-25 Given $X$: 2, 7, 6, 11, 0, calculate each of these quantities:

(a) $m_X$

(b) $\dfrac{\Sigma X^2 - \dfrac{(\Sigma X)^2}{n}}{n - 1}$

(c) $\dfrac{\Sigma(X - m)^2}{n - 1}$

2-26 Repeat the previous exercise with $X$: 4, 14, 12, 22, 0.

2-27 Make up a set of three numbers for which $(\Sigma X)^2 = \Sigma X^2$.

## 2.3 Measures of Variability

Suppose you are planning to go on a Caribbean cruise during your spring vacation. A travel agent tells you that there are three possible cruises, and that the mean ages of the passengers on each ship are 20, 29, and 41. Which cruise will you select? Which one will your mother select?

Did you pick the ship with a mean age of 20? 29? 41? After you have made your choice, look at Table 2-1 for a detailed listing of the passengers' ages.

Having seen the passenger lists, would you like to change your selection? You can see that the mean does not accurately reflect the distribution of ages in the Niña and the Pinta. We need a measure that will indicate whether the numbers in a distribution are close together or far apart. Such a measure is called a **measure of variability, scatter**, or **spread**. Ideally, such a measure should be large if the raw scores are spread out and small if they are close together.

**measure of variability**
**scatter**
**spread**

## 2.3-1 The Range

One simple measure of variability is the **range**.

**range**

> *Definition:* The **range** is the difference between the largest number in the distribution and the smallest number.

Thus, in the Niña the range is $62 - 2 = 60$ years, in the Pinta the range is $52 - 19 = 33$ years, and in the Santa Maria the range is $43 - 39 = 4$ years.

**Table 2-1 Ages of the Passengers on Each Cruise Ship**

| | Niña | Pinta | Santa Maria |
|---|---|---|---|
| | 2 | 19 | 39 |
| | 3 | 20 | 39 |
| | 4 | 20 | 39 |
| | 5 | 21 | 39 |
| | 8 | 22 | 41 |
| | 9 | 23 | 41 |
| | 9 | 23 | 41 |
| | 10 | 24 | 41 |
| | 40 | 25 | 43 |
| | 44 | 49 | 43 |
| | 44 | 50 | 43 |
| | 62 | 52 | 43 |
| Sum = | 240 | 348 | 492 |
| Mean = | $\frac{240}{12} = 20$ | $\frac{348}{12} = 29$ | $\frac{492}{12} = 41$ |

**midrange**

A simple measure of central tendency is the **midrange**. This is the arithmetic mean of the smallest value and the largest value in a distribution. Like the range, the midrange is extremely easy to compute. In the above example the midranges are:

| | |
|---|---|
| Niña: | $(2 + 62) \div 2 = 32$ |
| Pinta: | $(19 + 52) \div 2 = 35.5$ |
| Santa Maria: | $(39 + 43) \div 2 = 41$ |

## 2.3-2 The Standard Deviation

For most "everyday" problems, which you might have to solve on an intuitive basis, the range serves very well as the measure of variability. For more-technical problems, especially the type we will be doing later in this book, there is another measure of variability that is useful. It is called the **standard deviation** and is especially useful when used in conjunction with the mean.

To illustrate the concept of standard deviation, let us consider two small populations. Two students, David and Laura, have the same mean grade in algebra, 70. David's grades were 67, 70, 72, and 71. Laura's grades were 100, 62, and 48. Laura's grades are spread out, while David's are close together. One way this can be seen clearly is to look at the **deviations from the mean** (Table 2-2).

**deviations from the mean**

> *Definition:* The **deviation from the mean** of a score is found by subtracting the mean from the score.

A positive sign (+) on a deviation tells us that the grade is *above* the mean, while a negative sign (−) indicates that it is *below* the mean. A zero deviation indicates that a particular grade equals the mean. Note that the devi-

**Table 2-2**

| David's Grades $X$ | Mean $\mu$ | Deviations from the Mean $X - \mu$ | Laura's Grades $Y$ | Mean $\mu$ | Deviations from the Mean $Y - \mu$ |
|---|---|---|---|---|---|
| 67 | 70 | −3 | 100 | 70 | +30 |
| 70 | 70 | 0 | 62 | 70 | −8 |
| 72 | 70 | +2 | 48 | 70 | −22 |
| 71 | 70 | +1 | | | |
| | | $\Sigma(X - \mu) = 0$ | | | $\Sigma(Y - \mu) = 0$ |

ations in David's grades are closer to zero than those in Laura's grades. This is because David's grades are less scattered than Laura's.

If you compute the mean of the deviations for David's grades, you will find that it is zero: $(-3 + 0 + 2 + 1) \div 4 = 0 \div 4 = 0$. If you compute the mean of the deviations of Laura's grades, you will see that it also is zero. In fact, it is true for any distribution that the sum of the deviations is zero, and therefore the mean of the deviations is zero.

**Who Said It First?**

Karl Pearson (1857–1936) invented the term "standard deviation" around 1893.

Recall that we are trying to introduce a new measure of variability, called the standard deviation. We want the "standard" deviation to be representative of the deviations. You might think to use the mean of the deviations as the representative deviation, but we have just mentioned that it is always zero, no matter how much variability there is in the distribution. So statisticians have developed a procedure that is not immediately obvious. First, the squares of the deviations are obtained. None of these can be negative. Then their mean is found. For example, the deviations in David's grades were $-3$, 0, 2, and 1, and so the squared deviations are 9, 0, 4, and 1. The mean of the squared deviations is $(9 + 0 + 4 + 1) \div 4 = 14 \div 4 = 3.5$. This number, the mean of the squared deviations, can be used as a measure of variability. It is called the **variance** of David's grades. Chapter 15 deals with several situations where the variance is the easiest measure of variability to use.

variance

*Definition:* The **variance** is the mean of the squared deviations.

You will notice, though, that in this process we have squared all the original deviations, so that the variance is representative of the squares of the deviations. To get a number representative of the original deviations, we take the square root of the variance. This final number is called the **standard deviation**.

standard deviation

*Definition:* The **standard deviation** is the square root of the variance.

In the case of David's grades, since the variance was 3.5, the standard deviation was $\sqrt{3.5} \approx 1.9$. Since the deviations from the mean were between 0 and 3 units, 1.9 is reasonable as a representative of the deviations. The standard deviation is always in the same units as the original raw scores. In this case, the standard deviation is 1.9 grade points. If we have an example where the raw scores are in feet, then the standard deviation is also in feet.

Table 2-3

| Laura's Grades $X$ | Deviation from the Mean of 70 $(X-\mu)$ | Squared Deviation $(X-\mu)^2$ |
|---|---|---|
| 100 | 30 | 900 |
| 62 | −8 | 64 |
| 48 | −22 | 484 |
| $\Sigma X = 210$ | $\Sigma(X-\mu) = 0$ | $\Sigma(X-\mu)^2 = 1448$ |

Let us now calculate the standard deviation of Laura's grades (Table 2-3).

$$\text{Variance} = \text{mean of squared deviations}$$

$$= \frac{\Sigma(X-\mu)^2}{n} = \frac{1448}{3} \approx 482.67$$

$$\text{Standard deviation} = \text{square root of variance}$$

$$= \sqrt{\frac{\Sigma(X-\mu)^2}{n}} = \sqrt{482.67} \approx 21.97$$

Again, notice that the original deviations were 8, 22, and 30. So 21.97 is reasonable as a representative deviation.

## 2.3-3 Formulas

We use the Greek letter for lowercase *s*, which is $\sigma$ (read: sigma), to represent the *s*tandard deviation of a population. Therefore, $\sigma^2$ represents the variance.

Thus the formula for the variance of a population is

$$\sigma^2 = \frac{\Sigma(X-\mu)^2}{n}$$

and the formula for the standard deviation of a population is

$$\sigma = \sqrt{\frac{\Sigma(X-\mu)^2}{n}}$$

**Note** $\sigma$ and $\sigma^2$ are parameters, because they are based on the entire population.

**Symbols**

| Greek for Populations (Parameters) | English for Sample (Statistics) |
|---|---|
| $\mu$ mu | $m$ (mean) |
| $\sigma$ sigma | $s$ (standard deviation) |
| $\rho$ rho | $r$ (correlation coefficient) |

However, *when you wish to estimate $\sigma$ using only sample data, the formula must be adjusted.* This estimate is denoted by *s*.

The formula for $s$ is

$$s = \sqrt{\frac{\Sigma(X - m)^2}{n - 1}}$$

Notice that we use $m$ instead of $\mu$ and $n - 1$ instead of $n$. Dividing by $n-1$ instead of $n$ gives a larger value. This compensates for the fact that estimates of $\sigma$ which use $n$ in the formula tend to be too small because there is usually less variability in a sample distribution than there is in the whole population. Since $s$ is an estimate of $\sigma$ based on sample data, $s$ is a statistic, while $\sigma$ is a parameter. Statisticians refer to $s$ as the **sample standard deviation**, meaning that it is an estimate of the standard deviation of the population derived from *sample* data. The corresponding estimate for the variance is denoted by $s^2$.

### *The Biernaymé-Chebyshev Rule*

In 1867 the Russian mathematician Pafnuti Lvovich Chebyshev (1821–1894) rediscovered an inequality previously discovered by I. J. Biernaymé in 1853. They proved that in *any set* of data *at least* $(1 - 1/k^2) \cdot 100\%$ of the data fall within $k$ standard deviations of the mean.

For example, if $k = 2$, then $(1 - (1/2)^2) = 0.75$, which means that for *any* set of numbers you make up 75% (or more) of the data lie within 2 standard deviations of the mean. Consequently, less than 25% of the numbers in *any* set of data will lie more than 2 standard deviations from the mean.

The formula for $s^2$ is

$$s^2 = \frac{\Sigma(X - m)^2}{n - 1}$$

## A Computational Formula for $s$

It turns out that, in practice, the above formula for $s$ is often awkward to use because of the many subtractions. A second, more convenient formula is*

$$s = \sqrt{\frac{\Sigma X^2 - \frac{(\Sigma X)^2}{n}}{n - 1}}$$

This formula produces the same answer as the previous one. Let us illustrate this by computing $s$ both ways for the sample data 1, 8, 0, 3, and 9.

Using the first formula,

$$s = \sqrt{\frac{\Sigma(X - m)^2}{n - 1}}$$

*Some people prefer the following alternate version:

$$s = \sqrt{\frac{n\Sigma X^2 - (\Sigma X)^2}{n(n - 1)}}$$

we need $m$, $X - m$, and $\Sigma(X - m)^2$:

$$m = \frac{\Sigma X}{n} = \frac{21}{5} = 4.2$$

| $X$ | $X - 4.2$ | $(X - 4.2)^2$ |
|---|---|---|
| 1 | −3.2 | 10.24 |
| 8 | 3.8 | 14.44 |
| 0 | −4.2 | 17.64 |
| 3 | −1.2 | 1.44 |
| 9 | 4.8 | 23.04 |
| $\Sigma X = 21$ | | $\Sigma(X - 4.2)^2 = 66.80$ |

$$s = \sqrt{\frac{66.80}{4}} = \sqrt{16.70} \approx 4.1$$

Using the second formula,

$$s = \sqrt{\frac{\Sigma X^2 - \frac{(\Sigma X)^2}{n}}{n - 1}}$$

we need $\Sigma X$ and $\Sigma X^2$:

| $X$ | $X^2$ |
|---|---|
| 1 | 1 |
| 8 | 64 |
| 0 | 0 |
| 3 | 9 |
| 9 | 81 |
| $\Sigma X = 21$ | $\Sigma X^2 = 155$ |

**Helpful Hint**

Check your calculator to see if you can compute the value of $s$ automatically.

$$s = \sqrt{\frac{155 - \frac{21^2}{5}}{5 - 1}} = \sqrt{\frac{155 - \frac{441}{5}}{4}}$$

$$= \sqrt{\frac{155 - 88.20}{4}} = \sqrt{\frac{66.80}{4}}$$

$$= \sqrt{16.70} \approx 4.1$$

The corresponding computational formula for the variance is

$$s^2 = \frac{\Sigma X^2 - \frac{(\Sigma X)^2}{n}}{n - 1}$$

### Exercises

2-28 Last year the mean high temperature in two cities, Squaresville and Octothorpe, was 70° F. Explain how the two cities might be very different from one another in daily temperatures.

2-29 From each of these tables of life expectancies for people in northern African and northern European countries, find the median and the range. (Data are from *World Population Data Sheet* (1993) published by the Population Reference Bureau.) Note that there is no data for Western Sahara. It is not uncommon in studies to have *missing data.* When you give the results, you should also just note that Western Sahara was not included.

How would you compare these two sets of data with respect to centrality and variability?

| *Northern Africa* | | *Northern Europe* | |
|---|---|---|---|
| Country | Life Expectancy | Country | Life Expectancy |
| Algeria | 66 | Denmark | 75 |
| Egypt | 60 | Finland | 75 |
| Libya | 63 | Iceland | 78 |
| Morocco | 65 | Ireland | 75 |
| Sudan | 53 | Norway | 77 |
| Tunisia | 68 | Sweden | 78 |
| Western Sahara | — | United Kingdom | 76 |
| | | Estonia | 70 |
| | | Latvia | 70 |
| | | Lithuania | 71 |

2-30 A student observed cars entering the college parking lot, and reported these results:

20 cars had only a driver
10 cars had a driver and 1 passenger
5 cars had a driver and 2 passengers
1 car had a driver and 3 passengers

From this data, compute the total number of cars and the median and the range for the number of passengers per car.

2-31 An IQ test was given to two groups of fourth-grade students. One group was from a grade school in Nassau County; the other group was from a school for psychotic children. Both groups had a mean score of 100, but the standard deviation for the normal students was 14, and the standard deviation for the psychotic children was 23. Interpret the statistics.

2-32 A mine foreman is comparing the products of two manufacturers of blasting materials. Both companies' materials explode a mean time of 40 minutes after they are set off, but the standard deviation for KaBoom brand materials is 4 minutes, and the standard deviation for It's-a-Blast brand materials is 14 minutes. Which brand should he choose?

2-33 Two brands of yardsticks both have mean lengths of 36 inches. The standard deviation for the Euclidean brand is 0.002 inches, while the standard deviation for the Pythagorean brand is 0.001 inches. Which brand is better?

2-34 After a class test in statistics, you go to your teacher's office to find out your grade. The teacher is not there, but on the desk you see the following results:

Class I: mean = 80, standard deviation = 5

Class II: mean = 80, standard deviation = 10

(a) Which class would you rather be in? Why?
(b) If Mary is in class I and she is within 3 standard deviations of the mean, find the interval containing her grade.
(c) If Bill is in class II and he is more than 2 standard deviations below the mean, find the interval containing his grade.

2-35 Two experiments were done on different brands of artificial hearts. The first compared Cor brand with Valentine brand. The second compared brand X with brand Y. All of the artificial hearts wear out eventually and must be replaced. We show here the results of tests on 10 hearts of each brand. Explain why the results are conclusive in the first experiment, but not in the second.

| | *Experiment I* | | *Experiment II* | |
|---|---|---|---|---|
| | Cor | Valentine | Brand X | Brand Y |
| Average lifetime, days | 1000 | 1400 | 1000 | 1400 |
| Standard deviation | 22 | 22 | 300 | 320 |

2-36 An honesty test was given to a random sample of prisoners at Singsong State Prison. The scores went from $-31$ to $+9$, giving a range equal to 40. The average score was $-17$, and the standard deviation for the sample distribution was 7. Warden Warren Wardon estimates that the average score for the entire prison population is also about $-17$, but that the range is larger than 40 and the standard deviation is *larger than* 7. Explain *why* Warden Wardon is probably correct.

2-37 For the following two distributions, find $s$ first by the definition

$$s = \sqrt{\frac{\Sigma(X - m)^2}{n - 1}}$$

and then by the computational formula

$$s = \sqrt{\frac{\Sigma X^2 - \dfrac{(\Sigma X)^2}{n}}{n - 1}}$$

State which method is easier for each of the following distributions:
(a) 4, 6, and 8
(b) 3, 8, 9, 17, and 20

2-38 List all the raw scores in part (a) of the previous exercise which are 1 or more standard deviations below the mean. List all the raw scores in part (b) of the previous exercise which are more than 1 standard deviation away from the mean.

2-39 Consider the following grades on an arithmetic test: 70, 75, 80, 81, 82, 83, 85, 85, 86, 86, 89, 90, 90, 91, 92, 94, and 95. In this distribution, $\bar{X} = 85.53$ and $s = 6.7$.
(a) *How many* scores are within 1 standard deviation of the mean?
(b) What *percentage* of the scores are within 1 standard deviation of the mean?
(c) *How many* scores are within 2 standard deviations of the mean?
(d) What *percentage* of the scores are within 2 standard deviations of the mean?

2-40 The number of UFOs reported to the National UFO Reporting Center each month over the preceding 12 months was 30, 3, 27, 0, 15, 40, 37, 1, 1, 20, 10, and 5. Compute the mean and $s$ for this distribution.
(a) *How many* sightings are within 1 standard deviation of the mean?
(b) What *percentage* of the sightings are within 1 standard deviation of the mean?
(c) *How many* sightings are within 2 standard deviations of the mean?
(d) What *percentage* of the sightings are within 2 standard deviations of the mean?

*In Exercises 2-41 to 2-45, use the following ideas.*

*Some Properties of* s:

- If you add the same value to all the numbers in a distribution, the standard deviation is not changed at all. The same holds for subtraction.

  *Example* $X$: 1, 2, 3 gives $s = 1$. $X + 6$ gives 7, 8, 9, which still has $s = 1$.

- If you multiply all the numbers in a distribution by the same positive value, the standard deviation is multiplied by that value. The same holds for division.

  *Example* $X$: 1, 2, 3 gives $s = 1$. $6X$ gives 6, 12, 18, which has $s = 6$.

2-41 (a) Using either formula discussed in this chapter, find $m$, $s^2$, and $s$ for the distribution 2, 5, 6, and 7.
(b) In part (a) we found $m$, $s^2$, and $s$ for the distribution 2, 5, 6, and 7. Use the ideas above to find $m$, $s^2$, and $s$ for the distribution 12, 15, 16, and 17.

2-42 (a) Using either method, find $m$, $s^2$, and $s$ for the distribution 3, 4, 7, 9, and 11.
(b) Find $m$, $s^2$, and $s$ for the distribution 30, 40, 70, 90, and 110.

(c) For the distribution in part (a), list all the values that are more than 3 standard deviations away from the mean.

2-43 As the result of a strike, the International Brotherhood of Pogo Stick Workers obtained a contract in which the mean salary was 11,000 marks per year, with a standard deviation of 800 marks.

(a) This year each worker will receive an increase of 500 marks per year. What will the new mean and standard deviation be?

(b) Next year each worker will receive a 10% increase. Using the result from part (a), what will the new mean and standard deviation be?

2-44 (a) Without actually calculating, compare the mean and the standard deviation of the following ages:

*X:* 5, 2, 7, and 3

*Y:* 65, 62, 67, and 63

(b) Without actually calculating, compare the mean and the standard deviation of the following temperatures:

Kodiak, Alaska: 10, 8, 0, and −1

Coldfoot, Alaska: −10, −8, 0, and 1

2-45 To change an awkward Celsius temperature to a more convenient Fahrenheit temperature you multiply by 1.8 and then add 32°. For example, 20°C becomes 20(1.8) + 32 = 36.0 + 32 = a comfortable 68°F.

(a) If a distribution of temperatures has a mean of 25°C and a standard deviation of 10°C, express the mean and standard deviation in the more familiar Fahrenheit scale.

(b) Here's a challenge for you: If the range of temperatures is 18°C, find the range in degrees Fahrenheit.

2-46 Here are the bills and the tips that a waiter collected in one night.

The data to be treated as a sample

| Bill | Tip |
|---|---|
| $12.46 | $1.75 |
| 20.16 | 3.00 |
| 6.25 | .75 |
| 22.00 | 3.25 |
| 15.88 | 2.50 |
| 38.50 | 5.50 |

(a) What is the average tip?

(b) What are the variance and the standard deviation of these tips?

(c) Determine what percentage of the bill each tip represents. What is the mean percentage?

(d) What are the variance and the standard deviation of these percentages?

2-47 Gilberto follows the prices of spaghetti at the local supermarkets. Each Monday he goes to 20 markets and checks the price of a 1-pound box of Biskety brand spaghetti. Here are his results for the past 10 weeks.

| Week | Average Price, Cents | Standard Deviation, Cents |
|---|---|---|
| 1 | 79.3 | 4.01 |
| 2 | 80.0 | 3.79 |
| 3 | 78.7 | 4.00 |
| 4 | 84.1 | 4.61 |
| 5 | 83.0 | 3.98 |
| 6 | 80.0 | 3.90 |
| 7 | 79.7 | 3.82 |
| 8 | 83.0 | 3.00 |
| 9 | 83.2 | 4.45 |
| 10 | 83.4 | 4.30 |

(a) What is the mean price for the 10-week period?
(b) What is the mean standard deviation for the 10-week period?

2-48 Make a list of five numbers whose variance is:
(a) Larger than its standard deviation
(b) Smaller than its standard deviation
(c) Zero

2-49 Infant mortality refers to babies that are born alive but then die before they reach age 1. The infant mortality *rate* tells how many such deaths occur for every 1000 babies born. From each of these tables of infant mortality rates for eastern Europe and east Asia, find the mean and the standard deviation. [Data are from *World Population Data Sheet* (1993), published by the Population Reference Bureau.] To be treated as a sample

How would you compare these two sets of data with respect to centrality and variability?

| *Eastern Europe* | | *East Asia* | |
|---|---|---|---|
| Country | Infant Mortality Rate | Country | Infant Mortality Rate |
| Bulgaria | 16.7 | China | 53.0 |
| Czech Republic | 10.4 | Hong Kong | 6.7 |
| Slovakia | 13.2 | Japan | 4.4 |
| Hungary | 15.1 | North Korea | 30.0 |
| Poland | 14.4 | South Korea | 15.0 |
| Romania | 22.7 | Macao | 8.0 |
| | | Mongolia | 50.0 |
| | | Taiwan | 5.1 |

## 2.4 Some Measures of an Individual in a Population

The results of the last exam in this class were good: the class mean was 83, the median was 87, the range was 24, and the standard deviation was 5.

All of the above information is nice, but what you want to know is: How did *you* do on the examination?

### 2.4-1 *z*-Scores

We have already mentioned one measure of an individual score, the *raw score*. Another very important measure of an individual's rank in a population is the ***z*-score**, sometimes called the **standard score.**

*z*-score

> *Definition:* The ***z*-score** measures how many standard deviations a raw score is above or below the mean.

Basically, the *z*-score uses the standard deviation as a unit of measurement.

In the example above, $\mu = 83$ and $\sigma = 5$.

Therefore, the *z*-score corresponding to 88 (written $z_{88}$) is $+1.00$, since 88 is 1 standard deviation (5 units) above the mean; $z_{73} = -2.00$, since 73 is 2 standard deviations (10 units) below the mean.

We illustrate this below:

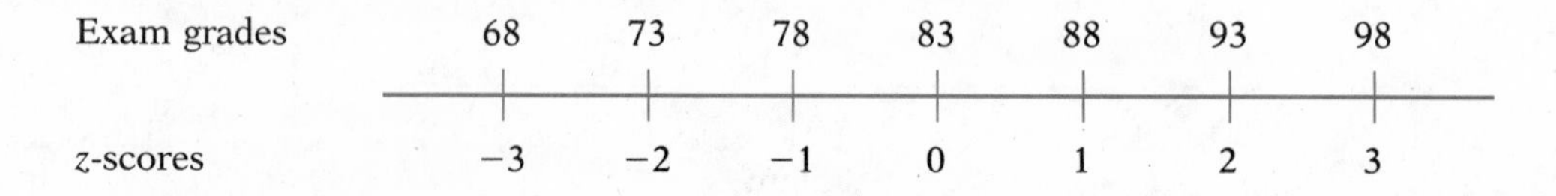


It is common in statistics to place an individual by how many standard deviations away from the mean he or she is.

---

### ❑ EXAMPLE 2.1 Raw Scores and *z*-Scores

A distribution of weights has a mean of 17 pounds and a standard deviation of 3 pounds. Fill in the missing *z*-scores and raw scores in the following table:

| Raw Score, Pounds | $z$-Score |
|---|---|
| 20 | |
| 11 | |
| 26 | |
| 5 | |
| 21.5 | |
| 17 | |
| | 2 |
| | −1 |
| | −3 |
| | −4 |
| | 1.8 |

**SOLUTION**

| Raw Score, Pounds | $z$-Score |
|---|---|
| 20 | 1 |
| 11 | −2 |
| 26 | 3 |
| 5 | −4 |
| 21.5 | 1.5 |
| 17 | 0 |
| 23 | 2 |
| 14 | −1 |
| 8 | −3 |
| 29 | 4 |
| 22.4 | 1.8 |

Can you see why each answer is correct? ❑

---

A formula for finding the $z$-score corresponding to a particular raw score $X$ in a population is

$$z_{\text{corresponding to a raw score}} = \frac{\text{raw score} - \text{mean}}{\text{standard deviation}}$$

Represented in symbols, the formula for the $z$-score is

$$z_X = \frac{X - \mu}{\sigma}$$

When there is no confusion, we often drop the subscript and simply write

$$z = \frac{X - \mu}{\sigma}$$

### ❑ EXAMPLE 2.2 Finding a z-Score

Micki is in the class described in the opening sentence of Section 2.4. Her exam grade was 69. Find the $z$-score for her grade.

SOLUTION

Recall that the class mean was 83, and the standard deviation was 5. We simply substitute into the formula $X = 69$, $\mu = 83$, and $\sigma = 5$:

$$z_{69} = \frac{69 - 83}{5} = \frac{-14}{5} = -2.80$$

This tells us that Micki's score was below average by almost 3 standard deviations. You will learn that in many situations this is very far below average. ❑

In the tables that we will be using, most of the $z$-scores will be rounded off to two decimal places, hence we will usually round off $z$-scores to the same precision.

If in the above class Beverly had a $z$-score of +2.00, what was her score on the test? Since the standard deviation was 5 and the mean 83, two standard deviations above the mean would be $83 + 2.00(5) = 93$.

A formula for the raw score corresponding to a particular $z$-score is

$$\text{Raw score}_{\text{corresponding to a } z\text{-score}} = \text{mean} + (z\text{-score})\ (\text{standard deviation})$$

In symbols, this is

$$X_z = \mu + z\sigma$$

or, simply,

$$X = \mu + z\sigma$$

If David, a classmate of Micki and Beverly, had a $z$-score of $-0.60$, then his raw score is

$$\begin{aligned} X &= 83 + (-0.60)(5) \\ &= 83 - 3.00 \\ &= 80 \end{aligned}$$

Notice that we multiply before we add.

If we are using $s$ as an estimate of $\sigma$, these two formulas become

$$z_X = \frac{X - \mu}{s} \quad \text{and} \quad X_z = \mu + zs$$

**Which Goes First?**

"Please Excuse My Dear Aunt Sally" is a mnemonic used by some to recall that the order of arithmetic operations is:
- parenthesis first
- then exponentiation
- followed by multiplication and division
- with addition and subtraction last

❑ **EXAMPLE 2.3 *z*-Scores and Raw Scores**

Suppose someone claimed that the mean depth of successful oil-well drillings is 2500 feet. If you estimate the standard deviation of these depths from some sample data and you get $s = 100$ feet, find the $z$-score corresponding to a depth of 2250 feet. Round your answer to hundredths.

SOLUTION

$$z_{2250} = \frac{X - \mu}{s} = \frac{2250 - 2500}{100} = \frac{-250}{100} = -2.50$$

In contrast, to find the depth that corresponds to a $z$-score of 1.65, we have

$$X = \mu + zs = 2500 + 1.65(100) = 2665 \text{ feet}$$ ❑

### Relative Standing via z-Score

**relative standings**

Let us consider two test grades you might receive. Suppose you scored an 85 in English and a 65 in physics. Clearly you would rather receive a raw score of 85 than a raw score of 65, but a second consideration is how well you did relative to the other students in the class. Suppose we tell you that the mean in English was 70 and the mean in physics was 50. Thus, in both classes you scored 15 points above the mean. Does this mean that, relatively speaking, you did the same in both classes? The answer is no; the number of points above or below the mean is insufficient information to give you a rating relative to your position in the class, as you can see from the class scores given in Table 2-4.

**Table 2-4**

| English | Physics |
|---|---|
| 100 | 65 (your score) |
| 99 | 57 |
| 98 | 55 |
| 85 (your score) | 53 |
| 73 | 50 |
| 67 | 49 |
| 60 (Alice's score) | 47 |
| 53 | 44 |
| 45 | 44 (Alice's score) |
| 20 | 36 |
| Mean = 70 | Mean = 50 |
| $\sigma$ = 25.1 | $\sigma$ = 7.7 |

We can see from this table that although you scored 15 points above the mean in both classes, when compared to the other students you did better in physics than in English, since your physics grade was the top in the class, while three students scored higher than you did on the English test.

In order to see how well you did compared to the rest of the class, you can use $z$-scores. In English class, your $z$-score is

$$z_{85} = \frac{85 - 70}{25.1} = \frac{15}{25.1} = 0.60$$

In physics class, your $z$-score is

$$z_{65} = \frac{65 - 50}{7.7} = \frac{15}{7.7} = 1.95$$

Thus you see that even though you scored 15 points above the class average in both subjects, your physics score was *relatively* better.

---

❑ **EXAMPLE 2.4 Relative Standing**

Using the information in Table 2-4, if Alice scored 60 in English and 44 in physics, which was a better grade relative to the other students in each class?

**SOLUTION**

In English, Alice's $z$-score is

$$z = \frac{60 - 70}{25.1} = \frac{-10}{25.1} = -0.40$$

In physics, Alice's $z$-score is

$$z = \frac{44 - 50}{7.7} = \frac{-6}{7.7} = -0.78$$

Since $-0.40$ is greater than $-0.78$, the 60 in English had the higher $z$-score than the 44 in physics, and Alice did better, relatively speaking, in the English class. ❑

---

## 2.4-2 Percentile Rank and Percentiles

**percentile rank**

Another measure of an individual's position in a population is the **percentile rank**. This is primarily used for large populations. In a small population, one would simply use ordinary rankings, such as "5th out of 9." Essentially, the percentile rank tells you the percentage of a distribution below a given value. Consider a person whose raw score has a percentile rank of 75. Approximately 75%, or three-fourths, of the population scored below this individual. In a large distribution, if a raw score of 72 has a percentile rank of 80, then approximately 80% of the scores are below 72 and 20% are above.

> *Definition:* A **percentile rank** of a raw score gives the percentage of scores less than the raw score.

Now consider the following example. In a distribution of weights of babies, 70% of the babies weighed less than Steven, 10% weighed the same as Steven, and 20% weighed more than Steven. Since 70% weighed less than Steven and 20% weighed more than Steven, his percentile rank should be between 70 and 80. We will use 75, the value that is halfway between 70 and 80. You can find the percentile rank of a raw score directly by finding the percent of scores that are below the given score and adding one-half of the percent of scores that are the same as that raw score. In our example, 70% weighed less than Steven and 10% weighed the same as Steven, so the percentile rank of Steven's weight is $70 + \frac{1}{2}(10) = 75$.

Suppose that, in a senior class of 500, Andrea's score on an aptitude test was 603. She finds that 16 grades are above hers, 3 (including Andrea's) are 603, and 481 are below her grade. Thus, $481/500 = 0.962 = 96.2\%$ scored below Andrea's grade and $3/500 = 0.6\%$ are at Andrea's grade. The percentile rank of Andrea's grade is $96.2 + \frac{1}{2}(0.6) = 96.2 + 0.3 = 96.5$, or 97.

**Note** We usually round off percentile ranks to the nearest whole number.

We can express the procedure for finding the percentile rank with a formula. Let $B$ equal the number of raw scores **b**elow a particular score $X$. Let $E$ be the number of scores **e**qual to $X$, including $X$ itself. Let $n$ be the total number of raw scores.

Then the percentile rank of $X$ is

$$\frac{B + \frac{1}{2}E}{n}(100)$$

In the preceding example, the percentile rank of Andrea's grade is

$$\frac{481 + \frac{1}{2}(3)}{500}(100) = \frac{481 + 1.5}{500}(100) = \frac{482.5}{500}(100) = 97$$

A grade which has a percentile rank of 97 is said to be "at the 97th percentile."

**percentile**

> *Definition:* A **percentile** is a raw score which has a particular percentile rank.

**We Know That . . .**

$P_{\text{percentile rank}} = \text{raw score}$

Some people use the inverse

$PR_{\text{raw score}} = \%$

Thus, if one quarter of the applicants for a job score below 17 on an ice cream-scooping aptitude exam, we have both

$P_{25} = 17$

and

$PR_{17} = 25\%$

We will write $P_{97}$ (read: *P* sub 97) to indicate the raw score that is at the 97th percentile. Thus, in the example that we just did, Andrea's grade would be at the 97th percentile and would be symbolized by $P_{97}$. Note the difference—the *percentile rank* is a percentage, a value from 0 to 100, while a *percentile* can be any raw score. Thus we say both:

1. The *percentile rank* of raw score 603 is 97.
2. The 97th *percentile* is raw score 603. In symbols, $P_{97} = 603$.

That is:

1. The *percentile rank* of a score is a percentage.
2. The *percentile* is the raw score.

## ❑ EXAMPLE 2.5 Percentile Ranks

Refer to the data in Table 2-4. Find the percentile ranks for your grades in English and in physics. Find the percentile ranks for Alice's grades.

**SOLUTION**

The percentile rank for your 85 in English is given by

$$\text{Percentile rank of 85} = \frac{6 + 0.5(1)}{10}(100) = 65$$

and the percentile rank for your 65 in physics is given by

$$\text{Percentile rank of 65} = \frac{9 + 0.5(1)}{10}(100) = 95$$

The percentile rank for Alice's 60 in English is given by

$$\text{Percentile rank of 60} = \frac{3 + 0.5(1)}{10}(100) = 35$$

and the percentile rank for Alice's 44 in physics is given by

$$\text{Percentile rank of 44} = \frac{1 + 0.5(2)}{10}(100) = 20$$

We summarize these results in Table 2-5.

Table 2-5

| | Grade | *z*-Score | Percentile Rank |
|---|---|---|---|
| English, you | 85 | 0.60 | 65 |
| Physics, you | 65 | 1.95 | 95 |
| English, Alice | 60 | −0.40 | 35 |
| Physics, Alice | 44 | −0.78 | 20 |

❑

## 2.4-3 Quartiles

$P_{25}$ is called the "1st quartile," $Q_1$.

$P_{50}$ is called the "median," $Q_2$.

$P_{75}$ is called the "3rd quartile," $Q_3$.

Computer statistics packages and statistics calculators often automatically produce five summary values for a distribution: The lowest value, $Q_1$, $Q_2$, $Q_3$, and the highest value. For example, for one set of data, the following 5 values appeared: 45, 76, 80, 91, and 98. Associated with this 5-number summary is a graph called a **box-and-whisker display**.

**box-and-whisker display**

A box-and-whisker display (also called a box plot) of these 5 summary values follows. It clearly indicates the range and the *middle half* of this distribution. The bottom quarter is represented by the left "whisker," and the top quarter is represented by the right "whisker."

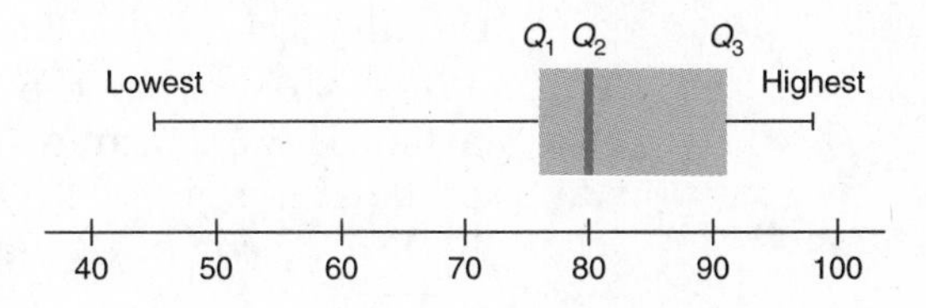


### Exercises

2-50 Complete the following table. (The first row is already done.)

| | Symbol | Pronunciation | Meaning |
|---|---|---|---|
| | $n$ | en | Number of items in a distribution |
| (a) | $m$ | mean | Σx/n of sample |
| (b) | $\bar{X}$ | X-bar | mean |
| (c) | μ | mu | Σx/n of population |
| (d) | + | plus | The command to add |
| (e) | $\sigma$ | sigma | Standard deviation of pop. |
| (f) | $z_{98}$ | z sub 98 | #'s of standard deviation. |
| (g) | PR | percentile rank | A score at the 78th percentile |
| (h) | | | An estimate of the standard deviation |

2-51 A distribution of temperatures has a mean of 98.6° and a standard deviation of 0.5°. Find the $z$-scores corresponding to the following temperatures:
(a) 99.1° (b) 97.6° (c) 98.6° (d) 100° (e) 98°

2-52 A distribution of air pressures has a mean of 32 pounds per square inch and a standard deviation of 1.2 pounds per square inch. Find the air pressures corresponding to the following $z$-scores:
(a) 2.00 (b) 0.00 (c) $-3.00$ (d) 1.35 (e) $-0.06$

2-53 A sensitivity test was given to 1000 people. Here are the $z$-scores for five of them: Adelbert, $-0.02$; Bastion, 1.27; Carmulon, 0.001; Joe, $-2.03$; and Elfremde, 0.48.
(a) Which of these five people scored above the mean for the test?
(b) Which of these five people scored below the mean for the test?
(c) Rank these five people from the highest scorer to the lowest.

2-54 In the previous exercise, suppose that the mean test score for the 1000 people who took the sensitivity test was 10 and the standard deviation was 5.
(a) Find the raw score for each of the five people named.
(b) In general, must negative $z$-scores always correspond to negative raw scores?
(c) In general, must positive $z$-scores always correspond to positive raw scores? (*Hint:* Consider temperatures at the North Pole.)
(d) Another six people scored as follows on the sensitivity test: Lou, 10; Nes, 9; Dierdre, 5; Inez, 15; Pat, 12; and Sue, 11. Find their $z$-scores.

2-55 Astronauts discover that the mean height of a Martian is 3.6 Martian inches with a standard deviation of 0.2 Martian inch. (*Note:* 12 Martian inches equals 1 Martian foot.)
(a) Given the heights of the following Martians, find their $z$-scores: Xgol, 3.8; Zib, 2.6; Mni, 2.6; and Rfd, 4.
(b) How tall is President Mil, whose $z$-score is $-0.5$?
(c) Mr. Zar is 3.9 Martian inches tall. Ms. Zar's height has a $z$-score of 1.6. Who is taller?

2-56 In a given year, the mean length of American-made cars was 171 inches and the standard deviation was 5 inches.
(a) Find the $z$-scores for cars with lengths 169 inches, 171 inches, and 180 inches.
(b) Three models from one manufacturer had $z$-scores of $-1.00$, 0.00, and 0.30. Find the lengths of these models.
(c) All Colonel Motors cars measured within two standard deviations of the mean. (1) James claims his CM car was 185 inches long. Find the $z$-score for 185 inches. Why is James's claim not possible? (2) What is the maximum possible length of James's car? (3) What is the minimum possible length of James's car?

2-57 Below are given student Pentak's scores on some standard exams. Also given are other statistics for the exams.

| Test | Mean | Standard Deviation | Pentak's Score |
|---|---|---|---|
| Math | 47.2 | 10.4 | 83 |
| Verbal | 64.6 | 8.3 | 71 |
| Geography | 74.5 | 11.7 | 72 |

(a) Transform each of Pentak's test scores to a $z$-score.
(b) On which test did Pentak stand relatively highest? Relatively lowest?

2-58 A doctor collects the heights, weights, and blood pressures of a large group of people, called a "control group," and then computes the three means and the three standard deviations. After taking your height, weight, and blood pressure, the doctor computes your $z$-scores with regard to the control group. They are height, $z = 2.10$; weight, $z = -1.30$; and blood pressure, $z = 0.003$. Interpret these results.

2-59 Danny Kazort is a demolition expert. He has been able to knock down 75 apartment houses at an average time of 3.5 weeks per house, with a standard deviation of 4 days. One tricky job took 5 weeks. What $z$-score would that job have? Assume he works 5-days per week.

2-60 Bess scored 87 on her typing exam and 78 on her physics exam, yet the $z$-score for the 87 was 0 and the $z$-score for the 78 was 2. Explain how this can be and what it means. With relation to her fellow students, is she a better typing student or a better physics student?

2-61 A test is scored +1 for each correct response and −1 for each wrong or omitted response. A group of people take the test. Half of them score $+a$ points and the other half score $-a$ points. Find
(a) $\mu$ (b) $\sigma$ (c) $z$-scores for $+a$ and $-a$.

2-62 The table gives the 1992 population (in thousands) of the world's 10 largest cities. Compute the mean and the standard deviation. Then find the $z$-score for Tokyo. The data appear in the *Statistical Abstract of the United States* (1993).

| City | Population (in thousands) |
|---|---|
| Tokyo-Yokohama, Japan | 27,540 |
| Mexico City, Mexico | 21,615 |
| São Paulo, Brazil | 19,373 |
| Seoul, South Korea | 17,334 |
| New York, United States | 14,628 |
| Osaka-Kobe-Kyoto, Japan | 13,919 |
| Bombay, India | 12,450 |
| Calcutta, India | 12,137 |
| Rio de Janeiro, Brazil | 12,009 |
| Buenos Aires, Argentina | 11,743 |

Treat data as a sample

2-63 The table shown gives the poverty rate in several countries in the mid-1980s. The table appears in the *Statistical Abstract of the United States* (1993). (Each entry is the percent of people with income below 40% of adjusted household median disposable income after taxes and transfers.) Find the mean and the standard deviation. Then find the $z$-score for the United States.

| Country | Poverty Rate |
|---|---|
| United States | 13.3 |
| Canada | 7.0 |
| Australia | 6.7 |
| United Kingdom | 5.2 |
| France | 4.5 |
| Sweden | 4.3 |
| Netherlands | 3.4 |
| West Germany | 2.8 |

2-64 In a large distribution of ages at Golden Vista Nursing Home, the percentile rank of age 72 is 50. Are the following true or false?
(a) The median age in this population is about 72.
(b) The mean age in this population is about 72.

2-65 On an important exam, Phil's $z$-score was negative. He claimed that his score had a percentile rank of 60. Can this happen?

2-66 If 50% of the scores in a population are below 70, then which of these is (are) correct?
(a) Percentile rank of 50 = 70 (b) $P_{50} = 70$ (c) Percentile rank of 70 = 50 (d) $P_{70} = 50$

Answers for percentile rank should be ← stated as percents (rounded to nearest whole percent.)
a) 84%
b) 15%

2-67 (a) Leda Swan took the National Safe Driver Test. Of the 120,000 people who took the test, there were 100,000 who scored lower than Leda and 2400 who scored the same as Leda. Find the percentile rank for Leda's score.
(b) Bob Byrde took the National Safe Driver Test. Of 120,000 people who took the test, there were 18,000 who scored lower than Bob and 1000 who scored the same as Bob. Find the percentile rank for Bob's score.

2-68 Here is a table of household incomes. Use it to find the approximate percentile rank of households with each of these incomes: \$10,000, \$25,000, \$50,000. Explain how you can tell that the category \$25,000 to \$34,999 contains the median household income. (In fact, the median income was \$30,126.)

**Money Income of Households—Distribution, by Income Level**

| | Under \$5,000 | \$5,000–9,999 | \$10,000–14,999 | \$15,000–24,999 | \$25,000–34,999 | \$35,000–49,999 | \$50,000–74,999 | \$75,000 & over |
|---|---|---|---|---|---|---|---|---|
| Number of households | 4,800 | 10,100 | 9,400 | 17,400 | 15,200 | 17,300 | 15,400 | 10,400 |

2-69 According to the statistics of the Quick and Easy Data Company, some percentiles for family incomes in Nowso County are as follows:

| Percentile | Family Income |
|---|---|
| 25 | $24,000 |
| 50 | 26,800 |
| 75 | 30,200 |
| 90 | 34,500 |

State approximately the *percentage* of the families that earn:
(a) Less than $24,000
(b) Less than $26,800
(c) Less than $30,200
(d) Less than $34,500
(e) More than $34,500
(f) Between $24,000 and $30,200

2-70 There are about 500,000 families in Nowso County. Using the figures from the previous exercise, state approximately *how many* families earn:
(a) Less than $24,000
(b) Less than $26,800
(c) Less than $30,200
(d) Less than $34,500
(e) More than $34,500
(f) Between $24,000 and $30,200

2-71 A test of susceptibility to photographic stimuli was given to 500 subscribers to *Sportfellow* magazine. A *few* results are tabled below.

| Test score | 68 | 84 | 100 | 116 | 132 | 148 |
|---|---|---|---|---|---|---|
| $z$-score | −2 | −1 | 0 | 1 | 2 | 3 |
| Percentile rank | 2 | 16 | 30 | 50 | 98 | 99 |

By inspection of the table, is it possible to tell:
(a) What is the value of $n$? Why?
(b) What was the mean test score of all 500 who were tested? Why?
(c) What was the median test score of all those who were tested? Why?
(d) What percentage of these subscribers scored below 68? Between 84 and 116?
(e) What is the standard deviation for the 500 test scores?
(f) What test score would be 1.5 standard deviations above the mean?
(g) What test score would be transformed to a $z$-score of $-1.20$?

2-72 Give an example, *where possible,* of a negative value for each of the following: mean, median, mode, standard deviation, variance, range, $z$-score, percentile rank, and 25th percentile.

2-73 In a study of waste disposal in Nosewer County, it was discovered that the mean amount of garbage was 30 pounds per day per family and the median was 35 pounds per day per family. Which one of the following is true?
(a) Exactly half the families produced 30 pounds or more of garbage.
(b) More than half the families produced 30 pounds or more of garbage.
(c) Less than half the families produced 30 pounds or more of garbage.

2-74 The annual contributions to charity of the employees of Pan-Europa Whistles and Samovars Ltd. are given below in pounds sterling.

| Employee | Contribution |
|---|---|
| Señor Francisco Castaño | 500 |
| Señora Ana | 750 |
| Señorita Maria | 100 |
| Tio Luis y | 25 |
| Hermano Aloysius | 1000 |
| Herr Franz von Braun | 3000 |
| Frau Anna | 450 |
| Onkel Ludwig und | 600 |
| Fräulein Maria | 33 |
| Monsieur François de Brun | 100 |
| Madame Anne | 2500 |
| Oncle Louis | 0 |
| Frère Henri et | 444 |
| Soeur Anne-Marie | 100 |
| Signore Francesco di Bruno | 2000 |
| Signora Anita | 900 |
| Zio Luigi e | 100 |
| Signorina Maria | 40 |
| Mijhneer Frans van der Bruin | 986 |
| Mevrouw Hannah en | 3345 |
| Oom Hendrik | 9800 |
| Zuster Hannah-Marysa | 700 |

(a) Find the mean, median, mode, and midrange.
(b) Find the range and the standard deviation.
(c) Find the maximum, and minimum, and the quartiles.

2-75 Here are box-and-whisker displays for the average weekly salaries of 100 high school graduates and 100 college graduates.

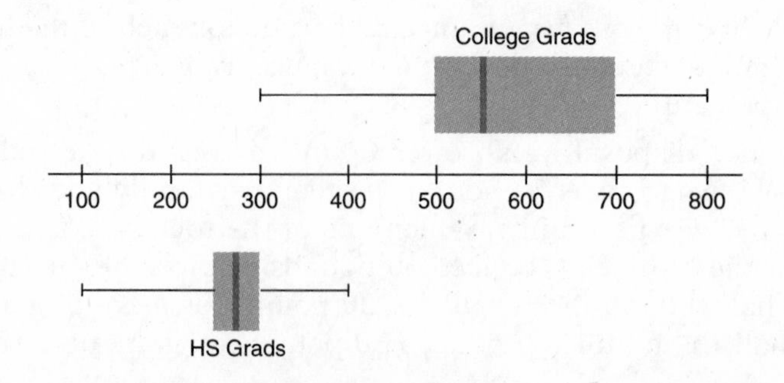

(a) What is the median salary for each group?
(b) What is the range of the salaries in each group?
(c) Do all graduates in the high school group earn less than all graduates in the college group?
(d) What significance is there to the fact that the two rectangles in the college box plot are unequal?
(e) What significance is there to the fact that the center box of the college salaries is 4 times as long as the center box of the high school salaries?

2-76 The age of the mayor of each of the following cities is given. (a) Find the mean, median, and modal ages. (b) Find the quartiles and draw a box-and-whisker display.

| City | Mayor's Age |
|---|---|
| Clinton, Ala. | 41 |
| Clinton, Ark. | 45 |
| Clinton, B.C., Canada | 59 |
| Clinton, Ont., Canada | 49 |
| Clinton, Conn. | 51 |
| Clinton, Ga. | 51 |
| Clinton, Ill. | 47 |
| Clinton, Ind. | 43 |
| Clinton, Iowa | 47 |
| Clinton, Ky. | 24 |
| Clinton, La. | 52 |
| Clinton, Me. | 53 |
| Clinton, Md. | 47 |
| Clinton, Mass. | 61 |
| Clinton, Mich. | 45 |
| Clinton, Minn. | 73 |
| Clinton, Miss. | 49 |
| Clinton, Mo. | 57 |
| Clinton, Mont. | 51 |
| Clinton, Neb. | 52 |
| Clinton, N.J. | 53 |
| Clinton, N.Y. | 47 |
| Clinton, N.C. | 47 |
| Clinton, Ohio | 49 |
| Clinton, Okla. | 57 |
| Clinton, S.C. | 51 |
| Clinton, Tenn. | 53 |
| Clinton, Utah | 57 |
| Clinton, Wis. | 49 |

## 2.5 Rates

In Chapter 1 we said that one major use of statistics is "descriptive"—the summarizing of large amounts of data. So far in this chapter we have discussed several quantities useful for such a description. For example, the *mean* summarizes the location of the "center" of the data; the *variance* summarizes the variability of the data. Another descriptive device often used in statistical reports is called a **rate**.

rate

> *Definition:* A **rate** is a fraction used for measuring relative values.

Rates are especially useful in the area of study called **vital statistics,** which is concerned largely with population problems such as birth (natality), death (mortality), and various social phenomena. No doubt you have heard statements like, "The birthrate is dropping," "The divorce rate is going up every year," and "The death rate due to lung cancer among women is rising rapidly."

A rate is basically a fraction, though it is usually expressed in a convenient decimal form. Let us consider birthrates as our first example. According to U.S. government data, during 1990 there were about 4,158,000 (4.158 million) live births in the United States. Also, during 1990 the population of the United States was about 249,900,000 (249.9 million) people.* We say that the birthrate for the year 1990 was 4.158 million births per 249.9 million population. As a fraction, this is

$$\frac{4,158,000}{249,900,000}$$

To express this as a decimal, we divide top by bottom and find

$$\frac{4,158,000}{249,900,000} = 0.0167$$

This represents the birthrate expressed as births per 1 member of the population. This is usually not a sensible or convenient way to think of the rate. It is common to want rates in terms of each 1000 (or 100,000) members of the population. So we multiply the decimal by 1000 (or 100,000). This gives, for example,

$$0.0167 \times 1000 = 16.7$$

*Technically, this is the Census Bureau's estimate of the population at midyear, July 1, 1990.

We say that the 1990 birthrate was 16.7 births **per thousand** population. This means that for every 1000 people in the United States, about 17 babies were added to the population. We can do the calculation in one step:

$$\begin{array}{l}\text{Annual birthrate} \\ \text{(per 1000 population)}\end{array} = \frac{\text{number of births during year}}{\text{size of population during year}}(1000)$$

$$= \frac{4{,}158{,}000}{249{,}900{,}000}(1000) = 16.7 \text{ per } 1000$$

We use rates when it is natural to refer to a base population. Suppose, for example, we are studying how people in various cultures cope with stress, and we believe it would be useful to study suicide in, say, the United States and Finland. Suppose we find that in 1990 there were about 1370 suicides in Finland and about 30,500 suicides in the United States. Does this tell us that people in the United States are more likely to commit suicide than people in Finland? No, because we have not used the information that there are many more people in the United States to begin with. We incorporate the base populations by using them as the denominators in the suicide-rate fractions:

$$\begin{array}{l}\text{Annual suicide rate} \\ \text{(per 100,000 population)}\end{array} = \frac{\text{number of suicides}}{\text{size of population}}(100{,}000)$$

Now we get

$$\text{Finland, 1990 suicide rate} = \frac{1370}{4{,}986{,}000}(100{,}000)$$

$$= 27.5 \text{ suicides per 100,000 people}$$

$$\text{United States, 1990 suicide rate} = \frac{30{,}500}{249{,}900{,}000}(100{,}000)$$

$$= 12.2 \text{ suicides per 100,000 people}$$

We see that the 1990 Finnish suicide rate was more than double that of the United States. People are more likely to commit suicide in Finland than in the United States.

Often rates are computed in the same location but at different times to see if there are trends in the rate. For example, here are some birthrates for different years in the United States (expressed per 1000 population).

| Year | 1970 | 1972 | 1974 | 1976 | 1978 | 1980 | 1982 | 1984 | 1986 | 1988 | 1990 |
|---|---|---|---|---|---|---|---|---|---|---|---|
| Birthrate | 18.4 | 15.6 | 14.9 | 14.8 | 15.3 | 16.3 | 15.9 | 15.6 | 15.6 | 16.0 | 16.7 |

It appears that the trend was decreasing, "bottomed out" about 1976, and then began to increase. Time trends in rates are often shown in graphs, as illustrated in Figure 2-1.

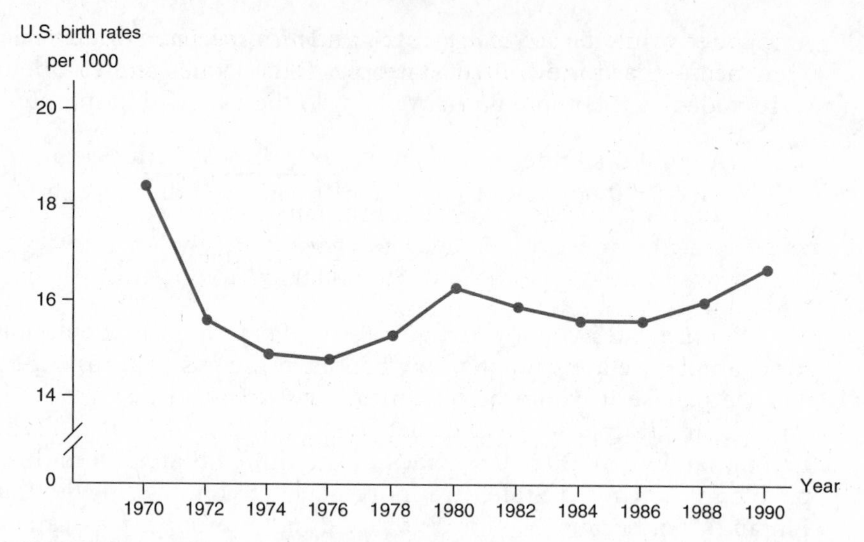


**Figure 2-1.** U.S. birthrates, births per 1000 population.

In summary, to understand a rate fully you must know:

1. The period of time it covers (1 year, 1 month, etc.).
2. What is being counted (births, deaths, etc.). This is the numerator.
3. The base population (U.S. citizens, males, etc.). This is the denominator.
4. The units in which the base population is expressed (e.g., "per 1000 population," "per 100,000 population").

Here are some examples.

1. The 1990 U.S. marriage rate was 9.8 marriages per 1000 population. This means that, in 1990, for every 1000 people there were about 9.8 marriages. So about 20 people out of every 1000 got married. Note that the 1000 includes children as well as people who were already married.
2. For the first quarter of 1991 the U.S. unemployment rate was 6.6%. This means that between January 1 and March 31 about 7 people out of 100 were unemployed. Unemployment rates are supposed to represent the fraction of people who want jobs but don't have them, so the reference population does not include students, retired people, or others not looking for work. For instance, it does not include parents who stay home to raise children, nor does it include people who have given up looking for a job. The population that makes up the denominator for the unemployment rate therefore consists of all those who are either employed or unemployed by these definitions. This population is called the "civilian labor force."
3. The 1991 death rate in the United States for males in the age group 15–24 years was 161 per 100,000. This means that, in 1991, about 161 out of 100,000 males in this age group died. Because age and sex are mentioned, this death rate is called *age-* and *sex-specific.* A further breakdown indicat-

ing race would be called *age-*, *sex-*, and *race-specific*. For example, the death rate in 1991 in the United States for Black males age 15–24 was 277 per 100,000.

### Percent Change

**percent change**

A very common calculation that accompanies values that change over time is called **percent change**. Percent change gives one way to answer a question like "How does the birthrate in 1990 compare to the birthrate in 1980?" Percent change is calculated as follows:

$$\text{Percent change} = \frac{\text{new value} - \text{old value}}{\text{old value}}(100)$$

We illustrate with the birthrate data given in the table above. Find the percent change in the U.S. birthrate from 1980 to 1990.

$$\text{Percent change} = \frac{16.7 - 16.3}{16.3} \times 100 = \frac{0.4}{16.3} \times 100 = 0.0245 \times 100 = 2.45$$

We conclude that the birthrate increased by 2.45% between 1980 and 1990.

Note that if the new value is smaller than the old value, the answer is negative and represents a percent *decrease*.

## STUDY AIDS

### Chapter Summary

In this chapter you have learned:

- How to calculate and interpret measures of central tendency such as the mean, median, and mode
- How to calculate and interpret measures of variability or spread such as the range, variance, and standard deviation
- How to calculate and interpret measures of individual scores such as $z$-scores and raw scores
- How to draw and interpret the box-and-whisker display (or box plot)
- How to calculate and interpret rates

### Vocabulary

You should be able to explain the meaning of each of these terms:

1. Measure of central tendency, or average
2. Mean
3. Mode
4. Median
5. Parameter
6. Statistics
7. Measure of variability
8. Range
9. Midrange
10. Deviations from the mean
11. Variance
12. Standard deviation
13. $z$-score
14. Relative standing
15. Percentile rank
16. Percentile
17. Box-and-whisker display
18. Rate
19. Percent change

### Symbols

You should understand the meaning of each of these symbols:

1. $X$, Y
2. $n, n_X, n_Y$
3. $\Sigma$
4. $\mu, m, \bar{X}$
5. $\sigma, s$
6. $z, z_X$
7. $P_{83}$
8. $B$
9. $E$

### Formulas

You should know when and how to use each of these formulas:

1. $\mu = \dfrac{\Sigma X}{n} \qquad m = \dfrac{\Sigma X}{n} \qquad \bar{X} = \dfrac{\Sigma X}{n}$

2. $s = \sqrt{\dfrac{\Sigma(X - \mu)^2}{n - 1}} \qquad s^2 = \dfrac{\Sigma(X - \mu)^2}{n - 1}$

3. $s = \sqrt{\dfrac{\Sigma X^2 - \dfrac{(\Sigma X)^2}{n}}{n - 1}} \qquad s^2 = \dfrac{\Sigma X^2 - \dfrac{(\Sigma X)^2}{n}}{n - 1}$

4. $z_X = \dfrac{X - \mu}{\sigma} \qquad z_X = \dfrac{X - \mu}{s} \qquad z_X = \dfrac{X - m}{s}$

5. $X = \mu + z_X\sigma \qquad X = \mu + z_Xs \qquad X = m + z_Xs$

6. Percentile rank of $X = \dfrac{B + \frac{1}{2}E}{n}(100)$

## EXERCISES

2-77 For each of these rates give (1) the time period, (2) the base population, and (3) what is being counted.

(a) The 1994 U.S. birthrate

(b) The death rate for people 65 and over in the United States in 1994

(c) The divorce rate in the United States for July 1995

(d) The apple consumption rate in the Garden of Eden in the year 1

2-78 Why are these reports incomplete or uninterpretable?

(a) The U.S. marriage rate is 45 per 1000 population.

(b) The annual motor vehicle death rate is 24.2.

(c) The death rate is 2.4 for disease A and 24.0 for disease B.

2-79 The population of Conception City in June 1992 was 400,000. During June there were 400 births. Find:

(a) The monthly birthrate per 1000 population

(b) The annual estimated birthrate per 1000 population (by multiplying the monthly rate by 12)

2-80 In 1991 the U.S. population was about 250 million. During that year, there were about 2,371,000 marriages, 1,187,000 divorces, 718,090 deaths due to heart disease, 514,310 deaths due to cancer, and 45,200 deaths due to motor vehicle accidents. Express the marriage and divorce rates as rates per 1000 population. Express the death rates as rates per 100,000 population.

2-81 In Amusement Land 3500 people ride the bumper cars each day, and there are about 68,000 collisions each day. Express the daily accident rate per person.

2-82 In 1980 there were 2,390,300 marriages in the United States, giving a rate of 10.6 per 1000 population. In 1985 there were 2,412,600 marriages, giving a rate of 10.1 per 1000 population. How can the rate go down if the number of marriages went up?

2-83 A government report stated that in 1988 the marriage rate in the United States was 54.6 per 1000 in the category "unmarried women 15 years and over," but was 91.0 per 1000 in the category "unmarried women 15–44 years." What do these two rates mean? Why are they so different?

2-84 Coward Hossel stated, "In 1924 Babe Ruth's batting average was three seventy-eight." What kind of an average is a batting average?

The data go up and the rate goes down.

2-85 *Epidemiology* is the study of who in a population gets sick and who does not. Epidemiologists are usually called in when there is a sudden outbreak of a disease. Their job is to help pinpoint the cause of the disease. A typical example would be to try to locate which food was the source of an outbreak of food poisoning. A useful rate in such cases is the *attack rate* for and against various foods:

$$\text{Attack rate "for" a food} = \frac{\text{number ill who ate the food}}{\text{number of people who ate the food}}$$

$$\text{Attack rate "against" a food} = \frac{\text{number ill who did not eat the food}}{\text{number of people who did not eat the food}}$$

The table below shows the information the epidemiologist collected about a picnic in Oswego, N.Y., in 1940, where there was an outbreak of food poisoning. Look over the data and decide which food is the likely source of infection. Calculate the various attack rates to confirm your intuition.

| | *Persons Who Ate Food* | | | *Persons Who Did Not Eat Food* | | |
|---|---|---|---|---|---|---|
| Food | Ill | Not Ill | Total | Ill | Not Ill | Total |
| Baked ham | 29 | 17 | 46 | 17 | 12 | 29 |
| Spinach | 26 | 17 | 43 | 20 | 12 | 32 |
| Mashed potatoes | 23 | 14 | 37 | 23 | 15 | 38 |
| Cabbage salad | 18 | 10 | 28 | 28 | 19 | 47 |
| Jello | 16 | 7 | 23 | 30 | 22 | 52 |
| Rolls | 21 | 16 | 37 | 25 | 13 | 38 |

*(continued on next page)*

| Food | Persons Who Ate Food | | | Persons Who Did Not Eat Food | | |
|---|---|---|---|---|---|---|
| | Ill | Not Ill | Total | Ill | Not Ill | Total |
| Bread | 18 | 9 | 27 | 28 | 20 | 48 |
| Milk | 2 | 2 | 4 | 44 | 27 | 71 |
| Coffee | 19 | 12 | 31 | 27 | 17 | 44 |
| Water | 13 | 11 | 24 | 33 | 18 | 51 |
| Cakes | 27 | 13 | 40 | 19 | 16 | 35 |
| Vanilla ice cream | 43 | 11 | 54 | 3 | 18 | 21 |
| Chocolate ice cream | 25 | 22 | 47 | 21 | 7 | 28 |
| Fruit salad | 4 | 2 | 6 | 42 | 27 | 69 |

*Source:* Handout for Centers for Disease Control summer course in epidemiology, based on article by M. B. Gross, "Oswego County Revisited," *Public Health Reports,* 1976, 91:160–170.

2-86 The population of Lagos, Nigeria, in 1991 was 7,998,000. In 1992 it was 8,487,000. Find the percent change between 1991 and 1992.

2-87 In 1991 the birthrate in China was 22.3 per 1000. In 1993 it was 18.3. Find the percent change between 1991 and 1993.

## CLASS SURVEY QUESTIONS

1. (a) Find the mean, median, and modal ages of this class.
   (b) Do you think the mean age of this class would be a good estimate of the mean age of the school? Why or why not?
   (c) Draw a box-and-whisker display for the distribution of ages in the class.
2. (a) Find the range and the standard deviation of the heights of the smokers.
   (b) Find the range and the standard deviation of the heights of the nonsmokers.
   (c) Which group has more variability?
   (d) Find the $z$-score and the percentile rank for your height in whichever is your group.

## FIELD PROJECTS

1. Calculate the mean, median, modal, and midrange averages for the ages you gathered as part of the field project in Chapter 1.
2. Find the variance and the standard deviation of your data.

## CLASS EXPERIMENTS

1. Record the pulse rate for all members of the class (in beats per minute). Compute the mean, median, range, and standard deviation. Find the $z$-score for the highest and lowest pulse rates.
2. Have each person record his or her own pulse rate 5 times and then compute the mean and standard deviation of their own 5 values.

# 3

# Frequency Tables and Graphs

One picture is worth a thousand words.

CONTENTS



GOALS

At the end of this chapter you will have learned:

- How to construct stem-and-leaf displays, frequency tables, bar graphs, and histograms
- How to interpret the data presented in stem-and-leaf displays, frequency tables, bar graphs, and histograms
- The relationship between the percent of area in a histogram and the percent of outcomes in a frequency table
- How to construct a histogram from percentile ranks and *z*-scores

# 3.1 Organizing the Data

Usually, when a collection of statistical data is gathered, it must be organized in some way before much sense can be made of it.

## 3.1-1 Stem-and-Leaf Display

One simple and convenient way to organize some data sets is by stem-and-leaf display. This method is not too cumbersome to do by hand for small data sets, and is done quickly by computer for large sets. Here is an illustration. It provides a quick way to group data into categories which give a visual impression of the distribution of the data.

### ❑ EXAMPLE 3.1 Heart Rates

A group of 34 volunteers sat quietly in a classroom and then recorded their heart rates (in beats per minute). The results are as follows:

Heart Rates (Raw Data)

| | | | | | | | | | |
|---|---|---|---|---|---|---|---|---|---|
| 71 | 74 | 63 | 81 | 71 | 70 | 66 | 72 | 60 | 82 |
| 70 | 83 | 57 | 66 | 125 | 67 | 71 | 100 | 71 | 96 |
| 73 | 74 | 78 | 82 | 82 | 88 | 53 | 83 | 58 | 89 |
| 72 | 96 | 71 | 109 | | | | | | |

To create the stem-and-leaf display, we simply sort the values into groups such as 50s, 60s, 70s, etc. The data above range from the 50s to the 120s. So we make **stems** from 5 (for the 50s) to 12 (for the 120s), as shown in Figure 3-1.

stems

5 |
6 |
7 |
8 |
9 |
10 |
11 |
12 |

**Figure 3-1** Stems for a stem-and-leaf display.

Proceeding through the data, we record the units digit of each value in the appropriate row. (Thus, the first entry would be a 1 next to the 7.) These are the **leaves**. This gives the stem-and-leaf display shown in Figure 3-2. From this display you get an immediate impression of the shape of the distribution; in particular, it is evident that many heart rates are in the 70s and 80s, and that a heart rate of 125 is unusual.

leaves

|  |  |
|---|---|
| 5 | 738 |
| 6 | 36067 |
| 7 | 1410201134821 |
| 8 | 12322839 |
| 9 | 66 |
| 10 | 09 |
| 11 | |
| 12 | 5 |

**Figure 3-2** Stem-and-leaf display of heart rates.

For some purposes, it is convenient to take one more step and arrange the leaves on each row in numerical order. Then the display would look like the one in Figure 3-3.

|  |  |
|---|---|
| 5 | 378 |
| 6 | 03667 |
| 7 | 0011111223448 |
| 8 | 12223389 |
| 9 | 66 |
| 10 | 09 |
| 11 | |
| 12 | 5 |

**Figure 3-3** An ordered stem-and-leaf display of heart rates.

From this version, you can see quickly that the lowest heart rate in the group was 53 and that the highest was 125. The median value is also easy to find. Since there are 34 values, it will be halfway between the 17th and 18th values. By counting (from either end) these are seen to be 72 and 73, so the median heart rate for this group is 72.5. ❑

## 3.1-2 Frequency Tables

Probably the most common way to organize data is to combine the individual raw scores into a few categories and then simply indicate how many values are in each category. The resulting table is called a "frequency table." The next example illustrates this approach.

### ❑ EXAMPLE 3.2 Late Buses

George Stephen, a famous mathematician, gets bored waiting for the 8 A.M. bus each morning. He decides to record to the nearest minute the length of time the bus is late each day. His raw data for the past 30 days look like this:

| Day | 1 | 2 | 3 | 4 | 5 | 6 | 7 | 8 | 9 | 10 | 11 | 12 | 13 | 14 | 15 |
|---|---|---|---|---|---|---|---|---|---|---|---|---|---|---|---|
| Minutes late | 9 | 7 | 3 | 4 | 2 | 6 | 3 | 7 | 2 | 6 | 6 | 3 | 10 | 1 | 10 |

| Day | 16 | 17 | 18 | 19 | 20 | 21 | 22 | 23 | 24 | 25 | 26 | 27 | 28 | 29 | 30 |
|---|---|---|---|---|---|---|---|---|---|---|---|---|---|---|---|
| Minutes late | 3 | 3 | 2 | 6 | 1 | 4 | 6 | 4 | 3 | 5 | 6 | 3 | 5 | 3 | 4 |

To organize the data, he first lists the values from smallest to largest:

Minutes late: 1 1 2 2 2 3 3 3 3 3 3 3 3 4 4 4 4 5 5 5 5 5 6 6 6 7 7 9 10 10

**frequency table**

Then he condenses the data into a **frequency table**. To do this, he must first decide the size of each category. *If he decides* that each category will be 1 minute, he gets the following frequency table (Table 3-1):

Table 3-1

| Minutes Late | Frequency $f$ |
|---|---|
| 10 | 2 |
| 9 | 1 |
| 8 | 0 |
| 7 | 2 |
| 6 | 3 |
| 5 | 5 |
| 4 | 4 |
| 3 | 8 |
| 2 | 3 |
| 1 | 2 |
| | $30 = \Sigma f = n$ |

**frequency**

*Definition:* The **frequency** of an outcome is the number of times it occurs.

The frequency table allows him to answer certain common questions easily. For example:

(a) What percentage of the 30 latenesses were more than 5 minutes late?
(b) Which lateness occurred most often? That is, find the mode of the distribution of latenesses.

## SOLUTIONS

(a) Add the frequencies for all latenesses of more than 5 minutes and divide by 30:

$$\frac{2 + 1 + 0 + 2 + 3}{30} = \frac{8}{30} = 0.2667$$

About 27% of the latenesses were more than 5 minutes.

(b) See which category has the highest frequency. The most common lateness was 3 minutes. It occurred 8 times. ❑

## 3.1-3 Bar Graphs and Histograms

Often it is a good idea to present a graphical display of the data. This gives shape to the data and may make certain trends or patterns in the data very clear. The simplest graph is the **bar graph**. We can translate a frequency table directly into a bar graph by labeling the horizontal axis according to our grouping categories and then drawing bars according to the corresponding frequencies. Each line of the frequency table becomes one bar of the graph. For the data in Table 3-1, we get the bar graph in Figure 3-4.

bar graph

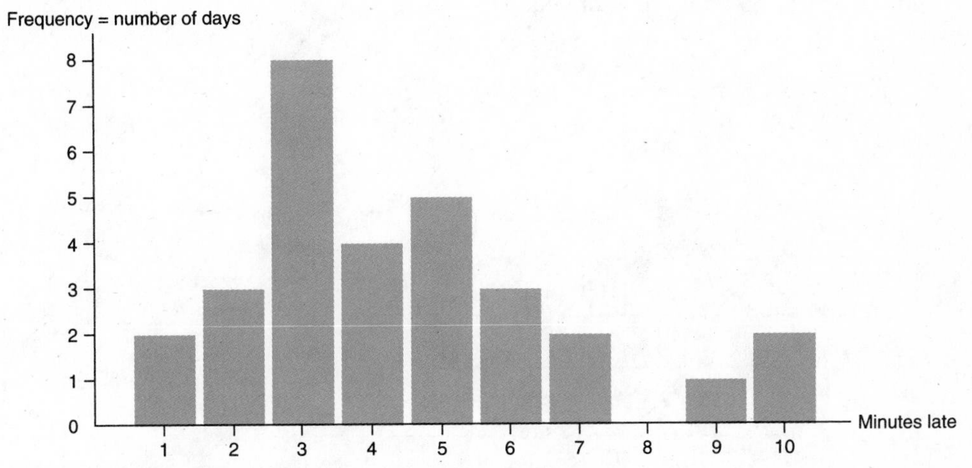


Figure 3-4 Bar graph for data in Table 3-1.

Now it stands out very clearly, for example, that a 3-minute wait occurred more often than any other.

histogram

Closely related to the bar graph is the **histogram**. This is a graph where we let the bars touch and then (usually) erase the inside vertical lines. If there is a bar missing, we imagine that it is there and that its height is zero. For the bar graph of Figure 3-4, we get the histogram in Figure 3-5.

A histogram makes sense when the horizontal axis describes some *increasing quantity*, such as time in our example, where the categories can be put in a natural numerical order. It does not make sense when the groups do not have any natural order. For example, if we were counting the number of blond-, brown-, red-, and black-haired people who made appointments during one week at a certain hair stylist, we might get a bar graph like Figure 3-6. It is not helpful to join these bars to make a histogram. There is no logical numerical order to the categories, they are just names of colors. It is best to just use the bar graph.

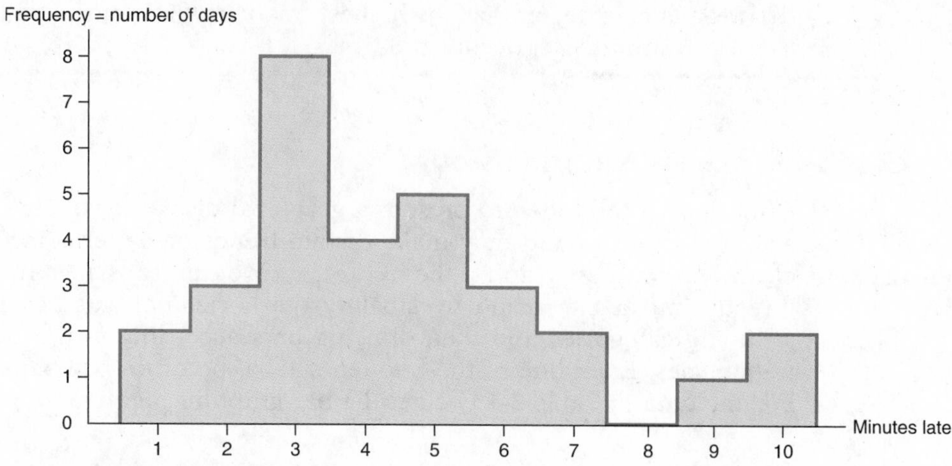


Figure 3-5 Histogram based on Figure 3-4.

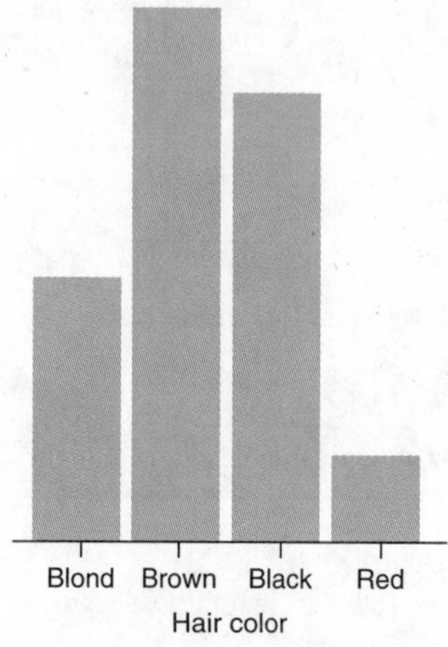


Figure 3-6 Bar graph with nonnumerical categories.

It is useful to learn some specific vocabulary for describing histograms.

**boundary**

*Definition:* A **boundary** is the value on the horizontal axis where two bars of a histogram meet.

For example, George measured bus waits to the nearest *minute*. So the waits which he recorded as 3 minutes actually go from 2.5 minutes to 3.5 minutes. Therefore, 2.5 and 3.5 are the boundaries for that category.

Often a histogram shows the labels for the boundaries. In Figure 3-7, we redraw the histogram of Figure 3-5, showing the boundaries.

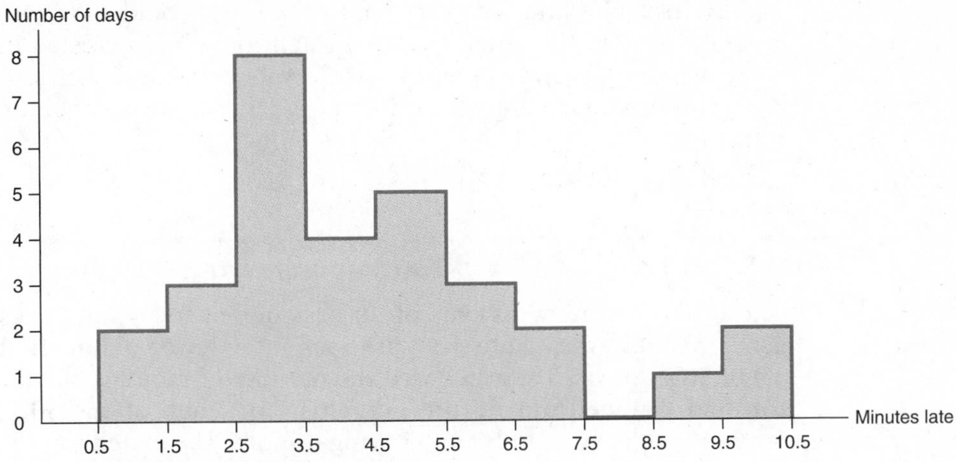


Figure 3-7 Histogram with labeled boundaries.

Alternatively, it is also common for the labels on a histogram to show just the midpoint value for the bars. In this case, the readers must determine the boundaries for themselves.

**interval**

> *Definition:* An **interval** is the segment on the horizontal axis between two consecutive boundaries.

For example, the piece from 0.5 to 1.5 is called the "interval from 0.5 to 1.5."

**interval width**

> *Definition:* The **width of an interval** is the difference between the upper boundary and the lower boundary of the interval.

For example, the width of the leftmost interval is 1.5 − 0.5 = 1.0. Notice that all the intervals in any one histogram are of **the same width.** The width of the intervals in a histogram is decided by the person who is making the graph. The width can be small or large, depending upon the purpose of the graph. For example, George Stephen could have measured how long he waited for the bus to the nearest half

### *Sir Francis Galton (1822–1911)*

A gentleman, a man of leisure, a world traveler, and one engaged in many disciplines, Galton became interested in meteorology and weather reports. These data led him to the field of statistics. He later became famous for his statistical work in heredity, particularly as a founder of eugenics, the goal of which was the improvement of mankind through controlled breeding. Galton was first cousin to Charles Darwin.

minute instead of the nearest minute; then he could have chosen intervals of width 0.5. On the other hand, he might only be interested in waits to the nearest 2 minutes and might set up intervals with width 2. The width can also be calculated by taking the difference between two consecutive midpoint values.

## ❑ EXAMPLE 3.3 A Histogram for a Distribution of Sample Means

One Friday morning, as part of an assignment for a statistics class, 30 students in a university statistics class each interviewed a different random sample of 50 students. They recorded the number of minutes of TV that each person had watched during the previous day. Each of the 30 students then reported the mean number of viewing minutes calculated from his or her own sample. The list of 30 means is shown, accompanied by a histogram (Figure 3-8) of the kind often produced by computer. If you turn the page sideways, you will see the histogram in the usual orientation.

| Student | Mean | Student | Mean |
|---|---|---|---|
| Augustina | 54.1 | Purvi | 60.3 |
| Blanca | 54.2 | Quincy | 60.7 |
| Charom | 55.9 | Redentor | 60.8 |
| Danilo | 56.4 | Shireen | 61.0 |
| Eartley | 57.1 | Theophile | 61.1 |
| Fiona | 58.0 | Ursula | 61.1 |
| Gonzalo | 58.5 | Vashali | 61.2 |
| Hariklia | 58.7 | Wanda | 62.2 |
| Ione | 58.8 | Xaveria | 62.2 |
| José | 58.9 | Yeong | 62.5 |
| Kristen | 59.0 | Zachariah | 62.9 |
| Linh | 59.0 | Annemarie | 63.2 |
| Mohammed | 59.2 | Kerianne | 63.6 |
| Neville | 59.5 | Roseanne | 66.0 |
| Osmund | 59.5 | Rosemarie | 67.4 |

❑

Notice the shape of this particular graph in Figure 3-8. It is typical that the histogram of a distribution of sample means is low on both ends and high in the middle. We will say more about distributions of sample means in Chapter 10. In this graph it is easy to see, by subtracting 56 − 54, that the interval width is 2 minutes. Also, since each boundary is exactly halfway between the two midpoint values next to it, the boundaries for this histogram are 53, 55, 57, . . . , 69.

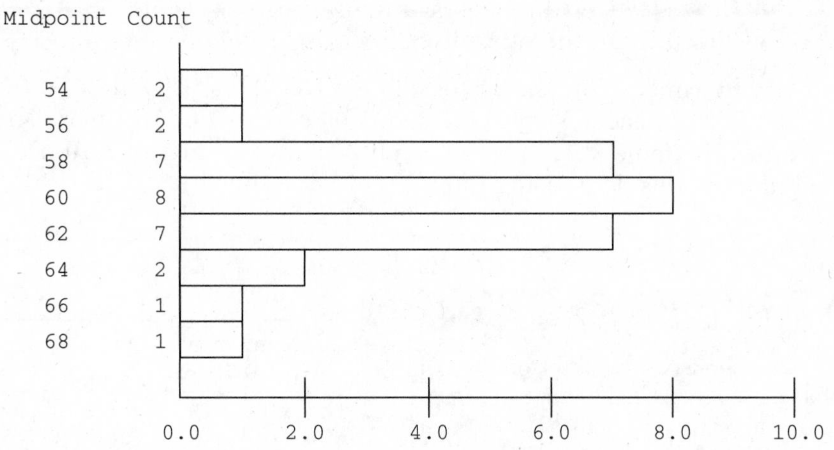


**Figure 3-8** A computer-generated histogram.

Some observations that can be made from this histogram are:

1. The median of the distribution is some number between 59 and 61. This is because the median is between the 15th and 16th values, which are both in the interval from 59 to 61.
2. A majority (73 percent) of the sample means were between 57 and 63 minutes. That is because the three tall bars account for 22 out of 30 of the values shown in the graph (22/30 = 0.7333).

***Note*** A stem-and-leaf display is also very similar to a sideways histogram. In fact, if we draw lines about the data in the stem-and-leaf display which we drew previously in Figure 3-2, we obtain the graph shown in Figure 3-9.

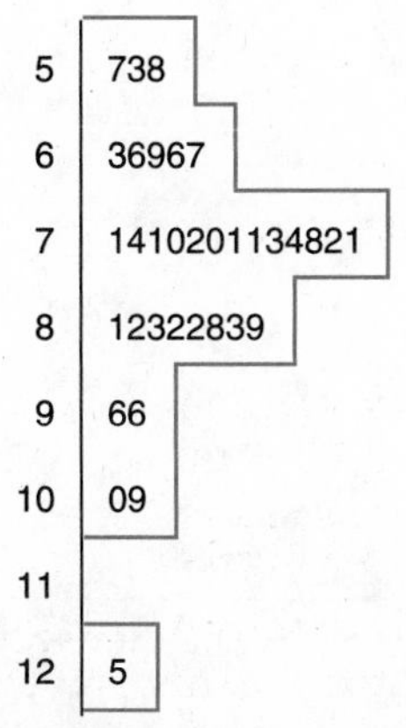


**Figure 3-9** A histogram from the stem-and-leaf display of heart rates.

## ❑ EXAMPLE 3.4 Tossing a Silver Dollar

An experiment consists of tossing a silver dollar 20 times and counting the number of heads. The experiment was carried out 30 times. The results are shown in Table 3-2. The bar graph is shown in Figure 3-10. Draw the histogram for this data, using intervals of width one.

### *Computer Simulation*

It is easy to program a computer to simulate the tossing of a coin 20 times and to have the computer repeat this experiment 30 times. In a matter of seconds, what would take a human much longer is finished and the results are tabulated.

In fact, the computer can just as easily repeat the experiment 300 or 3000 times, performing in a matter of seconds what might take you many hours or even days.

**Table 3-2**

| Number of Heads in 20 Tosses | Number of Experiments |
|---|---|
| 15 | 2 |
| 14 | 1 |
| 13 | 3 |
| 12 | 0 |
| 11 | 4 |
| 10 | 5 |
| 9 | 7 |
| 8 | 3 |
| 7 | 2 |
| 6 | 2 |
| 5 | 0 |
| 4 | 0 |
| 3 | 1 |
| Total number of experiments = 30 | |

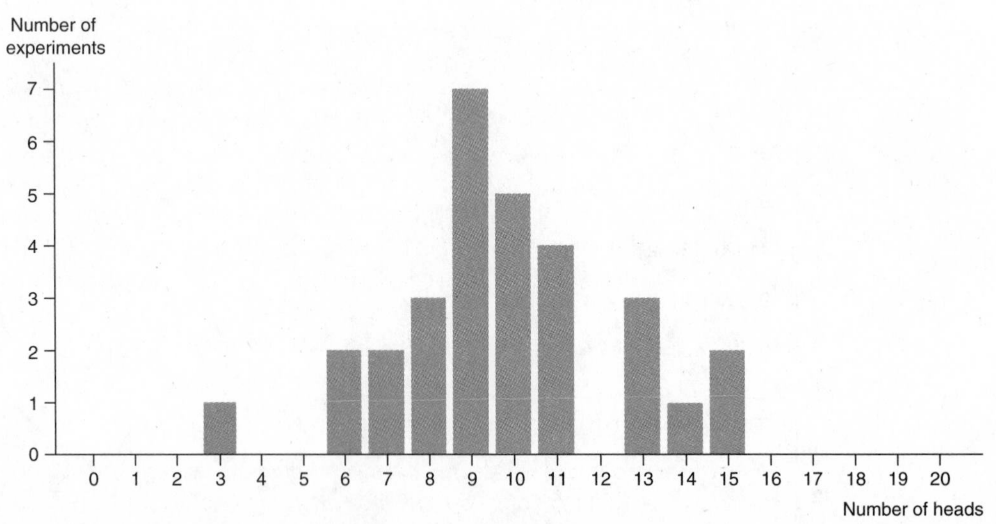


**Figure 3-10** Bar graph for data in Table 3-2.

## SOLUTION

We make the histogram by widening the bars and finding the boundaries of the intervals. Figure 3-11 shows the resulting histogram.

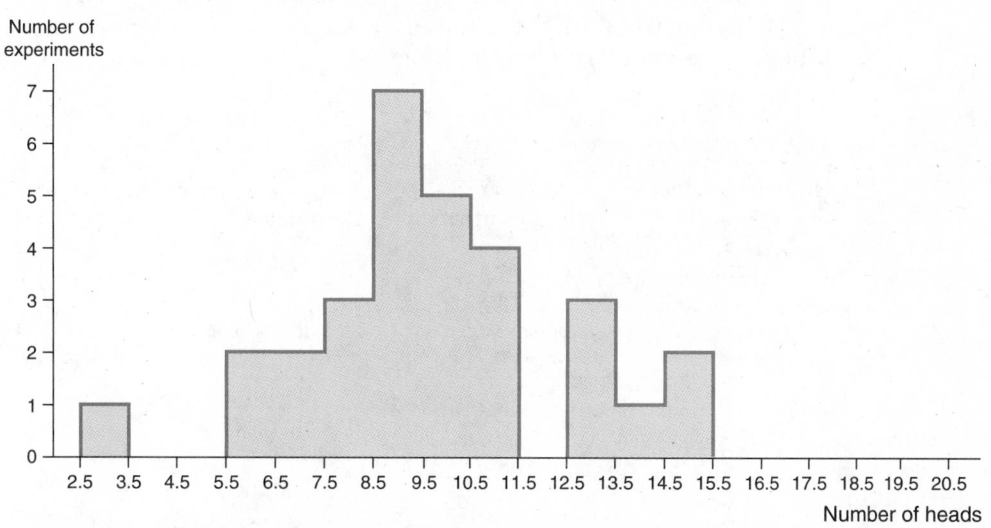


**Figure 3-11** Histogram for data in Table 3-2.

What is the effect of such a graph? It gives the impression that the number of heads is a continuously increasing quantity. It appears, for example, that in the 7 experiments represented by the interval from 8.5 to 9.5 the number of heads could have been anything from 8.5 to 9.5, when in fact the number of heads could only have been exactly 9. On the other hand, the graph still gives an accurate picture of the outcomes of the experiments and their frequencies. For example, you can still clearly see that 9 heads was the outcome that occurred more often than any other outcome. And you can still see that no experiment resulted in 12 heads. In other words, the basic *shape* of the distribution is preserved. We will see in Chapters 5 and 7 that it can be very useful to graph some noncontinuous quantities using histograms, as we will be asking questions about areas which will be easier to answer if we have one continuous graph rather than separate bars. As long as you keep in mind the kinds of data you have, continuous or not, no confusion will result.

Let us go back to the coin-tossing experiment and discuss drawing a histogram for those data. Two decisions have to be made first:

1. What width should the intervals be?
2. Where should we start the lowest interval?

There are no fixed rules for answering these questions. It is up to the person drawing the graph. You might try graphing the same data in histograms of different width intervals to see what the overall effect is. Generally speaking, if the intervals are too wide, you will have too much data lumped together, and the trends in the data will be hard to spot. At the other extreme, if the intervals are too narrow, the graph will be too spread out and is likely to become very spotty, and, once again, any trends will be hard to spot. Suppose that for the coin-tossing data we decide to use intervals of width 3, and suppose that we start the lowest interval with a boundary of 2.5. We could tabulate the data as shown in Table 3-3.

Table 3-3

| Number of Heads in 20 Tosses | Number of Experiments | Boundaries | Frequency (Number of Experiments) |
|---|---|---|---|
| 15 | 2 | 14.5 to 17.5 | 2 |
| 14 | 1 | | |
| 13 | 3 | 11.5 to 14.5 | 4 |
| 12 | 0 | | |
| 11 | 4 | | |
| 10 | 5 | 8.5 to 11.5 | 16 |
| 9 | 7 | | |
| 8 | 3 | | |
| 7 | 2 | 5.5 to 8.5 | 7 |
| 6 | 2 | | |
| 5 | 0 | | |
| 4 | 0 | 2.5 to 5.5 | 1 |
| 3 | 1 | | |

The histogram for these data would be as shown in Figure 3-12.

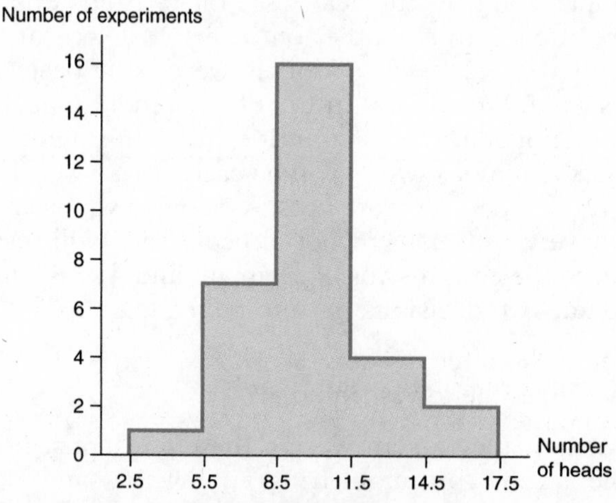


Figure 3-12 Histogram based on data in Table 3-3.

You can see that this histogram is similar to that in Figure 3-11, but it is easier to understand. It shows clearly that in this experiment we often get *around* 10 heads and that outcomes become more and more rare the farther away they are from 10 heads.

## 3.1-4 Reading Histograms

What information is contained in a histogram? We note that the percentage of area of the graph which is over any particular interval is equal to the percentage of the outcomes that are in that interval. Let us see why. In Figure 3-13 we redraw the histogram of Figure 3-12, and we add some vertical and horizontal lines to show how each of the 30 outcomes is represented by an equal amount of area. This is the key to histograms: Because each interval has the same width, each outcome is represented by an *equal amount of area.*

You see, for example, that there are 16 equal sections in the third interval. This corresponds to the frequency of 16 in Table 3-3. The next two examples let us answer some specific questions to illustrate the principles involved.

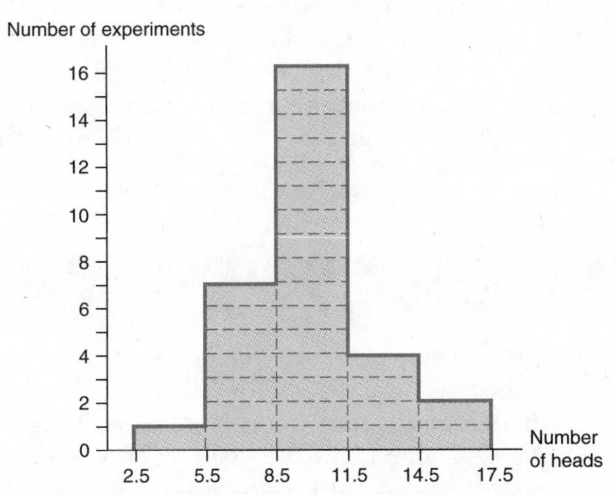


**Figure 3-13** Histogram with units of area indicated.

---

### ❑ EXAMPLE 3.5 Reading a Frequency Table

In what percentage of the experiments described in Table 3-3 was the result from 3 to 5 heads?

**SOLUTION**

From the last 2 columns of the **frequency table** in Table 3-3, we see that the frequency for these outcomes was 1. The total number of outcomes was 30. The percentage of outcomes in this interval therefore is $1/30 = 0.0333 = 3.33\%$. ❑

---

### ❑ EXAMPLE 3.6 Reading a Histogram

What percentage of the area of the histogram corresponds to experiments where the outcome was from 3 to 5 heads?

SOLUTION

We see in the **histogram** of Figure 3-13 that over this interval is 1 unit of area. The total number of units of area in the graph is 30. Therefore, the percentage of area in the graph over this interval is $1/30 = 0.0333 = 3.33\%$. ❑

You notice from Examples 3.5 and 3.6 that, as far as percentages of outcomes are concerned, the graph and the table contain the same information. In the next chapter you will see that percentages of outcomes are very important in discussing basic ideas of probability. This, in turn, means that we will be able to discuss questions of probability by looking at graphs and measuring percentages of area over various intervals. It will turn out that this is indeed a major tool in statistics, because many apparently different questions lead us to graphs of the same shape. So once we know how to measure percentages of area over the intervals of that graph, we will be in good shape to answer those statistical questions. All that remains in this chapter is to show how we can relate what is known about percentages of area over various intervals and the actual overall *shape* of the graph.

## 3.2 Percentile Rank, z-Scores, and Graphs

In different distributions, the same *z*-score may be associated with different percentile ranks. This is because the distributions have different *shapes*.

If we know which percentile ranks correspond to the *z*-scores at the boundaries of the intervals, we can draw the histogram. We demonstrate this relationship between *z*-scores, percentile ranks, and area in the following examples.

### ❑ EXAMPLE 3.7 Constructing a Histogram from z-Scores and Percentile Ranks

Draw the histogram for the following data:

| *z*-Score | Percentile Rank |
|---|---|
| 2 | 100 |
| 1 | 70 |
| 0 | 30 |
| −1 | 20 |
| −2 | 0 |

## SOLUTION

Since the $z$-scores of 0 and 1 have percentile ranks of 30 and 70, respectively, 40% of the distribution must lie between $z = 0$ and $z = 1$. In order to draw the histogram, we need to know the area corresponding to each interval. As we have just shown, you find the area corresponding to an interval by subtracting the percentile ranks of its boundaries.

| Boundaries | Percentage of Area |
|---|---|
| 1 to 2 | 100 − 70 = 30 |
| 0 to 1 | 70 − 30 = 40 |
| −1 to 0 | 30 − 20 = 10 |
| −2 to −1 | 20 − 0 = 20 |

We can now draw the histogram (Figure 3-14).

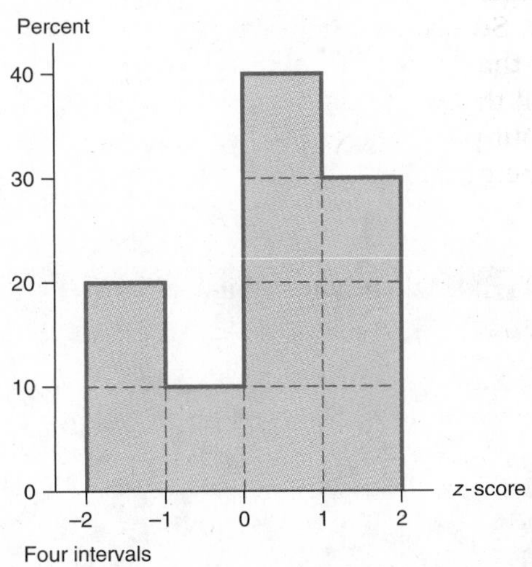


**Figure 3-14** Histogram based on $z$-scores and percentile ranks. ❑

If we had nine $z$-scores instead of five, we could draw a more accurate histogram, as indicated in the next example.

### ❑ EXAMPLE 3.8 Constructing a Histogram from z-Scores and Percentile Ranks

Draw a histogram similar to the one shown in Example 3.7, given these 9 $z$-scores and their corresponding percentile ranks:

| $z$-Score | Percentile Rank |
|---|---|
| 2.0 | 100 |
| 1.5 | 90 |
| 1.0 | 70 |
| 0.5 | 55 |
| 0.0 | 30 |
| −0.5 | 25 |
| −1.0 | 20 |
| −1.5 | 5 |
| −2.0 | 0 |

## SOLUTION

We find the percentage of area for each interval as we did in Example 3.7.

| Boundaries | Percentage of Area |
|---|---|
| 1.5 to 2.0 | 10 |
| 1.0 to 1.5 | 20 |
| 0.5 to 1.0 | 15 |
| 0.0 to 0.5 | 25 |
| −0.5 to 0.0 | 5 |
| −1.0 to −0.5 | 5 |
| −1.5 to −1.0 | 15 |
| −2.0 to −1.5 | 5 |

The histogram for these data is shown in Figure 3-15.

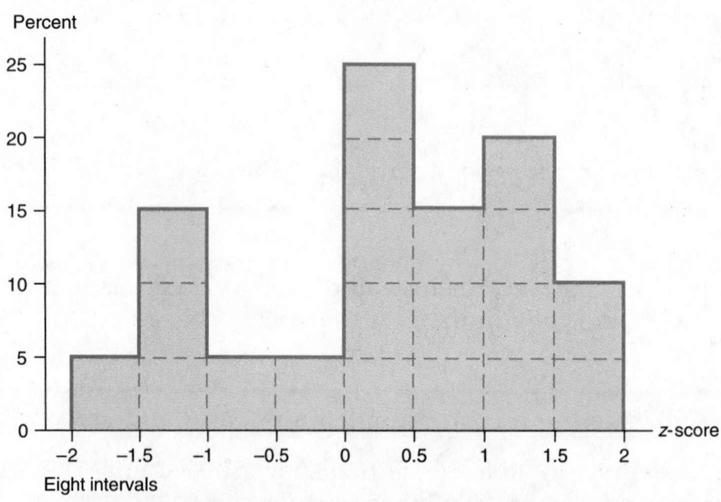


**Figure 3-15** Histogram based on $z$-scores and percentile ranks. ❑

If we had 100 $z$-scores, we would have a histogram with 99 intervals. The histogram would begin to resemble a smooth curve, as shown in Figure 3-16.

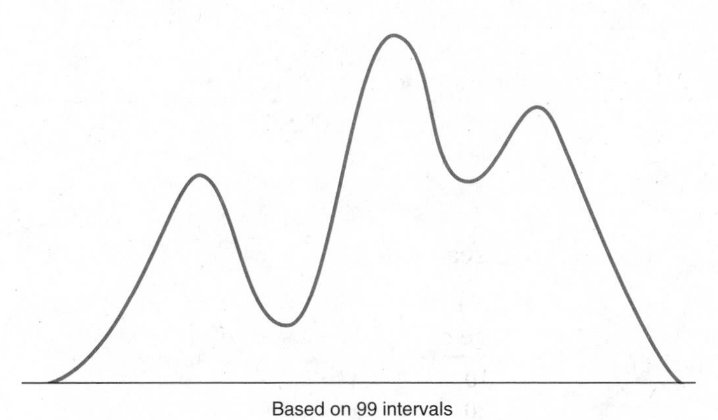


Figure 3-16 Smooth approximation of histogram with many intervals.

## STUDY AIDS

### Chapter Summary

In this chapter you have learned:

- To construct frequency tables, stem-and-leaf displays, bar graphs, and histograms
- To interpret the data presented in frequency tables and in graphs
- The relationship between the percent of area in a histogram and the percent of outcomes in a frequency table
- How to draw a histogram given the percentile ranks and the $z$-scores

### Vocabulary

You should be able to explain the meaning of each of these terms or phrases.

1. Stem-and-leaf displays
2. Frequency table
3. Frequency
4. Bar graph
5. Histogram
6. Boundary
7. Interval
8. Width of an interval

## EXERCISES

3-1 Here are the ages at inauguration of U.S. presidents from Washington to Clinton.

(a) Construct a frequency table for the ages.

(b) Draw a stem-and-leaf display.

(c) Draw the histogram using 5-year categories starting with age 40. (First category is 40–44.)

(d) Compute the mean and median ages.

(e) What percentage of U.S. presidents were in their 50s when inaugurated?

| Name | Age at Inauguration | Name | Age at Inauguration |
|---|---|---|---|
| 1. Washington | 57 | 22. Cleveland | 47 |
| 2. J. Adams | 61 | 23. B. Harrison | 55 |
| 3. Jefferson | 57 | 24. Cleveland | 55 |
| 4. Madison | 57 | 25. McKinley | 54 |
| 5. Monroe | 58 | 26. T. Roosevelt | 42 |
| 6. J. Q. Adams | 57 | 27. Taft | 51 |
| 7. Jackson | 61 | 28. Wilson | 56 |
| 8. Van Buren | 54 | 29. Harding | 55 |
| 9. W. H. Harrison | 68 | 30. Coolidge | 51 |
| 10. Tyler | 51 | 31. Hoover | 54 |
| 11. Polk | 49 | 32. F. D. Roosevelt | 51 |
| 12. Taylor | 64 | 33. Truman | 60 |
| 13. Fillmore | 50 | 34. Eisenhower | 62 |
| 14. Pierce | 48 | 35. Kennedy | 43 |
| 15. Buchanan | 65 | 36. L. B. Johnson | 55 |
| 16. Lincoln | 52 | 37. Nixon | 56 |
| 17. A. Johnson | 56 | 38. Ford | 61 |
| 18. Grant | 46 | 39. Carter | 52 |
| 19. Hayes | 54 | 40. Reagan | 69 |
| 20. Garfield | 49 | 41. Bush | 64 |
| 21. Arthur | 50 | 42. Clinton | 46 |

3-2 For the countries of Central and South America, the following birthrates were reported in the *World Population Data Sheet* (1993), published by the Population Reference Bureau. The birthrate is births per 1000 population. (The U.S. birthrate in this report is 16.)

(a) Complete a stem-and-leaf display and a frequency table for this data using stems of high teens, low twenties, high twenties, etc., as indicated.

1 |
2 |
2 |
3 |
3 |

(b) Construct a histogram using intervals of width 5, starting at 15.

(c) Find the mean and median birthrates for these countries.

(d) What percentage of these countries have a birthrate higher than 25 per 1000?

(e) Find the range and standard deviation for the birthrates.

(f) Find the $z$-score for Uruguay. Why will it have to be negative?

| Country | Birthrate | Country | Birthrate |
|---|---|---|---|
| Belize | 38 | Costa Rica | 27 |
| El Salvador | 34 | Guatemala | 39 |
| Honduras | 39 | Mexico | 29 |
| Nicaragua | 38 | Panama | 25 |
| Argentina | 21 | Bolivia | 37 |
| Brazil | 23 | Chile | 21 |
| Colombia | 26 | Ecuador | 31 |
| Guyana | 25 | Paraguay | 34 |
| Peru | 28 | Suriname | 24 |
| Uruguay | 19 | Venezuela | 30 |

3-3 A total of 100 different random samples were taken from a large population of grades on a national aptitude test. From each sample, the mean was recorded, giving the following distribution of 100 sample means.

| | | | | | | | | | |
|---|---|---|---|---|---|---|---|---|---|
| 105.16 | 98.48 | 104.80 | 103.41 | 95.59 | 99.59 | 95.25 | 98.64 | 100.34 | 100.50 |
| 100.42 | 96.58 | 100.40 | 104.67 | 101.96 | 97.56 | 97.25 | 93.94 | 98.63 | 98.74 |
| 100.33 | 99.07 | 100.17 | 95.74 | 95.35 | 97.02 | 101.37 | 101.07 | 101.99 | 94.91 |
| 100.24 | 99.04 | 99.48 | 99.41 | 100.90 | 96.49 | 101.43 | 98.18 | 102.62 | 101.01 |
| 98.85 | 101.26 | 97.95 | 100.16 | 103.60 | 99.80 | 96.87 | 98.98 | 102.08 | 103.24 |
| 100.15 | 97.50 | 94.47 | 98.17 | 97.86 | 97.35 | 101.38 | 99.08 | 103.72 | 98.51 |
| 96.66 | 101.52 | 102.40 | 104.98 | 98.77 | 97.28 | 102.14 | 102.59 | 99.39 | 97.15 |
| 105.47 | 96.84 | 101.87 | 99.21 | 105.33 | 99.09 | 100.36 | 102.53 | 99.37 | 104.61 |
| 98.00 | 99.51 | 98.27 | 105.26 | 98.17 | 101.59 | 95.62 | 102.53 | 103.76 | 97.95 |
| 96.42 | 100.01 | 96.64 | 100.74 | 104.10 | 98.82 | 96.37 | 98.65 | 104.42 | 95.19 |

(a) Use the data to make a frequency table and a histogram with interval width equal to 2.

(b) Draw a histogram with interval width equal to 1.

3-4 Here are the 1993 salaries for the Toronto Blue Jays, as published in *Street and Smith's Guide to Baseball* (Ballantine Books, 1994).

| | | | |
|---|---|---|---|
| Alomar | $4,833,333 | Borders | $2,500,000 |
| Butler | 109,000 | Canate | 109,000 |
| Carter | 5,550,000 | Castillo | 185,000 |
| Coles | 500,000 | Cox | 625,000 |
| Guzman | 500,000 | Hentgen | 182,500 |
| Knorr | 112,500 | Leiter | 252,500 |
| Molitar | 3,575,000 | Olerud | 1,537,000 |
| Schofield | 800,000 | Sprague | 182,500 |
| Stewart | 4,300,000 | Stottlemyre | 2,325,000 |
| Timlin | 262,000 | Ward | 3,300,000 |
| White | 3,608,333 | | |

(a) Construct a frequency table and draw a histogram for these data, where the first interval covers salaries from $0 to $499,999.

(b) Compute both the mean and the median salaries. What property of the graph would lead you to expect that the mean and the median might be quite different from one another? In salary negotiations, what would you expect the owners to claim as the "average" salary? What would you expect the players to claim as the "average" salary?

3-5 The following table shows the percentages of the U.S. population in different age groups according to censuses from 1860 to 1990. Notice that the categories are not of equal width, which makes it inappropriate to draw a histogram.

| | *Age* | | | | |
|---|---|---|---|---|---|
| Year | Under 5 | 5–19 | 20–44 | 45–64 | 65 and Over |
| 1860 | 15.4 | 35.8 | 35.7 | 10.4 | 2.7 |
| 1870 | 14.3 | 35.4 | 35.4 | 11.9 | 3.0 |
| 1880 | 13.8 | 34.3 | 35.9 | 12.6 | 3.4 |
| 1890 | 12.2 | 33.9 | 36.9 | 13.1 | 3.9 |
| 1900 | 12.1 | 32.3 | 37.8 | 13.7 | 4.1 |
| 1910 | 11.6 | 30.4 | 39.1 | 14.6 | 4.3 |
| 1920 | 11.0 | 29.8 | 38.4 | 16.1 | 4.7 |
| 1930 | 9.3 | 29.5 | 38.3 | 17.5 | 5.4 |
| 1940 | 8.0 | 26.4 | 38.9 | 19.8 | 6.9 |
| 1950 | 10.7 | 23.2 | 37.7 | 20.3 | 8.1 |
| 1960 | 11.3 | 27.1 | 32.4 | 20.0 | 9.2 |
| 1970 | 8.4 | 29.4 | 31.7 | 20.6 | 9.8 |
| 1980 | 7.2 | 24.8 | 37.2 | 19.7 | 11.3 |
| 1990 | 7.5 | 21.4 | 40.2 | 18.5 | 12.5 |

(a) Using these categories, construct a bar graph for the data from 1860, then do the same for 1920 and 1990. Describe in a few sentences what is going on.

(b) For each of the census years from 1860 to 1990, find the percentage of the population that is less than 20 years old. Draw a "time-series" graph like that in Figure 2-1 which is reprinted below for these percentages and describe the meaning of any trend that you see.

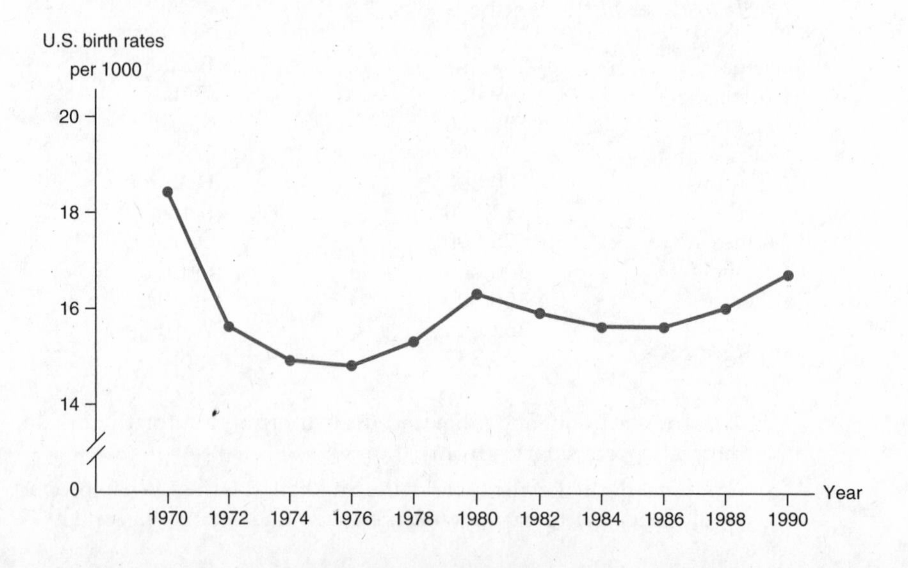

3-6 The table shown represents job performance ratings at a factory. (Assume that any individual rating is a whole number from 0 to 10.)

| Interval | Boundaries | Frequency |
|---|---|---|
| 10 | 9.5–10.5 | 7 |
| 9 | 8.5– 9.5 | 16 |
| 8 | 7.5– 8.5 | 15 |
| 7 | 6.5– 7.5 | 36 |
| 6 | 5.5– 6.5 | 99 |
| 5 | 4.5– 5.5 | 150 |
| 4 | 3.5– 4.5 | 216 |
| 3 | 2.5– 3.5 | 304 |
| 2 | 1.5– 2.5 | 401 |
| 1 | 0.5– 1.5 | 197 |
| 0 | −0.5– 0.5 | 253 |

(a) How many ratings are represented in this frequency table?

(b) Estimate the median rating.

(c) What percentage of the ratings are either 0 or 1?

(d) Collapse the data into 4 intervals of width 3 and draw the resulting histogram.

3-7 Here are some U.S. government estimates on income distribution from 1991. The data are given in the *Statistical Abstracts of the United States* (1993). All amounts are in dollars.

| Money Income | Percentage of Households |
|---|---|
| Under 5,000 | 4.8 |
| 5,000–9,999 | 10.1 |
| 10,000–14,999 | 9.4 |
| 15,000–24,999 | 17.4 |
| 25,000–34,999 | 15.2 |
| 35,000–49,999 | 17.3 |
| 50,000–74,999 | 15.4 |
| 75,000 and over | 10.4 |

(a) Regroup the table using these categories:

0–14,999
15,000–24,999
25,000–49,999
50,000 and over

(b) Draw a bar graph for the table you regrouped in part (a).

(c) What percentage of the households had money income less than \$15,000? Between \$15,000 and \$50,000?

(d) Estimate the median money income.

3-8 The table shows the breakdown of the expenses of one investment company for 3 consecutive years. We wish to compare the breakdowns to see if there is any pattern emerging.

All amounts are in thousands of dollars.

| Expenses | Year 1 | Year 2 | Year 3 |
|---|---|---|---|
| Management fee | 1147 | 3086 | 7478 |
| Shareholder service | 639 | 1541 | 2702 |
| Printing | 45 | 167 | 506 |
| Postage | 38 | 119 | 419 |
| Professional fees | 78 | 65 | 325 |
| Rent | 148 | 156 | 307 |
| Registration fees | 75 | 195 | 261 |
| Equipment maintenance | 72 | 102 | 151 |
| Custodian fees | 38 | 101 | 138 |
| Directors fees | 13 | 21 | 40 |
| State and local taxes | 11 | 8 | 7 |
| Total | 2304 | 5561 | 12,334 |

(a) For the first year, draw a bar graph corresponding to the dollar amounts spent in each category.

(b) For the first year, convert the dollar amounts to "fraction of total expense" (e.g., management fee = 1147/2304 = 0.50). Then draw a bar graph using the fractions to determine the heights of the bars. Compare the bar graphs from parts (a) and (b).

(c) For years 2 and 3, convert the amounts of fractions of total expense as in part (b). Then draw bar graphs for years 2 and 3.

(d) Compare the graphs for years 1, 2, and 3. Do you see any interesting patterns in the company's expenses?

3-9 The following histogram was the result of a study of the time it took to do a major engine repair job at a school of auto repair.

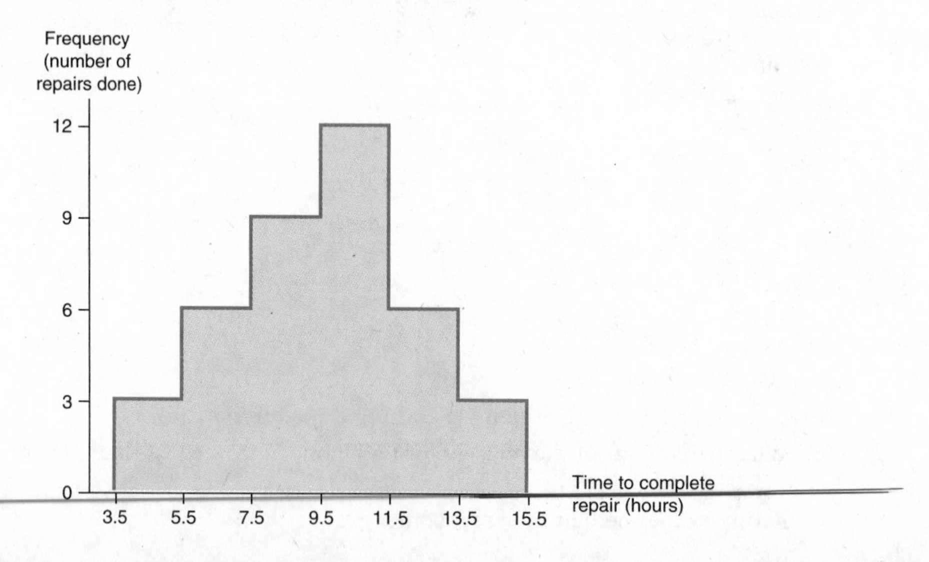

(a) How many repair jobs were included in this study?

(b) How many of the repair jobs took more than 11.5 hours? What percentage of the jobs took more than 11.5 hours?

(c) What percentage of the jobs took between 7.5 and 11.5 hours?

3-10 Using the histogram shown below, find:

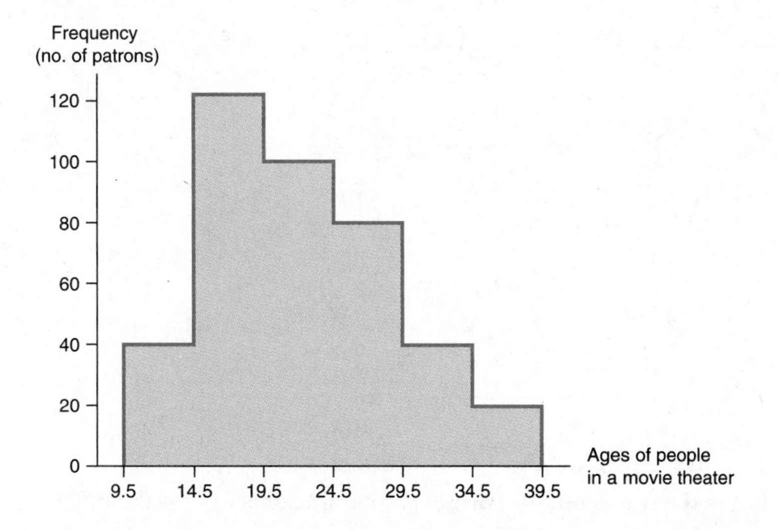


(a) The number of people represented in each interval and the total number of people represented in the graph.

(b) The percentage of people in each interval.

(c) The percentage of area of the graph which is in the space over each interval. Check that the total of these is 100%.

3-11 A computer was used to select 36 random weights from a large population of weights. The mean of these 36 weights was recorded.

This whole procedure was repeated 100 times, resulting in a distribution of 100 sample means. The resulting histogram of these 100 sample means is shown.

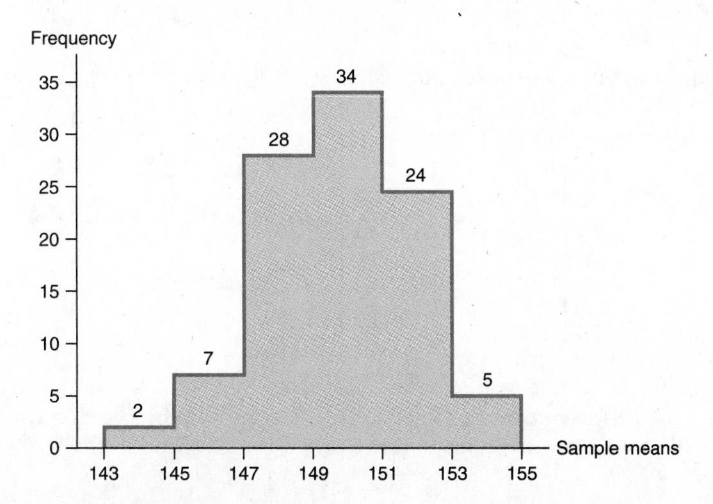

(a) What percentage of the sample means were between 147 and 151?

(b) What percentage of the sample means were more than 153?

3-12 An experiment is to roll a die 6 times. This experiment was repeated 20 times, and the number of times that a 2 appeared in each experiment was recorded. The results are tabulated by making 1 tally mark after each 6 tosses of the die.

| Number of Times a 2 Appeared | Tally | Frequency |
|---|---|---|
| 6 | | 0 |
| 5 | / | 1 |
| 4 | / | 1 |
| 3 | /// | 3 |
| 2 | // | 2 |
| 1 | ///// // | 7 |
| 0 | ///// / | 6 |
| Total | | 20 |

(a) Draw a histogram for the data in the table.

(b) What is the mean number of times that a 2 appeared on the die?

3-13 What visual impression can be created by lengthening or shortening the scale of numbers on the vertical axis of a bar graph? What happens if you do not start with zero? Suppose we are comparing sales of 3 competing brands of portable CD players. Here are the data.

| Brand | Sales for Last Year |
|---|---|
| Phonix | 30,000 units |
| Sanyaha | 25,000 units |
| Monimo | 15,000 units |

(a) Draw bar graphs for these results, using 3 different vertical scales, as shown. The sales are in thousands.

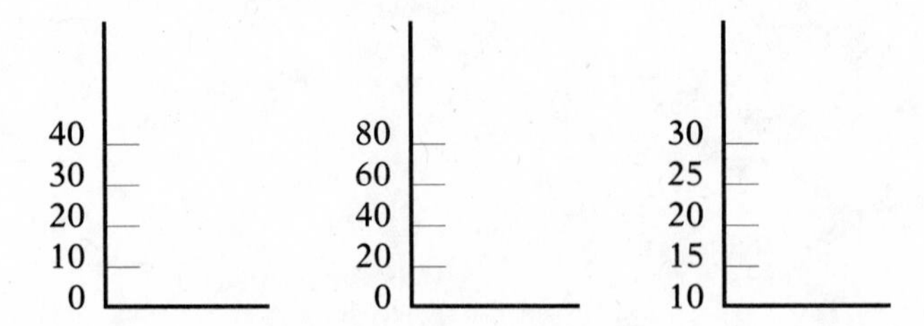


(b) Describe the different overall effects of the 3 graphs. Which graph do you prefer? Why? Which would you prefer if you worked for Monimo? Why?

3-14 The histogram illustrated below is symmetrical. The data represent the heights of children in Uncle Don's Nursery School. Some of the $z$-scores and percentile ranks are related as follows:

| $z$-Score | Percentile Rank |
|---|---|
| 3 | 100 |
| 2 | 98 |
| 1 | 84 |
| 0 | 50 |

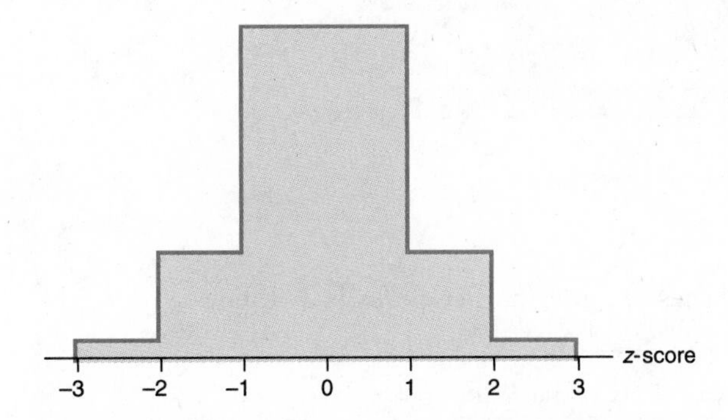


What percentage of the area of the graph is

(a) To the left of 0?

(b) To the left of 1?

(c) To the left of −1?

(d) To the right of 2?

(e) To the right of −2?

(f) Between 0 and 3?

(g) Between 0 and 2?

(h) Between −2 and 0?

(i) Between −1 and 1?

(j) Between −2 and 2?

3-15 Answer parts (a) to (j) from Exercise 3-14 for the following data and histogram which describe the grade distribution in Scuba Diving 2.

| $z$-Score | Percentile Rank |
|---|---|
| 3 | 100 |
| 2 | 80 |
| 1 | 50 |
| 0 | 40 |
| −1 | 20 |
| −2 | 10 |
| −3 | 0 |

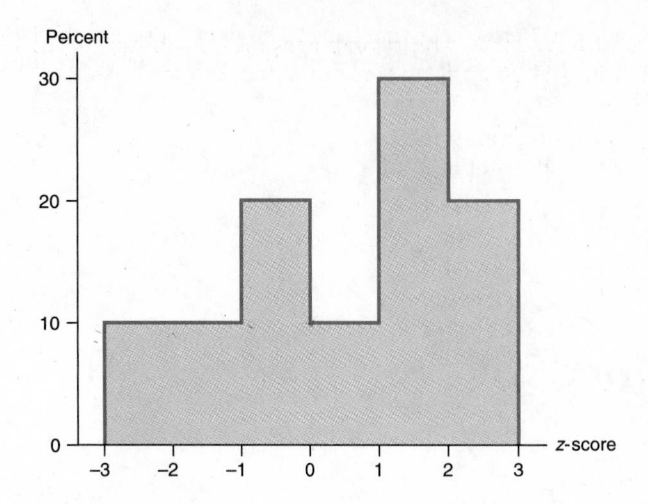


3-16 A distribution of salaries of supermarket clerks is as follows.

| $z$-Score | Percentile Rank |
|---|---|
| 3 | 100 |
| 2 | 85 |
| 1 | 45 |
| 0 | 35 |
| −1 | 30 |
| −2 | 10 |
| −3 | 0 |

(a) For the given data, what percentage of the salaries lie between $z = 1$ and $z = 2$?

(b) What percentage of the area of a histogram based on these data will be between $z = 1$ and $z = 2$?

(c) Complete the following table.

| Boundaries | Percentage of Salaries | Area of a Histogram |
|---|---|---|
| 2 to 3 | 15 | |
| 1 to 2 | | |
| . | | |
| . | | |
| . | | |

(d) Draw the histogram.

3-17 Construct a table of boundaries and percentages of area and draw the histogram for the following data derived from test scores in a prenursing course.

| $z$-Score | Percentile Rank |
|---|---|
| 3 | 100 |
| 2.5 | 75 |
| 2 | 65 |
| 1.5 | 50 |
| 1 | 45 |
| 0.5 | 40 |
| 0 | 35 |
| −0.5 | 33 |
| −1 | 30 |
| −1.5 | 25 |
| −2 | 10 |
| −2.5 | 5 |
| −3 | 0 |

3-18 The data given here represent amounts of money bet one night at the races. Construct a table of boundaries and percentages of area and draw a histogram.

| $z$-Score | Percentile Rank |
|---|---|
| 3 | 100 |
| 2 | 90 |
| 1 | 60 |
| 0 | 40 |
| −1 | 0 |

## CLASS SURVEY QUESTIONS

1. Draw 3 histograms for the *male heights*. Graph 1 has intervals of width 1 inch; graph 2 has intervals of width 2 inches; graph 3 has intervals of width 3 inches. Start all graphs at the same height.
2. Which of the three graphs do you prefer for showing the height? Why?
3. Draw a histogram for the *female heights* using intervals equal to those you selected in answering question number 2.
4. Compare the two histograms in questions 2 and 3, and comment on any differences.
5. What is the modal eye color?
6. Draw a stem-and-leaf display for the heights of the mothers.

## FIELD PROJECTS

1. Construct a stem-and-leaf display, a frequency table, a bar graph, and a histogram for the ages that you gathered in the project in Chapter 1.
2. Calculate the percentile rank of an age of 18 years.

## CLASS EXPERIMENT

1. (a) Toss a coin 5 times, and count the number of times that tails appears. Repeat this experiment 15 times. Record the data in the table below, and draw a histogram for the data. Make 1 tally mark after each 5 tosses of the coin.

| Number of Tails | Tally | Frequency |
|---|---|---|
| 5 | | |
| 4 | | |
| 3 | | |
| 2 | | |
| 1 | | |
| Total | | 15 |

(b) What is the mean number of tails that occurred in 5 tosses?

# Sample Test for Chapters 1, 2, and 3

1. For each of the following distributions decide which average you would use to measure the central tendency, and explain your choice:
   (a) The salaries of all the professors at your college.
   (b) The times that it takes 10 horses to run a measured mile.
   (c) The current time as shown on 10 different clocks.
   (d) The grades on the first test this class takes.
2. Which of the following are not measures of variability: range, $z$-score, variance, standard deviation, percentile rank, raw scores?
3. $s^2$ is a __________ (parameter or statistic) which estimates the value of $\sigma^2$.
4. A sample of people were asked how many brothers and sisters they had. Their responses were 3, 2, 1, 0, 2, 3, 2, 1, 6, and 1.
   (a) Find the mean, median, and mode of this distribution.
   (b) Find the range, variance, and standard deviation.
   (c) Find the largest $z$-score and the smallest $z$-score.
   (d) If somebody had a $z$-score equal to 0.54, how many brothers and sisters does the person have?
5. (a) Draw a histogram for these heights: 72, 50, 59, 60, 63, 71, 70, 35, 65, 63, 61, 75, 60, and 67 inches.
   (b) Find the percentile rank for a height of 61 inches.
6. A sociologist, Andrew Grievely, wants to determine the average number of children in the households of Apawlingburg. He goes to all the schools in the community and takes a random sample of schoolchildren. He asks each child how many children are in his or her household. He uses all of this data to estimate the average. Why will this be a biased sample for his study?
7. (a) If we treat your class as a population and calculate the average wrist size, what symbol should we use for this number?
   (b) If we treat your class as a sample of the entire campus, and calculate the average wrist size, what symbol should we use for this number?
8. (a) Do you think that the average from part (b) in the previous question would be a good estimate for the average wrist size of the entire campus? Explain your answer.
   (b) If we discovered that the wrist sizes extended from 5 inches to 9 inches, then the range for the class would be $9 - 5$ or 4 inches. Do you think that this would be a good estimate for the range of wrist measurements for the entire campus? Explain your answer.
9. Give an example where it would be necessary for a researcher to use information from a sample rather than an entire population.
10. A new medicine is tried on 400 patients who have disease X. In this study, distinguish clearly between the population and the sample.

11. One afternoon 100 listeners to a nationally broadcast radio talk show call in to give their opinion on Senator Smith's plan for health care reform. Of these callers, 85% are opposed to Smith's plan. Explain why this method of learning what the public feels about the Smith plan is probably biased. What would be a less biased method to ascertain public opinion?
12. Explain what is meant by "statistical inference."
13. Calculate the mean of these values and round the answer to the nearest hundredth: 11.66, 12.30, 13.58, 14.95, 15.66, 16.29, and 17.89.
14. Give an example of two small populations, $X$ and $Y$, each made up of five values, where $\mu_X = \mu_Y$, but $\sigma_X \neq \sigma_Y$.
15. In a 1-week study at a health center, the age, sex, and systolic blood pressure of each patient that week was recorded. What are the variables in this study? The mean blood pressure in this study was 122; what parameter does the 122 estimate?
16. In a certain school district, the mean salary for male teachers is $36,000 and the mean salary for female teachers is $28,000. Explain how it is possible that the mean salary for *all* the teachers is only $29,000.
17. Here are the ages of Stanley's pet mosquitos: 3, 17, 23, 18, 34, 52, 34, 21, 15, 34, 35, 45, 42, 41, 30, 28, 24, 25, 23, 43, 34, 13, 15, 43, 46, 65, 53, 34, 37, 23, 36, 37, 38, 23, 26, 26, 34, 54, and 39.
    (a) Draw a stem-and-leaf display for the above data, separating the data into the following groups: 0–4, 5–9, 10–14, 15–19, 20–24, 25–29, 30–34, 35–39, 40–44, 45–49, 50–54, 55–59, 60–64, and 65–69.

    Use these stems, placing the high sixties on the top line and the low sixties on the second line, etc.

    6 |
    6 |
    5 |
    5 |
    4 |
    4 |
    3 |
    3 |
    2 |
    2 |
    1 |
    1 |
    0 |
    0 |

    (b) Draw a stem-and-leaf display for the above data, separating the data into the following groups: 0–9, 10–19, 20–29, 30–39, etc.
    (c) Which of these displays is a better depiction of this data? Why do you think so?

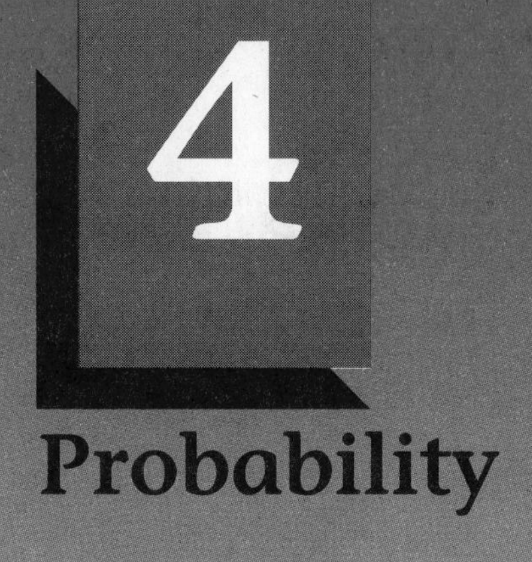

# Probability

**CONTENTS**



**GOALS**

At the end of this chapter you will have learned:

- The meaning of "probability" and how to apply it in a large number of trials
- How to calculate probabilities
- The relationship between probability, proportion, and the percentage of area in a histogram

# 4.1 Probability

The probability of drawing a heart out of an ordinary deck of 52 playing cards is 13/52. The probability of drawing a fifth heart after 4 have already been removed is 9/48.

The probability of tossing a head on a fair coin is 1/2.

The weather bureau announces that the probability of rain tomorrow is 40%.

What meaning does the word "probability" have? If I toss a fair coin 2 times, will I get exactly 1 head? If I toss it 10 times, will I get exactly 5 heads? If I toss it once, will I get one-half of a head? Does a 40% possibility of rain indicate that it will rain 40% of the time tomorrow? Or that it will almost rain tomorrow?

probability

**Probability** is usually interpreted in reference to a *large number* of trials. When we say that the probability that a fair coin will come up heads equals 1/2, we mean that *in a large number* of trials we expect about 50% of the outcomes to be heads.

"The probability of rain tomorrow is 40%" means that out of a large number of days with weather conditions like today's, about 40% of the time it will rain on the following day.

Consider the following illustration.

At a party Peter P. Parapsych professed to have powers of extrasensory perception. To probe this claim we placed 6 similar pieces of candy in a purple paper pouch; 2 of the pieces of candy were peppermint and the other 4 were peach. After the candies were well mixed, Peter was blindfolded and asked to reach into the pouch and pick a peppermint. Peter then proceeded to pick a peppermint. Does this mean he has ESP? What was the probability of picking a piece of peppermint candy just by luck?

## *A Little History*

If any particular questions can be said to have started the quest for a theory of probability it is the two that were put to Blaise Pascal in a Paris salon by the gambler Antoine Gombaud, the Chevalier de Méré, in about 1652:

1. How many tosses of two dice are necessary to have at least an even chance of getting a double 6?
2. How should the money on the table be divided if a betting game is interrupted before it is over?

These questions instigated the correspondence between Pascal and Pierre de Fermat concerning games of chance. However, the science of probability also developed from some very serious topics such as John Graunt's 1662 work on the causes of death in England. His work marks the origins of actuarial science.

**equally likely** In this experiment there were 6 ***equally likely*** possible outcomes, since he could have picked any of the 6 pieces of candy. Two of these outcomes, the peppermint candies, were favorable to Peter's claim. Thus we say that the probability of his picking a peppermint candy *just by luck* was 2 chances in 6; that is, the probability equals 2/6, or 1/3. This does not imply that he will pick peppermint precisely once every 3 trials, but, rather, that in a large number of trials, about one-third of the picks will be peppermint just by luck. We wouldn't even begin to believe his claim unless he could pick the proper candy much more often than 1/3 of the times he tried.

> *Definition:* When all the possible outcomes of an experiment are equally likely, then the **probability of any particular event** is the number of *favorable* outcomes of the experiment divided by the *total* number of possible outcomes.

### A Formula for Probability

The formula for the probability of one of several equally likely events is:

$$P(\text{an event will occur}) = \frac{\text{number of favorable outcomes}}{\text{total number of outcomes}} = \frac{F}{T}$$

> ***Three Outcomes***
>
> If we toss two coins, we obtain either:
>
> i two heads,
> ii two tails, or
> iii one head and one tail
>
> yet the probability of tossing two heads is less than 1/3. Why? (See Exercise 4-26.)

We read the symbol "$P(\ \ )$" as "the probability that."

In the above example,

$$P(\text{a piece of peppermint candy is picked}) = F/T = 2/6 = 1/3$$

You can see that the fraction of the population that makes up an event is the probability of that event.

**randomness** The idea of probability is related to the idea of ***randomness***. For example, when you play a game of cards, to ensure fairness you shuffle the cards. Why? To ensure that no card or cards have a *special place* in the deck; to ensure that *each card* has just the *same chance* of being in any given part of the deck. The statistician says you shuffle the deck to make certain that the cards are arranged *randomly*, or *in random order*. When an item (or number) is picked *at random* from some population, then that item has the same likelihood of being picked as any other item in the population; that is, it has no special, privileged position in the population and neither does any other item in the population.

> *Definition:* An item is selected **at random** if every item under consideration has an equally likely chance of being selected.

### *The Bernoullis*

The "Bernoullis" refers to eleven famous Swiss mathematicians produced in one family during the seventeenth, eighteenth, and nineteenth centuries; *four* of them did important work in the field of probability and statistics.

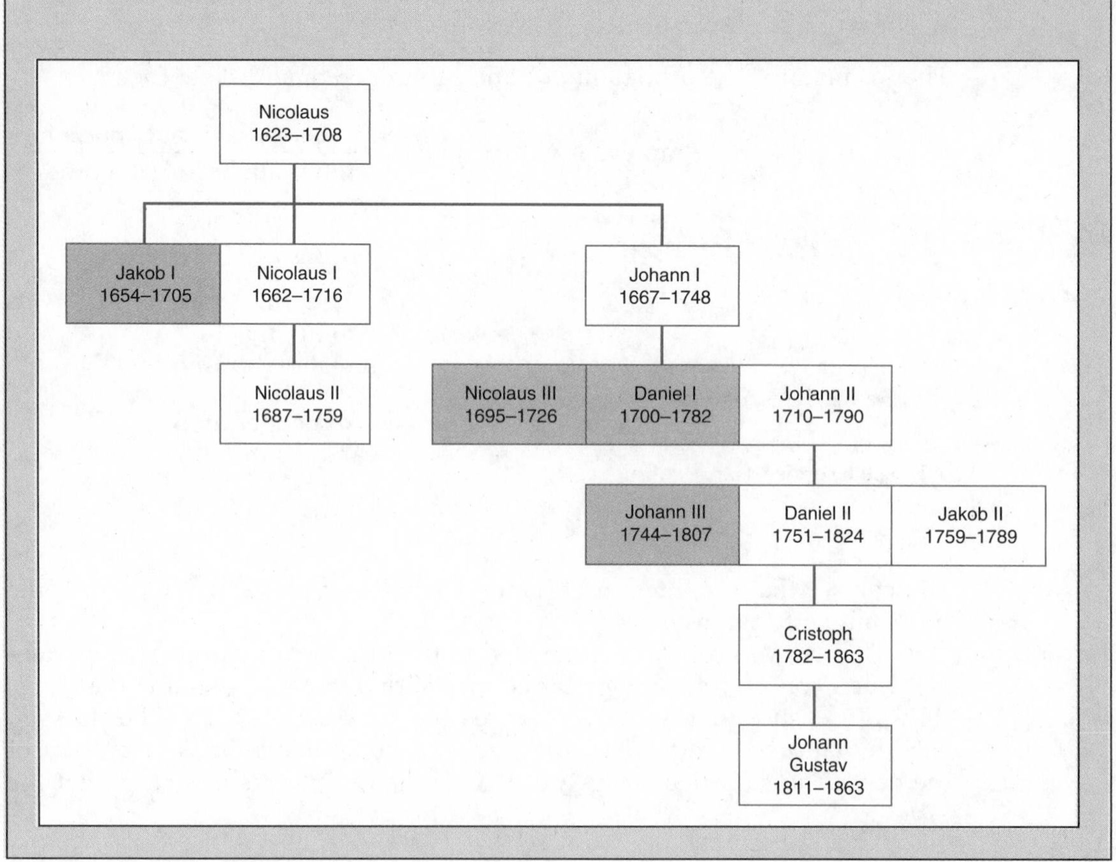

## *The Most Famous Probabilist in the Bernoulli Clan*

A major early contributor to and (an early student of) the theory of probability, Jakob I (also known as Jacques or James) wrote an early text on probability theory called *Ars Conjectandi* ("The Art of Guessing"), which was published posthumously by his nephew Nicolaus II. It treated permutations, combinations, and the binomial theorem. Bernoulli gave Pascal credit for the latter. (This is one reason Pascal's name is attached to the arithmetic triangle of the binomial coefficients.)

### Theoretical vs. Empirical Probability

While others, like Pascal and Fermat, had begun to develop the theory based on games of chance where it was possible to list all the possible, equally likely outcomes as the basis for computation, Jakob Bernoulli was concerned with calculating the probability of events where this could not be done. He proposed to assign probabilities by looking at the results observed in many similar instances, as indicated by the following quotation.

> "For example, if we have observed that out of 300 persons of the same age and with the same constitution as a certain *Titius*, 200 died within ten years while the rest survived, we can with reasonable certainty conclude that there are twice as many chances that Titius also will have to pay his debt to nature within the ensuing decade as there are chances that he will live beyond that time.
>
> —*Ars Conjectandi* according to Newman's *The World of Mathematics*.

Bernoulli's approach is called *empirical*, while Pascal's is called *theoretical*. Both approaches are still important.

It is in *Ars Conjectandi* that we have the first statement of the Law of Large Numbers, which is a fundamental principle of probability theory.

Jakob's brother Johann I (also known as Jean or John) was very touchy. Johann was envious, resenting his older brother and jealous of his own son Daniel. He did his major mathematical work in other fields, mostly differential equations. Johann invented the term "integral," which is widely used in calculus.

## Examples of Probability Calculations

### ❑ EXAMPLE 4.1 A Fair Coin

If we toss a fair coin, what is the probability that a head is tossed?

**A Birthday Question**

What is the probability that 2 people in this class have the same birthday, month and day?

Is it about 25%? 50%? 75%?

SOLUTION

There are 2 possible outcomes, heads or tails, and these 2 outcomes are equally likely. Therefore the probability of tossing heads is written

$$\begin{aligned} P(\text{a head is tossed}) &= F/T \\ &= 1/2 \\ &= 0.50 \\ &= 50\% \end{aligned}$$

❑

### ❑ EXAMPLE 4.2 A Wheel of Fortune

A carnival spinning wheel has the numbers 1 to 20 on it, all equally marked off. When it is spun, it will stop, at random, on 1 of the numbers.

(a) What is the probability that it will stop on the number 14?
(b) What is the probability it will stop on an even number?
(c) What is the probability it will stop on a number equal to 15 or higher?

SOLUTIONS

(a) The probability that it will stop on the number 14 is 1/20 (or 0.05) because (i) there are 20 numbers altogether, and (ii) exactly 1 of the numbers is a 14.
(b) The probability that it will stop on an even number is 10/20 (or 0.50) because (i) there are 20 numbers altogether, and (ii) 10 of them are even.
(c) The probability that it will stop on the number 15 or higher is 6/20 (or 0.30) because (i) there are 20 numbers altogether, and (ii) 6 of the numbers are 15 or higher (15, 16, 17, 18, 19, 20). ❑

### ❑ EXAMPLE 4.3 Rolling One Die

If we roll 1 die, there are 6 possible outcomes: 1, 2, 3, 4, 5, and 6. If we let $X$ represent the value of the outcome, then find each of these probabilities:

(a) $P(X = 3)$
(b) $P(X \text{ is even})$
(c) $P(X > 4)$
(d) $P(X \geq 4)$

SOLUTIONS

(a) $P(\text{the outcome is 3}) = P(X = 3) = 1/6$.
(b) $P(\text{the outcome is even}) = P(X \text{ is even}) = 3/6 = 1/2$.
(c) $P(\text{the outcome is more than 4}) = P(X > 4) = 2/6 = 1/3$.
(d) $P(\text{the outcome is 4 or more}) = P(X \geq 4) = 3/6 = 1/2$. ❑

### ❑ EXAMPLE 4.4 Young and Old

In the town of Juvena, 40% of the population are below 25 years old (the young folks); 60% of the population are 25 years old or older (the older folks). What is the probability that a person picked at random is young?

SOLUTION

If a person is picked at random from the population, the probability that the person will be young is 0.40, which is the fraction of young folks in the entire population. ❑

### *Pascal's Wager*

Historically, two aspects of probability have always existed side by side. One is the use of probability to describe what happens in repeated trials (the relative-frequency approach), and the other is to describe how strongly we believe in something (the subjective-probability approach). In the life of Pascal we see evidence of both. He was among the first to develop rules of probability for analyzing games of chance. But he also incorporated the second approach in thinking about some religious issues.

Pascal was an intensely religious man as well as a scientist and mathematician at a time when science was new and was challenging some aspects of religious doctrine. Pascal was deeply troubled by the fact that nature did not offer him a clear sign for either the existence or nonexistence of God. So he found it useful to apply probabilistic ways of thinking to the question. As one small part of his effort to resolve conflicts between science and religion, he proposed what is now known as "Pascal's wager," which may be summarized in a sketch:

| | God Exists | God Does Not Exist |
|---|---|---|
| We believe in God. | Eternal happiness | No reward after death |
| We disbelieve. | Eternal damnation | No penalty after death |

On the basis of this argument, Pascal concluded that the sensible choice was to believe in God, because even if the probability that God exists was small, the value of the prize of an eternity in heaven was immeasurably great.

### *Birthday Answer*

The following table gives the minimum probability that two or more have the same birthday:

| Class Size Over | Probability Exceeds |
|---|---|
| 14 | 25% |
| 22 | 50% |
| 31 | 75% |
| 40 | 90% |
| 52 | 98% |
| 56 | 99% |
| 69 | 99.9% |

The probability fraction $F/T$ is some value from 0 to 1. If an outcome is impossible, its probability is 0. Thus, the probability of rolling a 13 with two ordinary dice is 0. If an outcome is sure to happen, its probability is 1. Thus, if you draw five cards from a deck of playing cards, the probability that at least two are the same suit is 1, or 100%.

If the probability of your winning a raffle is 0.02, then the probability of your not winning is 0.98. In general, if $p$ is the probability of an event happening and $q$ is the probability of the event not happening, then

$$p + q = 1 \quad \text{or} \quad q = 1 - p$$

Consider a carnival's spinning wheel of fortune which has the numbers 1 to 36 marked off on it in equally spaced divisions. Let $X$ stand for the number at which the wheel stops. If we let $p = P(X < 13)$, then $q = P(X \geq 13)$. Since $p = 12/36 = 1/3$, then $q = 1 - 1/3 = 2/3$. Of course, we could have computed $q$ directly: $q = 24/36 = 2/3$.

**sample space**

When you list *all* the possible outcomes for some problem, that list is often called the **sample space** for the problem.

### *Helpful Hint*

Writing out the sample space often simplifies probability problems.

## ❑ EXAMPLE 4.5 Tossing a Fair Coin Twice

A fair coin is tossed twice. What is the probability of getting 2 tails?

SOLUTION

If on the first toss we get heads, the second outcome can be heads or tails. Similarly, if the first toss is tails, then the second toss can be heads or tails. Since each toss has 2 possible outcomes, we can see that there are $2 \times 2$, or 4, equally likely possible outcomes for the 2 tosses. The sample space consists of the 4 items in the middle column below:

| Outcome | First Toss, Second Toss | Number of Tails |
|---|---|---|
| 1 | Heads, heads | 0 |
| 2 | Heads, tails | 1 |
| 3 | Tails, heads | 1 |
| 4 | Tails, tails | 2 |

Only the last of the 4 possible outcomes was favorable. Therefore, $P(2 \text{ tails}) = 1/4$. ❑

**multiplication principle**

Example 4.5 above illustrates the **multiplication principle**: If there are $A$ possible outcomes for one event and $B$ possible outcomes for a second event, then there are $A \times B$ possible outcomes for the sequence of both events.

### ❑ EXAMPLE 4.6 A Penny and a Die

**_Words, Words, Words_**

As used in this chapter the words "proportion of," "percentage of," "fraction of," and "probability" have similar meanings and the *same value* as in:

The *proportion* of students passing is 9/10.

The *fraction* of students passing is 9/10.

The *percentage* of students passing is 90%.

The *probability* that a student picked at random passes is 0.90.

A penny is tossed once and a die is rolled once. List the possible outcomes.

SOLUTION

Since the penny can come up 2 ways and the die 6 ways, by the multiplication principle there are $2 \times 6$, or 12, possible outcomes—namely, H1, H2, H3, H4, H5, H6, T1, T2, T3, T4, T5, T6. ❑

### ❑ EXAMPLE 4.7 The Penny and the Die Revisited

Find the probabilities of the following outcomes for the previous example, where a die is rolled once and a coin is tossed once.

(a) Tossing a head **and** rolling an even number
(b) Tossing a head **or** rolling an even number
(c) Tossing a head and rolling a 5
(d) Tossing a head or rolling a 5
(e) Rolling either a 4 or a 6
(f) Rolling both a 4 and a 6
(g) Tossing a head or rolling a 7
(h) Tossing a head and rolling a 7

SOLUTIONS

(a) There are 3 favorable outcomes: H2, H4, H6. The answer is 3/12, or 1/4.
(b) There are 9 favorable outcomes: H1, H2, H3, H4, H5, H6, T2, T4, T6. The answer is 9/12 = 3/4.

***Note*** In probability calculations we interpret "$A$ or $B$" to mean the **inclusive or,** "$A$ or $B$ *or both*"

(c) There is 1 favorable outcome: H5. The answer is 1/12.

(d) There are 7 favorable outcomes: H1, H2, H3, H4, H5, H6, T5. the answer is 7/12.
(e) There are 4 favorable outcomes: H4, H6, T4, T6. The answer is 4/12 = 1/3.
(f) **None** of the outcomes are favorable. This is not possible in *one* roll of the die. The answer is 0/12 = 0.
(g) There are 6 favorable outcomes: H1, H2, H3, H4, H5, H6. The answer is 6/12 = 1/2.
(h) **None** of the outcomes are favorable. The answer is 0/12 = 0. ❑

### ❑ EXAMPLE 4.8 Sum of Two Dice

Find the probability of rolling a sum of 7 with 2 fair dice.

SOLUTION

There are 6 faces on each die, numbered 1 to 6. Thus, from the multiplication principle there are 6 × 6, or 36, possible outcomes for 2 dice. This sample space is shown below. Note that, given any way the first die lands, there are still 6 possible ways the second die can land.

(die 1, die 2)

| | | | | | |
|---|---|---|---|---|---|
| (1,1) | (2,1) | (3,1) | (4,1) | (5,1) | **(6,1)** |
| (1,2) | (2,2) | (3,2) | (4,2) | **(5,2)** | (6,2) |
| (1,3) | (2,3) | (3,3) | **(4,3)** | (5,3) | (6,3) |
| (1,4) | (2,4) | **(3,4)** | (4,4) | (5,4) | (6,4) |
| (1,5) | **(2,5)** | (3,5) | (4,5) | (5,5) | (6,5) |
| **(1,6)** | (2,6) | (3,6) | (4,6) | (5,6) | (6,6) |

Of the 36 possible outcomes, there are 6 in which the sum of the dice is 7, and so

$$P(\text{rolling a sum of 7}) = 6/36 = 1/6$$

❑

### ❑ EXAMPLE 4.9 Pamela Purloiner

Pamela Purloiner, a clerk in a jewelry store, has a passion for perfect purple pearls. In a tray containing 4 perfect purple pearls, she has replaced 2 gems with imitations. Her employer, Paul Perlemann, reaches into the tray and picks up 2 stones at random. What is the probability that Paul Perlemann picks 2 perfect purple pearls?

**SOLUTION**

We must determine the number of possible outcomes. To distinguish between the stones, let us label the perfect purple pearls P1 and P2, and the imitations I1 and I2. If the employer picks up a real pearl first, then any 1 of the 3 remaining stones can be picked second. Similarly, taking any one of the 4 stones first leaves a choice of 3 stones for the second selection. Therefore, there are 4 × 3 or 12 *equally likely* outcomes.

| | 1 | 2 | 3 | 4 | 5 | 6 | 7 | 8 | 9 | 10 | 11 | 12 |
|---|---|---|---|---|---|---|---|---|---|---|---|---|
| 1st stone | P1 | P1 | P1 | P2 | P2 | P2 | I1 | I1 | I1 | I2 | I2 | I2 |
| 2nd stone | P2 | I1 | I2 | P1 | I1 | I2 | P1 | P2 | I2 | P1 | P2 | I1 |

There are only 2 outcomes of the 12 that would be favorable, that is, selecting *both* real *perfect purple pearls*, and so the probability of taking the 2 perfect pearls is 2/12, or 1/6. ❑

## 4.2 Probability and Histograms

Since the fraction of the total population that corresponds to an event is the probability of that event, in a histogram the fraction of the area that represents the event is equal to the probability of that event.

### ❑ EXAMPLE 4.10 Fuel Mileage

A statistician has been studying fuel mileage on a particular model automobile by testing many such cars. The resulting graph is shown in Figure 4-1.

| Mileage | Percentile Rank |
|---|---|
| 15 | 0 |
| 18 | 2 |
| 21 | 16 |
| 24 | 50 |
| 27 | 84 |
| 30 | 98 |
| 33 | 100 |

What is the probability that 1 of these cars picked at random will get less than 27 miles per gallon?

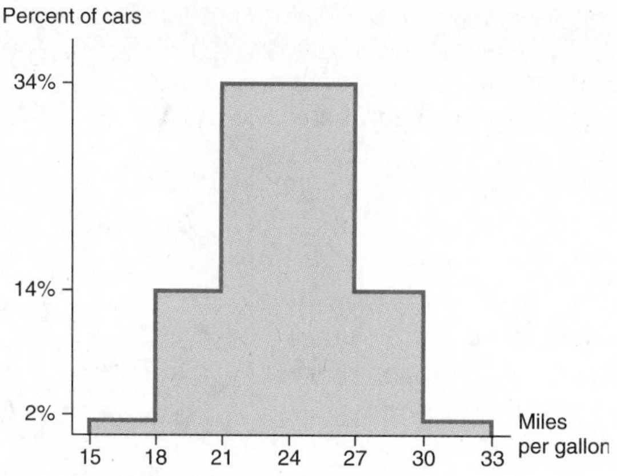


**Figure 4.1** Histogram of fuel mileage.

SOLUTION

We need to know the fraction of cars in this category. This is equal to the proportion of the *area* of the graph to the left of 27. The percentile rank of 27 is 84, and so to the left of 27 is 84% of the graph. Therefore,

$$P(\text{one of these automobiles gets less than 27 miles per gallon}) = 0.84$$ ❑

---

### *A Paradox*

Nicolaus Bernoulli III wrote on probability and proposed the famous *Petersburg Problem*, which deals with a situation such as the following:

A tosses a coin. If the first head appears on toss number:

1 B pays A \$1
2 B pays A \$2
3 B pays A \$4
4 B pays A \$8
⋮
$n$ B pays A $\$2^{n-1}$

What should A pay for the privilege of playing?

By mathematics, Nicolaus's brother Daniel (1700–1782) concluded that A should pay a huge amount, but common sense said the answer should be a small finite amount. This unresolved paradox still lacks a satisfactory explanation 200 years later. Technically, the problem hinges on the fact that if you conceive that the number of tosses is theoretically unlimited, then the calculation of the average payoff gives an answer which is not finite.

Georges Louis Leclerc, comte de Buffon (1707–1788), found by repeated trials that B would have paid \$10,057 in 2084 games, an average of about \$4.83 per game.

*In one computer simulation of 200,000,000 games, B paid A an average of \$9.69, with the payout ranging from a low of \$1 to a high of \$1,048,576 (which is $2^{20}$, indicating that A tossed 20 tails). A second run of the computer program produced an average of \$14.15, with a payout as high as \$67,108,864 ($2^{26}$).*

## STUDY AIDS

### Chapter Summary

In this chapter you have learned:

- The definition of probability and how to interpret it in a large number of trials
- How to calculate the probability of an event
- The relationship between probability, proportion, and the percentage of area in a histogram

### Vocabulary

You should be able to explain the meaning of each of these terms:

1. The probability of an event
2. Equally likely
3. Selected at random
4. Sample space
5. The multiplication principle

### Symbols

You should understand the meaning of each of these symbols:

1. $P$(event)
2. $p$
3. $q$

### Formulas

You should know when and how to use each of these formulas:

1. $P(\text{event}) = F/T$
2. $q = 1 - p$

## EXERCISES

4-1 From an ordinary playing deck of 52 cards, 1 card is selected at random. Find the probability:

(a) That a 3 of diamonds is selected

(b) That a heart is selected

(c) That a jack is selected

4-2 Find the following probabilities for the random selection of 1 card from an ordinary deck of 52:

(a) $P$(a queen)

(b) $P$(a heart)

(c) $P$(a queen or a heart)

(d) $P$(a queen and a heart)

(e) $P$(a king or a queen)

(f) $P$(a king and a queen)

(g) $P$(a heart or a diamond)

(h) $P$(a red card or a black card)

### *Bridge Hands*

The probability that all 4 players at bridge will be dealt a complete suit is 1/2,235,197,406,895,366,368,301,560,000.

4-3 A total of 365 capsules, each containing the date of 1 day of the year, are mixed, and 1 capsule is picked at random. What is the probability that the 1 capsule picked will be:

(a) Some day in January?

(b) March 2?

(c) Either in March or in April?

(d) Not in December?

4-4 Refer to the carnival wheel of Example 4.2. Let $X$ represent the number the wheel stops on. Find each of these probabilities:

(a) $P(X = 20)$

(b) $P(X \text{ is divisible by } 5)$

(c) $P(X \text{ is } 7 \text{ or } 11)$

(d) $P(X \leq 2)$

4-5 A tile floor is made up of squares as illustrated. The tiles are numbered, and the odd-numbered tiles are grey, while the even-numbered tiles are red.

| 1 | 2 | 3 | 4 | 5 |
|---|---|---|---|---|
| 6 | 7 | 8 | 9 | 10 |
| 11 | 12 | 13 | 14 | 15 |

A pin is dropped from a great height by a blindfolded person and is just as likely to fall in 1 square as in any other. If the pin falls on a crack between 2 tiles, it does not count, and the pin is dropped again. The outcome will yield both a color and a number. Find:

(a) $P$(an outcome is grey)

(b) $P$(an outcome is red)

(c) $P$(the number is less than 5)

(d) $P$(the number is greater than 5)

(e) $P$(an outcome is red and less than 5)

(f) $P$(an outcome is red or less than 5)

4-6 A standard roulette wheel in an American casino has 38 pockets, including 1 marked "zero" and 1 marked "double zero"; the others are numbered from 1 to 36. The pockets for the odd numbers from 1 to 35 are red, the pockets for the even numbers from 2 to 36 are black, and the other 2 pockets are green.

The wheel is spun, and the ball falls randomly into 1 of the pockets. Let $X$ represent that pocket. Find these probabilities:

(a) $P(X \text{ is red})$

(b) $P(X \text{ is not red})$

(c) $P(X \text{ is } 1)$

(d) $P(X \text{ is red and more than } 30)$

(e) $P(X \text{ is red or more than } 30)$

(f) $P(X \text{ is double zero})$

(g) $P(X \text{ is red or black})$

(h) $P(X \text{ is neither red nor black})$

4-7 What is the probability that 1 day selected at random from this month is

(a) A Monday?

(b) Not a Monday?

4-8 Assuming that births are just as likely to happen any one day as another, find the probability that a person picked at random has a birthday on:

(a) February 29 (*Hint:* Use a period of 4 years to make the calculation.)

(b) February 28

4-9 In a certain high school, 20% of the students are seniors, and 48% of the students are girls. Assuming that the students enter the cafeteria randomly, we stand at the door of the cafeteria and watch for the next pupil to enter. What is:

(a) $P(X \text{ is a senior})$?

(b) $P(X \text{ is not a senior})$?

(c) $P(X \text{ is a boy})$?

4-10 Here is a chart showing the racial breakdown of employees at a certain factory.

| Race | Percentage |
|---|---|
| Asian | 6 |
| Black | 46 |
| White | 43 |
| Other | 5 |

What is the probability that 1 employee picked at random will be:

(a) Asian?

(b) Not white?

(c) Either black or Asian?

(d) Not black?

4-11 There are 100 males and 100 females belonging to Klubbe Club; 80 of the males and 10 of the females are scientists.

(a) What is the probability that a *person* picked at random from the membership list is:

(i) Female?
(ii) A scientist?
(iii) A female scientist?

(b) What is the probability that a *scientist* picked at random in the organization is a female?

(c) What is the probability that a *female* picked at random from the organization is a scientist?

4-12 The *Statistical Abstract of the United States* (1993), published by the U.S. Census Bureau, projects the following breakdown of U.S. population by age for the year 2000:

| Age | Population |
|---|---|
| Under 5 | 18,908,000 |
| 5–13 | 36,051,000 |
| 14–17 | 15,734,000 |
| 18–24 | 26,117,000 |
| 25–34 | 37,416,000 |
| 35–44 | 44,662,000 |
| 45–54 | 37,054,000 |
| 55–64 | 23,988,000 |
| 65–74 | 18,258,000 |
| 75–84 | 12,339,000 |
| 85 and up | 4,289,000 |

Using these figures, if a person is picked at random from the U.S. population in the year 2000, what is the probability that person is;

(a) Under 18 years old?

(b) At least 85 years old?

(c) In the age group 25–44?

4-13 An experiment consists of tossing a penny and a dime.

(a) List the possible outcomes.

Find the probability of the following:

(b) $P$(a head on the penny)

(c) $P$(heads on both)

(d) $P$(exactly 1 head)

(e) $P$(at least 1 head)

4-14 Refer to Example 4.6 involving a penny and a fair die. Find the probabilities of each of the following outcomes:

(a) Tossing a tail and rolling a 6

(b) Tossing a tail and rolling a 5 or a 6

(c) Tossing a tail

(d) Tossing a head and rolling more than a 4

(e) Tossing a head or rolling more than a 4

4-15 In the game of Monopoly, Carol Capitalist needs to roll an 8 on the 2 dice to land on Park Place. Refer to the list of possible outcomes for 2 dice in Example 4-8.

(a) What is the probability that Carol will succeed in rolling an 8?

(b) Billy Badluck will go to jail if he rolls doubles 1 more time. What is the probability of that happening?

4-16 (a) Explain by referring to Example 4.8 why 7 is the most likely sum when you roll 2 fair dice.

(b) Is it true that you are twice as likely to throw a sum of 7 as a sum of 4? Explain your answer.

(c) What is the probability that the sum of 2 dice will be an even number?

(d) If you roll 2 dice over and over again and write down the sum each time, about what fraction of the results will be 7s?

4-17 A spinner is labeled as indicated in the diagram.

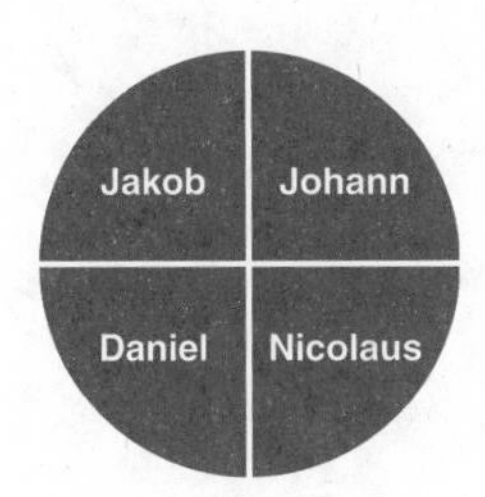


If we spin it once, find:

(a) $P$(the outcome is Jakob)

(b) $P$(the outcome begins with the letter "J")

If we play the game twice, there are 16 possible outcomes.

(c) List them.

Of the 16 outcomes in part (c), what is:

(d) $P$(the first spin is Jakob, and the second is Johann)?

(e) $P$(the first spin is Jakob, and the second is Jakob)?

(f) $P$(the outcome includes Jakob exactly once)?

(g) $P$(the outcome includes Jakob at least once)?

4-18 Refer to the spinner in the previous exercise.

(a) Suppose we spin it once. What is the probability that the outcome is Daniel or Jakob? What is the probability that Daniel and Jakob do not appear at all? Explain why these two answers must sum to 1.

(b) Suppose we spin it twice. What is the probability that at least 1 outcome is Daniel or Jakob? What is the probability that Daniel and Jakob do not appear at all? Explain why these 2 answers must sum to 1.

4-19 In a kitchen drawer are 5 AA batteries which look alike. Unfortunately, 2 of them are dead. You need 2 good ones for your camera's electronic flash. If you grab 2 at random, what is the probability that they will both be good? (*Hint.* Label the batteries as G1, G2, G3 for the good ones and B1 and B2 for the bad ones, and show the sample space of all 20 possible outcomes.)

4-20 A young girl has been picking mushrooms. She accidentally picks 2 toadstools which to her appear identical to the 3 mushrooms she picked. She will eat 2 of the 5 fungi picked.

(a) Using M1, M2, and M3 for the mushrooms and T1 and T2 for the toadstools, list all 20 possible outcomes.

What is the probability that she:

(b) Eats 2 toadstools?

(c) Eats at least 1 toadstool?

(d) Eats no toadstools?

4-21 **tree diagram** Sometimes it is easy to picture all the possible outcomes of an experiment by drawing a **tree diagram.** In Example 4.5, where we tossed a coin twice, the tree diagram looks like this.

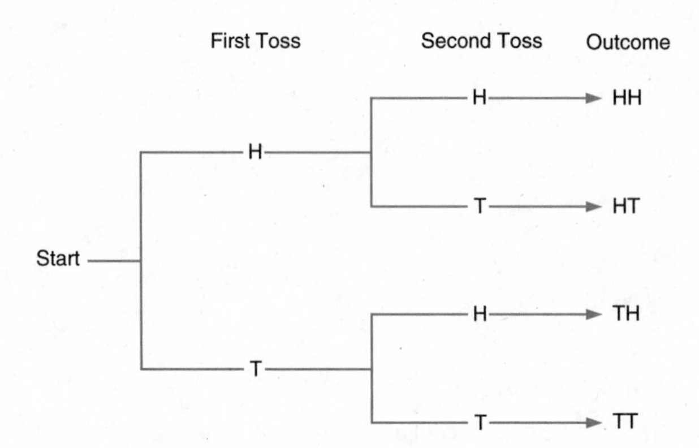


Each possible path from the left starting point to a right endpoint represents 1 possible outcome. It is easy to see from this diagram what the possible outcomes are.

(a) Draw a tree diagram for Example 4.6 (about tossing a penny and a die), and list the outcomes corresponding to each branch. Your answer should be the same as that given in the example.

(b) Four pennies are tossed. Draw a tree diagram for this experiment. Find the probability of tossing exactly 2 heads and of tossing more than 2 heads.

(c) If the experiment in part (b) consisted of tossing *4 dice*, would you want to solve it by use of a tree diagram? Why or why not?

4-22 A child's toy slot machine has 3 wheels, and each wheel has 1 banana, 1 cherry, and 1 lemon.

(a) How many outcomes are there? List them.

(b) Find $P(3 \text{ lemons})$.

(c) Find $P(3 \text{ the same})$.

(d) Find $P$(at least 2 lemons).

(e) If we play the slot machine 54 times, about how many times would you expect to see 2 or more lemons? Suppose this happened only once. What would your reaction be?

4-23 Prunella Fructus wants to select 2 items at random from a basket containing 2 figs, 2 dates, and 1 lime.

(a) If she selects 1 item and does not replace it before she selects the second item, find the probability that she selects 2 different fruits.

(b) Repeat part (a) if she replaces the first item before selecting the second.

(c) Suppose she picks 2 items without replacement. If you see the first item selected and it is a fig, what is the probability that she selected 2 different fruits?

(d) Repeat part (c) if the first item is a lime.

4-24 On a true-false test you have no idea of the answers to three questions, and so you decide to guess.

(a) List all the possible outcomes, using C to stand for "correct" and W to stand for "wrong."

(b) If $X$ = number of correct guesses, find
 (i) $P(X = 3)$
 (ii) $P(X \geq 2)$
 (iii) $P(X = 0)$

4-25 A teacher of ESP is thinking of a random letter of the alphabet. In front of a class of 260 students, the teacher asks them to concentrate very hard and to write down what letter comes to them. It turns out that 9 people have the correct letter. Have we discovered 9 people with ESP? Why or why not? Explain your conclusion.

4-26 If we toss 2 coins in the air, there are 3 possible outcomes: 2 heads, 1 head, or no heads. Why is the probability of obtaining 2 heads *not* equal to 1/3?

4-27 **Rates and Probability.** We can often interpret rates as probabilities. For example, in Chapter 2 we discussed death rates. Suppose the annual death rate for a particular disease is 2 per 100,000 in your community. One could say that the probability is about 2/100,000 that an individual picked at random in your community will die of this disease in the next year.

The highway patrol in Speed City predicts 20 traffic fatalities over the upcoming St. Swithin's day weekend. There will be about 5000 travelers.

(a) What is the predicted death rate per 1000 travelers?

(b) What is the probability that a traveler picked at random will be killed?

4-28 In 1993 the infant mortality rate in Cuba was 10.7 per 1000 live births. This means that between 10 and 11 out of every 1000 babies who were born alive died before their first birthday. Health officials often consider the infant mortality rate as a good indicator of health care in a population. (The corresponding figure for the United States was 8.6; for Canada it was 6.8.)

(a) What was the probability (in 1993) that a Cuban baby did not survive until its first birthday?

(b) There were about 176,000 babies born in Cuba in 1993. About how many of these babies died before reaching age 1?

4-29 When a baby is born, it is either male or female. If it is true that there is an approximately equal chance that a newborn baby will be male or female, then the histogram below shows the theoretical distribution of the relative frequencies of males in a family of 4 children.

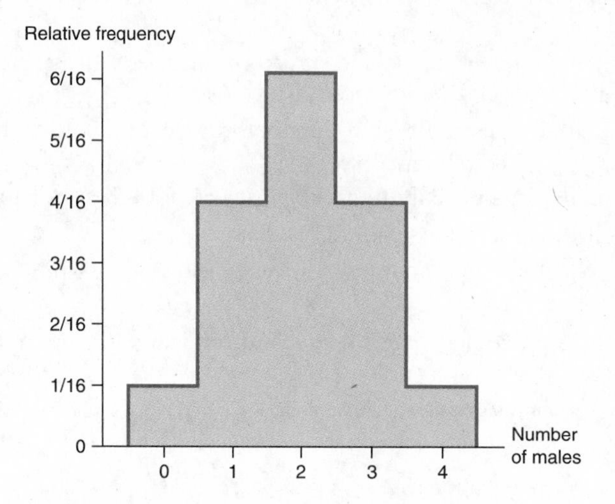


Use the histogram to find the probability that a family with 4 children will have:

(a) No male children.

(b) 2 female children.

(c) All female children.

(d) 2 or more male children.

(e) Either 1 male child or 3 male children.

4-30 The table and graph of relative frequencies for family incomes in Padsville are shown below:

| Income, Dollars | Percentile Rank |
|---|---|
| 0 | 0 |
| 10,000 | 35 |
| 20,000 | 50 |
| 30,000 | 67 |
| 40,000 | 100 |

0 10,000 20,000 30,000 40,000 Family income (dollars)

What is the probability that a family picked at random in Padsville has an income:

(a) Less than $20,000?

(b) Between $10,000 and $30,000?

(c) Either less than $10,000 or more than $30,000?

(d) Below $30,000?

4-31 We repeat the fuel mileage data from Example 4.10:

| Mileage | Percentile Rank |
|---|---|
| 15 | 0 |
| 18 | 2 |
| 21 | 16 |
| 24 | 50 |
| 27 | 84 |
| 30 | 98 |
| 33 | 100 |

Find the probability that an engine picked at random gets:

(a) Less than 21 miles per gallon

(b) Between 21 and 27 miles per gallon

(c) More than 24 miles per gallon

4-32 Use the data in Exercise 4-12 to draw a bar graph for the U.S. population in the year 2000. Explain why *with the intervals given* you cannot technically call this a histogram. If you use the given intervals *and make all the bars the same width*, then what percentage of the total area will fall in the bar corresponding to age "under 5 years?"

4-33 (a) There is an old saying, "Bread always falls buttered side down." Since there are 2 possible outcomes (buttered-side up and buttered-side down), why wouldn't the probability of the bread falling buttered side down equal 1/2?

(b) The professors in Good College are either male or female. We will pick 1 of them at random. Why can't we conclude that we have a 50% chance of choosing a female professor?

(c) On a fair die, the probability of getting a 2 is 1/6. Suppose a nasty person alters the die. There are still 6 possible outcomes, but why can't we be sure that the probability of getting a 2 is still 1/6?

4-34 (a) The following probability statements are either both true or both false. Which is it? Give a possible explanation.

(i) Insecurity Trust and Savings Bank employs 12 different classifications of workers, including tellers. Count de Munnee works there. Hence, the probability that the Count is a teller is 1/12.

(ii) Let $X$ be the middle digit in a 9-digit Social Security number. If a Social Security number is picked at random, then $P(X \text{ is even}) = 1/2$.

(b) Can you explain why (iii) is true and (iv) is false?

(iii) Let $X$ = the fourth digit in a 7-digit phone number without the area code, then $P(X \text{ is the digit } 2) = 1/10$.

(iv) Let $X$ = the fourth digit in a 10-digit area code + phone number, then $P(X \text{ is the digit } 2) = 1/10$.

4-35 A group of people are going to play Russian roulette continuously until only 1 is "left alive." A *statistical model* of this experiment can be constructed as follows. Each person will roll a die once. If the outcome is a 1, he or she is "dead." The die is passed around among the "living" until only 1 person is left. Find:

(a) $P$(first person "dies" on first roll).

(b) $P$(second person "dies" on first roll).

(c) Using the list of possible outcomes for the 2-dice experiment in Example 4.8, find $P$(a particular person "lives" through first 2 rolls). That is, find $P$(the first roll is not a 1, and the second roll is not a 1).

(d) Construct a tree diagram for the outcomes in a 2-roll game. Why are there only 31 branches and not $6 \times 6$, or 36?

4-36 A Bill of Mortality,* published in London in 1632, reads in part:

**The Diseases, and Casualties this year being 1632.**

| | |
|---|---|
| Affrighted | 1 |
| Aged | 628 |
| Bit with a mad dog | 1 |
| Cancer and Wolf | 10 |
| Dead in the street and starved | 6 |
| Kil'd by several accidents | 46 |
| Lunatique | 5 |
| Murthered | 7 |
| Over-laid, and starved at nurse | 7 |
| Plague | 8 |
| Purples, and spotted Feaver | 38 |
| Quinsie | 7 |
| Rising of the Lights | 98 |
| Suddenly | 62 |
| Teeth | 470 |
| Tissick | 34 |
| Worms | 27 |

Searchers ("ancient matrons") examined bodies and determined the cause of death. What is the probability that a body picked at random from the above sample was diagnosed as having died either by being bit by a mad dog or from the "rising of the lights"? (Source: *The World of Mathematics,* J. R. Newman, Simon and Schuster, 1956, p. 1425.)

4-37 A wooden cube is painted on all 6 faces. It is then cut into 27 smaller cubes, all of which are the same size.

*See Exercise 1-17(b) and the box on page 21.

(a) How many have paint on no face?
(b) On only 1 face?
(c) On 2 faces?
(d) On 3 faces?
(e) On 4 faces?
(f) The 27 cubes are placed in a bag. A blindfolded student pulls 1 cube out and rolls it on a table. What is the probability that it lands with a painted face up?

4-38 **The birthday problem.** "Sure Thing Jones" enters a room containing 21 other people whom she does not know. She announces, "I'll bet that at least 2 people here share the same birthday." Would you take the bet?

Calculate the probability that 2 or more people in a group of 22 people have the same birthday. (*Hint:* What is the probability that no 2 people in the room have the same birthday?)

## CLASS SURVEY QUESTIONS

1. What is the probability that someone picked at random in this class:
   (a) Is a smoker?
   (b) Is a nonsmoker?
   (c) Is left-handed?
   (d) Is right-handed?
   (e) Has had one or more broken bones?
2. What proportion of the mothers of people in your class were 5 feet, 6 inches tall or less?
3. What is the probability that the mother of 1 person picked at random from your class is 5 feet, 6 inches tall or shorter?
4. If you drew a histogram of the heights of the mothers of those in your class, what percent of the area under the graph would be to the left of 5 feet, 6 inches? To the right?

## FIELD PROJECTS

1. If you completed the field project described in Chapter 1, answer parts a, b, and c.
   (a) What proportion of the people were 19 years old?
   (b) What is the probability that 1 person picked at random from your sample is 19 years old?
   (c) Using the histogram you drew for this field project in Chapter 3, what percent of the area under the graph is above the interval representing 19-year-olds?
2. Go to the cafeteria or snack bar at lunchtime. Record 50 consecutive purchases at 1 register. Draw a graph of the distribution of purchases.

(a) What fraction of the purchases totaled \$2 or more?
(b) Based on your answer to part (a), what is the probability that a single purchase picked at random will be \$2 or more?
(c) Repeat the problem on a different day to see if there is much change in the answers to parts (a) and (b).
(d) Discuss whether these samples are random or not.
(e) Find the mean and variance for your 50 purchases.
(f) What proportion of the purchases are more than 1 standard deviation from the mean?

# 5

# The Binomial Distribution

A thing either is or it is not.
—Scholastic Philosophy's
Law of the
Excluded Middle

**CONTENTS**



**GOALS**

At the end of this chapter you will have learned:

- The definition and use of several types of variables, especially the binomial random variable
- How to use Pascal's triangle of binomial coefficients
- How to calculate probabilities for a binomial random variable

## 5.1 Variables

In Chapter 2 we spoke about variables such as ID, Name, Age, and Height. We now formalize the idea of a variable.

variable

> *Definition:* A **variable** is a characteristic of a population which can take on different values for different members of the population.

It is useful to realize that there are different types of variables, and that calculations that make sense for one type may not make sense for others. We consider two different types of variables.

quantitative variable

> *Definition:* A **quantitative variable** is one that uses a number which has a *measurable* interpretation.

An example of a quantitative variable is weight; when you record a subject's weight, you write down a specific number which measures something. For instance, a person who weighs 120 pounds is twice as heavy as a child who weighs 60 pounds. It makes sense to compute the mean weight of a group of people.

Another example of a quantitative variable is an integer test score ranging from 0 to 100. An individual score can be any 1 of 101 different values, and a 90 is 6 points better than an 84.

By contrast, sex is **not** a quantitative variable; a subject is either male or female. There is no obvious numerical interpretation of this. Even if we record the data numerically, such as 0 for male and 1 for female, 1 is not more than 0—the numbers do not measure anything. It makes no sense to speak of the mean sex of a group of people.

categorical variable

> *Definition:* A **categorical variable**, or qualitative variable, is one where each subject in a study is simply assigned a label.

Sex is a categorical variable. Some categorical variables can take on *many* values, such as the four marital states: *single, married, widowed, divorced*. An individual can be in any one of four categories.

Occasionally we may choose to treat even quantitative variables as categorical, as when we divide the heights of subjects into three categories: *short, average,* and *tall.*

In this chapter we will consider some problems that involve the simplest kind of categorical variable, one that is **dichotomous**, or has **only two** labels.

**dichotomous variable**

> *Definition:* A **dichotomous variable** is one which has exactly two possible values.

The word "dichotomous" comes from Greek words meaning "cut into two parts." A variable with exactly two possible outcomes is also known as a **Bernoulli variable** or, simply, a 2-valued variable, and the two outcomes are often referred to as "success" and "failure." Some instances of dichotomous or Bernoulli variables are:

**Bernoulli variable**

- Answers in a true-false test
- Responses to yes-no questions
- Results of a contest in terms of win-lose (but not in terms of win, lose, draw)
- Grades in a pass-fail grading system
- Consequences of tossing a coin (heads-tails)
- Outcomes in a dice game in terms of rolling a 7 or not rolling a 7

In some studies it makes sense to assign probabilities to the two labels of a dichotomous variable. For example, in tossing a coin repeatedly, the two outcomes, heads and tails, may each have a probability of 1/2. In a medical experiment, it may make sense to label the possible outcomes as "**successful**" and "not successful" and to assign probabilities to these outcomes allowing us to make statements like, "For any patient getting this treatment, there is an 80% chance of a successful outcome and a 20% chance of an unsuccessful outcome."

**successful outcome**

## 5.2 Binomial Random Variables

In this text we are interested only in those dichotomous variables in which the probabilities of these outcomes **remain the same** for *each trial* regardless of what has happened on previous trials.

We will not consider variables whose outcomes have changing probability from trial to trial, such as a variable whose two outcomes are *rain* and *no rain*—because the probability of rain **changes** each day; however, the result of tossing a *biased* coin is fair game (pun intended), since the probabilities for heads and tails remain the same for each toss.

### *A Binomial Random Variable*

*Note* Each "yes" or "no" is a value of a **dichotomous variable**. In contrast, the total 4 is a value of the **binomial random variable** $S$.

In most research involving dichotomous variables we are ultimately interested in answering questions about the **total number of successful outcomes** in a **set** of trials.

For instance, we ask not merely what is the probability of tossing heads on *1* toss of a coin, but what is the probability that in a *set of 10 tosses* of a coin we get a total of more than 6 heads? The **total number of successes** in such a set of trials is a random variable which is unknown before the trials begin. For instance, in 10 tosses of a coin we could end up with anything from zero to 10 heads, and we would like to be able to assign probabilities to the various possibilities. You will soon see that many applications in statistics can be modeled on coin-tossing experiments. The analysis of such experiments depends on the concept of a **binomial random variable**. For ease of reference, we shall often call any study that analyzes a binomial random variable a **binomial experiment**.

**Total number of successes**

**binomial experiment**

**binomial random variable**

> *Definition:* A **binomial random variable** $S$ is the total number of successes in the $n$ trials of a dichotomous variable, where on each trial the probability of success is $p$.

**binomial distribution**

When $S$ is a binomial random variable, we also say that $S$ has a **binomial distribution**.

### *Not Just Two*

A *bi*nomial random variable $S$ is named for the algebraic binomial $(p + q)$, where $p$ and $q$ are probabilities. $S$ takes on $n + 1$ values.

Since $S$ stands for the total number of successes, the possible values of $S$ are $0, 1, 2, \ldots, n$.

The definition will be easier to understand if you just imagine a coin-tossing experiment.

In the following illustrations of binomial random variables, we identify the various components of each experiment:

1. The total number of heads in a set of 10 tosses of a fair coin.

   One trial means one toss of the coin.

   Success means heads.

   $p = 0.5$ = probability of heads on any one toss.

   $q = 1 - 0.5 = 0.5$ = probability of tails on any one toss.

   $n = 10$ trials.

   $S$ represents the total number of heads and may take on any of the **11** whole-number values 0, 1, 2, 3, 4, 5, 6, 7, 8, 9, or 10.

2. The total number of lung cancer patients out of a group of 1000 who live at least 5 years after radiation treatment given that the probability of this survival for any one patient is 80%.

   One trial means treating one patient with radiation.

   Success means the patient survives at least 5 years.

   $p = 0.80$ = probability of surviving at least 5 years for any one patient.

   $q = 1 - 0.80 = 0.20$ = probability of death for any one patient before 5 years.

   $n = 1000$ patients treated.

   $S$ represents the total number who survive at least 5 years and may take on any of the 1001 whole-number values from 0 to 1000.

3. The total number of times Blaise Pascal rolls doubles in *30 tries* with two dice.

   One trial means rolling a pair of dice once.

   Success means getting doubles.

   $p = 1/6$ = probability that any one roll of the dice gives doubles.

   $q = 1 - 1/6 = 5/6$ = probability that any one roll of the dice does not give doubles.

   $n = 30$ trials.

   $S$ represents the total number of doubles and may take on any of the 31 whole-number values from 0 to 30.

4. Twelve students who have no idea of the proper answer are randomly guessing which 1 of the 11 Bernoullis is the correct response to a test question.

   One trial means guessing one of the 11 Bernoullis.

   Success means a student guessed the right one.

   $p = 1/11$ = probability the guess is correct.

   $q = 1 - 1/11 = 10/11$ = probability that the guess is wrong.

   $n = 12$ tries.

   $S$ represents the total number of correct guesses and may take on any of the 13 values 0, 1, 2, 3, . . . , 12.

## ❑ EXAMPLE 5.1 A Wheel of Fortune

The Knights of Columbus run an annual fund-raising bazaar for a local hospital. A game of chance at the bazaar is set up like this: A spinner has 3 equal areas colored red, green, and blue. A patron bets on 1 color. The wheel is spun, and if it stops with the pointer on the color the patron selected, the player wins. Marty decides to play the game twice, betting on red both times. The result of his betting is a binomial random variable with 2 trials since he will bet 2 times.

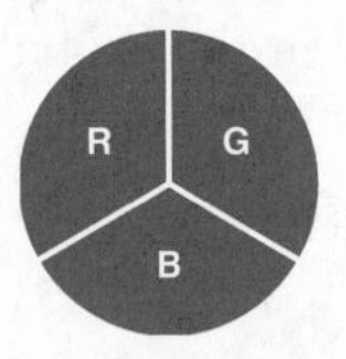


What is the probability that Marty will win:

(a) Both times?
(b) Just 1 time?
(c) Not at all?

We will solve this problem two ways; first, by looking at all the equally likely outcomes as before; and second, by a new approach we will call the "binomial" approach.

### SOLUTION Method 1: By Considering All of the Equally Likely Outcomes

Since there are 3 possible outcomes on each game, there are $3 \times 3 = 9$ possible outcomes for two games. These equally likely outcomes are listed below:

| Outcome | 1st Game, 2nd Game | $S$ = Total Number of Reds |
|---|---|---|
| 1 | Red, red | 2 reds |
| 2 | Red, green | |
| 3 | Red, blue | 1 red |
| 4 | Blue, red | |
| 5 | Green, red | |
| 6 | Blue, green | |
| 7 | Green, blue | No reds |
| 8 | Blue, blue | |
| 9 | Green, green | |

(a) The probability that Marty wins twice is $P$(Marty has 2 wins) $= P(S = 2)$ $= 1/9$, since there are 9 equally likely outcomes and only 1 of them consists of 2 reds.
(b) The probability that Marty wins exactly 1 time is $P$(Marty has 1 win) $=$ $P(S = 1) = 4/9$, because 4 of the outcomes consist of 1 red.
(c) Similarly, the probability that Marty does not win at all is $P$(Marty has 0 wins) $= P(S = 0) = 4/9$.

We can present these probabilities in a probability histogram, as shown in Figure 5-1. Note, since the probability histogram encompasses all possible cases, the total area is equal to $4/9 + 4/9 + 1/9 = 1$.

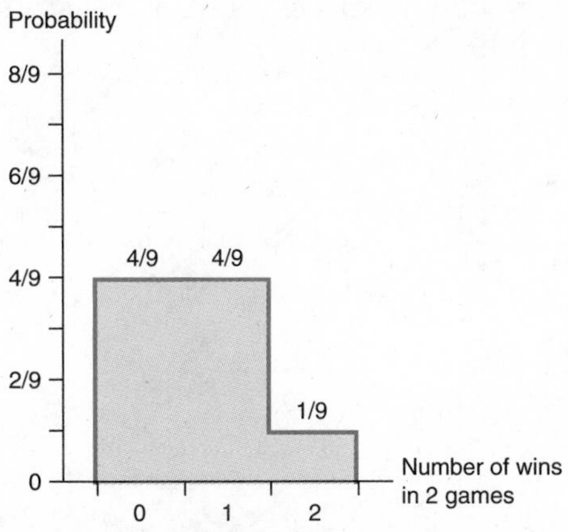


Figure 5-1 Probability histogram.

**SOLUTIONS Method 2: Using a Binomial Random Variable**

There is an alternative way to compute these probabilities (1/9, 4/9, 4/9), which is often easier. We can analyze the game directly in terms of win and lose, instead of in terms of color. As far as Marty is concerned, when he plays twice, he will win twice, once, or not at all. There is only one way he can win twice (he must win on both games); there are two ways he can win once (either win on the first game, or win on the second game); and there is only one way he can lose twice (he must lose on both games). We summarize this in Table 5-1.

Table 5-1

| Number of Wins | Ways the Wins Can Occur | Number of Different Ways the Wins Can Occur |
|---|---|---|
| 2 | WW | 1 |
| 1 | WL, LW | 2 |
| 0 | LL | 1 |

Recall that in a binomial experiment there are two possible outcomes on each trial. One of these is called a "success," and the other is called a "failure." We are now treating this problem as a binomial problem where "success" is win (W), "failure" is lose (L), and a "trial" is one spin. It is standard in this type of problem to let **n** stand for the **n**umber of trials, to let **S** stand for the number of **s**uccesses in $n$ trials, and to let the symbol $\binom{n}{S}$ stand for the number of different ways $S$ can occur.

**binomial coefficient**

The symbol $\binom{n}{S}$ is called a **binomial coefficient** and can be read as "the number of ways one can get exactly $S$ successes in $n$ trials." Thus, we can relabel Table 5-1 as shown in Table 5-2.

Table 5-2

| $S$ | Ways $S$ Can Occur | $\binom{2}{S}$ |
|---|---|---|
| 2 | WW | $\binom{2}{2} = 1$ |
| 1 | WL, LW | $\binom{2}{1} = 2$ |
| 0 | LL | $\binom{2}{0} = 1$ |

Now we have reformulated Marty's betting problem as a binomial problem.

One trial means 1 spin of the wheel.

Success means the spinner stops on red.

$p = 1/3 =$ probability that any 1 spin stops on red.

$q = 1 - 1/3 = 2/3 =$ probability that any 1 spin does not stop on red.

$n = 2$ spins.

$S$ represents the total number of times the spinner stops on red.

We want to find $P(S = 2)$, $P(S = 1)$, and $P(S = 0)$.

To compute the probabilities $P(S = 2)$, $P(S = 1)$, and $P(S = 0)$, we may apply the following rule. Multiply $\binom{2}{S}$ by $p$ for each success and by $q$ for each failure. This is shown in Table 5-3.

Table 5-3

| $S$ | $\binom{2}{S}$ Number of Ways $S$ Can Occur | $P(S)$ |
|---|---|---|
| 2 | 1 | $1pp = 1p^2 = (1)(1/3)^2 = 1/9$ |
| 1 | 2 | $2pq = 2pq = (2)(1/3)(2/3) = 4/9$ |
| 0 | 1 | $1qq = 1q^2 = (1)\quad(2/3)^2 = 4/9$ |
| | | $9/9 = 1$ |

You see that we can now read the probabilities directly from this table.

(a) The probability that Marty wins twice is $P(S = 2) = 1/9$.
(b) The probability that Marty wins once is $P(S = 1) = 4/9$.
(c) The probability that Marty wins not at all is $P(S = 0) = 4/9$.

The probability histogram is the same as before (Figure 5-1). ❑

The principle involved in binomial problems is that the probability of a sequence of events is found by *multiplying* the probabilities of the individual events. This multiplication rule was described in Chapter 4. For example, the probability of Marty first winning and then losing, $P$(WL occurs), is $1/3 \times 2/3 = 2/9$. The probability of his first losing and then winning, $P$(LW occurs), is $2/3 \times 1/3 = 2/9$. Marty wins exactly once by either of these 2 sequences, so there are 2 ways he can win once, each with probability 2/9, for a total probability of 4/9. This is exactly what we computed in Table 5-3.

## ❑ EXAMPLE 5.2 Four Spins

Suppose that Marty is going to play the spinner game 4 times. He can win 4, 3, 2, 1, or 0 times. Find the probability for each of these possible outcomes.

### SOLUTIONS

If we attempt to solve this problem by listing all the possible outcomes as we did in Method 1 in the previous example, we will have to list $3 \times 3 \times 3 \times 3 = 81$ different outcomes! Some of them would be (red, red, red, red), (red, red, blue, green), (red, red, green, blue). We can avoid having to list all 81 possible outcomes by solving the problem using Method 2 above. We treat it as a binomial problem and concern ourselves just with wins and losses. Since Marty is going to play 4 times, he will win either 4, 3, 2, 1, or 0 times. The possible ways he can win and lose are shown in Table 5-4. Recall that $p = 1/3$ and $q = 2/3$, because $p$ and $q$ are the probabilities of success and failure on any *one* trial.

**Table 5-4 The Number of Ways $S$ Can Occur When $n = 4$**

| $S$ | Ways $S$ Can Occur | $\binom{4}{S}$ Number of Ways $S$ Can Occur |
|---|---|---|
| 4 | WWWW | $\binom{4}{4} = 1$ |
| 3 | WWWL, WWLW, WLWW, LWWW | $\binom{4}{3} = 4$ |
| 2 | WWLL, WLWL, WLLW, LWWL, LWLW, LLWW | $\binom{4}{2} = 6$ |
| 1 | WLLL, LWLL, LLWL, LLLW | $\binom{4}{1} = 4$ |
| 0 | LLLL | $\binom{4}{0} = 1$ |

We compute the probabilities in Table 5-5.

**Calculators**

On many calculators, $\binom{n}{S}$ can be found by using the combination key, which is often labeled ${}_nC_R$. For example to find $\binom{4}{3}$, we press 4 ${}_nC_R$ 3.

**Table 5-5 Probabilities of the Binomial Outcomes When $n = 4$**

| $S$ | $\binom{4}{S}$ | $P(S)$ |
|---|---|---|
| 4 | 1 | $1pppp = 1p^4 = (1)(1/3)^4 = 1/81$ |
| 3 | 4 | $4pppq = 4p^3q = (4)(1/3)^3(2/3) = 8/81$ |
| 2 | 6 | $6ppqq = 6p^2q^2 = (6)(1/3)^2(2/3)^2 = 24/81$ |
| 1 | 4 | $4pqqq = 4pq^3 = (4)(1/3)\ (2/3)^3 = 32/81$ |
| 0 | 1 | $1qqqq = 1q^4 = (1)\ (2/3)^4 = 16/81$ |
| | | $81/81 = 1$ |

Now we can read the desired probabilities from the table.

(a) The probability that Marty wins all 4 times is $P(S = 4) = 1/81$.
(b) The probability that Marty wins 3 times is $P(S = 3) = 8/81$.
(c) $P(S = 2) = 24/81$.
(d) $P(S = 1) = 32/81$.
(e) $P(S = 0) = 16/81$.

The histogram for these probabilities is shown in Figure 5-2. This is the graph of the binomial distribution with $n = 4$ and $p = 1/3$.

You should also notice how you can interpret the expressions in the column labeled $P(S)$ in Table 5-5. Consider, for example, the $4p^3q$ in the line corresponding to $S = 3$. The exponent of $p$ is 3. This indicates that there were 3 successes. The exponent of $q$ is (understood to be) 1. This means that there was 1 failure. The 4 indicates that there are 4 ways that this can occur—namely, WWWL, WWLW, WLWW, and LWWW.

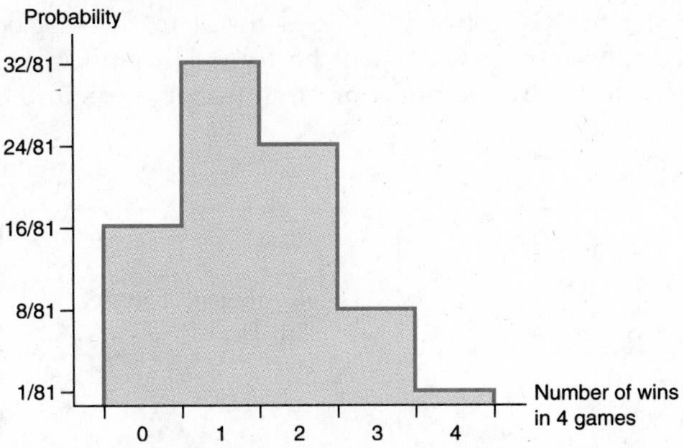


Figure 5-2 Probability histogram based on Table 5-5.

In general, the probability of $S$ successes in a binomial distribution with $n$ trials and probability $p$ of success is given by

$$P(S) = \binom{n}{S} p^S q^{n-S}$$

❑

### Pascal's Triangle

Two men are gambling by rolling a die. If the die shows a 1 or a 2, Jerome wins. Otherwise Ramon wins. Jerome will roll three times. Find the probability that Jerome wins:

(a) All three times
(b) Twice
(c) Once
(d) Not at all

Since Jerome will play three times, we have $n = 3$. We let $p = P$(Jerome wins on any one roll) $= 2/6 = 1/3$. Therefore, $q = P$(Jerome loses on any one roll) $= 2/3$. We set up a table similar to Tables 5-4 and 5-5 in the previous example.

| $S$ | Ways $S$ Can Occur | $\binom{3}{S}$ Number of Ways $S$ Can Occur | $P(S)$ |
|---|---|---|---|
| 3 | | | |
| 2 | | | |
| 1 | | | |
| 0 | | | |

To answer the four questions we need to fill in the last column in the table. We know that these answers will be found by multiplying the values from the second column by the appropriate number of $p$'s and $q$'s.

| $S$ | Ways $S$ Can Occur | $\binom{3}{S}$ Number of Ways $S$ Can Occur | $P(S)$ |
|---|---|---|---|
| 3 | | | $\binom{3}{3}p^3$ |
| 2 | | | $\binom{3}{2}p^2q$ |
| 1 | | | $\binom{3}{1}pq^2$ |
| 0 | | | $\binom{3}{0}q^3$ |

Previously we have found the entries in the $\binom{n}{S}$ column by listing all the ways $S$ can occur and then counting these ways. There is, however, an easier way to obtain these entries without counting. In this problem, for instance, the numbers are 1, 3, 3, and 1. Thus, the table can be filled out as follows:

| $S$ | Ways $S$ Can Occur | $\binom{3}{S}$ Number of Ways $S$ Can Occur | $P(S)$ |
|---|---|---|---|
| 3 | | **1** | $1(1/3)^3 = 1/27$ |
| 2 | | **3** | $3(1/3)^2(2/3) = 6/27$ |
| 1 | | **3** | $3(1/3)(2/3)^2 = 12/27$ |
| 0 | | **1** | $1(2/3)^3 = 8/27$ |

Therefore, the answers to questions (a), (b), (c), and (d) are:

(a) $P(S = 3) = 1/27$
(b) $P(S = 2) = 6/27$
(c) $P(S = 1) = 12/27$
(d) $P(S = 0) = 8/27$

The binomial coefficients $\binom{n}{S}$ in column 3 are the number of ways that $S$ can occur. We can get these numbers from the triangle of numbers below,* which is known as **Pascal's triangle**. Blaise Pascal, one of the founders of the science of probability, used the following triangle of the numbers $\binom{n}{S}$, where $n$ stands for the number of trials and $S$ for the number of successes.

**Pascal's triangle**

| | S Successes | | | | | |
|---|---|---|---|---|---|---|
| $n$ Trials | 0 | 1 | 2 | 3 | 4 | 5 |
| 0 | 1 | | | | | |
| 1 | 1 | 1 | | | | |
| 2 | 1 | 2 | 1 | | | |
| 3 | 1 | 3 | 3 | 1 | | |
| 4 | 1 | 4 + | 6 ↓ | 4 | 1 | |
| 5 | 1 | 5 | 10 | 10 | 5 | 1 |

### Pascal's Triangle

The French mathematician Blaise Pascal (1623–1662) investigated the properties of what has become known as "Pascal's triangle." Pascal himself referred to it as the "arithmetic triangle," but later scholars named it after him. This important array of numbers was known to the Chinese in the thirteenth century. Hang Hui, who wrote around 1261 to 1275, used it.

The coefficients obtained by raising the **binomial** $(p + q)$ to the $n$th power are exactly the entries in Pascal's triangle. For example,

$$(p + q)^3 = \mathbf{1}p^3 + \mathbf{3}p^2q + \mathbf{3}pq^2 + \mathbf{1}q^3$$

In the above problem Jerome rolled the die three times, therefore $n = 3$. Looking at the row corresponding to $n = 3$, we find the entries 1, 3, 3, and 1. Recall that in Example 5.2 Marty played a spinner game four times. Looking at the row of Pascal's triangle corresponding to $n = 4$, we see the entries 1, 4, 6, 4, and 1. These are the values that we had calculated ourselves in that example.

It is easy to construct the triangle. In order to get each number, we add two numbers, the number directly above the one we want and the one above and to the left. As indicated, the 10 in the row corresponding to $n = 5$ comes from adding the 6 and the 4 which are in the row above it. We can continue this process to construct more rows. Note that each row starts and ends with a 1. Try constructing the row corresponding to $n = 6$. You can check your answer by turning to Table D-1 (see Appendix D).

*Or by formula; see Exercise 5-31.

## EXAMPLE 5.3 Genetic Engineering

Let us use Pascal's triangle in one further example. A genetic engineer is experimenting in an attempt to produce a certain protein. The genetics of the situation is such that the probability of a successful experiment is 0.7. The engineer has enough equipment and funds for 6 repetitions of the experiment. Let $p = P(\text{one experiment succeeds}) = 0.7$ and $q = P(\text{one experiment fails}) = 0.3$. Find the following probabilities:

(a) $P(S = 3)$
(b) $P(S > 3)$
(c) $P(S \geq 3)$
(d) $P(S \leq 3)$
(e) $P(1 \leq S \leq 4)$
(f) $P$(there are at most 3 successes)
(g) $P$(there are at least 3 successes)

### SOLUTIONS

We first list the number of successes and the corresponding line from Pascal's triangle for $n = 6$ (because there are 6 repetitions of the experiment).

| $S$ | $\binom{6}{S}$ Number of Ways $S$ Can Occur | $P(S)$ |
|---|---|---|
| 6 | **1** | $1p^6 = 1(0.7)^6 = 0.12$ |
| 5 | **6** | $6p^5q = 6(0.7)^5(0.3) = 0.30$ |
| 4 | **15** | $15p^4q^2 = 15(0.7)^4(0.3)^2 = 0.32$ |
| 3 | **20** | $20p^3q^3 = 20(0.7)^3(0.3)^3 = 0.19$ |
| 2 | **15** | $15p^2q^4 = 15(0.7)^2(0.3)^4 = 0.06$ |
| 1 | **6** | $6pq^5 = 6(0.7)(0.3)^5 = 0.01$ |
| 0 | **1** | $1q^6 = 1(0.3)^6 = 0.00$ |

(a) $P(S = 3) = 0.19$. This is the probability that exactly 3 of the 6 experiments are successful.

(b) In this problem, $S > 3$ means $S = 4$, $S = 5$, or $S = 6$. To find $P(S > 3)$, we add $P(S = 4) + P(S = 5) + P(S = 6)$.

$$\begin{aligned} P(S > 3) &= P(S = 4) + P(S = 5) + P(S = 6) \\ &= 0.32 + 0.30 + 0.12 \\ &= 0.74 \end{aligned}$$

This is the probability that more than 3 of the 6 experiments are successful.

(c) $P(S \geq 3) = P(S = 3) + P(S > 3)$. We calculated $P(S > 3)$ in part (b), so we now have $P(S \geq 3) = 0.19 + 0.74 = 0.93$.

(d) $$P(S \leq 3) = P(S = 3) + P(S = 2) + P(S = 1) + P(S = 0)$$
$$= 1 - P(S > 3)$$

and so from part (b),

$$P(S \leq 3) = 1 - 0.74 = 0.26$$

(e) $$P(1 \leq S \leq 4) = P(S = 1) + P(S = 2) + P(S = 3) + P(S = 4)$$
$$= 0.01 + 0.06 + 0.19 + 0.32 = 0.58$$

This is the probability that either 1, 2, 3, or 4 of the 6 experiments are successful.

(f) "At most 3 successes" means 0, 1, 2, or 3 successes. So we can write $P(S \leq 3)$. This is the same as part (d).

(g) "At least 3 successes" means 3, 4, 5, or 6 successes. We can write $P(S \geq 3)$. See part (c). ❑

---

## STUDY AIDS

### Chapter Summary

In this chapter you have learned:

- To identify various kinds of variables
- What a binomial random variable is and how to calculate its probability using Pascal's triangle

### Vocabulary

You should be able to explain the meaning of each of these terms:

1. Variable
2. Quantitative variable
3. Categorical variable
4. Dichotomous random variable
5. Bernoulli random variable
6. Successful outcome
7. Total number of successes
8. Binomial experiment
9. Binomial random variable
10. Binomial distribution
11. Binomial coefficient
12. Pascal's triangle

### Symbols

You should understand the meaning of each of these symbols:

1. $n$
2. $S$
3. $\binom{n}{S}$
4. $p$
5. $q$

## EXERCISES

5-1 Explain the difference between a dichotomous random variable and a binomial random variable. How many values does each take on?

5-2 In every course Stuart Dente takes, he will either pass or not pass. Still, the question of whether or not Stu will pass exactly 3 of his 5 courses is not a binomial experiment. Why not?

*In the experiments of Exercises 5-3 to 5-8, describe success for 1 trial, and explain why* S *is or* is not *a binomial random variable. If you think that* S *is a binomial random variable, describe* n, p, *and* q, *and indicate their values. Indicate the possible values for* S.

5-3 In an experiment of rolling 1 die 25 times, Little Joe groups the 6 possible outcomes into 2 categories: "3" and "not 3." $S$ is the total number of times he rolls a 3.

5-4 Slim Jim is on a diet. He has decided to eat twice as much as he ordinarily ate at 1 meal every day. Each day for 1 week he will pick the meal by spinning a spinner divided into 3 equal parts labeled "breakfast," "lunch," and "dinner." $S$ is the number of days he picks dinner.

5-5 The Joker draws a card from an ordinary shuffled deck to see if it is a face card or not. The card is replaced, and the draw is repeated 5 times. $S$ is the total number of face cards drawn.

5-6 The experiment is similar to the previous exercise, except that the card is not replaced.

5-7 Mr. McGregor tags the ears of 60 rabbits in a colony of rabbits which at this time has 100 members. For 1 year, on the first day of each month, he selects 1 rabbit at random and notes if it has an ear tag. $S$ is the total number of times he selects a rabbit with an ear tag.

5-8 Sure Shot, the rising basketball star, practices foul shots every day in order to improve. During the course of a season, she attempts 1000 foul shots. $S$ is the total number of baskets scored.

5-9 Write the first 4 numbers of the row corresponding to $n = 21$ in Pascal's triangle. Use Table D-1 in Appendix D.

5-10 List the 32 possible outcomes of success $S$ or failure $F$ in a binomial experiment with 5 trials. Do your results agree with Pascal's triangle for $n = 5$?

5-11 In Example 5.2 we stated that there were 81 possible outcomes in terms of color. Have your kid brother or sister list them.

5-12 A fair coin is tossed 3 times.

(a) Complete the chart below:

| $S$ = Number of Heads | $\binom{3}{S}$ Number of Ways $S$ Can Occur from Pascal's Triangle | $P(S)$ |
|---|---|---|
| 3 | | |
| 2 | | |
| 1 | | |
| 0 | | |

(b) Draw the histogram depicting these results. What is the probability of getting heads:

(c) All three times?

(d) Exactly twice?

(e) At least once?

Find:

(f) $P(S = 0)$

(g) $P(S < 2)$

(h) $P(S \leq 2)$

5-13 (a) A true-false quiz is given with 5 questions. To pass, you need at least 4 right. You guess every answer. (1) what is the probability that you pass? (2) If the class is large and if everyone else in the class is also guessing, about what percentage of the class will probably pass?

(b) If in each pregnancy the probability of having a girl is 1/2, what is the probability that a childless couple planning to have 5 children will have at least 4 girls? What percentage of couples with 5 children would you expect to have 4 or 5 girls?

(c) In a certain binomial experiment, $p = 0.50$, $q = 0.50$, and $n = 5$. Find $P(S \geq 4)$. In what percentage of the results would you expect to have 4 or more successes? Discuss what you have discovered from parts (a) and (b).

5-14 (a) Eileen Dover needs a grade of 70 to pass an examination in abnormal physiology. The exam consists of 10 true-false questions. If she guesses blindly at all the questions, what is the probability that she will pass?

(b) Her brother Ben also needs a grade of 70 to pass an examination in nematode anatomy. However, Ben's test will consist of 10 multiple-choice questions where each question has 5 possible answers. If he guesses blindly at all the questions, what is the probability that he will pass?

5-15 A bent coin has a probability of falling heads equal to 0.40. This coin is tossed 5 times.

(a) What is the probability of getting at least 3 heads?

(b) What is the probability of getting at most 3 heads?

5-16 When a thumbtack is tossed in the air, it will land point up or not. The probability of landing point down is 0.21 (for a No. 35 tack). Three tacks are tossed. Find the probability of:

(a) All landing point up

(b) All landing point down

(c) At least 1 landing point up

5-17 According to a magazine article, if a child is conceived through artificial insemination, the probability that the child will be a boy is 0.80. Ms. Clark plans on having 2 children by this method. Find the probability that

(a) Both are girls

(b) At least 1 is a girl

5-18 (a) In the game of Montezuma's Revenge you must toss a run of more heads than your opponent using an ordinary coin. Michelle has scored a run of 3 heads in a row. Mike must now score a run of 4 heads to beat her. What is the probability that Mike will beat Michelle? That he will not beat her?

(b) In an alternative form of the game, a die is used. Felix rolled the same number 3 times in a row. Felicia must now score a run of 4 rolls of any 1 number. What is the probability that she will lose? (Why should you use $n = 3$ and not $n = 4$?)

5-19 A space hero, Luke Warmwater, has 4 functioning rockets left on his ship. The ship has been damaged, and the probability that any individual rocket will fire is 0.20. To escape sudden death (and survive to the next episode), he needs at least 1 of the rockets to fire. He pushes all 4 firing buttons. What is the probability that he will be around for the next movie?

5-20 **Medical Screening Test.** A screening test for a disease is a simple procedure which indicates if a patient is *likely* to have a disease. A person whose results on the screening test are "positive" is likely to have the disease and would be asked to submit to further tests in order that a proper diagnosis be made. This means that some people who screen "positive" do not, in fact, have the disease. These people are "false positives." The false-positive rate is the proportion of people who screen positive but who do not, in fact, have the disease. Suppose the false-positive rate for a certain screening test is 0.10.

(a) What does a false-positive rate of 0.10 mean? About how many false positives would you expect in 10 positive screenings?

(b) What is the probability that of 10 patients who screen positive, more than 1 is a false positive?

(c) Which is more likely, that there are 0 or 1 false positives in 10 positive screenings or that there is more than 1 false positive?

5-21 Consider the following spinner game. A wheel has 4 sections marked as shown. Each afternoon, Algernon, a student, spins the spinner. It stops at random on 1 of the sections. (If it stops on a line, it does not count.) Let us suppose that if it stops on the section P, then Algernon goes out to play that evening; if it stops on a section S, then Algernon stays home to study that evening. Algernon spins the spinner twice.

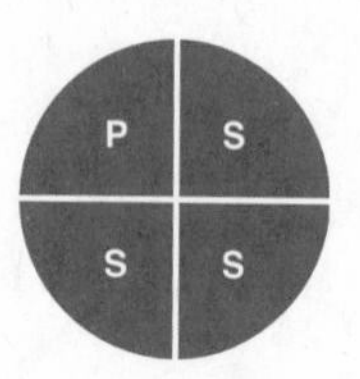


What is the probability that he:

(a) Plays both times?

(b) Does not play at all?

(c) Plays at least once?

(d) Plays fewer than 2 days?

5-22 Algernon's friend, Morgel, decides to use the spinner-game method to decide how to spend his evenings, too. He makes a spinner wheel as shown, where 4 of the sections are marked P for play and 1 section is marked S for study. Morgel decides to spin the wheel 7 times so that his entire week's schedule will be worked out.

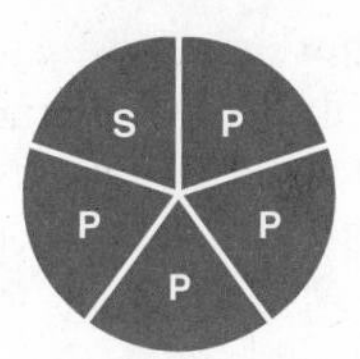


What is the probability that Morgel will end up:

(a) Going out to play all 7 evenings?

(b) Studying exactly 3 evenings?

(c) Not playing at all?

(d) Studying less than 4 evenings?

5-23 Morgel's friend, Chastain, wants to play the spinner game too. His spinner looks like the one shown below, where J is for Jay's Joint and K is for Ken's Kubicle, his two favorite nightspots. If the spinner lands on J, he spends the evening at Jay's. If the spinner lands on K, he spends the evening at Ken's. There is a 4-day holiday coming up, and so Chastain spins the spinner 4 times to determine his plans.

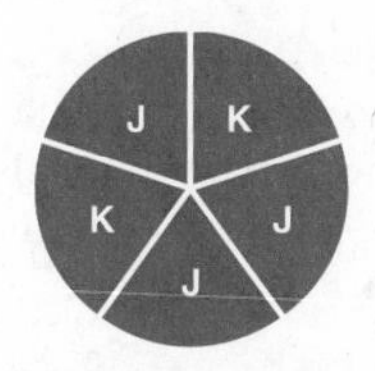


What is the probability that:

(a) He will spend 1 evening at Jay's and 3 evenings at Ken's?

(b) He will spend 2 evenings at Jay's and 2 at Ken's?

(c) He will spend all 4 evenings at the same nightspot?

5-24 Assistant Professor Ratso, a leading experimental psychologist, is in the habit of sending rats through mazes. She predicts that a rat reaching the end of a T-shaped maze is more likely to turn left than right. She believes that the proportion of rats which turn left is 0.65. If this is true and she sends 6 rats down a maze, what is the probability that:

(a) 3 will turn left and 3 will turn right?

(b) Fewer than 5 will turn left?

(c) All will turn in the same direction?

5-25 An emergency life-support system has 4 batteries. The probability of any battery failing is 0.01. What is the probability that:

(a) None fails?

(b) All fail?

(c) More than 2 fail?

5-26 A manufacturer claims that 4 of 5 dentists recommend sugarless gum for their patients who chew gum. Assuming that this claim is true, find the probability that in a randomly selected group of 20 dentists, 16 or more will recommend sugarless gum for their patients who chew gum.

5-27 Suppose that 2 baseball teams, the Yanquis and the Codgers, are equally matched. The outcome of each game between them can be considered as a random variable, with the probability of the Yanquis winning equal to 0.50.

(a) If they play 6 games, what is the probability that each team will win 3 games?

(b) If they play a series of games where the first team to win 4 games wins the series, what is the probability that the Yanquis will win the series on the seventh game? [*Hint:* Use your answer to part (a).]

5-28 Monica and Steffi are playing tennis. The probability that Monica wins a point is 0.80. The game is currently stopped, with "advantage Steffi" (she needs 1 point to win). What is the probability that Monica will win the next 3 points consecutively, and thus win the game?

5-29 Peter Peddler must bike 2 miles to get the Sunday newspaper. If he gets to the store too early, the papers have not arrived. If he gets there too late, they are all sold. He has learned that if he arrives at 8:30 he has an 85% chance of getting the paper. What is the probability that he will get the paper at least 6 of the next 8 Sundays if he shows up at 8:30?

5-30 Each morning Chef Victoir prepares 1 of his 3 yummy breakfast specialties: *Quiche Fillisse, Toast á la Naomi,* or *Eggs Ari-Bari.* Mme. Sharonne, a patron of his restaurant, has noticed that 63% of the time Victoir prepares *Eggs Ari-Bari.* If Mme. Sharonne enters the restaurant 6 times, what is the probability that she will get exactly 3 meals of *Eggs Ari-Bari?*

SKIP 5-31 An element of Pascal's triangle is referred to as $\binom{n}{S}$. For example, $\binom{6}{2} = 15$. This can be calculated with the formula

$$\binom{n}{S} = \frac{n!}{S!(n - S)!}$$

where $n! = 1 \times 2 \times 3 \times \cdots \times n$. (Your calculator may have an $x!$ button.)

(a) Verify that $\binom{6}{2} = \frac{6!}{2!(6 - 2)!} = 15$.

(b) Verify that $\binom{14}{9} = 2002$.

(c) Calculate the value of $\binom{35}{3}$.

(d) Calculate the value of $\binom{50}{47}$.

5-32 In his novel *Congo,* Michael Crichton informs us that the probability of a successful parachute jump is 0.7980, and he concludes that there is only "one chance in five that someone would be badly hurt." This would be accurate if only 1 person were about to jump, but 7 people are about to do so.

(a) Calculate the probability of at least 1 person being badly hurt.

(b) Comment on the probability figure 0.7980. Do you think it is reasonable that we could know this probability correct to 4 decimal places?

(c) Recalculate the probability of part (a) using 0.8.

## CLASS SURVEY QUESTIONS

1. Let $p = P$(a student in class is female).

   (a) Evaluate $p$.

   (b) Imagine that each student's name is printed on a piece of paper and that these papers are folded, placed in a large box, and thoroughly mixed. Withdraw 1 name at random, record it, and replace it in the box. Do this 2 more times. What is the probability that all 3 will be female? Male? The same sex?

2. (a) Which of the questions we asked in the survey could be used for binomial experiments?

   (b) Using one of your answers to part (a), construct an exercise similar to class survey exercise 1. That is, define $p$, evaluate $p$, and calculate the probability of some specific event of your choice.

## CLASS EXPERIMENTS

1. This experiment contains a quiz whose questions are so weird that we assume you will have to guess randomly at the answers. This means you can predict ahead of time what grade you are likely to get. There are 9 questions; each question has 3 possible choices for the correct answer. If we let $S$ equal the total number you get right, then $S$ is a binomial random variable, with possible values from 0 to 9. Answer the following questions **before** you take the quiz.

   (a) Explain why you have a 1/3 probability of getting question 1 right. Why does it follow that we expect about 1/3 of the students in your whole class to get question 1 right? Explain why the same conclusion holds true for question 2 and every other question on the exam.

   (b) What is the probability you get both question 1 and 2 right? About what fraction of the class do you expect to get both right?

   (c) What is the probability that you get exactly 3 questions right? About what fraction of the class do you expect to get exactly 3 questions right?

   (d) Find the probability associated with each possible value of $S$. Which value of $S$ is the most likely one? Are you more likely to get 3 questions right or none right?

   (e) Suppose a passing score is 6 or more right. What is the probability of passing? About what fraction of the class do you expect to pass?

   (f) Now have the whole class take the quiz. After you have answered all of the questions, check your answers with the correct answers at the end of the answer key in the back of the book. Let $S$ = the number of correct answers you obtained. Now compare the results of the class with what you predicted. Compute the percentage of the class that got each possible test score. Construct a histogram for $P(S)$, and compare it to the histogram for the class percentages.

**QUIZ**

1. The agamat is a unit of (a) volume (b) weight (c) length.
2. Scotopic vision is the function of (a) the cones (b) the rods (c) a fortune teller.
3. N. R. Finsen was (a) a Norwegian astronomer (b) a Swedish poet (c) a Danish scientist.
4. Mount Huancarhuas is located in (a) Ecuador (b) Colombia (c) Peru.
5. Blaise Pascal was born on June (a) 19th (b) 20th (c) 21st.
6. In 1968 the Pulitzer Prize for Feature Photography went to (a) Toshio Sakai (b) Moneta Sleet, Jr. (c) Dallas Kinney.
7. The 35th decimal digit of $\pi$ is (a) 8 (b) 9 (c) 2.
8. The average density of the sun is about (a) 2.4 grams per cubic centimeter (b) 1.4 grams per cubic centimeter (c) 3.4 grams per cubic centimeter.
9. A piece of meteorite contains seven Pb-206 atoms for every atom of U-238. Therefore, the age of the meteorite is about (a) 13 billion years (b) 500 million years (c) 2 million years.

2. Make up another test like the previous one and try it out in class. Change the number of questions and the number of choices. By the way, the "purest" version, to insure that everyone guesses, is to omit asking the questions altogether and just have each student fill out an answer sheet which you can compare to some "official" answer sheet made up randomly.

SKIP 5.3

## 5.3 Using Binomial Probability Tables

*(This section may be omitted without loss of continuity.)*

Now that you understand the theory, the ideas behind the binomial distribution, let us look at another approach. In Exercise 5-13 we saw that different problems have the exact same computational solution. This leads us to the idea of listing the results of such computations in tabular form. Thus, if you look at Table D-2 (Appendix D), you will find the solution to Exercise 5-13 by locating the two numbers for $S = 4$ and $S = 5$ corresponding to $n = 5$, $p = 0.50$. These numbers are 0.1563 and 0.0313. Since their sum is 0.1876, the solution to the problem is about 19%.

Statisticians often use tables of binomial probabilities to solve problems. Note that, as with any table, "round-off" errors do occur. For example, if we calculate the solution to Exercise 5-13 directly, we obtain $0.15625 + 0.03125 = 0.1875$, and not 0.1876. Also, since the table cannot possibly contain every feasible value of the probability $p$, it is often necessary to be satisfied with approximate tabular values. As an example, $\binom{6}{4}(0.59)^5(0.41)^2$ is approximately equal to 0.3055. However, $p = 0.59$ is not in our table. The closest value is 0.60.

Looking in our table for the entry corresponding to $p = 0.60$, $n = 6$, we find the solution for $S = 4$ to be 0.3110. If the researcher does not need extremely accurate results, then the approximate value can be found very quickly from the table. (Of course, the proliferation of calculators which determine exponents quickly has reduced the dependence of statisticians on such tables.)

❑ EXAMPLE 5.4

Let us redo parts (a), (b), and (c) of the genetics engineering experiment in Example 5.3, where $p = 0.7$ and $n = 6$.

SOLUTIONS

(a) To find $P(S = 3)$, we simply look in Table D-2 at the entry corresponding to $p = 0.70$, $n = 6$, $S = 3$. The solution is quickly found to be 0.1852, or 0.19.

(b) To find $P(S > 3)$, we add the entries corresponding to $S = 4$, $S = 5$, and $S = 6$. The solution is $0.3241 + 0.3025 + 0.1176 = 0.7442$, or about 0.74.

(c) To find $P(S \geq 3)$, we again add $P(S = 3)$ to the answer in part (b), obtaining $0.1852 + 0.7442 = 0.9294$, or 0.93. ❑

❑ EXAMPLE 5.5: Frimframs

Sundat Motors had to replace the frimfram on the left widget in 39% of last year's models. Acme Trucking bought 12 Sundats at various times last year. What is the probability that Acme will need to replace the frimfram in 6 trucks?

SOLUTIONS

(a) The solution is $P(S = 6) = \binom{12}{6}(0.39)^6(0.61)^6 = 0.1675$, or about 17%.

or

(b) Using Table D-2, we can approximate the solution. Since 0.39 is not in the table, we will use $p = 0.40$, $n = 12$, and $s = 6$. The solution is found to be 0.1766, or about 18%.

Depending on the accuracy that the statistician needs, this second, quicker solution may be close enough for practical purposes. ❑

## EXERCISES

*Solve any of the Exercises 5-15 to 5-30 by using the binomial probability table.*

# 6 The Normal Distribution

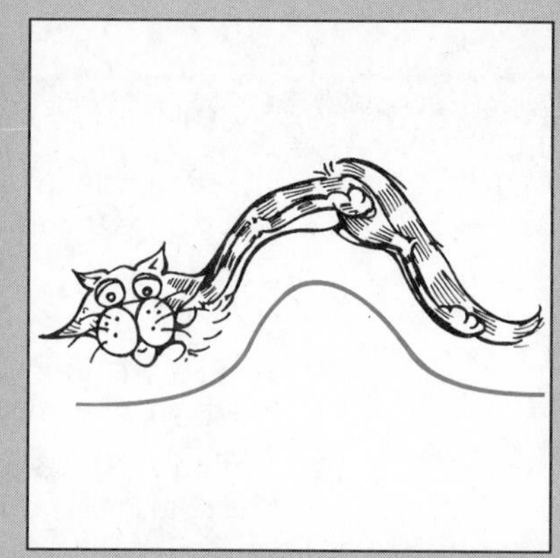

Does your professor mark on "the curve"?

CONTENTS



GOALS

At the end of this chapter you will have learned:

- The main features of the graph of the normal distribution
- How to read a table of normal curve areas and $z$-scores
- How to compute probabilities of normally distributed random variables

## 6.1 Binomial Histograms

Consider a problem based on the drawing of a card at random from a shuffled deck of cards. Suppose we have defined success as "drawing a club." Our experiment consists of drawing 1 card, seeing if it is a club, replacing it, reshuffling, drawing the next card, and so on, 5 times all told. This is a binomial experiment, with $n = 5$ and $p = 0.25$. Consequently, if we did a binomial analysis, we would get the probability histogram of Figure 6-1.

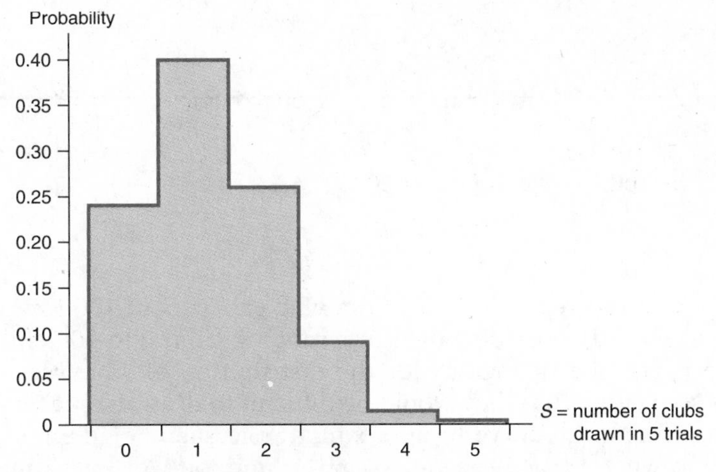


**Figure 6-1** Binomial probability histogram: $n = 5$, $p = 0.25$.

If we were interested in a similar experiment, but with $n = 10$ or $n = 20$, we would get the histograms shown in Figures 6-2 and 6-3.

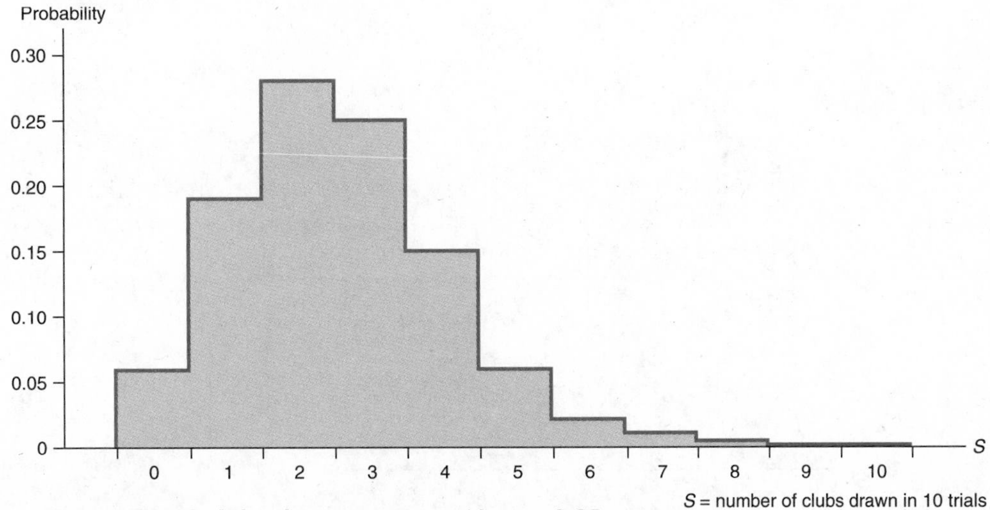


**Figure 6-2** Binomial probability histogram: $n = 10$, $p = 0.25$.

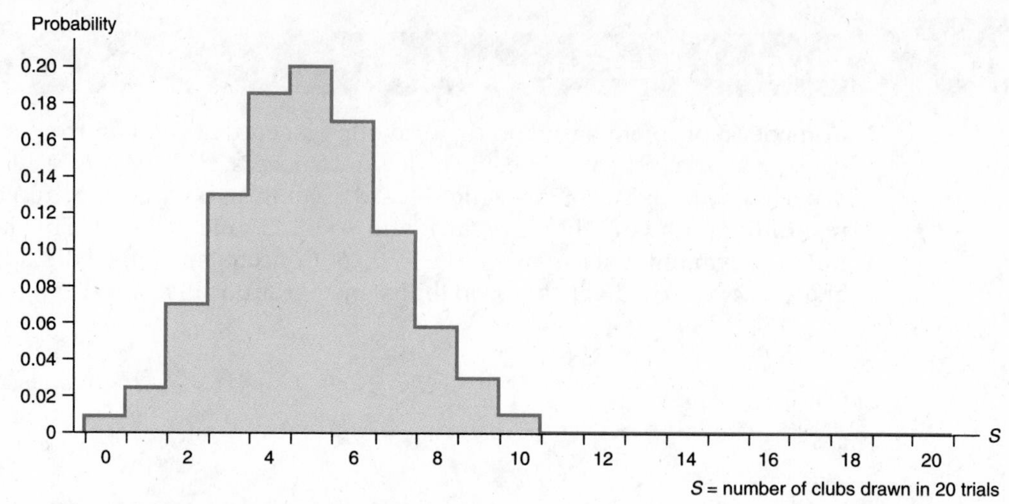


Figure 6-3 Binomial probability histogram: $n = 20$, $p = 0.25$.

If we were to repeat this card-drawing experiment 100 times ($n = 100$), there would be 101 possible outcomes, ranging from 0 to 100 clubs. On ordinary-size paper, the histogram for the distribution of successes would have intervals so narrow that they would be difficult to draw. In such cases, we will often draw a smooth curve to approximate the shape of the histogram. The curve is drawn by placing a dot over the midpoint of each interval at the proper height (the same height we would make the bar if we were drawing the histogram).

For example, if we are approximating the three histograms previously shown, we would get the curves that are shown in Figures 6-4, 6-5, and 6-6.

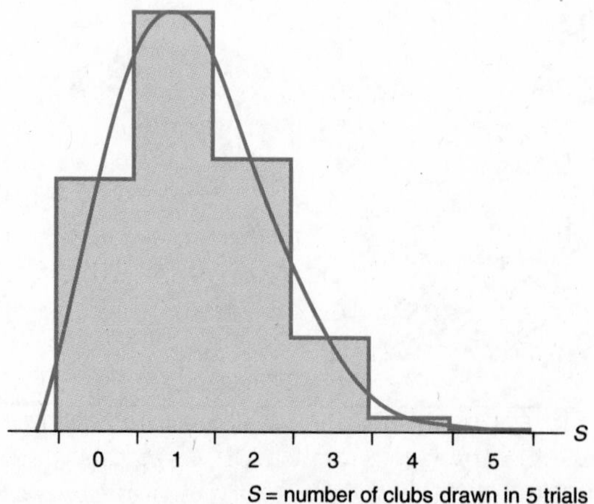


Figure 6-4 Smooth approximation of histogram in Figure 6-1.

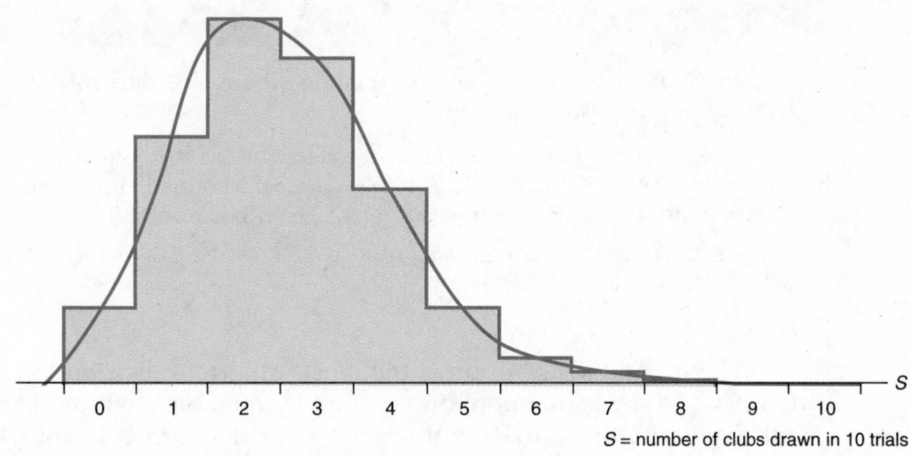


Figure 6-5 Smooth approximation of histogram in Figure 6-2.

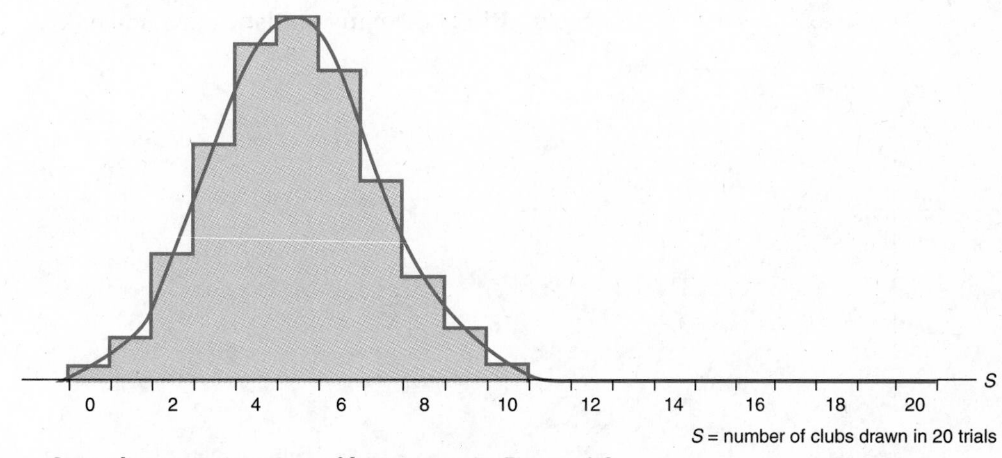


Figure 6-6 Smooth approximation of histogram in Figure 6-3.

Fortunately, according to mathematical theory, we know that under *certain common conditions* the curve used to approximate a binomial histogram will be **bell-shaped** and, more specifically, that it will be very close to a particular bell-shaped curve called the **normal curve**. (Note that not every bell-shaped curve is normal). The normal curve pictures a distribution of numbers called the **normal distribution**. Any set of observations whose frequency distribution has this shape is said to be "normally distributed."

normal curve

normal distribution

This close approximation of certain binomial distributions with a normal distribution means that, if we are faced with tedious binomial calculations, we are often able to approximate the solution quickly via a normal distribution instead. Because all the necessary values for the normal distribution have already been tabulated, using a normal distribution saves a great deal of time at the expense of only a little accuracy.

### Names for the Normal Curve

Our use of "normal curve" and "normal distribution" is due to Karl Pearson (1894), the English statistician.

The normal distribution is sometimes referred to as the "Gaussian distribution" in honor of Carl Friedrich Gauss (1777–1855), who made important use of it in his work on patterns of measurement error in astronomy.

SOURCE: T.M. Porter, *The Rise of Statistical Thinking 1820–1900,* Princeton, NJ: Princeton University Press, 1986.

The normal distribution is the most important distribution in statistics. Its earliest use was to approximate the binomial distribution, and this use is still very important. However, it has other applications, some of which we will study in this text. In fact, many collections of raw data have been found to be approximately normally distributed. Figure 6-7 illustrates an example from major league baseball records that looks approximately normal. In addition, we will see that the distributions of many statistics are normal.

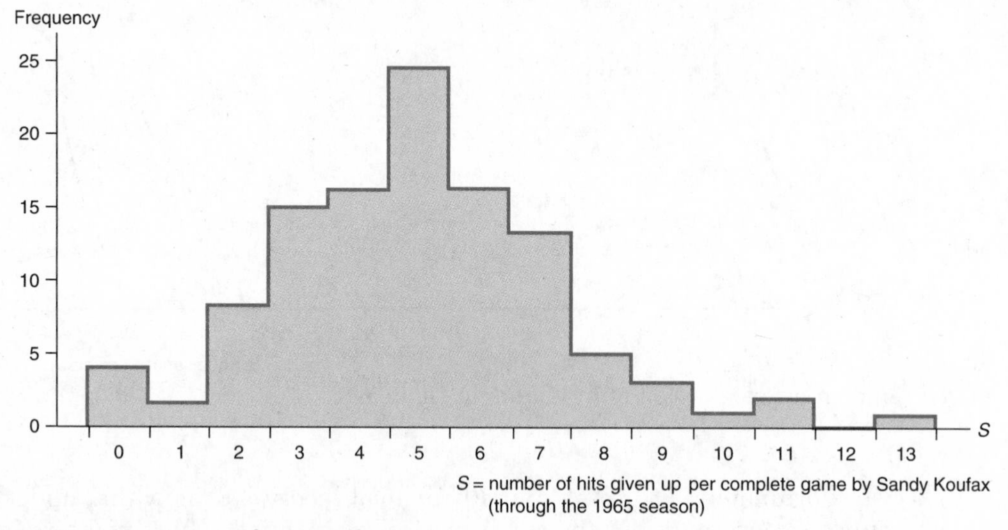


**Figure 6-7** Approximately normal raw data.

## 6.2 The Theoretical Normal Curve

The normal distribution and its graph (the normal curve, Figure 6-8) have the following important properties:

1. There is symmetry about the mean (i.e., the left and right halves of the curve are mirror images of each other). Therefore, the mean equals the median.

2. The scores that make up a normal distribution cluster about the middle.
3. The range of scores is unlimited, but only a very small fraction of the scores, fewer than 3 in 1000, are more than 3 standard deviations away from the mean.

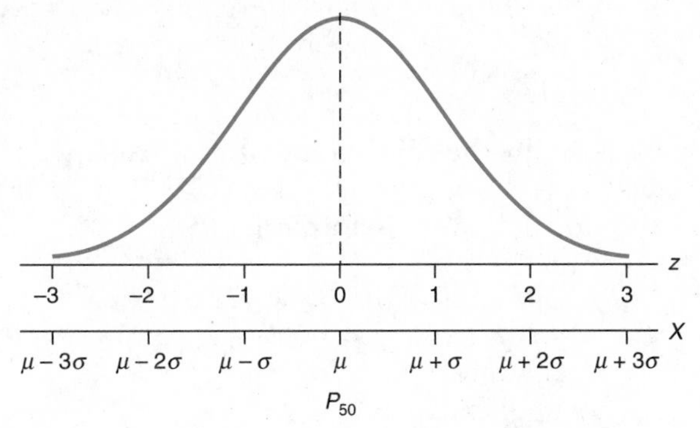


Figure 6-8 Normal curve.

**A Quick Rule of Thumb for a Normal Curve**

About 68% of the distribution lies between $z = -1$ and $z = +1$.

About 95% of the distribution lies between $z = -2$ and $z = +2$.

About 99.7% (or almost all) of the distribution lies between $z = -3$ and $z = +3$.

For our purposes the normal distribution will be defined by a table of $z$-scores and percentile ranks. We saw in Chapter 3 that a table of $z$-scores and percentile ranks does determine the shape of a distribution; the more $z$-scores and percentile ranks we have, the more accurate the representation of the distribution. In your own work it is not necessary to draw a precise graph for the normal curve. You can represent it by sketching any reasonably bell-shaped curve, because all the numbers used in calculations based on the normal curve are recorded accurately in Tables D-3 and D-4 (Appendix D).

## 6.2-1 Use of the Normal Curve Table

Table D-4 gives a list of $z$-scores from $-4$ to $+4$. For each $z$-score, the table also gives the area under the curve to the left of the $z$-score (i.e., its percentile rank). We shall illustrate the use of this table in the following examples.

## ❑ EXAMPLE 6.1 Area to the Left

In a normal curve, find the area to the left of $z = 1$.

**Statistical Calculators**

Some calculators have the capability to produce the area under a normal curve. Does yours?

SOLUTION

For $z = 1$, we look in the table under $z = 1$, and we read "Area = 0.8413." Sometimes we write "Area($z = 1$) = 0.8413." This means that 84.13% of the area under the normal curve is to the left of $z = 1$, as seen in Figure 6-9.

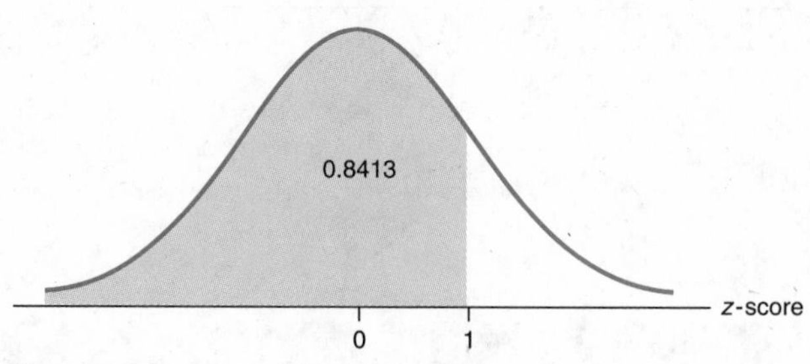


Figure 6-9

Some interpretations of this are the following:

1. 84.13% of the elements in any normal distribution have $z$-scores less than 1.
2. The element of a normal distribution which has a $z$-score of 1 also has a percentile rank of 84.
3. The probability of randomly selecting an element from any normal distribution with a $z$-score less than 1 is 0.8413. ❑

## ❑ EXAMPLE 6.2 Area to the Right

In a normal curve, find the area to the *right* of $z = -1.65$.

SOLUTION

Looking up $z = -1.65$, we read "Area($z = -1.65$) = 0.0495." Since this is the area to the *left* of $z$, and the total area is 100%, or 1, we subtract 0.0495 from 1.0000, to find what remains on the right. Thus, $1.0000 - 0.0495 = 0.9505$. This is shown in Figure 6-10.

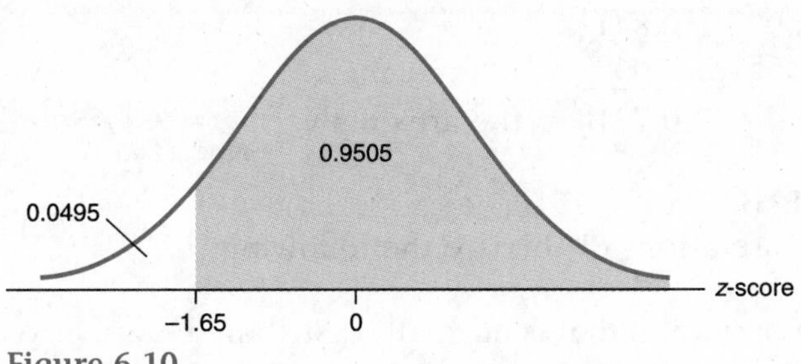


Figure 6-10

Some interpretations of this are the following:

1. 95.05% of the elements of any normal distribution have $z$-scores greater than $-1.65$.
2. The probability of randomly selecting an element from any normal distribution with a $z$-score greater than $-1.65$ is 0.9505. ❑

---

## ❑ EXAMPLE 6.3 Area Between Two z-Scores

Find the area under a normal curve between $z = -1.65$ and $z = 1.00$.

### SOLUTION

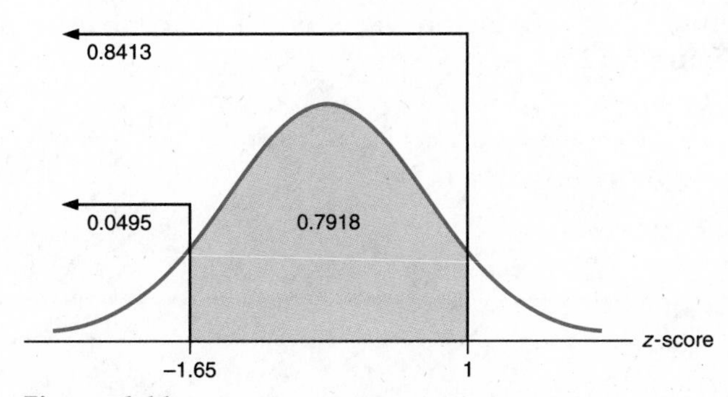


Figure 6-11

From Figure 6-11, we see that we want to subtract the area to the left of $z = -1.65$ from the area to the left of $z = 1.00$.

Looking up these two $z$-scores, we obtain the two areas to the left. We find the answer by subtracting them.

$$\begin{aligned} \text{Area}(z = 1.00) &= 0.8413 \\ \text{Area}(z = -1.65) &= \underline{0.0495} \\ & \quad 0.7918 = \text{the area between } z = -1.65 \text{ and } z = 1.00 \end{aligned}$$

Thus, the area between is 0.7918.

Some interpretations of this are the following:

1. 79.18% of a population that is normally distributed falls between $z = -1.65$ and $z = 1.00$.
2. The probability that a normally distributed random variable would fall between $z = -1.65$ and $z = 1.00$ is 0.7918; that is, $P(-1.65 < z < 1.00) = 0.7918$. ❑

In the preceding examples we have interpreted area as a percentage of raw scores, as a percentile rank, and as a probability. If we start with any one of these three quantities, we can reverse the preceding procedures and find the corresponding $z$-score.

## ❑ EXAMPLE 6.4 Given Percent Less Than

In a normal distribution, 10.2% of the members of the population have $z$-scores less than a certain $z$-score. Find that $z$-score.

### SOLUTION

We use the symbol $z_?$ for this unknown $z$-score. The problem is to find $z_?$, as illustrated in Figure 6-12.

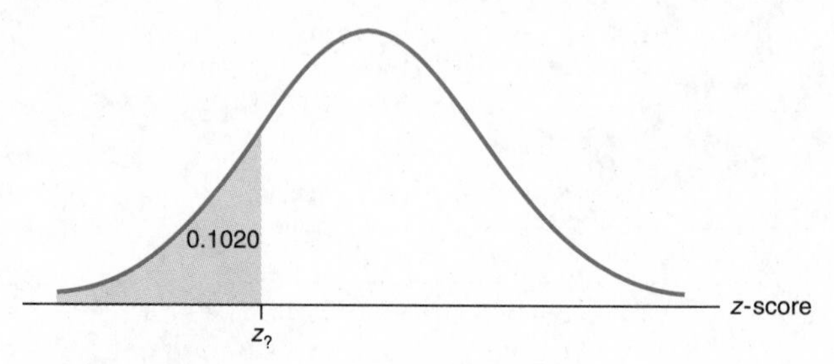


Figure 6-12

We write 10.2% as 0.1020, because the table records areas to four decimal places. Then, in the table we find the corresponding $z$-score, $z_? = -1.27$. ❑

### ❑ EXAMPLE 6.5 Given Probability Greater Than

$z_?$ has the property that there is a 95% probability that a $z$-score picked at random from a normal distribution is greater than $z_?$. Find $z_?$.

**SOLUTION**

This $z$-score is illustrated in Figure 6-13.

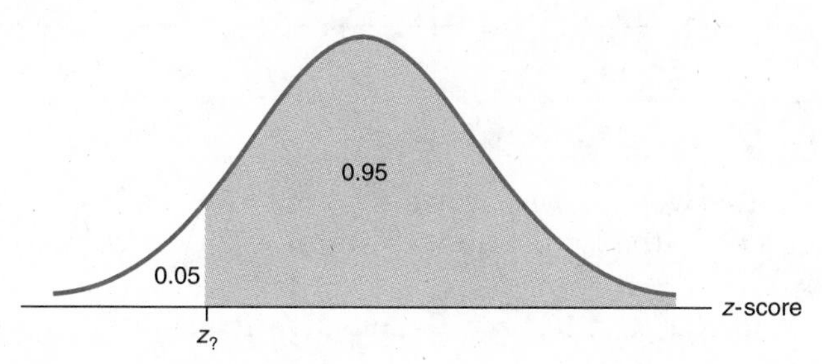


**Figure 6-13**

Since 95% of the $z$-scores must be greater than $z_?$, then 5% must be *less than* $z_?$. We look in the table for an area equal to 0.0500. This exact value is not there. However, we do find an area of 0.0495, which corresponds to a $z$-score of $-1.65$. Hence, $z_?$ is approximately $-1.65$. ❑

## 6.2-2 The Normal Curve Table and Raw Scores

In many problems involving normal distributions we will deal primarily with raw scores. Sometimes we will be given raw scores and will be asked to make a statement about percentage, percentile rank, or probability. In other problems we will be given the percentage, percentile rank, or probability and will be asked to find a raw score. Table D-4 does not contain raw scores, but *only z-scores* and *percentages*. In order to do such problems, we must convert raw scores to $z$-scores, and vice versa.

Outlines of these procedures are given below.

To find a percent or a probability when you know a raw score:

1. Change the raw score to a $z$-score by using the familiar formula

$$z\text{-score} = \frac{\text{raw score} - \text{mean}}{\text{standard deviation}}$$

2. Look up the $z$-score in the table to find the percent less than.
3. Use this percent to answer the question.

Conversely,

> To find a raw score when you know a percent or a probability:
>
> 1. Change the percent or probability into a percent less than.
> 2. Look up the percent less than to find the $z$-score.
> 3. Change the $z$-score into a raw score by using the familiar formula
>
> $$\text{Raw score} = \text{mean} + z\text{-score} \times \text{standard deviation}$$

## ❑ EXAMPLE 6.6 Normally Distributed Raw Scores

It was found that the weights of a certain population of laboratory rats are normally distributed* with $\mu = 14$ ounces and $\sigma = 2$ ounces. We denote such a distribution by **ND($\mu = 14$, $\sigma = 2$)**.

**ND symbol**

The expression ND($\mu = 14$, $\sigma = 2$) tells us three things:

1. The distribution is normal.
2. The mean is 14.
3. The standard deviation is 2.

(a) One of the rats weighs 12 ounces. What is the percentile rank for this weight?

(b) In such a population, what percentage of the rats would we expect to weigh between 10 and 15 ounces?

### SOLUTIONS

(a) Recall that the formula for converting raw scores to $z$-scores is

$$z = \frac{X - \mu}{\sigma}$$

Thus, the $z$-score for 12 ounces is

$$z_{12} = \frac{12 - 14}{2} = \frac{-2}{2} = -1$$

In Table D-4 we find Area($z = -1$) = 0.1587. This is shown in Figure 6-14, where we have drawn two horizontal axes, one for the $z$-scores and one for the corresponding raw scores.

*In reality, this collection of weights must be only *approximately* normal, because a "true" normal distribution contains an infinite set of values. But it is customary in much statistical literature to refer to such distributions as "normal."

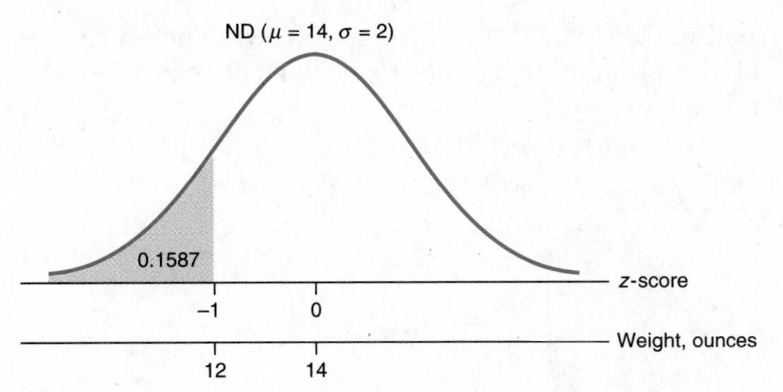

Figure 6-14

This means that approximately 15.87% of the rats weigh less than 12 ounces. Rounding off, we say that a weight of 12 ounces has a percentile rank of 16, and we write $P_{16} = 12$ ounces.

(b) We draw and label a rough sketch of this population (Figure 6-15).

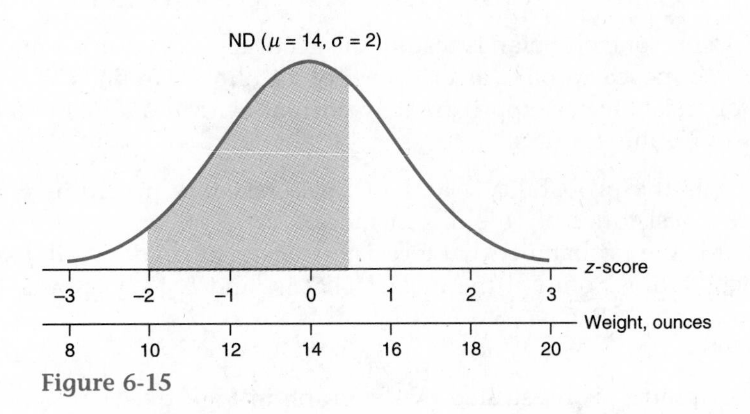

Figure 6-15

Since $\mu = 14$, 14 corresponds to $z = 0.00$. The standard deviation is 2 ounces. Therefore, the weight increases by 2 ounces each time the $z$-score increases by 1.

We want to compute the area between 10 and 15 ounces. Converting the raw scores to $z$-scores, we get

$$z_{15} = \frac{15 - 14}{2} = \frac{1}{2} = 0.5 \qquad \text{and} \qquad z_{10} = \frac{10 - 14}{2} = \frac{-4}{2} = -2$$

In Table D-4 we find

$$\text{Area}(z = 0.5) = 0.6915$$

$$\text{Area}(z = -2) = 0.0228$$

Subtracting 0.6915 − 0.0228, we get 0.6687. Therefore, we expect about 67% of the population of rats to weigh between 10 and 15 ounces, as shown in Figure 6-16.

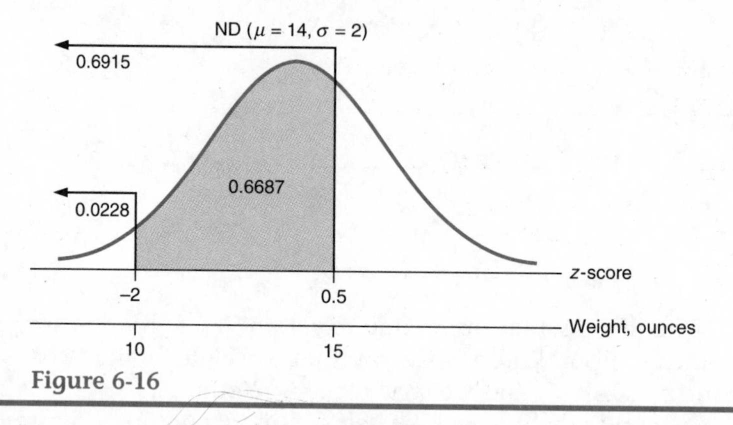


Figure 6-16

## EXAMPLE 6.7 Probabilities Involving Raw Scores in a Normal Distribution

A quality control technician is testing the accuracy of a certain type of resistor which is supposed to have a resistance of 14 ohms. He finds that the distribution of resistances is approximately normal, with $\mu$ about 14.06 ohms and $\sigma$ about 1.73 ohms.

(a) What is the probability that 1 of these resistors picked at random will have a resistance of 16 ohms or more?

(b) What is the probability that a resistor picked at random will have a resistance within 1 ohm of the mean? That is, find $P(13.06 \leq X \leq 15.06)$.

### SOLUTIONS

(a) This situation is illustrated by the graph in Figure 6-17.

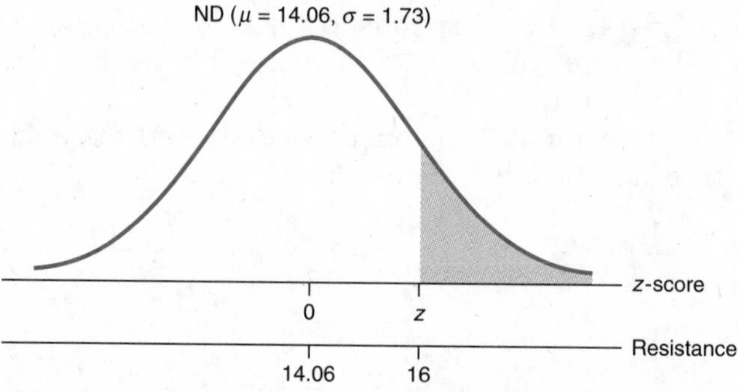


Figure 6-17

We first find the area to the left of 16 by converting 16 to a $z$-score:

$$z_{16} = \frac{16 - 14.06}{1.73} = \frac{1.94}{1.73} = 1.12$$

In Table D-4 we find Area($z$ = 1.12) = 0.8686. Our problem can now be represented as shown in Figure 6-18.

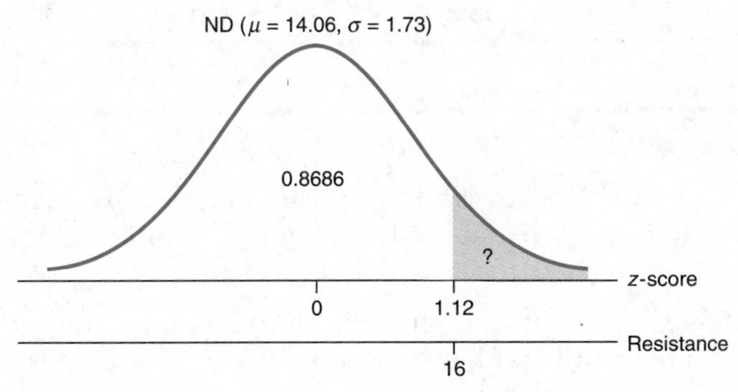


Figure 6-18

Hence, the probability that 1 of these resistors picked at random will have resistance 16 ohms or more is 1.0000 − 0.8686 = 0.1314, or about 0.13. If we use $X$ to stand for the resistance of any resistor that we pick at random, then we can write $P(X \geq 16) = 0.13$.

(b) This situation is illustrated in Figure 6-19.

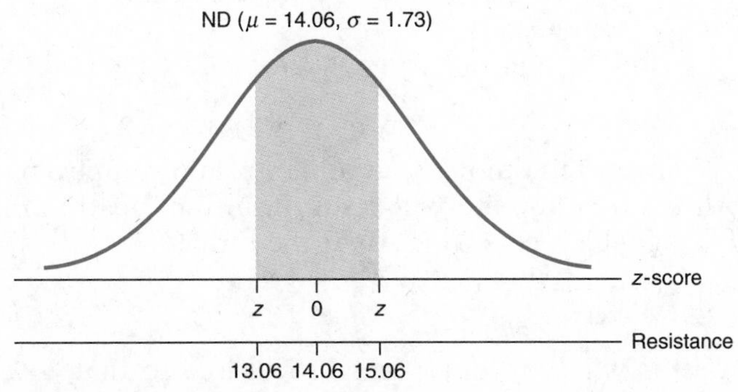


Figure 6-19

$$z_{15.06} = \frac{15.06 - 14.06}{1.73} = \frac{1.00}{1.73} = +0.58$$

$$z_{13.06} = \frac{13.06 - 14.06}{1.73} = \frac{-1.00}{1.73} = -0.58$$

$$\text{Area}(z = 0.58) = 0.7190$$

$$\text{Area}(z = -0.58) = 0.2810$$

Subtracting, we get 0.4380, or about 0.44. Therefore,

$$P(13.06 \leq X \leq 15.06) = 0.44.$$

❑

## ❑ EXAMPLE 6.8 Finding Raw Scores for Normal Vampire Bats

Professor Frankenstein raises laboratory vampire bats. The lengths of their left fangs are normally distributed with $\mu = 28$ millimeters and $\sigma = 4$ millimeters.

(a) The Professor's favorite vampire, Sheldon, has a left fang the length of which has a percentile rank of 84 (Figure 6-20). How long is Sheldon's left fang?

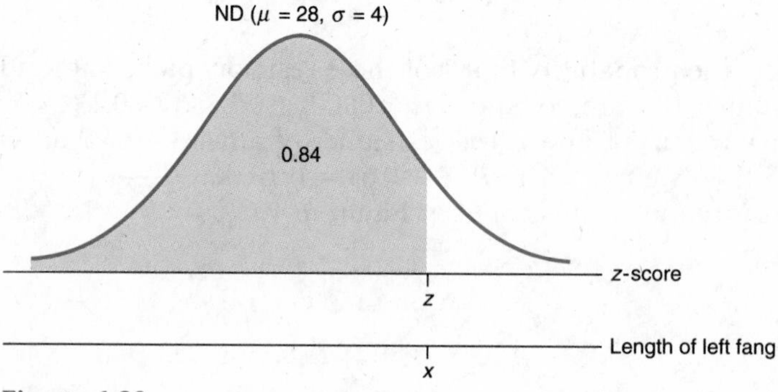


Figure 6-20

(b) Professor Frankenstein knows that a bite from a vampire bat whose left-fang length is in the top 5% will result in instant death. Find the length of the left fang which cuts (bites?) off the top 5%.

SOLUTIONS

(a) The problem is where to draw the vertical line so that 84% of the area will be to the left of that line. We search Table D-4 for an area entry equal (or close) to 0.8400. The closest entry is 0.8389, which corresponds to $z = 0.99$, and so the vertical line is drawn at $z = 0.99$. We then convert this $z$-score to a raw score. Recall that the formula for converting $z$-scores to raw scores is

$$X = \mu + z\sigma$$

Thus,

$$X = 28 + 0.99\ (4) = 28 + 3.96 = 31.96$$

Therefore, Sheldon's left fang is about 32 millimeters long.

(b) This situation is illustrated in Figure 6-21.

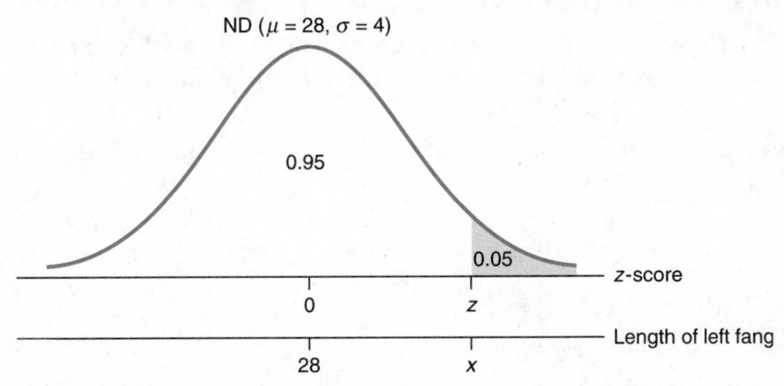


Figure 6-21

Since the area given in Table D-4 is to the left, we look up 1.0000 − 0.0500 = 0.9500. Since exactly 0.9500 is not in the table, we choose 0.9505, which corresponds to $z = 1.65$. Converting this $z$-score to a raw score, we have

$$X = \mu + z\sigma$$
$$= 28 + 1.65(4) = 34.6 \text{ millimeters, or about 35 millimeters}$$

Thus, a left fang of length 35 millimeters cuts off the top 5%; that is, about 5% of the vampire bats have left fangs more than 35 millimeters long. ❑

## STUDY AIDS

### Chapter Summary

In this chapter you have learned:

- Some properties of the normal distribution and its graph, the normal curve

### Vocabulary

You should be able to explain the meaning of each of these terms:

1. Normal curve
2. Normal distribution
3. ND symbol

### Symbols

You should understand the meaning of each of these symbols:

1. Area($z =$ )
2. ND($\mu =$ , $\sigma =$ )

## EXERCISES

6-1 Mention a variable from your experience (at work, at the mall, at the beach, etc.) which you believe is approximately normally distributed. What opinions lead you to this conclusion?

6-2 In a normal distribution of weights of third-grade boys, find the percentage of weights whose $z$-scores are:

(a) Less than $z = 2.33$.

(b) More than $z = 1.65$.

(c) Between the mean and $z = 2.5$.

(d) Between $z = 0$ and $z = -1.6$.

(e) Less than $z = -1.6$.

(f) More than $z = -1.6$.

(g) Between $z = 2$ and $z = 2.5$.

(h) More than 1 standard deviation from the mean.

(i) If these were the weights of fourth-grade girls, would the answers be different? Explain.

6-3 In a normal distribution of lengths of Venusian middle legs, find the percentage of lengths whose $z$-scores are:

(a) Above $z = -1.96$.

(b) Below $z = 0.23$.

(c) Between $z = -2.13$ and $z = -1.45$.

(d) Above $z = 3$.

(e) Within 2 standard deviations of the mean.

(f) Above $z = 6$.

For the same distribution, find:

(g) $P(z < 5)$

(h) $P(z > 1.5)$

(i) $P(0.3 < z < 0.7)$

(j) $P(z > 7)$

(k) If these were the lengths of Martian middle arms, would the answers be different? Explain.

6-4 In a normal distribution of diameters of bolts, find the percentage of diameters whose $z$-scores are:

(a) Above $z = 1.96$

(b) Below $z = -2$

(c) Not between $z = -1$ and $z = +1$

For the same distribution, find:

(d) $P(z < -3)$

(e) $P(z > 0.15)$

(f) $P(z > 1.5)$

(g) $P(2 < z < 3)$

(h) $P(-3 < z < -2)$

6-5 (a) For a normal distribution of bushels of corn per acre, find the sum of the percentile rank of $z = 2$ plus the percentile rank of $z = -2$.

(b) What happens if you replace the 2s in part (a) by 3s?

6-6 In a normal distribution of automobile gasoline mileage,

(a) Which $z$-score has a percentile rank of 2?

(b) Which $z$-score has a percentile rank of 95?

(c) Which $z$-score has a percentile rank of 50?

(d) Which $z$-scores cut off the middle 70% of the population?

(e) If parts (a) through (d) of this exercise had been about a normal distribution of weights of bags of sugar, what would the four answers have been? Why?

6-7 In a normal distribution:

(a) Which $z$-score cuts off the top 10%?

(b) Which $z$-score cuts off the bottom 20%?

(c) Which $z$-scores cut off the middle 30%?

(d) If $P(z < z_?) = 0.45$, find $z_?$.

(e) If $P(z > z_?) = 0.35$, find $z_?$.

(f) If $P(-z_? < z < z_?) = 0.40$, find $z_?$ and $-z_?$.

6-8 In a normal distribution, what $z$-score has a percentile rank equal to:

(a) 1

(b) 5

(c) 10

(d) 50

(e) 95

(f) 99

6-9 For a normal distribution, find the $z$-score that cuts off the top:

(a) 5%

(b) 2.5%

(c) 1%

(d) 0.5%

6-10 Label the raw scores corresponding to the $z$-scores given on the axis of this graph.

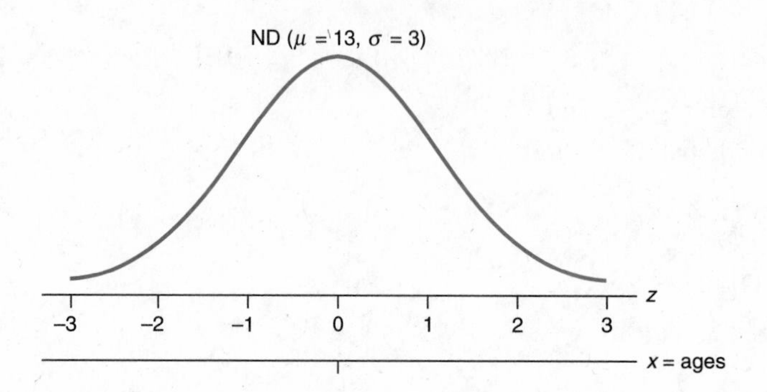


6-11 Label the $z$-scores corresponding to the raw scores given on this graph.

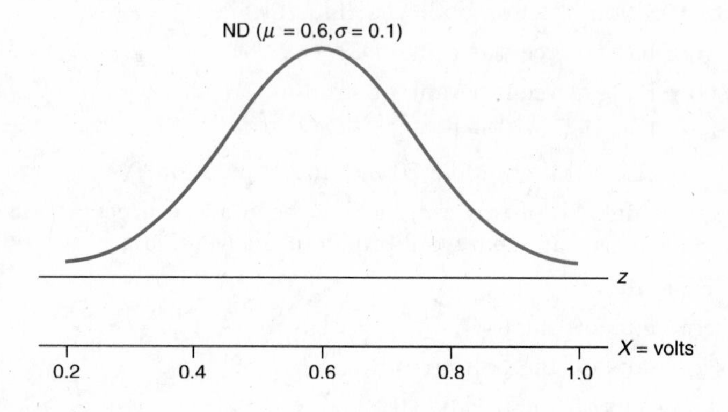


6-12 Boris Gudenuv entered a Russian literature contest. Last year's competition results are symbolized by ND($\mu$ = 230, $\sigma$ = 11). If the results are similar this year and Boris wants to score in the top 10%, what score will be good enough for Gudenuv?

6-13 The distribution of raw scores on a test of moral fitness is normal, with mean equal to 0 and $\sigma$ = 10. What raw score cuts off:

(a) The top 5%?

(b) The bottom 5%?

(c) The bottom 2.5%?

(d) The top 2.5%?

6-14 This year's distribution of the earnings of the Aunty Pasto franchises was ND($\mu$ = \$24,500, $\sigma$ = \$320). The earnings of the Aunt Chilada franchises were ND($\mu$ = \$23,900, $\sigma$ = \$600).

(a) What is the probability that the owner of an Aunty Pasto franchise earned more than \$25,200?

(b) What is the probability that the owner of an Aunt Chilada franchise earned over \$25,200?

6-15 An electronics technician repeats an experiment many times, each time recording a voltage reading. The technician finds that the collection of readings is approximately normally distributed, with $\mu$ about 74 volts and $\sigma$ about 6 volts.

(a) What percentage of the readings were between 70 and 80 volts?

(b) What is the probability that a reading taken at random will be over 86 volts?

(c) What is the probability that a random reading will be outside the range of 69 to 79?

6-16 The results on a certain blood test performed in a medical laboratory are known to be normally distributed, with $\mu = 60$ and $\sigma = 18$.

(a) What percentage of the results are between 40 and 80?

(b) What percentage of the results are between 76 and 78?

(c) What percentage of the results are above 100?

(d) What percentage of the results are below 60?

(e) What percentage of the results are between 78 and 80?

(f) What percentage of the results are outside the "healthy range" of 30 to 90?

(g) What is the probability that a blood sample picked at random will have results in the "healthy range" of 30 to 90?

(h) Which test result has a percentile rank of 5?

6-17 A study was done to see how many hours during the school day high school seniors spend thinking about sex. The results were normally distributed with $\mu = 5.7$ hours and $\sigma = 0.6$ hour. What percent of these students think about sex:

(a) More than 4 hours per school day?

(b) More than 6.5 hours?

(c) Between 5 and 6 hours?

6-18 At an urban hospital the weights of newborn infants are normally distributed, with $\mu = 7$ pounds, 2 ounces, and $\sigma = 15$ ounces. Let $X$ be the weight of a newborn infant picked at random. Find the following probabilities:

(a) $P(X \geq 8$ pounds$)$

(b) $P(X \leq 5$ pounds, 5 ounces$)$

(c) $P(6$ pounds $\leq X \leq 8$ pounds$)$

(d) What infant weight is at the 70th percentile? (That is, find $P_{70}$.)

(e) Let $W$ be a fixed weight. The probability is 0.70 that a baby picked at random weighs less than $W$. Find $W$. [That is, find $W$ such that $P(X < W) = 0.70$.]

(f) Find $W$ such that $P(X < W) = 0.10$.

6-19 The lifetimes of a certain brand of movie floodlights are normally distributed, with $\mu = 210$ hours and $\sigma = 56$ hours. Let $X$ be the lifetime of a light picked at random. Find the following probabilities:

(a) $P(X \geq 300)$

(b) $P(X \leq 100)$

(c) $P(100 \leq X \leq 300)$

(d) The company guarantees that its light will last at least 120 hours. What percentage of the bulbs does it expect to have to replace under this guarantee?

6-20 It happens that income for junior executives in a large retailing corporation is normally distributed, with $\mu$ = \$42,800 and $\sigma$ = \$4000.

(a) There is an unspoken agreement that you have "arrived" if you are in the top 15%. What salary must a junior executive earn in order to arrive?

(b) The highest 25% get keys to the executive washroom. Victoria earns \$45,590. Does she have a key?

(c) Because of a recession, the bottom 5% may be let go. What salary cuts off the bottom 5%?

(d) The top 20% go out to lunch. The bottom 30% bring their lunches in interoffice envelopes. The remaining 50% bring their lunches in attaché cases. Find the 2 salaries that separate these 3 categories.

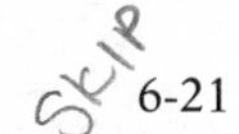

6-21 Perhaps the most famous set of naturally collected data that is approximately normal was used by Adolfe Quetelet in 1846 to demonstrate the presence of normality in "real life." He used figures which had been previously published in 1817 in a Scottish medical journal. The figures from the medical journal are given below. The values represent the chest circumference of Scottish soldiers and were collected in connection with providing uniforms for the soldiers.

| Circumference, Inches | Frequency |
|---|---|
| 33 | 3 |
| 34 | 19 |
| 35 | 81 |
| 36 | 189 |
| 37 | 409 |
| 38 | 753 |
| 39 | 1062 |
| 40 | 1082 |
| 41 | 935 |
| 42 | 646 |
| 43 | 313 |
| 44 | 168 |
| 45 | 50 |
| 46 | 18 |
| 47 | 3 |
| 48 | 1 |
| Total | 5732 |

*Source*: Stephen Stigler, *The History of Statistics*, Cambridge, MA: Harvard University Press p. 208.

(a) Use these data to make a histogram, and confirm that the distribution does appear bell-shaped.

(b) Confirm that the mean and the standard deviation for these 5732 observations are 39.85 inches and 2.07 inches, respectively.

(c) Find the $z$-scores for the largest and smallest chests.

(d) In theory, according to the normal curve, approximately 68 percent of the observations should be between $z = -1$ and $z = 1$. Find the chest measurements that correspond to $z = -1$ and $z = 1$, and determine the percentage of these observations that actually have $z$-scores between $-1$ and 1.

6-22 Cat Moran, the sailing instructor at Marny's Marina, has noticed that on a beginner's first "solo" the boat invariably tips over. She has observed that the time it takes them to capsize is normally distributed, with the mean equal to 21 minutes and the variance equal to 256 minutes.

(a) What is the percentile rank of someone who manages to stay afloat 1/2 hour?

(b) Is it more likely that a beginner will tip over during the first 15 minutes or during the second 15 minutes?

6-23 True or false? For a normal curve:

(a) The area between $z = 0$ and $z = 1$ equals the area between $z = 0$ and $z = -1$.

(b) The area between $z = 0$ and $z = 1$ equals the area between $z = 1$ and $z = 2$.

(c) The percentile rank of $z = 1$ equals the percentile rank of $z = -1$.

6-24 The graphs below represent the distribution of incomes in 2 populations. Discuss any differences between the 2 distributions.

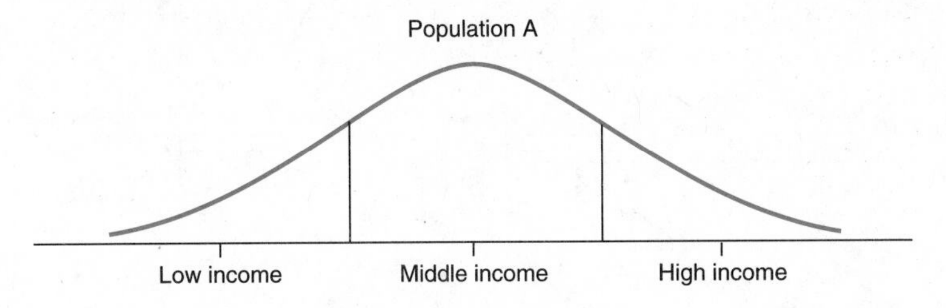


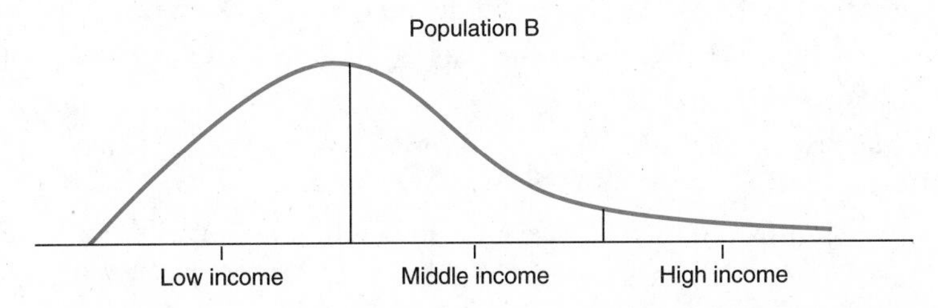


6-25 The graphs below represent the distribution of grades at a large university. One graph is for freshman-level courses. The other is for advanced courses. Discuss any differences between the 2 distributions of grades.

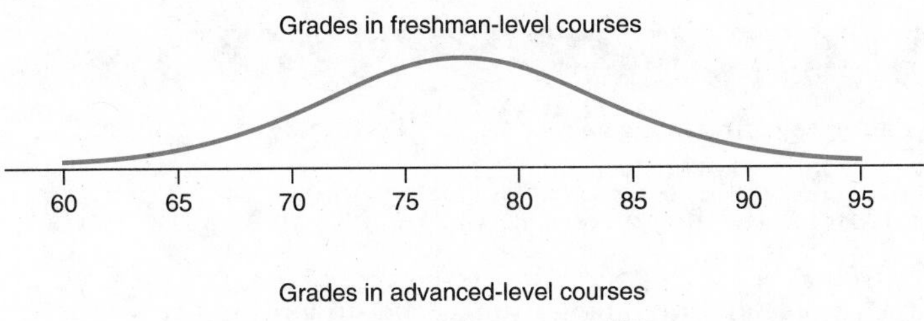


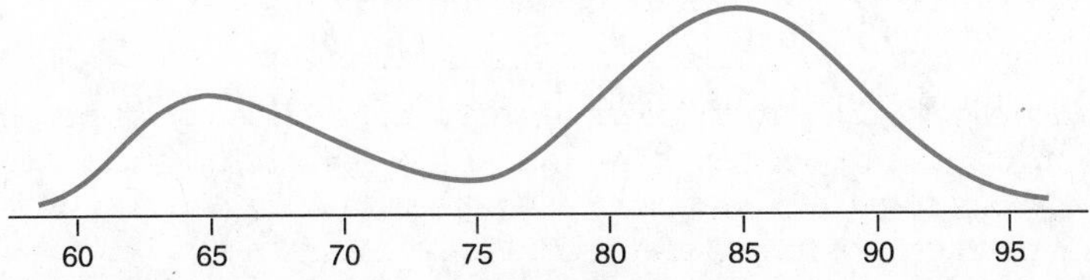

6-26 In a distribution of grades, $\mu = 80$ and $\sigma = 10$. Yet it is not true that 34% of the distribution is between 80 and 90. In fact, we cannot find what percentage of the grades lie between 80 and 90 from the above information. Why?

6-27 A distribution of professors' salaries at a large college has $\mu = \$37{,}000$ and $\sigma = \$3400$. What percentage of the professors earn between $37,000 and $41,000?

6-28 A technical writer wants to write a brochure describing a new steam iron. He finds that 15 irons, when filled with 7 ounces of distilled water, will operate for an average of 17 minutes before all the water is gone. He also finds that the distribution of times is approximately normal, with $s = 2$ minutes. He decides to write that the irons will operate for "about 20 minutes." What percentage of the people who use this model of iron will find that their iron does not live up to the description in the brochure? (That is, their iron operates less than 20 minutes before running out of water.)

Skip

6-29 Here are the data for the Sandy Koufax histogram (See Figure 6-7):

| Number of hits per game | 0 | 1 | 2 | 3 | 4 | 5 | 6 | 7 | 8 | 9 | 10 | 11 | 12 | 13 |
|---|---|---|---|---|---|---|---|---|---|---|---|---|---|---|
| Frequency | 4 | 2 | 8 | 15 | 16 | 24 | 16 | 13 | 5 | 3 | 1 | 2 | 0 | 1 |

We have found the mean and the standard deviation for these 110 games. Sandy gave up an average of 5.0 hits per game with a standard deviation of 2.3.

(a) From the data, calculate both the $z$-score for 6 hits and the percentile rank for 6 hits.

(b) Find the percentile rank in Table D-4 for the $z$-score you just calculated in part (a). Compare the 2 percentile ranks you have found.

(c) Repeat parts (a) and (b) for several other numbers of hits. How does this confirm or deny the visual impression that the distribution looked approximately normal?

## CLASS SURVEY QUESTIONS

1. (a) Draw a histogram for the heights of all the females in your sample. Do they appear to be normally distributed?

   (b) Do you think the heights of all the females in your school are approximately normal?

2. (a) Draw a histogram for the heights of all the students in your sample. Do they appear to be normally distributed?

   (b) Do you think the heights of all the students in your school are approximately normal?

## CLASS EXPERIMENTS

1. (a) Give everyone in the class a foot-long-ruler, and have each person use it to measure the length of the classroom to the nearest 1/8th of an inch. Discuss ahead of time why it is not likely that everyone will get the same answer. Draw a histogram of the answers, and see if it appears to be approximately normal.
   (b) Get a more accurate measuring device, and *carefully* measure the length of the classroom. Convert your measurements from part (a) into errors. Draw the histogram of these errors. Does it appear symmetrical? Does it appear to be approximately normal?
2. Many calculators have a key which gives a random decimal between 0 and 1. All possible decimals from .000 to .999 are equally likely to appear. These values are **NOT** normally distributed; they are evenly, or uniformly, distributed. They have a mean value of 0.5.

   Use the calculator to find 12 of these random values. Then find their sum and save it. Repeat this experiment and find another sum.

   Repeat this until you have 100 sums. These 100 values are your data for this problem. Draw a histogram for these 100 sums. The distribution of these sums should be approximately normal with mean 6 and standard deviation 1.
3. In the 1870s Sir Francis Galton (1822–1911) invented a device for illustrating the normal distribution which he called a "quincunx."* Build one as described below.

   A "quincunx" usually refers to 5 dots arranged in a square with one in the center:

   •   •
     •
   •   •

   Galton used this arrangement in his design.

### *Errors*

In the mid-18th century a current problem in astronomy was how to decide the best value for a reading through an instrument when every time you made the reading with this instrument you got a different value. Some mathematical justification was needed for saying that the mean observation was best. In about 1740 Thomas Simpson (1710–1761) wrote the first paper which focused not on the distribution of observations but on the distribution of errors. He made some assumptions about this distribution. He assumed it was symmetric and that the probability of small errors was greater than the probability of large errors. Each size error had its own probability; this was a discrete distribution—the possible errors were of size 1, 2, 3, etc. By 1757 his error distribution was refined and, for the first time, representable by a continuous smooth graph something like a normal curve. In some fields of research today, the normal curve is called the "error curve." Further important contributions to the treatment of errors were made by Laplace in the 1770s and Gauss in about 1800.

---

*See *The History of Statistics* by Stephen M. Stigler, Harvard University Press, Belknap Press, 1986, pages 276–281.

Two examples are illustrated below.

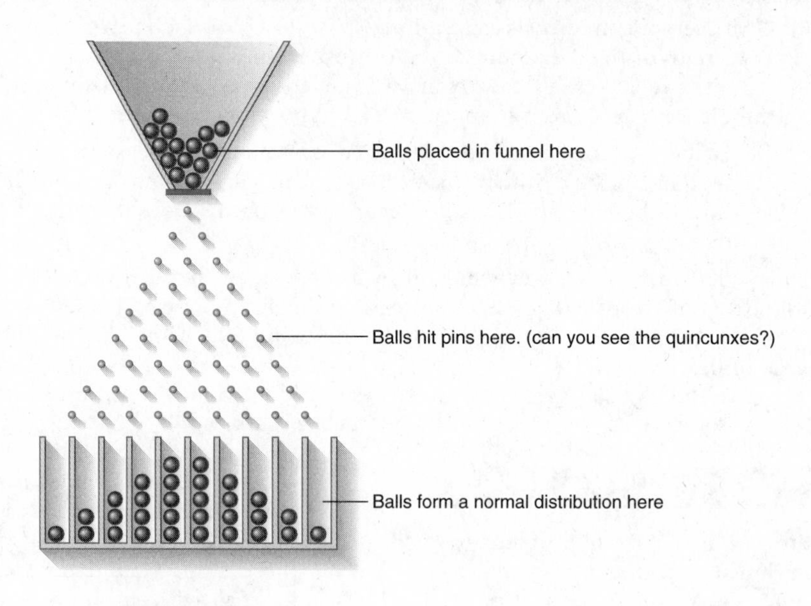


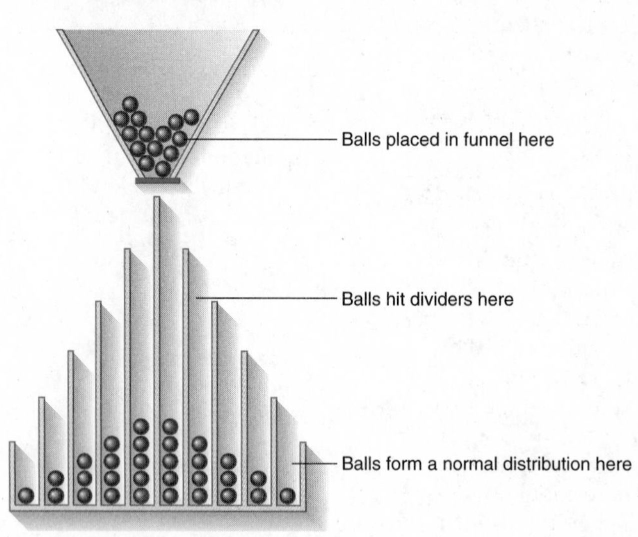

# 7

# Approximation of the Binomial Distribution by Use of the Normal Distribution

This is how we *normally* solve binomial problems.

**CONTENTS**



**GOALS**

At the end of this chapter you will have learned:

- How to answer binomial distribution questions by quicker normal distribution methods
- When such approximations can be used

## 7.1 The Normal Approximation to the Binomial Distribution

Dr. Y. B. Normal, the noted research psychologist, is doing a study with 20 of her amazing rats. Each rat is sent through a T-shaped maze. She assumes that due to chance factors alone it is equally likely that any one that will go left or right at the T, and so she expects about 10 rats to go left and 10 to go right. But since she would like some better description of what usually happens when 20 rats go through the maze, she plans to repeat this experiment many times.

She asks her assistant to send the whole group of 20 rats through the maze 200 times. If you think of one rat going through the maze as one trial, and turning left as "success," then each time 20 rats go through the maze, this is a binomial experiment with $n = 20$, and $p = P(\text{a rat turns left}) = \frac{1}{2}$. In this binomial experiment $S$ represents the total number of rats out of 20 that turned left. Thus, Dr. Normal's whole study consists of 200 repetitions of a binomial experiment. The list of the results obtained by her assistant is given in the following table:

Number out of 20 rats who turned left, $S$

| | | | | | | | | | | | | | | | | | | | | | |
|---|---|---|---|---|---|---|---|---|---|---|---|---|---|---|---|---|---|---|---|---|---|
| 8 | 11 | 11 | 8 | 12 | 10 | 11 | 10 | 12 | 12 | 7 | 8 | 9 | 8 | 10 | 12 | 13 | 9 | 9 | 10 | 10 | 12 |
| 9 | 11 | 7 | 10 | 14 | 10 | 10 | 8 | 11 | 12 | 12 | 9 | 8 | 8 | 8 | 11 | 7 | 8 | 10 | 8 | 11 | 9 |
| 13 | 2 | 7 | 16 | 11 | 9 | 10 | 11 | 15 | 13 | 9 | 11 | 12 | 6 | 11 | 13 | 8 | 9 | 6 | 10 | 7 | 9 |
| 14 | 11 | 14 | 9 | 6 | 12 | 11 | 10 | 15 | 10 | 11 | 11 | 12 | 11 | 10 | 11 | 10 | 8 | 16 | 8 | 6 | 9 |
| 9 | 4 | 12 | 8 | 9 | 10 | 10 | 12 | 12 | 7 | 8 | 8 | 11 | 6 | 9 | 9 | 11 | 6 | 11 | 12 | 8 | 14 |
| 8 | 11 | 8 | 7 | 9 | 10 | 9 | 10 | 8 | 11 | 10 | 10 | 8 | 8 | 10 | 8 | 13 | 9 | 7 | 10 | 11 | 10 |
| 10 | 14 | 8 | 12 | 5 | 10 | 9 | 9 | 14 | 11 | 11 | 11 | 12 | 10 | 11 | 7 | 15 | 9 | 9 | 9 | 10 | 8 |
| 12 | 9 | 6 | 11 | 10 | 10 | 12 | 15 | 11 | 13 | 13 | 12 | 8 | 12 | 11 | 11 | 11 | 12 | 9 | 12 | 15 | 6 |
| 10 | 10 | 6 | 12 | 9 | 9 | 12 | 6 | 5 | 9 | 6 | 9 | 10 | 7 | 13 | 11 | 14 | 7 | 9 | 13 | 11 | 11 |
| 10 | 8 | | | | | | | | | | | | | | | | | | | | |

The histogram in Figure 7-1 shows the overall shape of this distribution.

This graph shows which values $S$ tends to take on in this experiment and gives an idea of how consistent the results are. To create a summary of this sample data, the assistant computes the mean and the standard deviation for these 200 numbers. He finds that the mean is 9.9 and the standard deviation is 2.3.

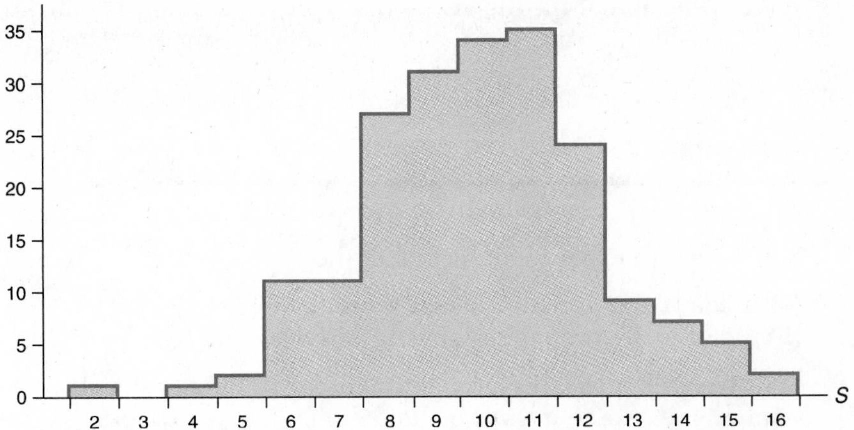


Figure 7-1 A histogram for 200 values of $S$, the number of rats that turned left in each experiment of 20 rats.

Notice that the distribution of $S$ in Figure 7-1 looks something like a normal curve. This is no accident and could have been predicted ahead of time, because under certain circumstances (which we will specify later) the distribution of $S$ from a binomial experiment is very close to a normal distribution. Now consider sending 20 rats through a T-shaped maze over and over again—much more than merely 200 times—where the probability of a left turn is 1/2. You might expect the mean number of left turns to be about 10, and outcomes near 10 to be more common than outcomes far from 10. This intuitive feeling is correct. In fact, statisticians have shown mathematically that, in theory, the mean number of successes in **many** repetitions of a binomial experiment is equal to the product $n$ times $p$. They have also shown, though it is not intuitively evident, that, in theory, the standard deviation for the number of successes is equal to $\sqrt{npq}$. These theoretical findings are summarized in the box below.

> Let $S$ be a binomial random variable, with $n$ trials and $P(\text{a success}) = p$. Then, if we repeat the binomial experiment many times (theoretically, forever) and record all the values of $S$, we will find
>
> $$\mu_S = np \quad \text{and} \quad \sigma_S = \sqrt{npq}$$

In this case,

$$n = 20$$

$$p = 1/2 \text{ and } q = 1/2$$

$$\mu_S = np = 20(1/2) = 10$$

and

$$\sigma_S = \sqrt{npq} = \sqrt{(20)(0.5)(0.5)} = \sqrt{5} = 2.24$$

**Note** Recall that the mean and the standard deviation obtained above using only 200 trials were approximately the same, 9.9 and 2.3.

## ❑ EXAMPLE 7.1 Rats

For the above experiment, if in fact the probability of a left turn equals 1/2:

(a) What is the probability that more than 15 rats turn left?
(b) What is the probability that 15 or fewer rats turn left?

We will solve this problem two ways. First, using Pascal's triangle to obtain the exact solution, and then, more quickly, by the normal approximation.

### SOLUTIONS Method 1: Exact Binomial Solution Using Pascal's Triangle

| $S$ | $\binom{20}{S}$ Number of Ways $S$ Can Occur | $P(S)$ |
|---|---|---|
| 20 | 1 | $1(0.5)^{20} = 0.000001$ |
| 19 | 20 | $20(0.5)^{19}(0.5) = 0.000019$ |
| 18 | 190 | $190(0.5)^{18}(0.5)^2 = 0.000181$ |
| 17 | 1140 | $1140(0.5)^{17}(0.5)^3 = 0.001087$ |
| 16 | 4845 | $4845(0.5)^{16}(0.5)^4 = 0.004621$ |

(a) Adding these probabilities, we get $P(S > 15) = 0.005909$, which rounds to 0.006, or about 6 chances in 1000.
(b) To find the probabilities that 15 or fewer rats turn left, we could continue the above table for values of $S = 15, 14, 13, \ldots, 0$ and compute the corresponding probabilities. If we added these probabilities together, the result would be 0.9941. Of course, we would most likely have solved the problem by simply subtracting $1 - 0.0059 = 0.9941$, or about 0.99. The histogram for the theoretical probabilities of $S$ is shown in Figure 7-2.

**Note** The theoretical graph (Figure 7-2) predicts that $S$ will equal 10 about 17.6% of the time and that almost every observed value of $S$ will lie between 4 and 16, inclusive. Our experiment, which consisted of only 200 trials, yielded $S = 10$ a total of 35 times (which is 17.5% of the time) and 199 of the outcomes (99.5%) were from 4 to 16.

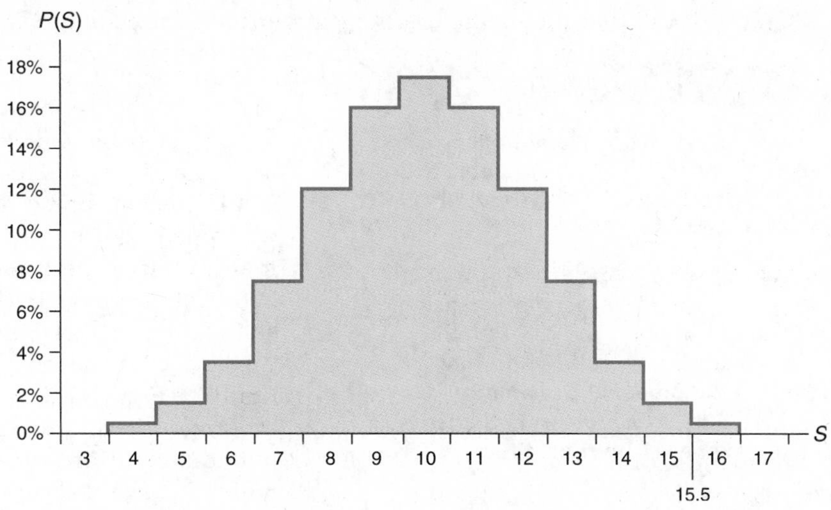


Figure 7-2 A histogram for the theoretical probabilities of $S$, the number of rats that turned left in each experiment of 20 rats.

In our second solution to Example 7.1 we will show how using the normal distribution can provide a quick way to approximate the answers to questions about a binomial random variable, but first we point out an essential difference between the binomial distribution and the normal distribution which explains why the two approaches do not give precisely the same answers.

### Discrete vs. Continuous Variables

continuous

discrete

The times given on a watch with a sweep hand are **continuous**. Every value from 0.0 to 12.0 is theoretically possible. The times given on a digital watch—no matter how accurate—are **discrete**. They come in steps, jumping over values in between the smallest adjoining steps.

Some examples of discrete variables are:

The 11 possible outcomes from rolling two dice: 2, 3, 4, 5, 6, 7, 8, 9, 10, 11, and 12.

The 2 possible results of a coin toss: heads or tails

The 201 possible grades on a 200-question true-false test: $0, \frac{1}{2}, 1, 1\frac{1}{2}, \ldots, 100$

The number of people at a football game

Any binomial random variable, $S$.

**Remember**

Age is continuous, while sex is discrete.

Some examples of continuous variables are:

Weights

Heights

Car velocities

## SOLUTIONS Method 2: Approximate Solution Using the Normal Curve

(a) The histogram in Figure 7-2 displays the theoretical probabilities for the possible values of $S$ when $n = 20$ and $p = \frac{1}{2}$. This is the same shape histogram we would get if we repeated the rat experiment many, many times, not merely the 200 trials we have performed. The shape of the histogram for $S$ is very close to the normal distribution (see Figure 7-3). Note that the binomial distribution and the normal distribution which approximates it have the same mean and the same standard deviation.

We seek an area under the normal curve that closely matches the area in the histogram taken up by the bars for $S = 16, 17, 18, 19$, and 20. The question is, Where do we draw the line that cuts off this area? We draw it at exactly the same point we use on the histogram, at the boundary between the bar for $S = 15$ and that for $S = 16$, namely at $S = 15.5$.

We use 15.5 as the cutoff point for "more than 15 rats" because the number of rats is a whole number, a discrete variable. By contrast, if $W$ is the weight of a dog, to find the probability of a dog weighing "more

### History of the Normal Distribution

The first attempt to approximate binomial probabilities by use of the normal curve was published in 1718 in the *Doctrine of Chance* by Abraham DeMoivre (1667–1754).

By 1721 DeMoivre "began to make progress in approximating the terms of a binomial expansion, in work that was to culminate in 1733 with the publication of what we now call the normal approximation to the binomial distribution." (from Stephen M. Stigler, *The History of Statistics*, Cambridge, MA: Harvard University Press, p. 71.)

In 1733 he published his work in Latin, and in 1738 he translated it into English. This was the first appearance of a work on what was essentially a normal distribution in English.

Part of the work was improved on by an approximation by James Stirling (1692–1770), following which DeMoivre used Stirling's work to further simplify his own calculations.

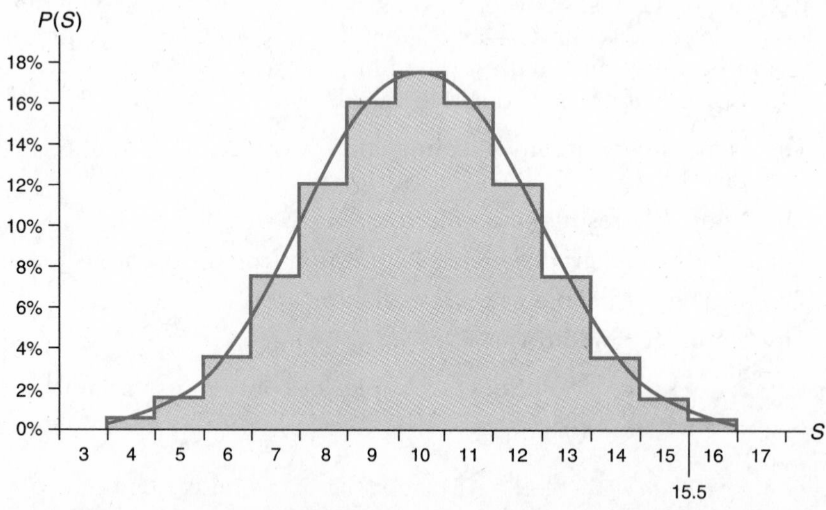


Figure 7-3 The histogram for $S$, the number of rats that turned left in each experiment of 20 rats, with a normal curve superimposed.

than 15 pounds," we would use 15 as our cutoff point, since weights are continuous. A dog can weigh 15.3 pounds, but we cannot count 15.3 rats turning left.

To find the area under the normal curve, we need the mean and standard deviation. Recall that since $n = 20$ and $p = \frac{1}{2}$, we have $\mu_S = np = 10$ and $\sigma_S = \sqrt{npq} = \sqrt{5} = 2.24$.

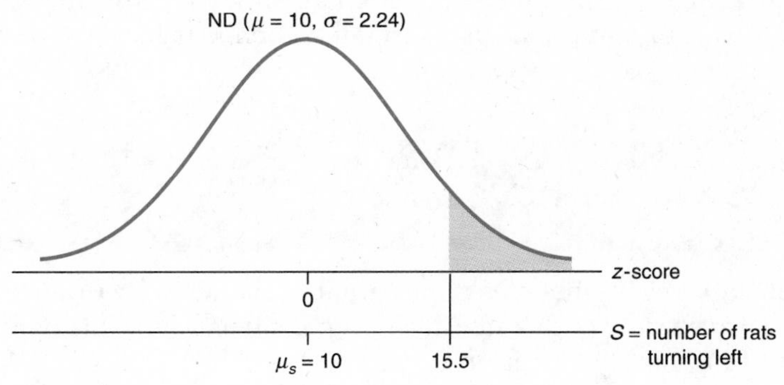


Figure 7-4

We convert the raw score $S = 15.5$ to a $z$-score, as follows:

$$z_{15.5} = \frac{S - \mu_S}{\sigma_S} = \frac{15.5 - 10}{2.24} = 2.46$$

In Table D-4 (Appendix D) we find that the area corresponding to $z = 2.46$ is 0.9931 (see Figure 7-5).

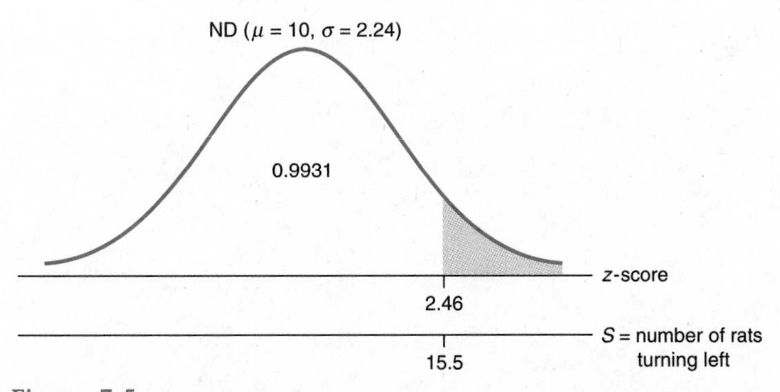


Figure 7-5

Therefore, $P(S > 15)$ from the binomial histogram is approximately $P(S > 15.5)$ on the normal curve. This area is $1 - 0.9931 = 0.0069$, or about 0.007.

(b) As shown in Method 1, part (b), we have

$$P(S \leq 15) = 1 - 0.007 = 0.993.$$

❑

You can see that the answers to Example 7.1 by Solution Method 1 and Solution Method 2 differ by only 0.001! The second solution gives a result which is accurate enough for most practical problems and which is *much easier* to calculate. What little we lose in accuracy is more than compensated for by the ease in computation, especially when $n$ is large.

## 7.2 Conditions for Approximating a Binomial Distribution by Use of a Normal Distribution

When can we say that this approximation is fairly accurate? In Examples 7.2 and 7.3 we address this question by considering the histograms for two different binomial distributions.

### ❑ EXAMPLE 7.2 A Close Approximation

Sketch the probability histogram for the binomial distribution with $n = 11$ and $p = \frac{1}{2}$, and draw a smooth curve through it.

SOLUTION

For $n = 11$, $p = \frac{1}{2}$, we get

| $S$ | $\binom{11}{S}$ Number of Ways $S$ Can Occur | $P(S)$ |
|---|---|---|
| 11 | 1 | $1(\frac{1}{2})^{11} = 1/2048$ |
| 10 | 11 | $11(\frac{1}{2})^{10}(\frac{1}{2}) = 11/2048$ |
| 9 | 55 | $55(\frac{1}{2})^9 (\frac{1}{2})^2 = 55/2048$ |
| 8 | 165 | $165(\frac{1}{2})^8 (\frac{1}{2})^3 = 165/2048$ |
| 7 | 330 | $330(\frac{1}{2})^7 (\frac{1}{2})^4 = 330/2048$ |
| 6 | 462 | $462(\frac{1}{2})^6 (\frac{1}{2})^5 = 462/2048$ |
| 5 | 462 | $462(\frac{1}{2})^5 (\frac{1}{2})^6 = 462/2048$ |
| 4 | 330 | $330(\frac{1}{2})^4 (\frac{1}{2})^7 = 330/2048$ |
| 3 | 165 | $165(\frac{1}{2})^3 (\frac{1}{2})^8 = 165/2048$ |
| 2 | 55 | $55(\frac{1}{2})^2 (\frac{1}{2})^9 = 55/2048$ |
| 1 | 11 | $11(\frac{1}{2}) (\frac{1}{2})^{10} = 11/2048$ |
| 0 | 1 | $1 (\frac{1}{2})^{11} = 1/2048$ |

The histogram for these probabilities appears in Figure 7-6.

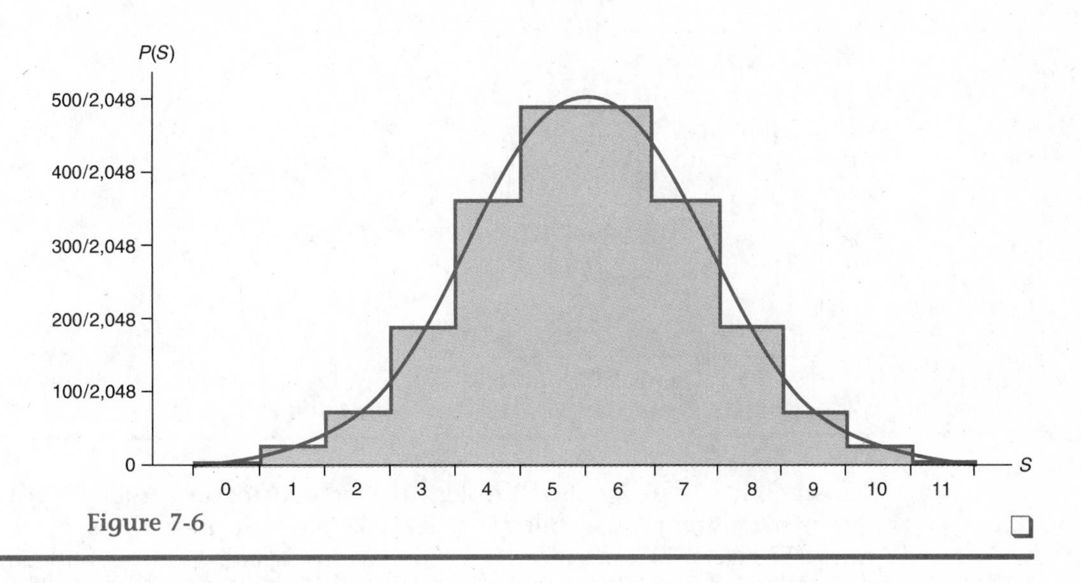


Figure 7-6

### EXAMPLE 7.3 A Dubious Approximation

Sketch the probability histogram for the binomial distribution with $n = 3$ and $p = \frac{1}{4}$, and draw a smooth curve through it.

**SOLUTION**

For $n = 3$, $p = \frac{1}{4}$, we get

| $S$ | $\binom{3}{S}$ Number of Ways $S$ Can Occur | $P(S)$ |
|---|---|---|
| 3 | 1 | $1(\frac{1}{4})^3 = \frac{1}{64}$ |
| 2 | 3 | $3(\frac{1}{4})^2(\frac{3}{4}) = \frac{9}{64}$ |
| 1 | 3 | $3(\frac{1}{4})(\frac{3}{4})^2 = \frac{27}{64}$ |
| 0 | 1 | $1\ (\frac{3}{4})^3 = \frac{27}{64}$ |

The graph for these probabilities is shown in Figure 7-7.

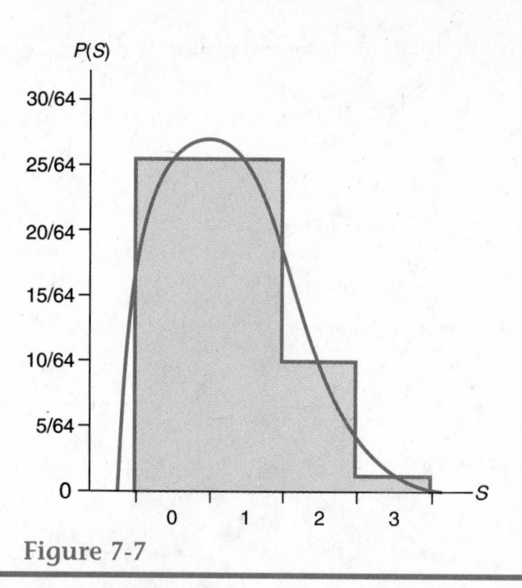

Figure 7-7

The smooth curve sketched through the first histogram (Figure 7-6) looks approximately normal, while the curve sketched through the second histogram (Figure 7-7) does not. It is not symmetrical. This second curve is chopped off on the left because the mean = $np = 3(\frac{1}{4}) = \frac{3}{4}$ is too close to zero, the smallest possible outcome. Whenever $np$ is too small this occurs. On the other hand, if $np$ is too large (or equivalently, when $nq$ is too small), the mean will be too close to $n$, the largest possible outcome. In this case, the curve will appear chopped off on the right.

rule of thumb

A good **rule of thumb** that will assure us that the binomial histogram is approximately normal is that both $np$ and $nq$ are *greater than* 5. Otherwise, the approximation may not be close enough, and so we would solve the problem by use of Pascal's triangle. In Example 7.2, $np = 5.5$ and $nq = 5.5$. Both of these are greater than 5, while in Example 7.3, $np = \frac{3}{4}$, which is too small.

## ❑ EXAMPLE 7.4 The Railroad Crossing

Let us consider the following example which further illustrates how we can handle the problem of approximating a binomial distribution by use of a normal distribution. Some car-pooling commuters drive to work during the morning rush hour. They must drive through a heavily traveled railroad crossing. From the railroad timetable, they have figured that the gate is closed 30% of the time. Because of traffic conditions, their time of arrival at the crossing is random.

(a) Find the probability that, on any given day, they arrive when the gate is open.

Next month they will drive to work 19 times. Let $S$ equal the number of times that they will be successful in arriving at the crossing when the gate is open.

(b) Find the probability that the gate is open fewer than 12 of the 19 days; that is, find $P(S < 12)$.
(c) Find the probability that $S$ is at least 15; that is, find $P(S \geq 15)$.
(d) Find $P(14 \leq S \leq 18)$.
(e) Find $P(S = 16)$, the probability that the gate is open exactly 16 out of 19 times.
(f) Certain numbers of successes are so high, and therefore, so unlikely, that there is only about a 0.05 probability that they will occur. Which numbers of successes are these? Using $S_?$ to represent the unknown number of successes, we symbolize this question as:

Find $S_?$ such that $P(S > S_?)$ is approximately 0.05.

## SOLUTIONS

(a) $p = P(\text{on one morning the gate is open}) = 1 - 0.30 = 0.70$
(b) $q = 0.30$
$n = 19$
$np = 19(0.70) = 13.30$
$nq = 19(0.30) = 5.70$

Because both 13.30 and 5.70 are greater than 5, we can use the normal distribution as an approximation to the binomial.

$$\mu = np = 13.30$$
$$\sigma = \sqrt{npq} = \sqrt{(13.3)(0.3)} = \sqrt{3.99} = 2.00$$

Because the outcomes are whole numbers, we find $P(S < 12)$ by using the boundary of 11.5, which separates the favorable outcomes 0, 1, 2, . . . , 11 from the unfavorable outcomes 12, 13, . . . , 19. this is shown in Figure 7-8.

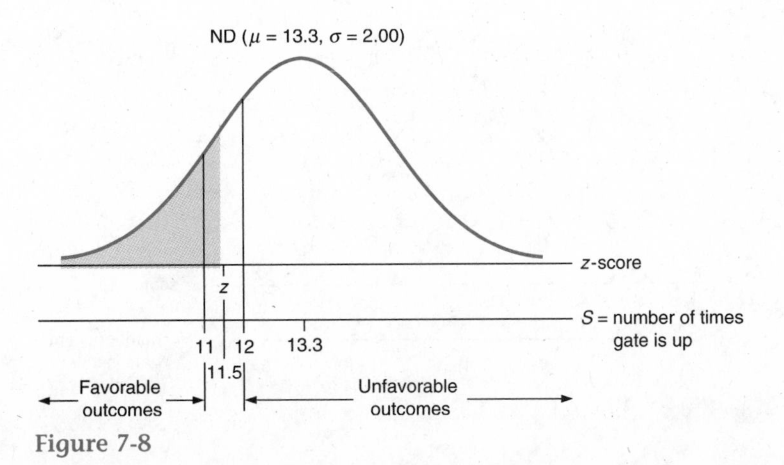


Figure 7-8

$$z_{11.5} = \frac{11.5 - 13.3}{2.00} = \frac{-1.8}{2.00} = -0.90$$

The area to the left of $-0.90$ is Area($z = -0.90$) = 0.1841, or about 0.18. Therefore, $P(S < 12)$ is approximately $P(S \leq 11.5) = 0.18$.

(c) To find $P(S \geq 15)$, we use the boundary 14.5, which separates the favorable outcomes from the unfavorable ones (Figure 7-9).

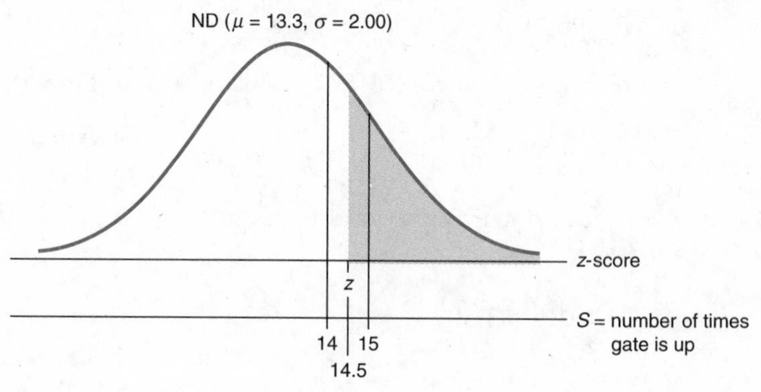


Figure 7-9

$$z_{14.5} = \frac{14.5 - 13.3}{2.00} = \frac{1.2}{2.00} = 0.60$$

$$\text{Area}(z = 0.60) = 0.7257$$

Therefore, $P(S \geq 15)$ is approximately $P(S \geq 14.5) = 1 - 0.7257 = 0.2743$, or about 0.27.

(d) To find $P(14 \leq S \leq 18)$, we will compute $P(13.5 \leq S \leq 18.5)$. This area is shown in Figure 7-10.

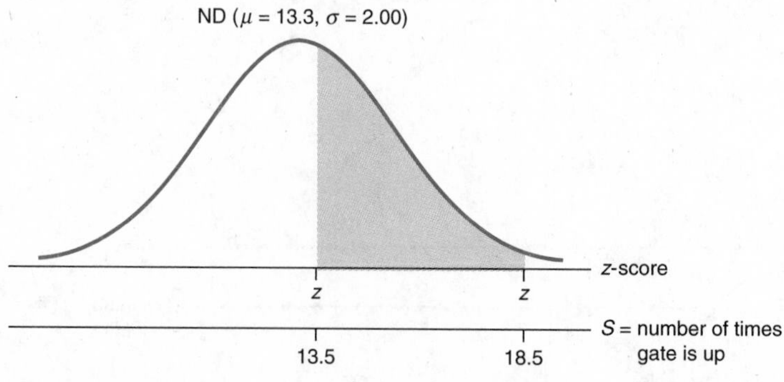


Figure 7-10

$$z_{18.5} = \frac{18.5 - 13.3}{2.00} = 2.60$$

$$z_{13.5} = \frac{13.5 - 13.3}{2.00} = 0.10$$

$$\text{Area}(z = 2.60) = 0.9953$$

$$\text{Area}(z = 0.10) = 0.5398$$

The area between 13.5 and 18.5 is found by subtracting these two areas, and thus $P(14 \leq S \leq 18)$ is approximately 0.4555, or about 0.46 (Figure 7-11).

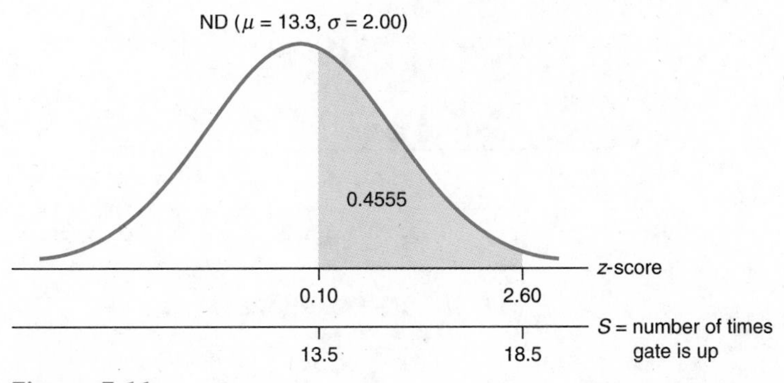


Figure 7-11

(e) To find $P(S = 16)$, we use the boundaries which cut off 16, namely, 15.5 and 16.5 (Figure 7-12).

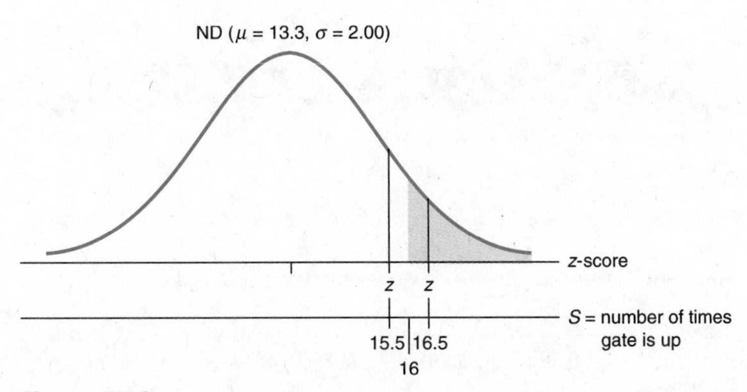


Figure 7-12

$$z_{16.5} = 1.60 \text{ and } z_{15.5} = 1.10$$

$$\text{Area}(z = 1.60) = 0.9452$$

$$\text{Area}(z = 1.10) = 0.8643$$

By subtraction we find that $P(S = 16)$ is approximately 0.0809, or about 0.08.

(f) To find the number of successes $S_?$ such that $P(S > S_?) = 0.05$, we look for area equal to 0.9500 in Table D-4. This area is depicted in Figure 7-13.

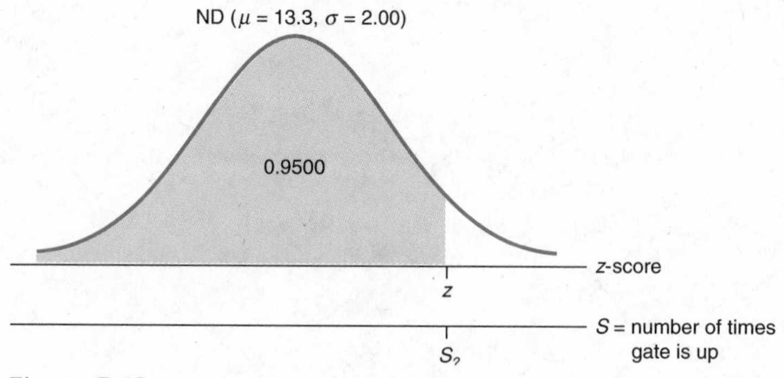


Figure 7-13

The area 0.9500 leads to a $z$-score of 1.65. To convert this $z$-score to a raw score of a number of successes, use the formula $S_? = \mu + z\sigma_S$. This yields

$$S_? = 13.3 + 1.65(2.00) = 13.3 + 3.30 = 16.6$$

Thus, outcomes 17, 18, and 19 occur only about 5% of the time. ❑

## STUDY AIDS

### Chapter Summary

In this chapter you have learned:

- How to solve binomial distribution problems quickly by using a normal distribution as an approximation
- When such approximations can be safely employed

### Vocabulary

You should be able to explain the meaning of each of these terms:

1. Continuous variable
2. Discrete variable

**Symbols**

You should understand the meaning of each of these symbols:

1. $n$
2. $S$

**Formulas**

You should know when and how to use each of these formulas:

1. $\mu = np$
2. $\sigma_S = \sqrt{npq}$
3. $S_? = \mu_S + z\sigma_S$

## EXERCISES

7-1 In using a normal curve to approximate a binomial distribution with $n = 20$ and $p = \frac{1}{2}$, where would the boundary be drawn for each of these events?
(a) $P(S > 12)$
(b) $P(S \geq 12)$
(c) $P(S < 12)$
(d) $P(S \leq 12)$

7-2 Consider the following binomial problem. Over the years, the percentage of freshmen who pass Calc I with Prof. Tomchick is 0.43. If 12 freshmen are in her class, what is the probability that 10 or more will pass? If we solve by Pascal's triangle, the answer is 0.005. The normal approximation yields 0.006. Which answer is more accurate? Comment.

7-3 Do this problem 2 ways. First, by the binomial techniques, and second, by normal approximation. Letitia tosses a coin to decide whether she will eat in a Chinese restaurant (heads) or a Greek restaurant (tails). She wishes to make plans now for her next 12 meals out. What is the probability that she will go to a Chinese restaurant more than 9 times? Comment on your answers.

7-4 Kent, a clerk in a supermarket, can work with either of 2 part-time accountants, Lois or Lane. Clerk Kent decides with whom he will work each day by using the spinner illustrated.

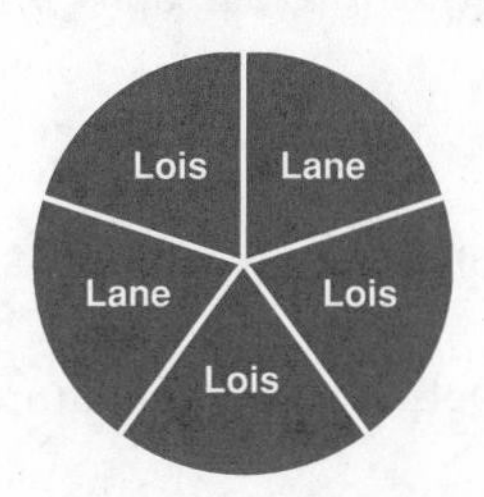


(a) Find $p = P(\text{he works with Lois})$.
(b) Find $q$.

If the supermarket man uses the spinner 14 times to determine which accountant he will work with each day for the next 2 weeks:

(c) Justify the use of the normal approximation to the binomial distribution.

(d) Find $\mu_S$.

(e) Find $\sigma_S$.

(f) Find the probability that Lois works with the superman 7 or fewer times.

7-5 A pair of loaded dice has $p = P(\text{rolling a } 7) = 0.3$. If the dice are rolled 20 times:

(a) Justify the use of the normal approximation to the binomial distribution.

(b) Find $\mu_S$.

(c) find $\sigma_S$.

(d) Find the probability of rolling more than three 7s.

(e) Find $P(S > 9)$, where $S$ is the number of 7s rolled.

(f) If $P(S > S_?)$ is approximately equal to 0.99, find the whole number $S_?$.

7-6 A game is played by two persons who each show 2, 1, or 0 fingers at the same time. If the total number of fingers is odd, then the person predesignated as "odd" wins. If the total number of fingers is even, then the other person, designated "even," wins.

(a) List the 9 possible outcomes.

(b) Show $P(\text{odd}) = 4/9$.

If Mauro and Gina play the game 18 times, find the probability that odds occur:

(c) More than 14 times.

(d) Exactly 9 times.

(e) Either 7, 8, or 9 times.

7-7 It is known that 30% of the autos produced by the Necromate Auto Company are defective when they come off the assembly line. (So we say that the probability that a car off the line is defective is 0.30.)

(a) What is the probability that if 900 of these cars are picked at random, more than 40% of the 900 cars will be defective? (This means that more than 360 cars will be defective.)

(b) What is the probability that more than 33% of the 900 cars will be defective?

(c) What is the probability that less than 28% of the 900 cars will be defective?

7-8 The probability that a 40-year-old man will have died before age 60 is 0.17. An insurance company insures 1200 men of age 40.

(a) What is the probability that fewer than 15% of these men will have died before age 60?

(b) What is the probability that more than 20% of the men will have died before age 60?

7-9 Under certain conditions, the probability that a tadpole survives to mature into a frog is 0.10. If we have 100 tadpoles:

(a) What is the probability that more than 14 survive?

(b) What is the probability that none survive? (*Hint:* You may interpret "none" to mean "less than 1.")

(c) If the probability of at least $S_?$ tadpoles maturing is about 0.95, find the whole number $S_?$.

7-10 The morbid *Journal of Morbidity* indicates that the probability of a teenager contracting terminal acne is 0.12. If 100 teenagers are checked at random, what is the probability that more than 20 will have contracted this disease?

7-11 For a certain disease half the victims recover in 1 week with no treatment. Dr. Quack invents a treatment for this disease. (The probability of recovering with his treatment in 1 week is still 0.5. It is no better than doing nothing.) He treats 20 patients. Of them, 12 recover in a week. He claims that this proves his treatment is good. What is the probability that at least 12 people out of 20 would have recovered with no treatment?

7-12 Using patients suffering from the disease of the previous problem, Dr. Chi wants to evaluate the effectiveness of mind over matter. He gives 40 patients a placebo, telling them it is a powerful new cure. After 1 week, only 11 of them still suffer from the disease. Under ordinary circumstances, with no treatment, about 20 would not recover in 1 week. What is the probability that 11 or fewer would not recover? That is, if Success $S$ is defined to be recovery, find $P(S \geq 29)$. Comment on these results.

7-13 Rosencrantz tossed a sixpence 10 times. It fell heads up each time. He claims that the coin is a fair one. Is that possible? Do you believe him? (Justify your answer mathematically.)

7-14 Refer to the rat problem at the beginning of this chapter. What would your reaction be if you performed this experiment 3 times, and each time more than 16 rats turned left?

7-15 Bob and Gene play a game by drawing 1 card from an ordinary deck and then replacing it. If the card is a spade, Bob wins. If not, then Gene wins. On the last 24 draws, Bob won 18 times.

(a) Using $p = P(\text{Bob wins}) = 1/4$, find the probability that Bob would win 18 or more times in 24 plays.

(b) Using $p = P(\text{Gene wins}) = 3/4$, find the probability that Gene would win 6 or fewer times in 24 plays.

7-16 Do Exercise 5-13(b) again by using the normal approximation. Explain any differences in your answers to the problem by the binomial method and by the normal approximation. We restate the problem.

If in each pregnancy the probability of having a girl is 1/2, what is the probability that a childless couple planning to have 5 children will have at least 4 girls? What percentage of couples with 5 children would you expect to have 4 or 5 girls? (The answer by the binomial method was 0.1875.)

7-17 Do Exercise 5-26 again by using the normal approximation. Explain any differences in your answers to the problem by the binomial method and by the normal approximation. We restate the problem.

A manufacturer claims that 4 of 5 dentists recommend sugarless gum for their patients who chew gum. Assuming that this claim is true, find the probability that, in a randomly selected group of 20 dentists, 16 or more will recommend sugarless gum for their patients who chew gum. (The answer by the binomial method was 0.63.)

7-18 A large distribution of the number of peanuts in a package of Sower's peanuts is approximately normal, with $\mu = 120$ and $\sigma = 15$. Suppose that 1 package is selected at random, and $X$ represents the number of peanuts in the package.

(a) Find $P(X < 100)$.

(b) Find the number $A$ for which $P(X > A)$ is approximately equal to 0.35. (*Hint:* The number of peanuts is a whole number.)

(c) Repeat parts (a) and (b) above if the distribution is the weight in grams of a package of Sower's peanuts. (*Hint:* Weight is continuous.)

7-19 Suppose that you have data from a continuous distribution. The distribution is approximately normal, with a mean of 14 and a standard deviation of 3.

(a) If a number $X$ is selected at random from the distribution, find $P(X > 18)$.

(b) The value of $A$ in this distribution is such that $P(X < A) = 0.4$. Find $A$.

(c) Repeat parts (a) and (b) above if the distribution consists of whole numbers only.

7-20 It is known that 20% of EZ brand zippers are defective. If many random samples of 100 zippers are examined:

(a) Find the mean number of defectives in these samples.

(b) Find the standard deviation of the number of defectives.

(c) Find the probability that more than 25 zippers in a random sample of 100 zippers will be defective.

(d) The probability that more than $S_?$ zippers in a random sample of 100 zippers will be defective is approximately equal to 0.30. Find the whole number $S_?$.

7-21 A carnival packages prizes in identical unmarked boxes in such a way that 9 of 10 boxes contain a prize worth 10 cents and 1 of 10 contains a prize worth 50 cents. There are hundreds of such boxes all mixed up in bins at the prize booth. For 30 cents, Sylvia may pick out any box and keep the prize.

(a) If she takes 1 chance, what is the probability that her prize will be worth less than what she paid to play?

(b) If 500 people each play once on a Saturday evening, what is the probability that more than 12% will win "good" prizes?

(c) How many times does Sylvia have to play to have better than a 50% chance of getting *at least one* 50-cent prize? (*Hint:* This means that the probability of no "good" prizes must be less than 50%.)

7-22 G. Whilakers runs an amusement parlor. One popular game there is called "Skee-Boll," in which a player can win by getting over 200 points. Mr. Whilakers has noticed that, overall, in 80% of the games the player wins. On the average night, 600 games are played. What is the probability that there will be over 500 winners?

7-23 (a) The probability that a person in a certain community is a carrier for the fatal Tay-Sachs disease is 1 in 30. If 300 people are examined in this community, what is the probability that more than 10 will be found to have the disease?

(b) If 2 people from this community marry, what is the probability that they will both be carriers?

(c) If both parents are carriers, there is a 1/4 probability that their baby will have the fatal disease. What is the probability if 2 people in this community marry that their first child will have the disease? (*Hint:* This is a sequence of events. First the 2 people would have to both be carriers, then they would have to produce a diseased baby. Use the multiplication rule mentioned in Chapter 4 for the probability of a sequence of events.)

7-24 A recent election in the city of Gotham was contested. The facts were these: 10,000 citizens voted; 4900 voted for Elizabeth Goodsoul, while 5100 voted for James Badguy. The election was challenged, and 1000 voters were found to have voted illegally. Ms. Goodsoul asked for a reelection, but Judge Blank ruled against her request. He reasoned that it was not likely that a reelection among the legal voters would change the final outcome.

Using $p = P(\text{a voter votes for Goodsoul}) = 4900/10{,}000 = 0.49$, and assuming that $n = 1000$, perform the following calculations. If we randomly throw out 1000 votes, Goodsoul would need at least 4501 of the remaining 9000 to win. This means that of the 1000 votes being randomly thrown out less than 400 of hers would have to be thrown out. Find the probability that less than 400 of the 1000 thrown out would be her votes.

## CLASS SURVEY QUESTION

1. If you wanted to use the class survey results to help decide whether or not your *entire* student body is about half male and half female, you could proceed as follows.
   (a) Count the number of students surveyed ($n$). Then count the number of males and females.
   (b) Calculate $p_m = P(\text{a student is male})$ and $p_f = P(\text{a student is female})$. Select the smaller of these two numbers as your test statistic.
   (c) Assuming that this survey is a random sample by sex, and assuming that your school equal opportunity officer claims that the school is $\frac{1}{2}$ male and $\frac{1}{2}$ female, find the probability of obtaining a value of $S$ equal to or less than the one you calculated. Does this probability tend to make you believe that your school may be about $\frac{1}{2}$ male and $\frac{1}{2}$ female, or does it tend to make you reject that assumption?

## FIELD PROJECT

If you did the field project assigned in Chapter 1, use that data for the Class Survey Question above.

## CLASS EXPERIMENT

*Coin Toss.* Have each person toss a penny 100 times and report the number of heads that occurred. Make a frequency table and a histogram of the results. In theory, what percent of the class should get from 45 to 55 heads? Compare this result to what actually happened.

# Sample Test for Chapters 4, 5, 6, and 7

1. Translate the symbol "$P$(one toss is heads)" into English.
2. Inferential statistics makes inferences about (a) a population or (b) a sample. [Select answer (a) or (b).]
3. An important question in statistical studies is whether a sample is a ______ sample or not. (Fill in the blank.)
4. People are to be classified according to Age (under 21, 21–39, or 40 and up), Sex (male or female), and Religion (Christian, Jewish, Muslim, Buddhist, or other). Using the multiplication principle, calculate how many categories there will be.
5. Write out the sample space of possible outcomes for the previous question.
6. What's wrong here? "Since the probability of rolling a 7 with two dice is 1/6, I will roll about two 7s in my next 12 rolls of two dice."
7. What is the probability that:
   (a) A fair coin tossed 3 times will come up heads each time?
   (b) 3 fair coins tossed together once will all come up heads?
8. In a history class $P(X \text{ is female}) = 0.43$.
   (a) What percentage of the class are females?
   (b) What percentage are males?
9. In a histogram 62% of the area is to the left of a grade of 75.
   (a) What percentage of the class got below 75?
   (b) What is the value of $P(X \geq 75)$?
10. A biased coin has $P(\text{heads}) = 70\%$. If it is tossed 50 times, how many heads will you expect to appear?
11. (a) The weights of the costumes worn at Disneyland are approximately normally distributed. What percentage weighs more than $1\frac{1}{2}$ standard deviations above the average?
    (b) If the mean weight is 6.7 pounds and $\sigma = 2.30$ pounds, how heavy are the top 10% of the costumes?
    (c) What weight cuts off the lightest 15% of the costumes?
    (d) What is the probability that a costume selected at random weighs between 7 and 8 pounds?
12. If employees at Friendly's Emporium are paid on average \$5.10 per smile with a standard deviation of \$1.29, is it possible that anyone in the highest-paid 6% of the employees is paid less than \$6.00 per smile?
13. One line of Pascal's triangle shows 1, 4, 6, 4, 1. What do the 4s signify?

14. Evaluate $x$:

(a) $\binom{10}{3} = x$ (b) $\binom{5}{x} = 10x$ (c) $\binom{125}{0} = x$

(d) $\binom{119}{119} = x$ (e) $\binom{117}{1} = x$ (f) $\binom{22}{2} = x$

Solve both Problem 15 and Problem 16 two ways: (a) By using Pascal's triangle. (b) By the normal approximation. Comment on your solutions.

15. A fair coin is tossed 18 times. Find the probability of getting either 8, 9, or 10 heads.

16. The probability that 1 landing light burns out on the XL-12 rocket to the moon is 0.42. Find the probability that all 12 landing lights are operating.

17. The publisher of *Honesty* magazine claims that 70% of its subscribers are under age 30. A random sample of 500 subscribers shows that only 63% of the sample are under 30. The publisher says that you only sampled some of the subscribers and that a second sample might yield 77%.
    (a) Presuming that his original claim is true, what is the probability of getting 63% or less under age 30 in a sample of 500 subscribers?
    (b) Is it possible that the publisher's claim is correct?
    (c) Would you believe the publisher?
    (d) It is possible that the sample was not random, but had some hidden bias. Discuss some ways to determine whether or not this sample was representative.

# 8

# Hypothesis Testing: One-Sample Tests of Percentages in Binomial Experiments

"The first precept was *never to accept* a thing as true until I knew it as such without a single doubt."
—René Descartes
(emphasis added)

**CONTENTS**



**GOALS**

At the end of this chapter you will have learned:

- How to formulate two types of hypotheses—null and alternative
- To identify and describe one-tail and two-tail hypothesis tests
- To distinguish between Type I and Type II errors
- To interpret the significance level of a hypothesis test
- How to perform a one-sample hypothesis test of percentages

This chapter brings together many of the ideas we have discussed so far. The material is of the utmost importance for success in understanding the remaining chapters. Hypothesis testing is one of the most widely used procedures in statistics. At the conclusion of this chapter you will be in a position to understand hypothesis testing and to perform your own experiments.

In this chapter we will illustrate the basic ideas of hypothesis testing by applying them to binomial experiments. In later chapters we will apply these basic ideas to other situations.

## 8.1 Statistical Hypotheses

Consider the following five questions:

1. What percentage of coupons printed in the newspapers are redeemed?
2. How much more effective is prescription A than prescription B?
3. Is it true that 30% of shoppers buy their favorite brand of toothpaste, regardless of price?
4. Is this die biased in favor of 3?
5. Do boys and girls score differently on the verbal portion of the SAT?

The five questions above fall into two categories. The first two ask for a *numerical* response, while the last three request a *yes-no* response. Statisticians approach these two types of questions differently. For now, we will discuss the *yes-no* response type of question, putting off the numerical-response type until Chapter 12.

In analyzing a study which was designed to decide which answer, "yes" or "no," is probably correct for some population, statisticians begin by first formulating a pair of opposing statements, called **hypotheses**.

**hypotheses**

**statistical hypothesis**

> *Definition:* A **statistical hypothesis** is a statement about one or more populations which is either true or false.

### *Statistical Hypothesis*

A hypothesis is a statement, a declarative sentence, that ends with a period (not a query that ends with a question mark). It can be labeled either "true" or "false."

For the purposes of statistical analysis, these hypotheses are in the form of numerical statements about one or more population parameters. Then sample statistics can be used to make inferences about these parameters.

hypothesis test

> *Definition:* A **hypothesis test** is a procedure used to make inferences about the truth or falsity of a hypothesis.

When we sample *one* population, we call the test a "one-sample test." When we sample two populations, we call the test a "two-sample test." Some illustrations based on questions 3, 4, and 5 above follow.

## ❑ EXAMPLE 8.1 Illustrations of Statistical Hypotheses

Formulate pairs of opposite hypotheses for questions 3, 4, and 5 above.

SOLUTION

*Question 3.* Is it true that 30% of shoppers buy their favorite brand of toothpaste regardless of price? The two opposite hypotheses are:

$H_1$: The percentage of shoppers who buy their favorite toothpaste regardless of price is 30%.

$H_2$: The percentage of shoppers who buy their favorite toothpaste regardless of price is not 30%.

Suppose that for this study the population is all the shoppers in some market. The parameter being considered is $p$, the percentage who buy their favorite toothpaste regardless of price.

$H_1$ claims that $p = 0.30$; $H_2$ claims that $p \neq 0.30$.

### *Two Yes-No Questions*

Every time we formulate a hypothesis there is a yes-no issue raised. Is the hypothesis true? But when the hypothesis is about $p$, there is also a second yes-no issue. In question 3 we have:

(a) Is $H_1$ true? Did 30% of the entire population buy their favorite toothpaste regardless of price?
(b) Does this individual being polled buy his or her favorite toothpaste regardless of price?

*Question 4.* Is this die biased in favor of 3? The two opposite hypotheses are:

$H_1$: The die is biased for 3s.

$H_2$: The die is not biased for 3s; it is fair.

For this study the population of interest is all the potential tosses of this die, and the parameter being considered is $p$, the probability of getting a 3 with this die, $P$(a 3 is rolled). We can rewrite our hypotheses so that they are mathematical statements about the numerical values of the population parameter $p$.

### *Again, Two Yes-No Questions*

Notice that the hypotheses in question 4 concern the parameter $p$. Thus, there are two different yes-no issues.

(a) Is $H_1$ true? When we consider all possible rolls of this die, do we get too many 3s?
(b) Did this particular roll of the die result in a 3?

$H_1$: $p > 1/6$.
$H_2$: $p = 1/6$.

For clarity, we will usually write each hypothesis *both* ways—in words and as a numerical statement about some population parameter:

$H_1$: The die is biased for 3s; $p > 1/6$.
$H_2$: The die is not biased for 3s; it is fair; $p = 1/6$.

**Note** By using just an = sign in $H_2$ (and not $\leq$), we have decided not to consider the possibility that the die gets too few 3s. Thus, $H_1$ and $H_2$ are not diametrically opposed. Sometimes a logical possibility just happens to not be of interest in a particular study, and so the hypotheses can be phrased accordingly. *The important point is that if one of the hypotheses is true, then the other hypothesis is false.*

*Question 5.* Do boys and girls score differently on the verbal portion of the SAT tests?

The question as given does not refer to any particular way to compare the boys' and girls' scores. But one reasonable approach would be to compare the average scores for the two groups. This would then become a question about averages rather than about percentages. Notice that this is not a binomial experiment—there are more than two answers to the question, "What is your SAT verbal score?"

**Only One Yes-No Question**

Notice that the hypothesis in question 5 concerns the parameter $\mu$. In tests about $\mu$ there is only one yes-no issue: Is $H_1$ true? Is the average score of all boys taking the test the same as the average score of all girls taking the test? In hypothesis tests about $\mu$, there are many possible numerical responses for each individual, not just yes or no.

For this two-sample comparison of averages the two populations are all the boys who took the SAT exam in a given year and all the girls who did so. The hypotheses are:

$H_1$: Boys and girls *have the same* average scores on the verbal portion of the SAT exam.

$H_2$: Boys and girls *do not have the same* average scores on the verbal portion of the SAT exam.

There are two parameters being compared, the mean score $\mu_B$ for all the boys and the mean score $\mu_G$ for all the girls:

$H_1$ claims that $\mu_B = \mu_G$, while $H_2$ claims that $\mu_B \neq \mu_G$, so we write:

$H_1$: The average verbal scores are the same; $\mu_B = \mu_G$.
$H_2$: The average verbal scores are not the same; $\mu_B \neq \mu_G$. ❑

## 8.1-1 Testing the Null Hypothesis

### The Logic of a Hypothesis Test

**null hypothesis**

For each example above we have presented two opposing hypotheses. For the purposes of statistical analysis, one of them is chosen as the basis for interpreting the data. This one is called the **null hypothesis**. In this section we explain how to decide which of your two hypotheses to use as the null hypothesis.

Suppose you were interested in the question, Is this a fair coin? A fair coin is one for which the probability of heads on a single toss is 0.50. Therefore, in this study, if we let $p = P$(a head appears on one toss), then our two hypotheses are:

$H_1$: The coin is fair; $p = 0.50$.

$H_2$: The coin is biased; $p \neq 0.50$.

We could attempt to establish the truth of either one of these two hypotheses, for if one is true, the other is false, and vice versa.

It turns out that it is much easier to test the first one: $p = 0.50$. This is because if we assume that $H_1$ is true, and then toss the coin, say, 80 times, it follows that we know what to expect. We would expect to get about 40 heads. We wouldn't be suspicious if the outcome were 38 heads, but an outcome of 72 heads would certainly arouse our suspicions. On the other hand, if we tried to test $H_2$ by tossing the coin 80 times, we would not know what to expect, because $H_2$ does not give us one specific value of $p$ with which to work. Would we expect 50 heads? 15 heads?

In general, when we perform a hypothesis test we call the hypothesis which gives us a specific value to work with the **null hypothesis**, and symbolize it as $\boldsymbol{H_0}$.

> *Definition:* A **null hypothesis** is one that leads to an expected value which can be confirmed or denied by analyzing the data from an appropriate sample.

The null hypothesis is often the one that implies fairness and honesty. It Pseems to look at the world through rose-colored glasses:

This die is fair.

This newspaper claim is true.

This theory is correct.

**motivated hypothesis** **alternative hypothesis**

The opposite hypothesis is called the **motivated**, or **alternative**, **hypothesis** and is symbolized by $H_a$. This *a*lternative hypothesis, however, is often the one that is of original interest, the one that motivated the experimenter to perform the hypothesis test.

We suspect:

That the die is biased

That the newspaper erred

That the theory is wrong

It is often this suspicion that motivates us to investigate the question in the first place.

| *Statistical Hypotheses* | |
|---|---|
| ***Alternative Hypothesis*** $H_a$ | ***Null Hypothesis*** $H_0$ |
| Is usually the statement of interest to the researcher | Views the world through rose-colored glasses |
| Usually does not lead to *one* specific expected value | Leads to *one* specific expected value |
| Usually contains an inequality symbol | Contains the equality symbol |
| | Is assumed to be true for testing purposes |

Go back to questions 3, 4, and 5 in Example 8.1. In each question, can you pick out which of the two hypotheses is the null hypothesis? That is, which one, by giving you a specific value to work with, would tell you most clearly *what to expect* if it were true?

*For question 3*, if you interviewed 200 shoppers, how many would you expect to respond, "Yes, I buy my favorite brand of toothpaste regardless of price"? If $H_1$ is true, you would expect about 30% of 200, or 60 shoppers to say yes. If $H_2$ is true, you do not know what to expect, except that it would not be 60. Thus, $H_1$ is the null hypothesis.

*For question 4*, if you rolled the die 60 times, how many 3s would you expect to occur? The answer depends upon whether the die is fair or not. If $H_2$ were true, we would know what to expect—about 1/6 of 60, or ten 3s to occur in 60 tosses of a fair die. However, if $H_1$ were assumed to be true, we would not know what to expect. Thus, $H_2$ would be the null hypothesis.

*For question 5*, suppose that you tested a group of 50 boys and a group of 50 girls and found the difference between the average score of each group. What should you expect the difference to be? Again, it depends upon which hypothesis is true. If $H_1$ were true, we would expect the difference between the two means to be near 0. $H_2$ gives no hint as to what we should expect. Therefore, $H_1$ is the null hypothesis.

To summarize, the null hypothesis gives us a specific value on which to base our expectations. Thus, when symbolized, the null hypothesis contains an equal sign (=), as in $p = 0.6$.

The alternative hypothesis, on the other hand, usually contains an inequality sign ($<$, $>$, or $\neq$) when symbolized.

We sum up by comparing the pattern of a hypothesis test in both a general and a specific way:

| **General Case** | **Specific Case** |
|---|---|
| Based on some prior experience or idea, you are motivated to perform a statistical experiment on a *specific* population. | Suppose we are suspicious about the behavior of a die, and we decide to study the number of 3s that appear in 60 tosses of the die. The hypotheses are: |
| If $H_0$ is true, the sample data will most probably come out as predicted. | $H_0$: This die is fair for 3s; $p = P(\text{a 3 is rolled}) = 1/6$. $H_a$: This die is not fair for 3s; $p \neq 1/6$. If the $H_0$ is true, then we expect to get *about* 10 threes in 60 tosses. |
| (a) If the data come out far from what is predicted by assuming $H_0$ is true we conclude that $H_0$ is most probably false, and we reject it. Our motivated hypothesis is probably true. | (a) If the number of 3s we roll is far from the 10 predicted by assuming that $H_0$ is true, we will reject the $H_0$ as probably false. We will conclude that $p$ is probably not 1/6 and that the die is most likely biased, as the motivated hypothesis states. |
| (b) If the data do come out close to what is predicted by assuming $H_0$ is true we will conclude that there is insufficient evidence to reject $H_0$. We will have failed to prove that the motivated hypothesis is true. | (b) If we do get close to 10 threes we will conclude that there is insufficient evidence to prove there is something wrong with the die. It might be true, $p$ might equal 1/6. We have failed to prove the die is not fair for 3s. |

When, for testing purposes, the null hypothesis is assumed to be true, the statistician often assumes the opposite of what he or she actually hopes to establish. The reasoning goes like this—How to prove indirectly that a statement is probably true:

Assume its opposite is true.

Show that this assumption is highly unlikely, untenable.

Hence, conclude that original statement *is* probably true.

This type of indirect logic is similar to what we use when we say such things as

"If that is true, I'll eat my hat" or "If that's true, then I'm a monkey's uncle."

Since we don't expect to eat our chapeau, we are implying that the statement under consideration is highly unlikely.

It follows from this description of a hypothesis test that no statistical "proof" is 100% infallible. Instead, we basically decide that a statement is false if it has only a very small chance of being true. A researcher attempts to prove or disprove the statement "beyond a reasonable doubt" by analyzing a sample from some population of interest.

## 8.1-2 One-Tail and Two-Tail Alternative Hypotheses

In each of the illustrations of hypotheses in Example 8.1 we have identified the null hypothesis. Now we take a closer look at the alternative hypotheses because they directly influence how the statistical analysis will be done. Here are the symbolic versions of each illustration.

*Question 3:* Null hypothesis: $p = 0.30$
Alternative hypothesis: $p \neq 0.30$.

*Question 4:* Null hypothesis: $p = 1/6$
Alternative hypothesis: $p > 1/6$.

*Question 5:* Null hypothesis: $\mu_B = \mu_G$.
Alternative hypothesis: $\mu_B \neq \mu_G$.

Notice that in questions 3 and 5 the alternative hypothesis contains a not-equal sign ($\neq$), while the alternative hypothesis in question 4 contains a greater-than sign ($>$). Alternative hypotheses which contain a not-equal sign are called *two-tail* alternatives, and a hypothesis test with such an alternative is called a **two-tail test**. By contrast alternatives which contain a greater-than sign (or a less-than sign) are called *one-tail* alternatives, and a hypothesis test with such an alternative is called a **one-tail test**.

two-tail test

one-tail test

If you suspect that a certain null hypothesis is false, you can formulate three different alternatives. For example, suppose you are suspicious of a claim made in a newspaper article. You therefore let this claim be the null hypothesis and assume it is true. Then you could choose any one of the following three alternatives:

(a) The figure in the newspaper is too large. This alternative leads to a one-tail test.
(b) The figure in the paper is too small. This alternative also leads to a one-tail test.
(c) The figure is simply suspicious and may be either too large or too small. This last alternative leads to a two-tail test.

### The Use of Symbols in Hypotheses

Note that we have formulated our hypotheses so that the equal sign (=) always appears in the null hypothesis.

The alternative hypothesis for a one-tail test contains either the less-than ($<$) or the greater-than ($>$) sign.

The alternative hypothesis for a two-tail test always contains the not-equal sign ($\neq$).

The choice of either a one-tail or a two-tail test is determined by what the statistician is interested in finding out.

The motives, purposes, and/or suspicions of the researcher would determine which alternative hypothesis is to be chosen.

The distinction between one- and two-tail tests is illustrated further in Example 8.2.

### ❑ EXAMPLE 8.2 Formulating Hypotheses

Formulate the null hypothesis and the alternative hypothesis for each of the following:

(a) Is the average life span of a dog more than 13 years?
(b) Is the proportion of 18-year-old drivers who have accidents the same as the proportion of 26-year-old drivers who have accidents?
(c) What percentage of people born with Down's syndrome can be taught to read?
(d) Do teenage girls receive a smaller weekly allowance than teenage boys?

SOLUTIONS

(a) Suppose this is a study based on the population of all dogs in Little Rock, Arkansas. Let $\mu$ = the average life span of dogs, then the two hypotheses are:

$H_0$: The average life span is 13; $\mu = 13$.
$H_a$: The average life span is greater than 13; $\mu > 13$ (a one-tail test).

(b) Suppose this study is set up in Wildwood, New Jersey. It compares two populations there: all 18-year-old drivers and all 26-year-old drivers.

Let $p_1 = P$(an 18-year-old driver has an accident)
$p_2 = P$(a 26-year-old driver has an accident)

This gives as hypotheses:

$H_0$: The proportion of accidents for the two age groups are the same; $p_1 = p_2$ or $p_1 - p_2 = 0$.
$H_a$: The proportion of accidents for the two age groups are different; $p_1 \neq p_2$ or $p_1 - p_2 \neq 0$ (a two-tail test).

(c) This question calls for a *numerical* response. Since there are no claims to test, the question does not lead to a hypothesis test.
(d) This is a two-sample hypothesis test comparing two averages. Suppose the populations are all the teenage girls living in London, England, and all the teenage boys living in London, England. Define $\mu_G$ = average weekly allowance for teenage girls and $\mu_B$ = average weekly allowance for teenage boys; then, we have:

$H_0$: The average allowances are the same; $\mu_G = \mu_B$ or $\mu_G - \mu_B = 0$.
$H_A$: The average allowance of girls is less than that of boys; $\mu_G < \mu_B$ or $\mu_G - \mu_B < 0$ (a one-tail test). ❑

## 8.1-3 Decision Rules

**decision rule**

**rejection region**

Once you have established the two opposite hypotheses for a study, you must then decide how to evaluate the data. This is the function of the **decision rule** for the study. For any hypothesis test some appropriate statistic is calculated from the sample data. The decision rule describes all values of that statistic which lead to the rejection of the null hypothesis. This set of values is called the **rejection region** for the hypothesis test. This is illustrated in the next example.

---

### ❑ EXAMPLE 8.3 Making a Decision Rule

Guildenstern suspects that a certain coin is biased for heads. What kind of statistical test can she do to confirm her suspicions? Describe the parameter of interest, the null and the alternative hypotheses, the sample test statistic, and the decision rule.

#### SOLUTION

Any experiment must involve tossing the coin, and the more times she tosses it the surer she can be about her conclusion. Suppose she decides to test the coin by tossing it 40 times. She reasons that *if* it is a fair coin, then she should get *about* 20 heads; but if it is biased *for heads*, then she should get substantially more than 20 heads. She needs to specify, though, what she means by "substantially more" than 20 heads. So she makes the following decision rule: if the 40 tosses produce 25 or more heads, then conclude that this coin is biased for heads.

Notice all the components of Guildenstern's one-sample hypothesis test. The population consists of all possible tosses of this coin. The parameter under investigation is $p = P(\text{a head is tossed})$. The null hypothesis says that the coin is fair, $p = 0.50$; the alternative hypothesis says that the coin is biased for heads, $p > 0.50$. The sample statistic is $S$, the total number of heads which appear in 40 tosses. The decision rule says that if $S$ is 25 or more, then reject the null hypothesis. Note that $S$ is a statistic on which we base a conclusion about the parameter $p$. Because $S$ is a binomial random variable and Guildenstern will make her decision based on the data in 1 sample, we say that she is carrying out a one-sample binomial hypothesis test of a percentage. ❑

---

### Exercises

*In Exercises 8-1 to 8-7:*

*(a) Indicate which questions are not formulated as hypothesis tests.*

*For the rest:*

*(b) Indicate whether the questions lead to one- or two-sample tests.*

*(c) Indicate whether the parameter is $\mu$ or $p$.*

*(d) Formulate the two hypotheses in both words and symbols, indicating whether the situation calls for a one-tail or a two-tail test. In each case explain the meaning of the symbols that you use, such as $p$, $p_1$, $p_2$, $\mu$, $\mu_1$, or $\mu_2$.*

8-1 Is the average workweek in Centerville less than 40 hours?

8-2 Do more than 10% of pet owners own goldfish?

8-3 What is the average jail sentence for bank robbery?

8-4 Do 12% of the students in your school major in mathematics?

8-5 Is the average height of 6-year-old male horses the same as the average height of 6-year-old female horses?

8-6 What is the average weight of adult baboons?

8-7 Do a greater percentage of 20-year-old females diet than do 20-year-old males?

8-8 A claim is made that 40% of all viewers watch the TV show "Mathematics and Humor." If we define $p = P$(a viewer watches "Mathematics and Humor"), the null hypothesis to test this claim would be $H_0$: The claim is true; $p = 0.40$. What would the motivated hypothesis be for the following 3 situations? In each case explain your choice.
(a) A statistics student who was considering testing the truth of the claim
(b) Prospective sponsors who were considering advertising on the show
(c) Al Jebra, the show's star, who is considering asking for a raise

## 8.2 Statistical Errors

One basic idea which is inseparable from hypothesis testing is that you can almost never have *absolute* proof as to which of the two hypotheses is the true one. The possibility always remains that a study may come up with the wrong conclusion. However, we can often make the probability of such an error very, very small indeed.

For example, in the case of testing a coin to see if it is fair, you must realize that the very definition of "fair coin" makes it impossible to completely test a coin. Recall that a fair coin is one for which the probability of heads on a single toss equals 0.50. But, "a probability equal to 0.50" means that the coin comes up heads one-half the time *in the long run,* and "in the long run" means you have to keep tossing the coin *forever!* So any time you test a coin, no matter how many times you toss it, it's only a small portion of how many times it *could* be tossed.

Suppose our friend Guildenstern tosses her coin 40 times and gets 30 heads. Since her decision rule was to reject fairness if she obtained 25 or more heads, she will say the coin is not fair. Who knows what *might* happen if she *kept* on tossing the coin? It's possible that Guildenstern does in fact have a fair coin, but an unlucky run of extra heads fools her. If so, the data have led her to make a mistake through *no fault of her own.* This is an unavoidable possibility in all hypothesis testing, because we look at only a sample and not the whole population. We can reduce the probability of this happening by collecting more data, but we have to stop sometime, and so there will always remain some possibility of a sampling error.

When you are testing a null hypothesis, you are trying to decide between two options: (1) true or (2) false. However, since statistical hypothesis testing is based on sample information and you cannot be absolutely sure that your decision is correct, you really are faced not with two, but with four possible situations:

4

2

1. $H_0$ is *true* and the sample data lead you to *correctly* decide that it is true.
2. $H_0$ is *true* but by bad luck the sample data lead you to *mistakenly* decide that it is false.
3. $H_0$ is *false* and the sample data lead you to *correctly* decide that it is false.
4. $H_0$ is *false* but by bad luck the sample data lead you to *mistakenly* decide that it is true.

Type I error

In the first and third situations above, you have been led to make a correct decision. In the second situation, some unlucky tosses misled you into making a **Type I error**

*Definition:* A **Type I error** is rejecting a true null hypothesis as a result of the luck of the sample.

Type II error

In the fourth situation, luck caused you to make a **Type II error**

*Definition:* A **Type II error** is failing to reject a false null hypothesis as a result of the luck of the sample.

This can be summarized in Table 8-1.

**Table 8-1**

| | You Fail to Reject $H_0$ | You Reject $H_0$ |
|---|---|---|
| $H_0$ is True | Correct | Type I error |
| $H_0$ is False | Type II error | Correct |

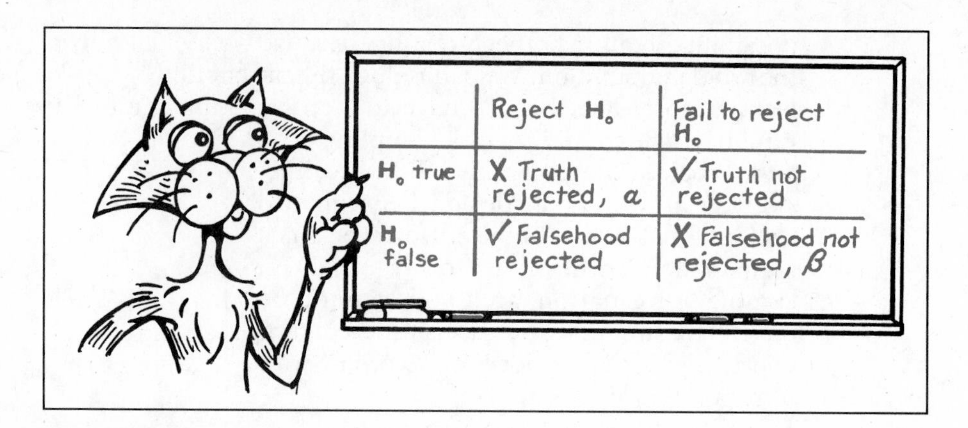


**Error Probabilities**

$\alpha = P$(a Type I error is made about a *true* $H_0$)

$\beta = P$(a Type II error is made about a *false* $H_0$)

We use the first letter of the Greek alphabet, alpha ($\alpha$), to represent the probability of making an error of Type I. Similarly, the second letter, beta ($\beta$), represents the probability of an error of Type II.

In a well-designed study, the researchers will try to have both $\alpha$ and $\beta$ as small as possible.

---

❑ **EXAMPLE 8.4 Guildenstern Revisited**

For Guildenstern's experiment of Example 8.3 describe what is meant by a "Type I error" and a "Type II error."

SOLUTION

First, we restate the null and alternative hypotheses and the decision rule.

$H_0$: The coin is fair; $p = 0.50$.

$H_a$: The coin is biased for heads; $p > 0.50$.

*Decision rule*: Decide that the coin is biased if we get 25 or more heads.

A Type I error is a mistake about a *true* null hypothesis. Thus a Type I error would happen if the coin were really fair but, by luck, came up heads 25 or more times. Guildenstern would decide it was biased, and she would be wrong.

A Type II error is a mistake about a *false* null hypothesis. This would happen if the coin were really biased but, by luck, came up heads less than 25 times. Guildenstern would decide it was fair, and she would be wrong. ❑

---

### Exercises

*In Exercises 8-9 to 8-16 think of statements such as "this person is innocent," "this drug is safe," or "this product is priced fairly" as typical null hypotheses. In each exercise state the null and alternative hypotheses and describe what is meant by a Type I and a Type II error.*

8-9 A penny and a centavo, which are fair coins, and a sen and a pfennig, which are biased for tails, are all tossed 40 times. The results are listed below.

| | Number of Heads | Conclusion |
|---|---|---|
| i penny | 21 | Fair |
| ii centavo | 33 | Biased for heads |
| iii sen | 18 | Fair |
| iv pfennig | 6 | Biased for tails |

(a) Which situation is a Type I error?
(b) Which situation is a Type II error?

8-10 Detective Sgt. Wednesday of the rackets squad confiscated 4 spinners, each divided into 7 equal areas. Of them, 2 are honest and 2 are biased. Bored while waiting for the grand jury hearings to begin, Wednesday begins spinning each one 70 times. For each spinner, the number of times area 6 wins is recorded. In table form, list the 4 situations that might occur and indicate the Type I and Type II errors.

8-11 A basketball coach screens new candidates by having them shoot 30 shots from a distance of 20 feet. A player is rejected if fewer than 20 shots are good. If the coach decides to increase the distance to 25 feet, will $\alpha$ increase or will it decrease?

8-12 If a person's results from a medical screening test for a disease are positive, he or she is given further tests to determine if he or she has the disease. Typically, these tests are more difficult, time-consuming, and expensive to administer than the screening test.

In such a screening test, 2 "errors" similar to Type I and Type II errors can occur. The test may indicate that a well person is sick, or it may indicate that a sick person is well.
(a) If you were designing the screening test, which error would you be more concerned about? Does it depend on the disease?
(b) A screening test for glaucoma measures the pressure of some fluid in the eye. If the pressure is higher than a certain value—say, 1000—that counts as "positive," and the person is given further tests. Discuss what would happen if the test were changed so that the new cutoff was 900. The figure shown will be helpful in solving the problem.

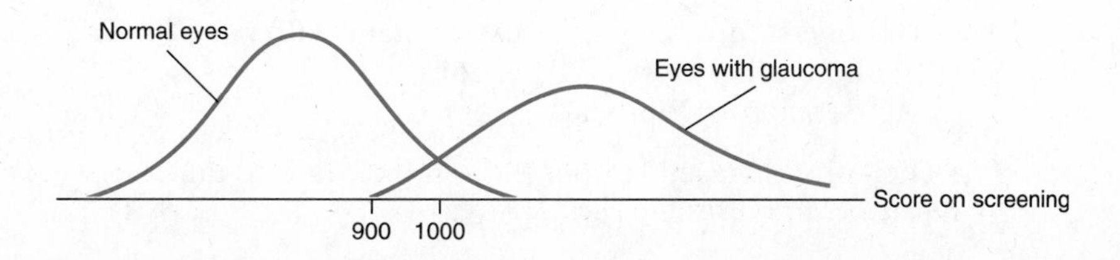

8-13 The sheriff in a gambling town prefers to believe that the roulette wheel is honest rather than take the chance of arresting the croupier and being charged with false arrest. If we think of the statement, "the wheel is honest" as a null hypothesis, does the sheriff prefer to make a Type I or a Type II error?

8-14 If a school principal prefers to believe that a student is guilty unless the accused student can prove his or her innocence, does the principal prefer to make a Type I or a Type II error?

8-15 If a pharmaceutical firm prefers to not sell a drug rather than risk selling 1 with bad side effects, does the firm prefer to make Type I or Type II errors?

8-16 A biology teacher is performing an experiment to see whether or not a new curriculum is superior to what had been used in the past. The hypotheses are:

$H_0$: The new curriculum is equal in value to the former curriculum.

$H_a$: The new curriculum is superior in value to the former curriculum.

In the eyes of the school board, how would the value of $\alpha$ depend on whether the new curriculum utilizes the present laboratory equipment or demands a large expense for new equipment?

8-17 Mary did a statistical analysis for her avionics instructor and ended up rejecting her null hypothesis. Mart says: "You could not possibly have made a Type II error." Marv interjects: "Therefore you must have made a Type I error." Are the boys correct? Explain.

## 8.3 More about Type I Errors and Decision Rules

The probability of a Type I error ($\alpha$) and the decision rule are interdependent, one upon the other. Knowing either one, we can calculate the other.

### ❑ EXAMPLE 8.5 Calculating $\alpha$ from a Given Decision Rule

In Example 8.3 Guildenstern used the following decision rule: I will conclude that the coin is biased if I get 25 or more heads in 40 tosses. Of course, it is possible for even a fair coin to fall heads 25 or more times just by chance, but it is not highly probable. *If the coin were fair* and fell heads more than 25 times, she would mistakenly call the coin biased and thereby make a Type I error. Find the probability that, using her decision rule, Guildenstern will mistakenly make a Type I error. That is, find $P(S \geq 25)$ if the coin is fair.

**SOLUTION**

This is a one-sample binomial experiment with $n = 40$. The population consists of all possible tosses of this coin.

We define $p = P(\text{a head is tossed})$.

$H_a$: The coin is biased for heads; $p > 0.50$ (a one-tail test).

$H_0$: The coin is fair; $p = P(\text{heads}) = 0.50$.

We take the *value* of $p$ from our assumption that the null hypothesis $H_0$ is true.

$$p = P(\text{heads}) = 0.50 \qquad \text{and} \qquad q = 0.50$$

### An Imaginary Distribution of Values of *S*

We now consider what *would happen* if we repeatedly tossed a fair coin 40 times and each time calculated $S$, the total number of heads that we obtained. The distribution of $S$ would look like the *theoretical* histogram of $S$. Since $np = 20 > 5$ and $nq = 20 > 5$, this distribution of values of $S$ would be close to a normal distribution, with

$$\mu = np = 20 \qquad \text{and} \qquad \sigma = \sqrt{npq} = \sqrt{10} \approx 3.16$$

We want to find the probability that we would toss 25 or more heads in 40 tries; that is, $P(S \geq 25)$. Since we are using the normal approximation, we will, as usual, use the boundary between 24 and 25 and find $P(S > 24.5)$. This is shown in Figure 8-1.

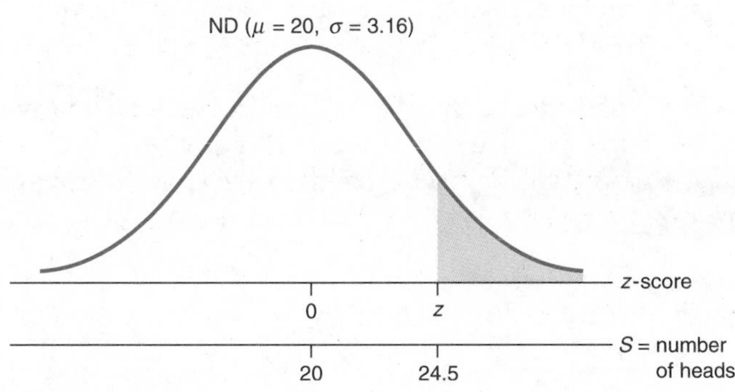


Figure 8-1

$$z_{24.5} = \frac{24.5 - 20}{3.16} = \frac{4.5}{3.16} = 1.42$$

$$\text{Area}(z = 1.42) = 0.9222$$

> **Imaginary vs. Real**
>
> *Note:* The theoretical distribution of many values of $S$ that we are considering exists only in our head.
>
> In the actual experiment, we have only one value of $S$.

Therefore, $P(S \geq 25) = 1 - 0.9222 = 0.0778$, or only about 8/100. In other words, the probability of a Type I error is about 0.08, and we would write $\alpha = 0.08$. Now, if Guildenstern actually tosses the coin 40 times and gets 25 or more heads, she will reject the null hypothesis. In doing so, she can be pretty confident that the outcome occurred because the coin is biased and that the result is not just the chance outcome of a fair coin. ❑

Remember that any time we reject a null hypothesis, no matter what decision rule we use, there is always some probability (hopefully, very small) that because we make our decision based on sample data the rejection is in error as a result of bad luck.

### Finding the Decision Rule That Corresponds to a Given Value of $\alpha$

In the previous example we found the value of $\alpha$ corresponding to a given decision rule. However, it is often the case that we prefer to find the *decision rule* corresponding to a given value of $\alpha$.

In Example 8.5, Guildenstern's decision rule led to $\alpha = 0.08$. Now perhaps she would not be satisfied with a probability of error as large as that. She might wish to do the problem in reverse. That is, instead of setting the decision rule and then calculating $\alpha$, she could first state how large a risk of a Type I error she is willing to accept—say, 0.01—and then calculate the decision rule for that value of $\alpha$. This is called "setting the **significance level** of the test" (at 0.01).

**significance level**

> *Definition:* The **significance level** of a hypothesis test is the probability that the decision rule will result in a Type I error—that is, $\alpha$.

Guildenstern's new decision rule will indicate how many heads must appear in order to support rejection of the null hypothesis such that the probability of a Type I error will be no larger than 0.01. We illustrate that procedure in Figure 8-2.

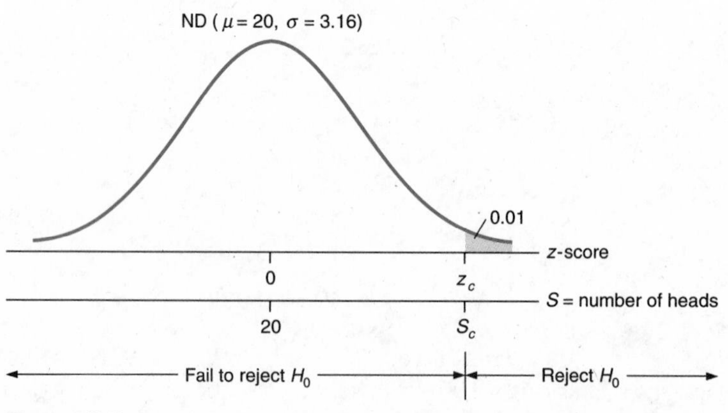


Figure 8-2

Since we wish $\alpha$ to be 0.01, and since this is a one-tail test on the right, we look up the area of $1.00 - 0.01$, or 0.99. This area corresponds to $z_c = 2.33$. We use the subscript $c$ to indicate that this is a $c$ritical value.

critical value

> *Definition:* The **critical value** is the value that marks the start or end of the rejection region.

We transform this $z$-score into a specific number of heads, as follows:

$$S_c = \mu + z_c\sigma$$
$$= 20 + 2.33(3.16) = 20 + 7.4 = 27.4$$

Therefore, if Guildenstern sets $\alpha = 0.01$, she will have to get more than 27.4 heads, that is, at least 28 heads, to reject the null hypothesis and conclude that the coin is biased for heads.

---

### ❑ EXAMPLE 8.6 Calculating a Decision Rule from a Given $\alpha$

Tenacious Tom, while recovering from an unfortunate accident to his hands, is once more trying to make a pair of loaded dice. He has altered one die in various secret ways. He thinks that the probability of getting a 6 on this die is now changed, but, because of his inexperience, he is not sure if it will now get too many 6s or not enough 6s. He decides to roll it 60 times and count the number of 6s that appear. What can he use for the decision rule if he decides to test the die at the 0.05 significance level?

#### SOLUTION

In order to determine the decision rule for this one-sample binomial hypothesis test, he clarifies the population of interest and the sample.

The population is all the possible tosses of this die, and the sample is 60 tosses.

He next states the hypotheses in terms of $p = P(\text{a 6 occurs})$:

$H_a$: The die is biased; $p \neq 1/6$ (a two-tail test).

$H_0$: The die is fair; $p = 1/6$.

Assuming that the null hypothesis is true,

$$p = 1/6$$
$$q = 5/6$$
$$n = 60$$

since $np = 10 > 5$ and $nq = 50 > 5$, the normal approximation may be used:

$$\mu = 10 \qquad \text{and} \qquad \sigma = \sqrt{8.33} \approx 2.89$$

Because he suspects that the die could be biased either for or against 6, he might end up with either too few or too many 6s. He divides the significance level of 5% into the two tails of the normal distribution, as shown in Figure 8-3.

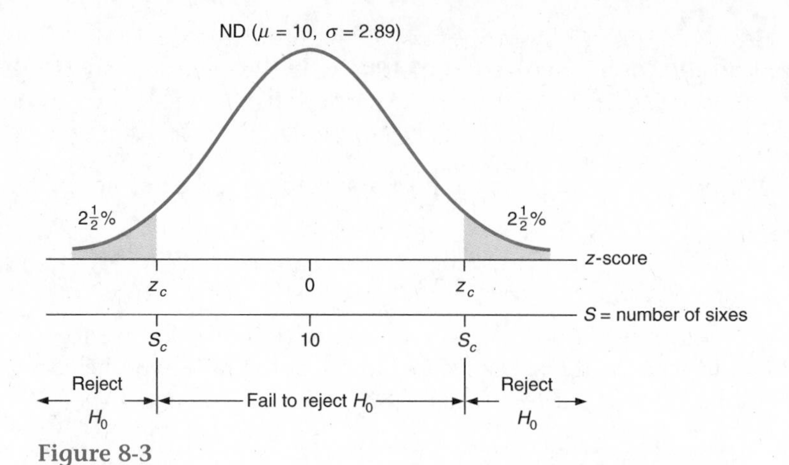

Figure 8-3

Looking up areas of 0.025 and 0.975, he gets the critical $z$-scores of $\pm1.96$.

To convert these into numbers of 6s he uses

$$S_c = \mu + z_c\sigma$$
$$= 10 + (\pm1.96)(2.89) = 10 \pm 5.7 = 15.7 \text{ and } 4.3$$

Therefore his decision rule will be: I will reject the null hypothesis if I get less than 4.3 sixes or more than 15.7 sixes. If this happens, he can conclude that the die is biased. The probability of making a Type I error is 0.05 or less. ❑

---

A researcher can set the significance level at any value, but in most current published reports it is more or less standard to use 0.01 or 0.05 as the significance level. These probabilities of error are small enough to be considered reasonable in most circumstances.

In the table below we show some commonly used significance levels and their critical values $z_c$:

Critical Values of $z$

| | One-Tail Test | Two-Tail Test |
|---|---|---|
| $\alpha = 0.01$ | 2.33 or −2.33 | ±2.58 |
| $\alpha = 0.05$ | 1.65 or −1.65 | ±1.96 |

By choosing $\alpha$ so small, we have nearly eliminated the possibility of a Type I error occurring. We now have to consider only *the remaining* three possible outcomes:

$H_0$ is false, and we correctly reject it.

$H_0$ is false, and we fail to reject it (a Type II error).

$H_0$ is true, and we correctly fail to reject it.

Note that simply by proper construction of the decision rule we control the probability of a Type I error in tests where we end up rejecting $H_0$. However, in cases where we do not end up rejecting $H_0$, we run the risk of a Type II error. As we show in the next section, it is usually harder in a typical research situation to control the probability of a Type II error, and often it is not known precisely. Thus, without more information, it is common in cases where we do *not* reject $H_0$ to say "We *fail to reject* $H_0$" or "The evidence is *not statistically significant*" rather than "We accept $H_0$." We adopt this convention throughout this text.

### A Note about Statistical Significance

When evidence results in rejecting the null hypothesis, it is common to say that the study has *statistical significance*. Conversely, when the evidence is not strong enough to reject the null hypothesis, we say that the study did not lead to a statistically significant result. It is important to realize the limited meaning of these expressions, particularly since media reports about scientific research often are misleading. Most importantly, "statistical" significance is not the same as "practical" significance. It only means that the data indicated that the null hypothesis is probably incorrect. It does not tell you whether you have learned anything valuable or important.

For instance, if we toss a U.S. penny millions of times, we may prove that the probability of getting a head is not exactly 0.50. However, it would be so close to 0.50 that for any practical purpose it wouldn't matter.

On the other hand, a study which compares two medical treatments may hint that treatment A is better than treatment B—without reaching statistical significance, particularly if the samples are small. In this case, maybe it is more important to continue the research until larger samples can clarify the situation.

### Exercises

8-18 A coin is suspected of being biased.

(a) Define the population, and state the two hypotheses. Will this be a one-tail or a two-tail test?

(b) The coin is tossed 40 times. How many or few heads are needed to establish statistically that the coin is biased if $\alpha = 0.05$? That is, calculate the decision rule.

(c) Repeat part (b) with $n = 100$.

8-19 A newspaper article states that 60% of children in the age bracket 1–4 who die do so as a result of motor accidents. Doubting that this is true, a public health official gathered information on the cause of death of 30 randomly selected children.

(a) Define the population, and state the motivated hypothesis.

(b) State the null hypothesis.
(c) How many or how few of the 30 deaths should be due to motor vehicle accidents in order to reject the claim of the article at the 0.01 significance level? That is, calculate the decision rule.

8-20 A national survey organization says that 40% of school-age children live with only 1 parent. A student wishes to see if this is a reasonable percentage for the children in his city. He plans to use a random sample of 200 school-age children for the study.
(a) For the student's study, define the population, and state the null and alternative hypotheses.
(b) If the student does the analysis at the 0.05 significance level, what is his decision rule?
(c) If the student does the analysis at the 0.01 significance level, what is his decision rule?

8-21 Some years ago a study by the U.S. Bureau of Justice Statistics concluded that 61% of jail inmates had not completed high school. A sociologist believes that now, in her state, the percentage is lower than 61%. She plans to use a random sample of 400 inmates for her study.
(a) For the sociologist's study, define the population, and state the null and alternative hypotheses.
(b) If she does the analysis at the 0.05 significance level, what is her decision rule?
(c) If she does the analysis at the 0.01 significance level, what is her decision rule?

8-22 A researcher is beginning to suspect that less than 3% of all genitz are pibled.
(a) State the 2 hypotheses in both words and symbols.
(b) If 1000 genitz are sampled, how few must be pibled to convince us that the suspicion is correct at the 0.01 significance level?

8-23 The population of Smalltown is 42% female. A statistician is trying to establish the claim that more than 42% of the Republicans in Smalltown are women. Assuming that sex has nothing to do with political affiliation, then $p = P(\text{a Republican is female}) = 0.42$. We select 100 Republicans at random. Using the 0.01 significance level, find the decision rule for the number of women that must occur in the sample before we reject the assumption.

8-24 The local chapter of Women Against the Exploitation of Females (WAEF) is picketing an "art" theater. They claim that at least 75% of the people who view such films are men. A newspaper reporter believes that this figure is too high. He gathers a random sample of 100 people from the audiences at the theater. He will use as his decision rule: I will reject the claim made by WAEF if the sample contains fewer than 60 men.
(a) Define the population, and state the motivated hypothesis.
(b) State the null hypothesis.
(c) Draw the normal curve including the $z$-score line, the line for the number of successes, and a shaded rejection region. Calculate the critical value, and state the decision rule.
(d) Find the probability of a Type I error.

8-25 A company official of the Ill Railroad claims that only 4% of its trains arrive more than 5 minutes late. A statistician hired by an irate commuter group gathers a random sample of 500 train arrivals. According to the company figures he would expect about 20 trains to be late. He is willing to give the company the benefit of the doubt, and so his decision rule is: reject the company's claim if more than 40 trains are late.

(a) State the motivated hypothesis.
(b) State the null hypothesis.
(c) Draw the normal curve including the $z$-score line, the line for the number of successes, and a shaded rejection region.
(d) Find the probability of a Type I error.
(e) If the decision were: reject $H_0$ if more than 30 trains are late, find $\alpha$.
(f) If in part (e) you were to change the number 30 to 50, would $\alpha$ increase or decrease?
(g) What decision rule gives $\alpha = 0.05$?

8-26 A statistician is testing the claim that a certain population is one-fourth male, three-fourths female. She is going to select 80 persons at random from the population. Her decision rule is: if she gets fewer than 15 or more than 25 males, she will reject the claim that $p = P(\text{a person is male}) = 0.25$. Suppose it is true that $p$ does equal 0.25. What is the probability that her results will lead her to mistakenly reject that fact? That is, what is $\alpha$? Draw a clearly labeled curve.

8-27 A manufacturer claims that the mixed nuts that he sells have only 30% peanuts. We open a large bag, select 100 nuts at random, and find that 36 of them are peanuts. If $p$ does equal 0.30, what is the probability that of 100 nuts picked at random 36 or more are peanuts? Would you be willing to accuse the manufacturer of a false claim?

8-28 A gambler tosses a fair coin 10 times and gets 8 heads. He makes a Type I error in stating that the coin is biased. Find the probability that a fair coin tossed 10 times will produce 8 or more heads.

8-29 Danny Dropout and Sally Student disbelieve a newspaper claim that says 30% of the students at Happy High cut class at least once a week. They think 30% is too high. They decide to test this using a random sample of 200 students. Danny says, "I'll reject the claim if our sample contains fewer than 50 cutters." Sally, who studied statistics, says, "I'll use $\alpha = 0.05$." Who has the larger risk of making a Type I error?

8-30 You decide. Which conclusion would you choose? Why?

(a) A coin is tossed 100 times, resulting in 97 heads.

| Conclusion 1 | Conclusion 2 |
|---|---|
| It is a fair coin, and I have just witnessed a very extraordinary event. | It is not a fair coin, and I have just witnessed an ordinary event for the tosses of a biased coin. |

(b) A random sample of 5000 U.S. teenagers shows that 25% more girls smoke than boys do.

| Conclusion 1 | Conclusion 2 |
|---|---|
| There really is no difference in the smoking habits of boys and girls; this random sample is very unusual. | This random sample is representative of all U.S. teenagers. I conclude that it is highly unlikely that the same percentage of boys and girls smoke. |

(c) A random sample is taken from a population, and the sample statistic is nowhere near some hypothetical population parameter.

| Conclusion 1 | Conclusion 2 |
|---|---|
| This was a highly unlikely event, and so I have just witnessed a miracle, but I see no reason to reject the proposed parameter. | This is a representative sample, an ordinary event, from a population with a different parameter. The proposed parameter is most likely not correct. |

8-31 Find the critical $z$-scores for $\alpha = 0.10$ for both a one- and a two-tail test.

| $z_c$ | One-Tail Test | Two-Tail Test |
|---|---|---|
| $\alpha = 0.10$ | | |

# 8.4 More about Type II Errors (optional)

### Historical Note

In the 1920s, Jerzy Neyman and Egon Pearson (son of the famous statistician Karl Pearson) developed a general theory of hypothesis testing (called "Neyman-Pearson Tests" in their honor) which tells what kind of decision rule to use for testing a given parameter so that the test has minimum $\beta$ for any given $\alpha$. These tests tell you what statistic to use and what the decision rule should be to test a given parameter. All the hypothesis-testing procedures for parameters in this text have been developed using Neyman-Pearson criteria.

(This section may be omitted without loss of continuity.)

We have just discussed the Type I error, which occurs when statistical evidence leads us to reject a null hypothesis when in reality the null hypothesis is true. Recall that the Type II error occurs when the null hypothesis is false but the statistical evidence is not strong enough to indicate it. In other words, even though the null hypothesis is false, the data are not statistically significant.

## 8.4-1 Calculating the Value of $\beta$ for a Given $p$

If the value of $p$ given in the null hypothesis is wrong, then some other particular value of $p$ is correct. Recall that the alternative hypothesis did not give us one specific value of $p$. Therefore, for each possible value of $p$, there is a corresponding value of $\beta$, the probability of a Type II error. We illustrate this in the following example.

---

### ❑ EXAMPLE 8.7 Calculating a Type II Error

Suppose you are asked to test a coin. You decide to toss it 60 times, reasoning that if it is fair, you will get about 30 heads. You choose for your decision rule: the coin is not fair if the number of heads is less than 26 or more than 34.

If, unknown to you, the coin is biased and $p = P(\text{heads}) = 0.60$, then what is the probability that your experiment will yield between 26 and 34 heads anyway? That is, what is the probability that you commit a Type II error and end up concluding that the coin is fair?

**SOLUTION**

$$p = 0.60$$

$$q = 0.40$$

$$n = 60$$

$$np = 36 > 5 \quad \text{and} \quad nq = 24 > 5$$

Therefore, the normal approximation may be used:

$$\mu = 36$$

$$\sigma = \sqrt{14.40} \approx 3.79$$

$P(26 \le S \le 34)$ is approximately equal to $P(25.5 < S < 34.5)$; see Figure 8-4.

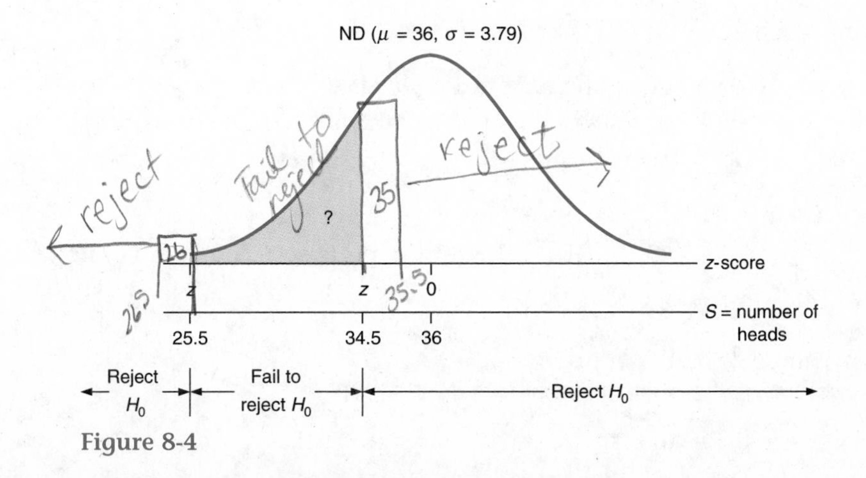


Figure 8-4

$$z_{34.5} = \frac{34.5 - 36}{3.79} = -0.40$$

$$z_{25.5} = \frac{25.5 - 36}{3.79} = -2.77$$

$$\text{Area}(z = -0.40) = 0.3446$$

$$\text{Area}(z = -2.77) = 0.0028$$

Therefore, $\beta = P(26 \leq S \leq 34) = 0.3446 - 0.0028 = 0.3418$, or about 0.34. ❑

---

In the previous example we found that the probability is 0.34, or about 1/3, that we would accept the statement "This coin is fair," when in fact the coin is biased with $p = 0.60$. We write $\beta = 0.34$. You see that our decision rule is not very powerful for distinguishing between fair coins and coins biased with $p = 0.60$. You will notice that we needed the probability $p = 0.60$ in order to compute $\beta$. We cannot compute a $\beta$ until we choose a specific value of $p$ from the alternative hypothesis. We could repeat the above calculation for other values of $p$, such as 0.70, 0.80, etc., in order to get an idea of the range of possible values for the Type II error. This contrasts with the case of the Type I error, where, as soon as the decision rule is made, $\alpha$ is known.

---

### ❑ EXAMPLE 8.8 Finding a Decision Rule

An experimenter believes that a coin is biased in favor of heads. He wants to test the claim "The coin is fair" at the 0.05 significance level by tossing the coin 40 times.

(a) Find $\alpha$.
(b) Find the decision rule.
(c) Find $\beta$.

**SOLUTIONS**

(a) $\alpha = 0.05$ because the experimenter chose it that way.
(b) $p = P(\text{a head is tossed})$.

$H_a$: The coin is biased; $p > 0.5$ (a one-tail test).
$H_0$: The coin is fair; $p = 0.5$.

$p = 0.5$ (from the null hypothesis "The coin is fair")

$q = 0.5$

$n = 40$

$np = 20 > 5$

$nq = 20 > 5$

$$\mu = 20$$
$$\sigma = \sqrt{10} \approx 3.16$$
$$\alpha = 0.05$$

The normal approximation may now be used, as shown in Figure 8-5.

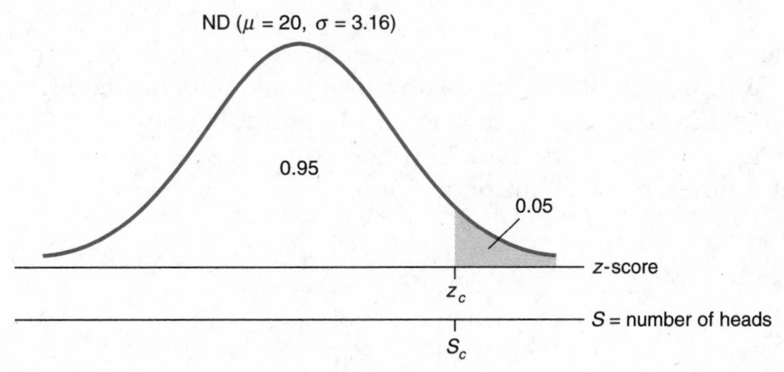


Figure 8-5

Looking up the area of 0.9500, we find $z_c = 1.65$. Converting this $z$-score to a raw score, we have

$$S_c = 20 + 1.65(3.16) = 25.2$$

Therefore, his decision rule is: I will reject the null hypothesis that the coin is fair if the coin turns up heads more than 25.2 times. In this problem, since we are using the continuous curve to approximate a distribution of whole numbers, we interpret "more than 25.2 heads" as any number of heads from 26 up. See Figure 8-6.

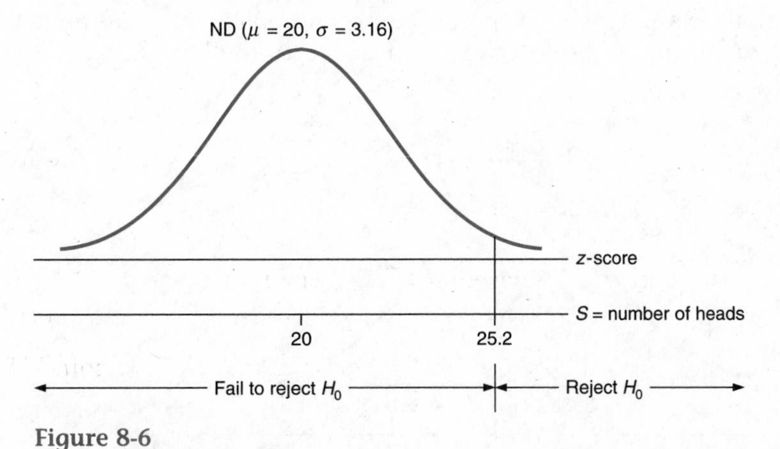


Figure 8-6

(c) A specific value of $\beta$ cannot be computed until we have decided upon a specific value of $p$ from the alternative hypothesis. ❑

## ❑ EXAMPLE 8.9 Continuing the Previous Example

In Example 8.8, suppose that in reality $p$ was approximately 0.70. Now we can compute $\beta$, the probability that the evidence will indicate that this biased coin is fair.

### SOLUTION

The decision rule comes from our choice of $\alpha$ and was computed in part (b) of Example 8.8 under the assumption that $p = 0.50$ (see Figure 8-6). The decision rule was: I will reject the null hypothesis that the coin is fair if the coin turns up heads more than 25 times. But if we now suppose that $p = 0.70$, we get a different distribution as follows:

$$p = 0.70$$

$$q = 0.30$$

$$n = 40$$

$$np = 28 > 5$$

$$nq = 12 > 5$$

$$\mu = 28$$

$$\sigma = \sqrt{8.40} \approx 2.90$$

Recall that in this problem $\beta$ measures the probability that a biased coin will produce results that are in support of the null hypothesis. Therefore, we want to measure the area of the graph that corresponds to those outcomes that support the null hypothesis that the coin is fair. Recall that in the previous example we interpreted "more than 25.2 heads" as any number of heads from 26 up. Therefore we use the area to the left of 25.5 (see Figure 8-7):

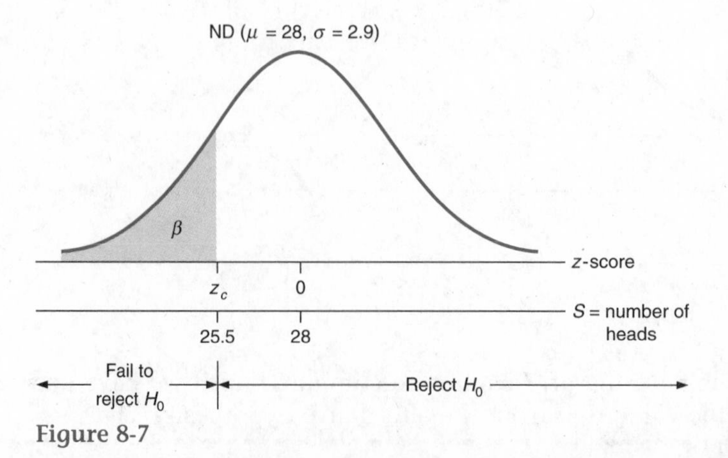


Figure 8-7

Converting 25.5 to a $z$-score in order to measure the area, we have

$$z = \frac{25.5 - 28}{2.90} = \frac{-2.5}{2.90} = -0.86$$

$$\text{Area}(z = -0.86) = 0.1949 = 0.19$$

Therefore $\beta = 0.19$.

Using this decision rule, there is a probability of approximately 0.19 that the experimenter will mistakenly decide that the coin is fair if in fact it is biased with $p = 0.70$. ❑

---

In the example above, if we compute $\beta$ for $p = 0.60$ and $p = 0.80$, we get values of $\beta$ equal to 0.68 and 0.005, respectively. Further, if we start over again with a $\alpha = 0.01$, then we find that $S_c = 20 + 2.33(3.16) = 27.4$. For the corresponding decision rule, the values of $\beta$ for $p = 0.60$, 0.70, and 0.80 turn out to be 0.87, 0.43, and 0.04, respectively. We can summarize these results in Table 8-2.

**Table 8-2 Values of $\beta$ for $n = 40$**

| True Value of $p$ | $\alpha = 0.05$ | $\alpha = 0.01$ |
|---|---|---|
| 0.60 | 0.68 | 0.87 |
| 0.70 | 0.19 | 0.43 |
| 0.80 | 0.005 | 0.04 |

Thus you can see by reading across each row of this table that for a fixed value of $n$ if we decrease $\alpha$, we will increase $\beta$, and vice versa.

Now let us see what happens to $\beta$ when we increase sample size. If we keep $\alpha = 0.05$ and let $n = 60$, the values of $\beta$ can be found as follows. First, we establish the decision rule assuming that the null hypothesis ($p = 0.50$) is true. We get

$$S_c = 30 + 1.65(3.87) = 36.4$$

Therefore, our decision rule is: we will reject $H_0$ if we see more than 36 heads. Now that we have a decision rule, we can compute $\beta$. The calculation for $p = 0.70$ is

$$np = (60)(0.70) = 42 > 5$$

$$nq = (60)(0.30) = 18 > 5$$

$$\mu = np = 42$$

$$\sigma = \sqrt{npq} = 3.55$$

$$\text{ND}_S(\mu = 42, \sigma = 3.55)$$

$$z_{36.5} = \frac{36.5 - 42}{3.55} = -1.55$$

$$\beta = 0.06$$

Similarly, values of $\beta$ for $p = 0.60$ and 0.80 can be found. We summarize these results in Table 8-3.

Table 8-3 Values of $\beta$ for $\alpha = 0.05$

| True Value of $p$ | $n = 40$ | $n = 60$ |
|---|---|---|
| 0.60 | 0.68 | 0.55 |
| 0.70 | 0.19 | 0.06 |
| 0.80 | 0.005 | 0.0001 |

You can see that increasing the sample size decreases the value of $\beta$ without increasing the value of $\alpha$. In performing a statistical experiment, if you have the time and the money, you can often make both $\alpha$ and $\beta$ as small as you want by using a sufficiently large value of $n$. For example, suppose you have $H_0$: $p = 0.50$ in a one-tail test, and you want to be reasonably sure to reject $H_0$ if the true value of $p$ is anything higher than 0.60. Then by making the sample large enough, you can have, for instance, both $\alpha$ and the maximum $\beta$ equal to 0.05. (You might try to find how big a sample is needed to do this. From Table 8-3 you can see that $n$ must certainly be bigger than 60.) These relationships among $\alpha$, $\beta$, and $n$ are the reason that statistical consultants always encourage researchers to use the largest samples they can afford.

## 8.4-2 The Power of a Test

power

Statisticians refer to the value of $1 - \beta$ as the **power** of a test. The power of a test is a measure of how good the test is at rejecting a false null hypothesis. The more "powerful" a test is (the closer the value of $1 - \beta$ is to 1), the more likely the test is to reject a false null hypothesis. An important part of statistical theory deals with the problem of finding a decision rule that will make a hypothesis test as powerful as possible for a given value of $\alpha$.

### ❑ EXAMPLE 8.10 The Power of a Test for Different Values of $p$

In this example, we again show that a test has different powers for different values of $p$. Suppose we have a hypothesis test where the 2 hypotheses are:

$H_0$: $p = 0.60$.

$H_a$: $p < 0.60$ (a one-tail test).

Suppose further that we set $\alpha = 0.05$ and $n = 50$.

Find the power of this test (a) when $p = 0.50$ and (b) when $p = 0.40$.

SOLUTION

Using the method of Example 8.9, we first calculate the decision rule and then we compute $\beta$ and the power $(1 - \beta)$ for each value of $p$. The results are given in Table 8-4. In Exercise 8-38 you are asked to confirm these results.

Table 8-4

| True Value of $p$ | $\beta$ | Power of Test |
|---|---|---|
| 0.50 | 0.56 | 0.44 |
| 0.40 | 0.10 | 0.90 |

Notice that the test has more power when the alternative is further from the null hypothesis. Not surprisingly, the further a null hypothesis is from the truth, the more likely it is that the test will reject it. ❑

### Exercises

8-32 In Exercise 8-29, Danny Dropout said, "I'll reject the claim if our sample contains fewer than 50 cutters." The problem essentially asked you to find $\alpha$. Why can we not simply ask you to find $\beta$?

8-33 Verify in Table 8-2 the value of $\beta$ for:
(a) $p = 0.60$, $\alpha = 0.05$.
(b) $p = 0.80$, $\alpha = 0.01$.

8-34 Verify in Table 8-3 the value of $\beta$ for:
(a) $p = 0.60$, $n = 40$
(b) $p = 0.80$, $n = 60$

8-35 Connie Consumer claims that 30% of Never Fail brand spark plugs are defective. Sparky the mechanic argues that the rate of defectives is lower. They decide to examine a random sample of 100 NF brand plugs.
(a) Using $\alpha = 0.05$, find their decision rule.
(b) If in fact 20% of the NF brand plugs are defective, what is the probability that the evidence will mistakenly lead them to believe that Connie is correct?

8-36 Repeat Exercise 8-35 if:
(a) $H_a$: $p > 0.30$.
(b) $H_a$: $p \neq 0.30$.

8-37 Refer to the story of Tenacious Tom in Example 8.6. If $p$ is really 0.35, what is the probability that Tom will be misled into believing the die is fair?

8-38 Verify the powers of the test in Example 8.10.

8-39 *AGRI* magazine recently reported the results of 2 studies on fertilizer effectiveness on corn. Dr. Bulschmidt reported that he "failed to reject" his null hypothesis at the 0.05 level. In the second article, Dr. Senserd reported that using a test with significance level 0.05 and a power better than 96%, he "accepted the null hypothesis."
(a) Why have they both reported their results correctly?
(b) What is the difference in what they have learned?

8-40 In Example 8.10, change $H_a$ to $p > 0.60$, and find the power of this new test for the indicated values of $p$.

| $p$ | $\beta$ | Power |
|---|---|---|
| 0.50 | | |
| 0.60 | | |
| 0.70 | | |
| 0.80 | | |
| 0.90 | | |

8-41 Change $H_a$ to $p \neq 0.60$ in Example 8.10, and find the power of this new test for the indicated values of $p$.

| $p$ | $\beta$ | Power |
|---|---|---|
| 0.40 | | |
| 0.50 | | |
| 0.70 | | |
| 0.80 | | |

8-42 In Middlesex, Massachusetts, 100 extraterrestrial humanoids were sampled to test whether more than 20% were neither male nor female.
(a) Find the critical number of successes for $\alpha = 0.01$.
(b) If the true fraction is 25%, find $\beta$.
(c) If the true fraction is 18%, find $\beta$.
(d) If the true fraction is 15%, find $\beta$.

8-43 A coin is tossed 50 times to test the null hypothesis that the coin is fair.
(a) If $\alpha$ is set at 0.05, find the 2 critical values for the decision rule.
If it happens to be true that $p = P(\text{heads}) = 0.80$, find $\beta$.
(c) If the coin is tossed 100 times instead of 50 times, repeat parts (a) and (b).
(d) How is $\beta$ affected by the increase in the sample size?

8-44 Let us reconsider Exercise 8-18(b). We decided that if we got more than 26 heads or fewer than 14 heads in 40 tosses, we would call the coin biased. If the true proportion of heads was really $p = 0.25$, find the probability of a Type II error.

8-45 A cancer research group wished to test the hypothesis that 25% of all college students smoke cigarettes. Using a 0.05 significance level, it was found that the critical points of the decision rule were 6.5 and 18.5 smokers for a random sample of 50 students. If the true proportion of smokers was 20%, find the probability of a Type II error.

8-46 Repeat the previous exercise, but this time find the probability of a Type II error if the true proportion of smokers was 15%.

8-47 Max, the bartender at Bernie's Paragon Cafe, claims that only one-fourth of the customers can differentiate between Northern Comfort and Rot Gut. Don, the barfly, makes a bet that the figure is higher. They gather a random sample of 25 customers. Each customer is given a glass of each liquor and asked to distinguish between them.

(a) Using $\alpha = 0.05$, how many customers would have to be correct for Don to win the bet?

(b) After the experiment is over, Don feels that $p$ is probably between 0.40 and 0.50. Assuming that this is true, what was the largest value of $\beta$ in the hypothesis test of part (a)?

8-48 A manufacturer of magic kits makes 2 types of "coins." One has $p = P(\text{heads}) = 0.80$, and the other is a fair coin with $p = P(\text{heads}) = 0.50$. An employee inadvertently mixed 100 of the fair coins with 200 of the biased coins. His employer instructs him to sort them. Since the coins look alike, the employee decides to toss each coin 30 times. The fair coin he expects to come up heads about 15 times. On the other coin he expects about 24 heads. He decides that if he tosses 19 or fewer heads, he will call the coin fair, but if he tosses 20 or more heads, he will call the coin biased. When he has finished sorting the coins into 2 piles:

(a) About how many coins are in each pile?

(b) About how many fair coins are in the biased pile? (How many Type I errors?)

(c) About how many biased coins are in the fair pile? (How many Type II errors?)

(d) How could he improve on his results?

8-49 A counterfeiter has a pile of mixed counterfeit and real coins that look alike. Detective Bridget Gallagher is on his trail. Needing to buy some food, the counterfeiter wishes to use only real coins so that he will not get caught. He decides to toss each coin 20 times. On the real coins he expects about 10 heads, but the counterfeit coins are not well balanced and do not have $p = P(\text{heads}) = 0.50$. He decides to consider a coin real if he gets 9, 10, or 11 heads and counterfeit otherwise.

(a) If he sorts through 100 coins this way, about what percentage of the real coins will he reject?

(b) What percentage of the counterfeit coins will he accept?

(c) He decides to move his hideout, and he realizes that he cannot transport all of the coins. He decides to continue tossing each coin 20 times, and he prefers to take real coins if possible. He reasons that by increasing the acceptance region from 9 to 11 heads to 5 to 15 heads he will obtain more real coins. His partner disagrees, stating that they should decrease the acceptance region to 10 heads only. That way they will accept fewer counterfeit coins. The counterfeiter's wife says that they are both right, but his partner's husband claims that her statement is stupid. With whom do you agree, and why?

## 8.5 Hypothesis-Testing Procedures

**hypothesis test**

The procedure we have described, by which statisticians analyze data in order to decide when the evidence is strong enough to support the motivated hypothesis, is called a **hypothesis test**

We will now summarize and formalize such a test. To test the validity of a statistical claim, the statistician decides on a population of interest and then formulates two opposite hypotheses: the alternative (or motivated) hypothesis and the null hypothesis. Often the statistician is motivated by a suspicion that the alternative hypothesis may be true. However, it is important not to let any preconceived ideas interfere with carrying out the experiment in an objective and unbiased manner. The experiment is set up as described in the following outline to see if there is sufficient evidence to prove beyond a reasonable doubt that the motivated (or alternative) hypothesis is true because the assumption of the null hypothesis strains one's credulity.

| **Outline** | **Example** |
|---|---|
| Based on some prior experience or idea, you are motivated to perform a statistical experiment on a *specific* population. | Your school paper reports that 63% of sophomores drive to school. Is this also true of freshmen? |
| 1. Identify the procedure that you will use. (So far you have only learned one: A *one-sample hypothesis test of percentages*.) | A one-sample binomial hypothesis test. |
| 2. Clearly describe the population(s) being investigated. | Although the paper's population was all sophomores at your school, the *population being tested* for this report is all freshmen at your school. |
| 3. Make a claim which you desire to investigate—that is, state the motivated hypothesis $H_a$ about the population(s) in which you are interested. This is stated both in words and with mathematical symbols (some of which may need to be defined) for the *parameter(s)* involved. It will contain an inequality symbol: $<$, $>$, or $\neq$. Indicate whether it is a one- or a two-tail test. | Let $p = P$(a freshman drives to school).<br>$H_a$: The percentage of freshmen who drive to school is different from the percentage of all sophomores who do so; $p \neq 0.63$ (a two-tail test). |
| 4. For statistical purposes, you test the opposite hypothesis. Therefore, state the null hypothesis, $H_0$ in both words and symbols. It will contain the = sign. | $H_0$: The percentage of freshmen who drive to school is the same as the percentage of all sophomores who do so; $p = 0.63$. |
| 5. Select the significance level $\alpha$ (the probability of a Type I error). That is, state how large a risk you are willing to run of | $\alpha = 0.05$ |

| Outline | Example |
|---|---|
| making an error if you claim at the end of the study that your motivated hypothesis is true. Select any significance level you wish; in most published statistics, the authors have used $\alpha = 0.05$ or $\alpha = 0.01$. These are generally accepted standards. | |
| 6. Choose the size of the random sample. This choice is often determined by the amount of time and/or money that you have to perform the experiment, and the availability of subjects. Ease of computation might also be a factor in the selection of $n$. | $n = 150$ |
| 7. Decide how you will gather a random sample from the appropriate population, and obtain your data. | I will stand near the entrance to the library at noon and ask people if they are freshmen at this school. For those who answer "yes," I will ask if they drive to school, and I will record each response. I will continue until I obtain 150 responses from freshmen. |
| 8. Assume for the sake of the test that the null hypothesis is true. Calculate the critical value(s). You need to justify the use of any tables used. Sketch a graph of the distribution being used, and indicate key values thereon. | Since $np$ and $nq$ are both > 5, I can use the normal approximation to the binomial distribution,* ND($\mu = 94.5$, $\sigma = 5.913$) and $S_c = 82.9$ and $106.1$. |

9. A hypothesis test terminates with a three-step ending:

| Outline | Example |
|---|---|
| A. State your decision rule. "I will reject $H_0$ if the sample statistic is . . ." | I will reject $H_0$ if $S < 82.9$ or $S > 106.1$. |
| B. State your experimental outcome derived from the data collected in step 7 above. | $S = ?$ (State the calculated statistic here—in this case, the number of freshmen who drive to school.) |

*For brevity, the calculations and the normal distribution curve have been omitted from the chart above; however, they should be shown in your report.

| Outline | Example |
|---|---|
| C. Applying the decision rule calculated in step 9A to your experimental *statistic* from step 9B, make one of two possible decisions: | Depending on the outcome, either: |
| i. *Reject the null hypothesis* and claim that the *motivated hypothesis* is *correct*. | i.<br>a. I reject the $H_0$.<br>b. The percentage of freshmen who drive to school *is* more than or less than the percentage of sophomores. |
| | or |
| ii. *Fail to reject the null hypothesis*. You have been unable to prove that the motivated hypothesis is correct; its opposite, the null hypothesis, *may be* true. | ii.<br>a. I have insufficient evidence to reject the $H_0$.<br>b. The percentage of freshmen who drive to school **may be** the same as the percentage of sophomores who do so. There is not enough evidence to prove otherwise. |
| In this second case, we do not wish to state that the null hypothesis is certainly true since we have not determined the value of $\beta$. If $\beta$ is large, there is a large probability of a Type II error. Hence, we use the phrase "fail to reject the $H_0$" rather than "accept the $H_0$." For example, we can never get (in a hypothesis test) sufficient proof to show absolutely that a null hypothesis—say, $p$ equals 0.25—is true. "Equals" is a very strong word. Perhaps $p$ really equals 0.2501 or 0.2499703. We could hardly expect to distinguish these from *exactly* 0.25. | |

| Outline | Example |
|---|---|
| 10. Comment on and clarify your results. What have you shown? Did you make any changes from your original design? Did people cooperate or not? (Etc.) Did you notice anything that might cause you to think your sample is not representative of the population? | I originally planned on asking 300 people, but so many were not freshmen or were not even students that I lowered $n$ to 150. This was large enough, since $np$ and $nq$ were $> 5$. Most people were very cooperative, especially when I told them that I was doing this for my Stat course.<br>(*If* S = *110 and you rejected the* $H_0$):<br>I was surprised that the percentage of freshmen who drive to school was higher. I would have guessed that it would be the same or lower. But I have no reason to believe the freshmen I spoke to are an unrepresentative sample. |

## More Hypothesis-Testing Examples

### ❑ EXAMPLE 8.11 Side Effects from the Pill

A manufacturer claims that no more than 2% of the women who use his birth control pill suffer from side effects. We have a feeling that this estimate is too low. We decide to test his claim at the 0.01 significance level using a sample of 900 randomly selected women. Find the decision rule.

#### SOLUTION

Our solution follows the steps outlined above:

1. Identify the procedure.
2. State the population(s).
3. Define the parameter(s) (if needed) and state $H_a$ two ways.
4. State $H_0$ two ways.
5. Select $\alpha$.
6. Choose $n$.
7. Gather sample data.
8. Calculate the critical value(s).
9. Terminate the test with the three-step ending:
   A. State the decision rule.
   B. State the outcome.
   C. State the conclusion two ways.

10. Comment on and clarify your results.

1. Because this is a claim about a percentage of successes in one population, we will use a one-sample hypothesis test of percentages.
2. The population being considered is all women who use this pill.
3. Let $p$ equal the probability that a randomly selected user of the pill suffers from a side effect. That is, $p = P$(a user suffers a side effect).
   $H_a$: More than 2% of the users suffer side effects; $p > 0.02$ (a one-tail test).
4. $H_0$: The percentage suffering side effects is only 2%; $p = 0.02$.
5. $\alpha = 0.01$.
6. $n = 900$.
7. We gather our sample data.
8. We can use the normal approximation to the binomial because $np = 900(0.02) = 18 > 5$ and $nq = 900(0.98) = 882 > 5$.

$$\mu = 18 \qquad \text{and} \qquad \sigma = \sqrt{(18)(0.98)} \approx 4.2$$

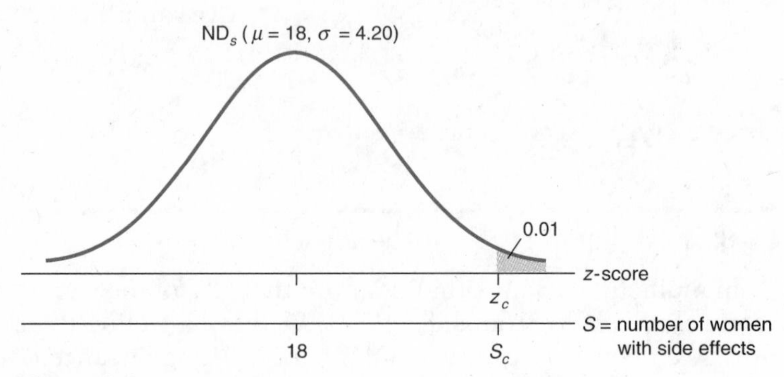


**Figure 8-8**

Looking up the $z$-score corresponding to Area = 0.99, we find $z_c = 2.33$. Using the formula for converting $z$-scores to raw scores, we have

$$\begin{aligned} S_c &= \mu + z_c\sigma \\ &= 18 + 2.33(4.2) = 18 + 9.8 = 27.8 \end{aligned}$$

We now have the normal distribution shown in Figure 8-9.

9A. With 27.8 as the critical value, our decision rule is as follows: *We will reject the null hypothesis if in our sample we find more than 27.8 women (that is, 28 or more) who have side effects.* (If we find fewer than 27.8 women with side effects, we will fail to reject the null hypothesis.)

9B. PSuppose that in step 7 we selected the sample and found 23 women with side effects. Thus, our statistic is $S = 23$.

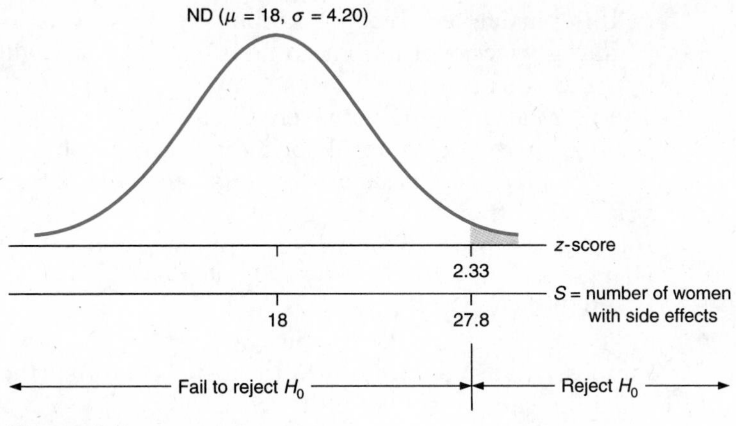


Figure 8-9

> ***Two Big If's***
>
> *Note:* There are two assumptions made in a hypothesis test:
>
> 1. *If* the null hypothesis is true, then the theoretical distribution of the sample statistic is known.
> 2. *If* we took many random samples, then the observed set of statistics would match the theoretical distribution.

9C. i. Since 23 is less than 27.8, according to our decision rule, we fail to reject the null hypothesis.

ii. We have been unable to prove the manufacturer wrong at the 0.01 significance level. We do not have sufficient evidence to say that the manufacturer's claim is too low. Our results are not statistically significant.

10. *Comments:* 23/900 is about $2\frac{1}{2}\%$ of our sample. Note that this is more than 2% maximum claimed for the population. Perhaps more testing should be done. ❑

---

## ❑ EXAMPLE 8.12 Rabbit Coloring

Dr. Bunny Hassenpfeffer, a noted biologist, is trying to change the coloring of rabbit offspring by diet. It is known that in a particular breed of rabbits 30% are pure white and the rest are spotted. A large group of this breed of rabbits is fed a special diet. Subsequently, a random sample of 100 offspring are selected, and their coloring is noted. Dr. Hassenpfeffer decides to use the 0.05 significance level. Find her decision rule.

### SOLUTION

1. She hopes that diet will affect the coloring of the offspring, but she is not sure whether it will produce more white or fewer white than usual

in this particular breed of rabbits. Because this experiment involves counting success of a dichotomous variable, she will use a one-sample hypothesis test of percentages.

2. The population being considered is all rabbits of this particular breed.
3. Defining $p = P$(a white offspring is born), we have $H_a$: Using the new diet, the probability that a white offspring will be born is no longer 30%; $p \neq 0.30$ (a two-tail test).
4. $H_0$: The proportion of white rabbits among the offspring will not change; $p = 0.30$.
5. $\alpha = 0.05$.
6. $n = 100$.
7. Suppose that she found only 15 of the 100 offspring to be pure white. Then the sample statistic $S$ is 15.
8. Assuming that the null hypothesis is true, we have $p = 0.30$, $q = 0.70$. Since $np = 30 > 5$ and $nq = 70 > 5$, we may use the normal approximation, with $\mu = 30$ and $\sigma = \sqrt{21} \approx 4.58$, as shown in Figure 8-10.

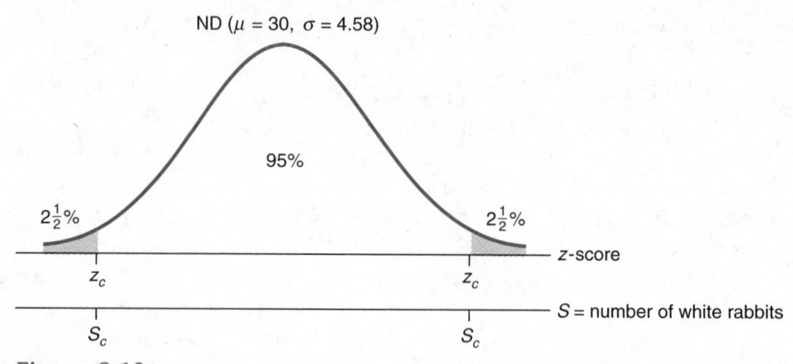


**Figure 8-10**

Looking up areas 0.025 and 0.975, we find that the corresponding $z$-scores are $-1.96$ and $+1.96$. Converting these $z$-scores to raw scores, we get

$$S_c = \mu + z_c\sigma$$
$$= 30 + (\pm 1.96)(4.58) = 30 \pm 8.98 = 21.02 \text{ and } 38.98$$

9A. We now have the situation shown in Figure 8-11.

Thus, *she will reject the null hypothesis if she gets fewer than 21.02 or more than 38.98 white rabbit offspring*, and she will claim that her idea that diet affects the coloring is true for this particular breed. Otherwise, she will say that she has been unable to prove such a relationship at the 0.05 significance level.

9B. Recall that she found only 15 of the 100 offspring to be pure white.

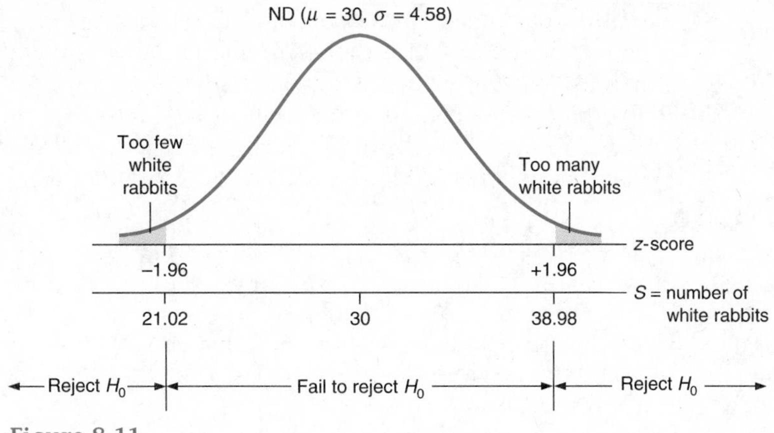


Figure 8-11

9C. i. Since 15 is less than 21.02, by the decision rule she will reject the null hypothesis.

ii. She will say that she has proved beyond a reasonable doubt—she has statistically significant evidence—that diet does affect the coloring of this breed of rabbit. It apparently reduces the number of white rabbits.

10. *Comments:* Because 15 is relatively far away from the critical value of 21.02, we have very strong evidence that the null hypothesis is probably false. ❑

---

## Two Other Approaches

1. *Converting Raw Scores to z-Scores*

In Example 8.12, in order to come to a conclusion, we had to compare our sample outcome, a raw score (of 15 rabbits), with the critical $z$-scores (+1.96). We chose to convert the $z$-scores to raw scores.

An equivalent approach, used by many researchers, would instead convert raw scores to $z$-scores. In this example, the researchers would get

$$z = \frac{X - \mu}{\sigma} = \frac{15.5 - 30}{4.58} = \frac{-14.5}{4.58} = -3.2$$

Since this is to the left of $z = -1.96$, they would reject the null hypothesis, just as we did.

2. *Reporting* p-*Values*

Some researchers prefer to reverse the approach used in Example 8.12 in another way. What we have done is:

Set $\alpha$ to a given value (say, 0.05).

Calculate a critical value for this $\alpha$.

Compare the experimental outcome to this critical value.

## *Common Sense and Statistics*[†]

When the American statesman Henry Clay (1777–1850) uttered the words "Statistics are no substitute for common sense," he may not have been referring to the conclusion of a hypothesis test, but his words certainly do apply. A statistical test of a hypothesis is merely a model—and often a very simplified model—of a real-world situation. Just as the real problem must be translated into some mathematical model, so, in turn, the model conclusion must be interpreted in light of the real world, and perhaps in conjunction with factors that were not included in the model. This process is often ignored by students.

Suppose, for example, a test of some statistical hypothesis leads to the decision rule "I will reject the null hypothesis at the 0.01 significance level if my outcome is greater than 61.1." Suppose further that your sample yields an outcome of 61.0. The textbooks say that you cannot reject the null hypothesis on the basis of this evidence; but some students react differently.

Alice: "Well I think that it's close enough."

Berry: "Can't you change the significance level to 0.05?"

Stosh: "Suppose we do a one-tail test instead of a two-tail test?"

Debbie: "Can we repeat the experiment with another sample?"

Annette: "I think we should gather a larger sample."

These students are looking for an exact numerical procedure which will make decisions for them, but it's not that simple. Here is an analogy which may clarify what's involved.

Suppose you want to paint your bedroom. You compute the area: 721 square feet. You read the label on the can of paint: "The contents of this can covers 350 square feet." Being a mathematical genius, you calculate that you need 2.06 cans. Now, how many cans do you buy—2 or 3? The mathematical model of your real-world problem provides an exact answer—namely, 2.06 cans. Clearly, 2 cans will not be enough; you must get 3 cans.

But the real world intrudes with other issues: Is the paint on sale? Is the paint store around the corner or 12 miles down the road? Are you a sloppy painter or a neat one? Are you a pessimist or an optimist? Do you believe that the number—350 square feet—on the label is an average or a minimum? Will you be satisfied if the last 20 square feet are covered only by a thin coat? Are you putting a dark color over a light color, or vice versa?

The mathematical model provides guidelines to the ultimate decision, but the risks involved in a wrong decision must be considered. If you buy only 2 cans of paint, you may have to go back to the store; they could be out of your color; the price may have gone up. If you buy 3 cans of paint, you still may have to go back to the store, to return an unused can of paint; and they might not make refunds.

In the same way, a statistical test of a hypothesis is only a model of a real-world situation. It can, at best, only provide guidelines for decisions about actual problems. Thus our response to Alice, Berry, Stosh, Debbie, and Annette is, "You know, you have something there; you may be right. Let's consider the consequences of the possible errors involved in a mistaken decision."

[†]Reprinted with permission from MATYC Journal, vol. 13, no. 2.

They, instead, do not specify a value of $\alpha$ before the data is collected. They wait until after the data are in and then just calculate the probability of obtaining an outcome as extreme as the one that occurred. This *prob*-ability is reported as the ***p*-value** of the test. In this approach, a $p$-value less than 0.05 is equivalent to rejecting $H_0$ at the 0.05 significance level. The researchers perform two steps:

***p*-value**

They calculate the $z$-score of the experimental outcome.

They determine the probability corresponding to this $z$-score.

For the rabbit example above, where we calculated $z = -3.2$, we use Table D-4 (Appendix D) to determine that the $p$-value for this experiment is 0.0007 (which is much less than 0.05). This indicates, once more, that it is correct to reject the null hypothesis.

### *Replication*

Many important scientific issues are subjected to repeated tests to insure that the results are accurate and not due to some unforeseen variation. If an experiment cannot be replicated (repeated) by other researchers in the same field, the results are not usually considered trustworthy.

## STUDY AIDS

### Chapter Summary

In this chapter, you have studied the hypothesis-testing procedure in general and a one-sample hypothesis test of percentages procedure in particular. Specifically, you learned:

- How to formulate two types of hypotheses, motivated and null
- To recognize one-tail and two-tail alternative hypotheses
- To interpret Type I and Type II errors and significance levels
- How to perform a one-sample hypothesis test of percentages
- Optionally, to compute the power of a test

### Vocabulary

You should be able to explain the meaning of each of these terms:

1. Hypotheses
2. Statistical hypothesis
3. Null hypothesis
4. Motivated hypothesis
5. Alternative hypothesis
6. Two-tail test
7. One-tail test
8. Decision rule
9. Rejection region
10. Type I error
11. Type II error
12. Significance level
13. Critical value
14. Power (optional)
15. Hypothesis test
16. $p$-value
17. Statistically significant

### Symbols

You should understand the meaning of each of these symbols:

1. $H_0$
2. $H_a$
3. $\alpha$
4. $\beta$
5. $S_c, z_c$

### Formulas

You should know when and how to use each of these formulas:

1. $\mu_S = np$
2. $\sigma_S = \sqrt{npq}$
3. $S_c = \mu_S + z_c \sigma_S$
4. Power $= 1 - \beta$

### A Brief Summary of a Hypothesis Test

1. Identify the procedure.
2. State the population(s).
3. Define the parameter(s) (if needed) and state $H_a$ in words and in symbols.
4. State $H_0$ in words and in symbols.
5. Select $\alpha$.
6. Choose $n$.
7. Gather sample data.
8. Calculate the critical value(s).
9. Terminate the test with the three-step ending:
   A. State the decision rule
   B. State the outcome
   C. State the conclusion twice: with respect to rejection of the null hypothesis and in the words of the particular application.
10. Comment on and clarify your results.

## EXERCISES

8-50 Complete the following table.

| | Symbol | Pronunciation | Meaning |
|---|---|---|---|
| a | $H_a$ | | |
| b | | *H* sub zero | |
| c | | Alpha | |
| d | | | Probability of a Type II error |

8-51 In the magazine article, "Common Sense and Statistics," which was reprinted above, consider the null hypothesis: Two cans of paint are sufficient.

(a) If you mistakenly buy 2 cans and you needed 3, what type of error did you make?

(b) If you mistakenly buy 3 cans and 2 were sufficient, what type of error did you make?

8-52 Lucky Larry notices that 1 of his coins comes up tails almost all the time. He decides to do a hypothesis test. He uses $p = P(\text{a tail is tossed})$, $H_0: p = 0.80$, and $H_a: p > 0.80$. He rejects $H_0$ at the 0.01 significance level and is therefore convinced that his coin comes up tails more than 80% of the time. Using this information, he makes several wagers and loses $500. Was this just bad luck with a biased coin, or could it be that the coin was fair after all?

8-53 A quality-control engineer at the Acme Bindery in South Jersey inspects every 25th book coming off the assembly line. His decision rule is: if more than 1% of the sample inspected is defective, the run is rejected. A recent run passed inspection and was shipped to Ketchum Book distributors in Philadelphia. There it was discovered that most of the books were defective. How could this have happened?

8-54 For the following hypotheses, decide if you would use a one-tail or a two-tail test. For each test, state a null and an alternative hypothesis. In the case of a one-tail test, state in which tail the rejection region would be.

(a) The coin is biased.

(b) The coin is biased for heads.

(c) The new procedure will reduce the number of defective parts produced on the assembly line.

(d) The manufacturer's claim about the percentage of people with iron deficiency is false.

(e) The manufacturer's claim about the percentage of people with iron deficiency is too high.

(f) This vaccine will reduce the number of cases of measles.

8-55 Follow the instructions given in the previous exercise.

(a) Less than 3% of the rattlesnakes found on Park Avenue have broken fangs.

(b) More than 18% of the teachers at the On-Time Railroad Dispatchers School are tardy.

(c) Joining the Sandworm Pickers Union will change the average take-home pay of a sandworm picker.

(d) Dr. Meany's practice of surprise tests in his course, "Do It Yourself Open-Heart Surgery," increases his students' grades.

(e) Coach Aquanut's policy of having his basketball players wear flippers during practice sessions improves their average number of points per game.

8-56 Lorenzo Jones has invented a new process for manufacturing computer chips. He claims it is a better process than the current one. A manufacturer decides to compare the two processes in a pilot study and to analyze the results by a statistical hypothesis test. The manufacturer wants to use a two-tail test, but Lorenzo wants to use a one-tail test. Why would Lorenzo prefer a one-tail test?

*For Exercises 8-57 to 8-71, perform hypothesis tests—that is, follow the 10 steps outlined in the Study Aids above.*

8-57 At a large university, the dean of desks and chairs has always assumed that 10% of the students are left-handed. Martinique Sinistra, a left-handed student, has been having trouble finding left-handed desks. She suspects that more than 10% of the students are left-handed. She takes a survey of 100 students picked randomly and finds 16 left-handers. Is this good evidence at $\alpha = 0.05$ to indicate that the dean is wrong?

8-58 Fargo North, decoder and cryptographer, stated that 40% of the letters in most messages are vowels. Using the first sentence of this exercise as a random sample, test the hypothesis that Fargo has overestimated the percentage of vowels. Use the 0.01 significance level.

8-59 A manufacturer of automobiles finds that 20% of the cars are unfit for delivery when they come off the assembly line. A worker proposes a new assembly technique which he claims will reduce the percentage of unfit cars. His technique is tried on 80 randomly chosen cars. It turns out that 3 are defective. Test with $\alpha = 0.05$.

8-60 For the disease Dandruffia Terminata, usually 68% of the victims recover without any treatment. The rest die within a short time. Dr. Ubaldo has a new drug which he hopes is a cure for the disease. He administers his drug to 64 patients picked at random from among victims of the disease, and 50 recover. Is this enough evidence to establish that the drug is effective? (Or is this result likely to have occurred just by chance?) Test at the 0.05 significance level.

8-61 The usual dropout rate in the freshman class at Wealth College is 50%. A new dean of admissions claims that recent policies have lowered the dropout rate, because in this year's class of 600 freshmen only 260 dropped out. Test at the 0.05 significance level. Do the statistics support the claim?

8-62 Using a particular type of bombing mechanism usually results in 70% of the bombs being on target. An engineer claims to have invented a more accurate mechanism. The new mechanism is used to drop 100 test bombs on a target; 75 hit the target. Is this enough evidence to establish that the new mechanism is better than the old one? Use $\alpha = 0.05$.

8-63 According to a nationwide poll, only 40% of the voters favor a health care bill that would benefit the poor. Senator Ted Gladkowski believes that the percentage of voters who favor the bill is higher in his district. A random sample of 30 voters in his district is taken, and it is found that 14 support the health care bill. Is this sufficient evidence to support Senator Gladkowski's belief at the 0.01 significance level?

8-64 A certain medicine is said to be at least 90% effective in giving relief to people with allergic reactions to cats and dogs. Dr. Kay Nyne believes that this claim is incorrect. A random sample of 60 people with such allergies is selected from patients at an allergy clinic. What would you say about the claim at the 0.05 significance level if 58 people got relief?

8-65 Ralph complains that on the Saturday morning programs on WW-TV about 25% of the time is devoted to commercials. A student tests this one Saturday by switching on the TV at 50 randomly chosen times between 7 A.M. and noon. She finds that at 9 of these times she sees a commercial. Does this support Ralph's claim? (Use $\alpha = 0.05$.)

8-66 It is known that in July on a certain stretch of the northeast coast of the United States 60% of the seagulls are Franklin's gulls. A birdwatcher goes out 1 morning and spots 80 seagulls. He finds that 75 of them are Franklin's gulls. Give 2 possible explanations for this. Use $\alpha = 0.01$.

8-67 Two professors read in the paper that 60% of all college freshmen are more interested in being popular than in doing well at school. They think this is too high, so they go around and interview 100 freshmen. They find that 10 say they are more interested in being popular, while 90 say they are more interested in doing well. Test their results at the 0.05 significance level. Interpret your findings.

8-68 An economics student, Darby Walbert, reads that in his county 35% of the employed earn more than $15,000 per year. He wants to see if the claim is accurate, so he mails out 500 questionnaires to people chosen at random from the phone book. He gets back 100 replies; 80 of them report incomes of more than $15,000. Interpret his results. He plans to test at the 0.05 significance level.

8-69 Under certain conditions the probability is 0.10 that a tadpole survives to mature into a frog. Now scientist Ann Juston believes that she has found a way to place vitamins in the frog pond so that more tadpoles will survive. Using this new approach, we take a random sample of 98 tadpoles and test using the 0.05 significance level. If 12 tadpoles survive, state whether we reject or fail to reject the null hypothesis, and explain what that means. If 27 tadpoles survive, state whether we reject or fail to reject the null hypothesis, and explain what that means.

8-70 Stan Sly claims that he can control the tosses of a fair coin. To see if he is correct, you take 2 ordinary coins and give him 1. You toss 1 and ask him to try to toss the same thing with his coin; that is, if your toss results in a tail, then he is to try to toss a tail also. You repeat this experiment 18 times, and Stan succeeds in tossing the same as you 15 times. Is this unusual at the 0.05 significance level?

8-71 With a fair pair of dice, doubles (2 numbers the same) should come up about 1/6 of the time. Marsha, who is losing at Monopoly, has rolled 15 doubles in the past 60 rolls. She claims that the dice are loaded.

(a) Find the probability that a pair of honest dice could produce 15 or more doubles in 60 rolls.

(b) Is the probability that you found in part (a) $\alpha$, $\beta$, or neither?

8-72 How large a sample size would be needed to use a normal curve approximation to test the hypothesis that 3% of left-handed Carpathians have at least one green eye?

## CLASS SURVEY QUESTIONS

1. It seems reasonable to assume that the last digit of Social Security numbers is odd half the time and even half the time. (*Note:* Zero is even.) Test this assumption with $\alpha$ = 0.05, using the data for this class.

2. Perform a similar hypothesis test on the fifth digit of your Social Security numbers. What happened?

## FIELD PROJECTS

Here are some suggestions for field projects that involve hypothesis tests. First, we suggest that you conduct field projects in two stages, as outlined below:

(a) ***Stage 1:*** Outline clearly what you *intend* to do. State the population or populations that you wish to sample. Describe your intended sampling procedure. Comment on its strengths and weaknesses. State your null and alternative hypotheses, the significance level, and the sample size. If you are going to ask questions of people in your sample, state these questions now exactly as you will ask them. If you are going to count something in your sample, state exactly what you are looking for. In any case, state how you will handle responses that do not fit into your predetermined categories. Give all this information to your instructor. After the instructor has approved it, then proceed with stage 2.

(b) ***Stage 2:*** Perform the experiment as approved. Do the calculations. Submit your results, with comments as to the strengths and weaknesses of the project as you actually carried it out.

### Example of a Field Project

A claim appears in a newspaper that 60% of Americans feel that the President is "doing a good job." A student doubts that this percentage is correct in his neighborhood, and so he sets up the following field project.

#### *Stage 1*

(a) *Population* All people 16 years old or older who live within 3 blocks of my house.

(b) *Sampling Procedure* There are 30 blocks in this neighborhood. I will pick 3 houses at random on each block. I will question only 1 person at each house. If no one is home, I will select another house on that block. Thus, n = 90.

(c) *Questions to Be Asked:*

(1) I am doing a survey for my college class in statistics. Would you please answer the following questions?

(2) Are you 16 years old or older?

(3) Do you think that the President is doing a good job?

I will continue until I get 90 yes responses to both questions 1 and 2, and either a yes or a no response to question 3. Since $np = 90(0.60) = 54 > 5$, and $nq = 90(0.40) = 36 > 5$, we can use the normal distribution.

(d) *Hypotheses* Defining $p = P$(a person thinks that the President is doing a good job), we have

$H_0$: The fraction in my neighborhood is the same as the fraction of all Americans, $p = 0.60$.

$H_a$: The fraction in my neighborhood is not the same as the fraction of all Americans, $p \neq 0.60$ (two-tail test).

I will use $\alpha = 0.05$.

(e) *Comments on Strengths and Weaknesses* Depending on when I am able to cover the houses, I may get more or fewer working people in my sample.

#### *Stage 2 (Conducted after Stage 1 was approved)*

I went to 123 homes. In 12 homes no one answered the door. In 3 homes there was no one over 16 present. In 2 homes the answer to question 1 was no. In 4 homes people first answered yes to question 1 but changed their minds after they heard question 3. In 12 homes the person was undecided on question 3. The remaining 90 responses were

| Yes | No |
|---|---|
| 48 | 42 |

### *Solution for This Project*

$H_0$: The fraction in my neighborhood is the same as the fraction of all Americans, $p = 0.60$.

$H_a$: The fraction in my neighborhood is not the same as the fraction of all Americans, $p \neq 0.60$ (a two-tail test).

$$p = 0.60 \qquad n = 90 \qquad \alpha = 0.05$$

Since $np = 54 > 5$ and $nq = 36 > 5$, we can use the normal approximation with $\mu = 54$ and $\sigma = \sqrt{21.6} \approx 4.65$.

$$z_c = \pm 1.96$$

The curve for this distribution is shown in Figure 8-12.

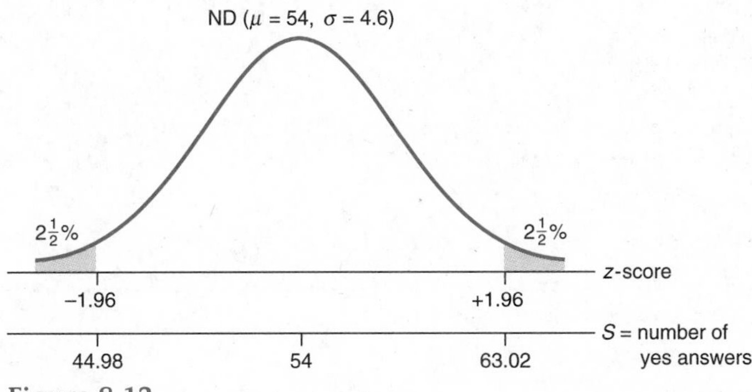


Figure 8-12

$$S_c = 54 + (\pm 1.96)(4.65) = 54 \pm 9.11 = 63.11 \text{ and } 44.89$$

Therefore, my decision rule is to reject the null hypothesis if I get more than 63 or fewer than 45 yes answers. Since the experimental outcome was 48 yes answers, I have failed to reject the null hypothesis. I was unable to show that the null hypothesis was false. Of the 33 nonresponses I do not have enough evidence to indicate that they would differ markedly from the 90 who did respond. If there were a marked difference, it might change my conclusion.

### Suggested Projects

Outline a one-sample hypothesis test of a percentage to be carried out on some population of your choice. After your instructor has approved your outline, gather your data and perform the hypothesis test.

1. Select a reported fact from a newspaper or some other source as in the preceding example and test it on some population of your own choice.
2. Test a theoretical hypothesis concerning coins, dice, cards, etc. For example, does the American penny have a probability of coming up heads equal to 0.50? If you want $n = 1000$, get 100 pennies and toss them from a large container 10 times to test this hypothesis.
3. Perform any one-sample hypothesis test of a percentage that interests you.

# 9

# Hypothesis Testing: Two-Sample Tests of Percentages in Binomial Experiments

Vive la difference!

CONTENTS



GOALS

At the end of this chapter you will have learned:

- Symbols for estimates of parameters in two-sample tests
- How to compute and use a pooled estimate of a parameter
- How to perform a two-sample binomial hypothesis test

## 9.1 Symbols for Estimates

In the previous chapter we have asked whether or not a sample could reasonably be expected to have been chosen at random from one population having a *given parameter.* For example, if a population is 62% female, what is the probability that we could draw a random sample of 40 people and find less than 48% of the sample to be female?

In this chapter we will discuss a related statistical test used for answering the question: Is it likely that two different populations have the *same* parameter? For example, is it likely that the *same percentage* of high school boys and girls in Giddyville commute via in-line skates? Many studies involve this kind of comparison, and they involve selecting random samples from two populations. The associated hypothesis tests are called **two-sample tests**.

**two-sample tests**

In conducting such studies we need to refer to *parameters* from both populations and to *statistics* from both samples. We use $p_1$, $p_2$, $\hat{p}_1$, and $\hat{p}_2$ for these symbols. The symbols for these values are listed in the table below. This system of symbols, where the statistic has a caret (^) over it and the parameter does not, is a common one for distinguishing between parameters and statistics. The symbol $\hat{p}$ is read as ***p*-hat**.

***p*-hat**

**Percentage Symbols for Parameters and Statistics**

| | Population Parameter | Sample Statistic |
|---|---|---|
| Population 1 | $p_1$ | $\hat{p}_1$ |
| Population 2 | $p_2$ | $\hat{p}_2$ |

Here are two illustrations:

1. Suppose populations 1 and 2 are the male and female teenagers in Giddyville.

   If 25 percent of all male teenagers in Giddyville used in-line skates to commute, we would write $p_1 = 0.25$.

   If 25 percent of a random sample of some female teenagers in Giddyville used in-line skates to commute, we would write $\hat{p}_2 = 0.25$.

2. If the population we are studying is all the people in this classroom, and 37½% of them are female, we write $p = 0.375$.

   If we use this figure to estimate that the proportion of females in the entire school is probably close to 0.375, we write $\hat{p} = 0.375$.

## 9.2 A Distribution of Differences

Imagine that, as a project, a class of 23 students wishes to compare two nearby campuses with regard to the percentage of students who support Senator Foghorn's policy on sales tax. Suppose that, unknown to anyone, the true figure on campus 1 is 63%; that is, $p_1 = 0.63$. On campus 1, each of the 23 students gathers a random sample of 100 people and computes the best estimate of $p_1$ which we call $\hat{p}_1$ (read: $p$ sub one hat). Their results are as follows.

| Student | $\hat{p}_1$ |
|---|---|
| 1 | 0.62 |
| 2 | 0.61 |
| 3 | 0.64 |
| 4 | 0.63 |
| 5 | 0.63 |
| 6 | 0.60 |
| 7 | 0.55 |
| 8 | 0.70 |
| 9 | 0.63 |
| . | . |
| . | . |
| . | . |
| 23 | 0.62 |

If we added these 23 values of $\hat{p}_1$ and divided by 23, we would probably get a value very close to the actual value of $p$, which is 0.63.

Next, each student visits campus 2 and gathers a second random sample of 100 persons from campus 2. (Unknown to anybody, the actual figure on campus 2 is 60%; that is, $p_2 = 0.60$.)

| Student | $\hat{p}_1$ | $\hat{p}_2$ |
|---|---|---|
| 1 | 0.62 | 0.57 |
| 2 | 0.61 | 0.62 |
| 3 | 0.64 | 0.60 |
| 4 | 0.63 | 0.59 |
| 5 | 0.63 | 0.61 |
| 6 | 0.60 | 0.57 |
| 7 | 0.55 | 0.58 |
| 8 | 0.70 | 0.60 |
| 9 | 0.63 | 0.61 |
| . | . | . |
| . | . | . |
| . | . | . |
| 23 | 0.62 | 0.64 |

Finally the students report the differences found between the percentages:

| Student | $\hat{p}_1$ | $\hat{p}_2$ | Differences, $\hat{p}_1 - \hat{p}_2$ |
|---|---|---|---|
| 1 | 0.62 | 0.57 | 0.05 |
| 2 | 0.61 | 0.62 | −0.01 |
| 3 | 0.64 | 0.60 | 0.04 |
| 4 | 0.63 | 0.59 | 0.04 |
| 5 | 0.63 | 0.61 | 0.02 |
| 6 | 0.60 | 0.57 | 0.03 |
| 7 | 0.55 | 0.58 | −0.03 |
| 8 | 0.70 | 0.60 | 0.10 |
| 9 | 0.63 | 0.61 | 0.02 |
| . | . | . | . |
| . | . | . | . |
| . | . | . | . |
| 23 | 0.62 | 0.64 | −0.02 |

**distribution of differences**
**sampling distribution**

The distribution of numbers in the last column is called a **distribution of differences** of sample proportions. The distribution of *all* such possible differences is often referred to as the **sampling distribution** of the differences. In general, when you repeatedly take random samples of the same size and compute the same statistic each time, the distribution of these values is called the "sampling distribution" of that statistic.

The symbol for the ***d***ifferences of the sample ***p***ercentages is $\boldsymbol{d\hat{p}}$. Thus, $d\hat{p}$ equals $\hat{p}_1 - \hat{p}_2$. In theory, this sampling distribution of $d\hat{p}$ has three important properties which are summarized in Table 9-1. (The first two properties you may have already guessed.)

The distribution described in Table 9-1 is shown in Figure 9-1.

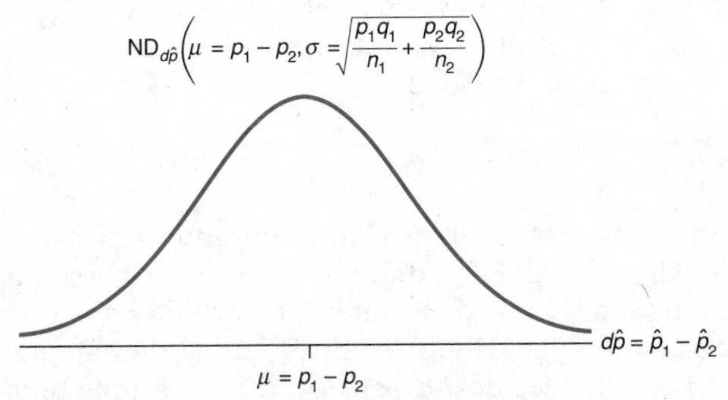


**Figure 9-1**

**Table 9-1 Theoretical Properties of the Sampling Distribution of $d\hat{p} = \hat{p}_1 - \hat{p}_2$**

1. If $n_1p_1$, $n_1q_1$, $n_2p_2$, and $n_2q_2$ are all larger than 5, then the distribution of sample differences will be approximately normal.

2. The mean of the differences will equal the true population difference, $p_1 - p_2$. We write

$$\mu_{d\hat{p}} = p_1 - p_2$$

3a. The standard deviation of the differences is given by

$$\sigma_{d\hat{p}} = \sqrt{\frac{p_1q_1}{n_1} + \frac{p_2q_2}{n_2}}$$

where $n_1$ = the size of each sample taken from first population (which population is denoted "first" and which is denoted "second" is arbitrary)
$p_1$ = true proportion* in the first population
$q_1 = 1 - p_1$
$n_2$ = the size of each sample taken from second population
$p_2$ = true proportion in the second population
$q_2 = 1 - p_2$

3b. In the particular case where $p_1$ and $p_2$ have the same value (say, $p$), the formula for $\sigma_{d\hat{p}}$ becomes

$$\sigma_{d\hat{p}} = \sqrt{\frac{pq}{n_1} + \frac{pq}{n_2}} \quad \text{or} \quad \sqrt{pq\left(\frac{1}{n_1} + \frac{1}{n_2}\right)}^{\dagger}$$

where $p$ is the common value of $p_1$ and $p_2$ and $q = 1 - p$.

*Recall that *proportion, percentage,* and *probability* all have the same meaning.
†This latter form is easier on most calculators.

## 9.3 Two-Sample Binomial Hypotheses Tests

Putting together this new material with what you learned in Chapter 8, you now know enough to perform a two-sample binomial hypothesis test for comparing percentages in two populations.

### ❑ EXAMPLE 9.1 Dorm Students

Leonard has been accepted at Adelphi University and at Hofstra University. He decides that he will go to the one that has the greater percentage of dorm students. A friend tells him that it does not matter which he chooses, because the proportions are about equal. Doubting his friend and unable to obtain the exact information, Leonard decides to perform a two-sample binomial test.

## SOLUTION

Leonard's motivated hypothesis is that the percentage of dorm students at Adelphi is different from the perçentage at Hofstra. His null hypothesis is that these proportions are the same. If we let $p_1 = P$(a randomly selected student at Adelphi is a dorm student) and $p_2 = P$(a randomly selected student at Hofstra is a dorm student), then the hypotheses are:

$H_0$: The percentage of dorm students at Adelphi is the same as the percentage of dorm students at Hofstra; $p_1 = p_2$ or $p_1 - p_2 = 0$.

$H_a$: The percentage of dorm students at Adelphi is different from the percentage of dorm students at Hofstra; $p_1 \neq p_2$ or $p_1 - p_2 \neq 0$ (a two-tail test).

Leonard's strategy is as follows. He will take a random sample on each campus and compute the percentage of dorm students in each. Then he will compute the difference. The size of this difference will suggest whether or not to reject the null hypothesis. He will be aided in this decision by the properties of the sampling distribution of $\hat{p}_1 - \hat{p}_2$ given in Table 9-1, because the particular difference that Leonard will find is one of the possible differences described by this distribution.

### A Prolific Mathematician

It had been noticed that the statistical tables of the positions of the planets and the theoretical positions did not agree (Jupiter went too fast, Saturn too slow). A treatment of this problem by the Swiss mathematician Leonhard Euler (1707–1783) won the 1748 Paris Academy of Sciences prize. He was the most productive mathematician of all time: his collected works exceed six dozen published volumes and more are still being edited for publication. He fathered 14 children.

The general approach, as in all hypothesis testing, is to assume that the null hypothesis is true, and then see if the sample evidence goes against this assumption. Now, according to Table 9-1, the mean value of all the possible sample differences is $p_1 - p_2$, and according to Leonard's null hypothesis $p_1 = p_2$, or $p_1 - p_2$ equals zero. Furthermore, if the sample sizes are large enough, then the distribution of possible differences is normal, which implies that most of the possible values are near the mean. Thus, Leonard's analysis will consist of deciding whether his particular difference is far enough from zero to indicate that the null hypothesis is probably wrong.

Once the strategy of the test is worked out, it remains to collect the data and interpret it. He is able to interview 100 randomly selected Adelphi students and 110 randomly selected Hofstra students. His data are given in Table 9-2.

We would like to use these data together with the information in Table 9-1 to come to a conclusion. To do this, we need an accurate description of the theoretical normal curve that would be generated by listing all the differences that are possible if the null hypothesis is true. This is called the *sampling distribution of differences* under the null hypothesis, and its properties are given in Table 9-1.

**Table 9-2 Leonard's Data**

| | Sample 1 (Adelphi) | Sample 2 (Hofstra) |
|---|---|---|
| Sample size | $n_1 = 100$ | $n_2 = 110$ |
| Number of dorm students | $S_1 = 60$ | $S_2 = 50$ |
| Percentage of dorm students | $\hat{p}_1 = \frac{S_1}{n_1} = 60/100 = 0.60$ | $\hat{p}_2 = \frac{S_2}{n_2} = 50/110 = 0.45$ |
| Percentage of non-dorm students | $\hat{q}_1 = 1 - \hat{p}_1 = 0.40$ | $\hat{q}_2 = 1 - \hat{p}_2 = 0.55$ |
| Difference in percentage of dorm students: $\hat{p}_1 - \hat{p}_2 = 0.60 - 0.45 = 0.15$ | | |

Notice that for property 3b in Table 9-1, the standard deviation depends on the values of $n_1$, $n_2$, $p$, and $q$. Since we do not know the true values of $p$ and $q$ in this study, we use their estimates, $\hat{p}$ and $\hat{q}$, in their places.

We begin by checking each of the three properties of the distribution of differences.

1. Check that the shape of the sampling distribution is normal: The sampling distribution of differences will be normal if all four numbers $n_1\hat{p}_1$, $n_1\hat{q}_1$, $n_2\hat{p}_2$, and $n_2\hat{q}_2$ are greater than 5.

$$n_1\hat{p}_1 = 100(60/100) = 60 > 5 \qquad n_1\hat{q}_1 = 100(40/100) = 40 > 5$$

$$n_2\hat{p}_2 = 110(50/110) = 50 > 5 \qquad n_2\hat{q}_2 = 110(60/110) = 60 > 5$$

***Note:*** These four numbers are the same as the four sample outcomes: at Adelphi, 60 dorm and 40 non-dorm students; at Hofstra, 50 dorm and 60 non-dorm students. *Clearly, this will always be the case and thus no calculation will be necessary.*

Since the numbers of successes and failures in each sample are 60, 40, 50, and 60, and they are all larger than 5, the distribution of the differences will be approximately normal.

2. Determine the mean of the sampling distribution. Assuming that our null hypothesis is true, we have the mean $\mu_{d\hat{p}} = 0$.
3. Determine the standard deviation of the sampling distribution. But what value can we use to estimate $\sigma_{d\hat{p}}$?

According to the null hypothesis, $p_1$ and $p_2$ are equal. But what is their actual common numerical value? In light of the fact that our null hypothesis claims that there is no difference between the two populations we are sampling, it makes sense to consider that the data come from one large sample rather than two smaller samples. Thus, we merge the data and form a **pooled estimate** of $p$, called $\hat{p}$.

**pooled estimate**

We know that Leonard looked at 210 students ($n_1 + n_2$) and found that 110 were dorm students ($S_1 + S_2$). So the pooled estimate for the common numerical value of the proportion of dorm students would be

$$\hat{p} = \frac{S_1 + S_2}{n_1 + n_2} = \frac{60 + 50}{100 + 110} = \frac{110}{210} = 0.52 \quad \text{and} \quad \hat{q} = 1 - 0.52 = 0.48$$

**estimate of $\sigma_{d\hat{p}}$**

We write $\hat{p} = 0.52$, and use this in the formula for $\sigma_{d\hat{p}}$. The result will be an **estimate of $\sigma_{d\hat{p}}$**. We call it $\hat{\sigma}_{d\hat{p}}$. Thus,

$$\hat{\sigma}_{d\hat{p}} = \sqrt{\hat{p}\hat{q}\left(\frac{1}{n_1} + \frac{1}{n_2}\right)}$$

$$= \sqrt{0.52(0.48)\left(\frac{1}{100} + \frac{1}{110}\right)} = \sqrt{0.004765} \approx 0.069$$

This distribution is shown in Figure 9-2.

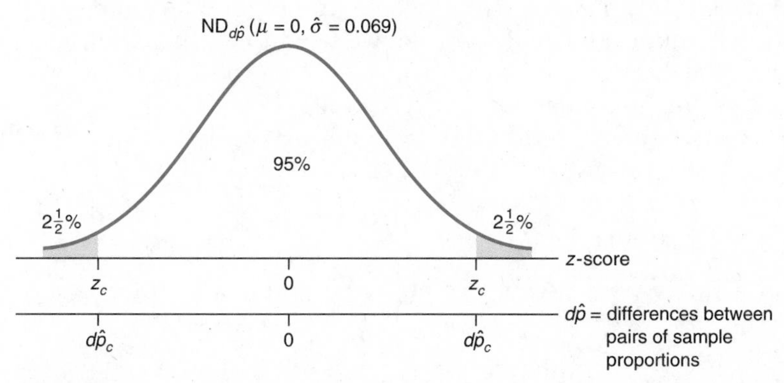


Figure 9-2

We can now use the curve in Figure 9-2 to make a decision rule. We already know that Leonard's study calls for a two-tail test. We need only to establish the significance level for the test. Let us suppose for the sake of illustration that Leonard chooses $\alpha = 0.05$, which gives $\pm 1.96$ as the critical $z$-scores.

**critical difference**

Next, we convert these critical $z$-scores to percentages **critical differences, $d\hat{p}_c$**:

$$d\hat{p}_c = \mu_{d\hat{p}} + \hat{\sigma}_{d\hat{p}} z_c$$

$$= 0 + 0.069(\pm 1.96) \approx \pm\, 0.14$$

His decision will be to reject the null hypothesis if his sample difference $d\hat{p} = \hat{p}_1 - \hat{p}_2$ is either less than $-0.14$ or greater than $+0.14$.

Since his sample difference, $d\hat{p} = \hat{p}_1 - \hat{p}_2 = 0.60 - 0.45 = 0.15$, is greater than 0.14, he rejects the null hypothesis and claims that the percentage of dorm students at Adelphi is different from the percentage of dorm students at Hofstra. Since $\hat{p}_1 > \hat{p}_2$, there is a *higher* percentage of dorm students at Adelphi. ❑

## Summary of Two-Sample Binomial Tests for Comparing Percentages

The two-sample binomial test is used when we compare two populations. We let $p_1$ be the true proportion of successes in the first population, that is, $p_1 = P(\text{a success in population 1})$; and $p_2$ the true proportion of successes in the second population, that is, $p_2 = P(\text{a success in population 2})$.

The null hypothesis states that the difference between these true proportions is a fixed number. For most of the problems in this book our null hypothesis is that $p_1 = p_2$, which implies that $p_1 - p_2 = 0$.

As before, the alternative will have a $<$, $>$, or $\neq$ symbol, which will tell us whether we are doing a one-tail or a two-tail test.

The distribution of differences will be normal if the four sample outcomes (or $n_1\hat{p}_1$, $n_1\hat{q}_1$, $n_2\hat{p}_2$, and $n_2\hat{q}_2$) are all greater than 5.

To compute the decision rule, we need the mean and the standard deviation of the distribution of differences. We get the mean from the null hypothesis, and thus the mean will usually be 0.

We estimate the standard deviation using the formula

$$\hat{\sigma}_{d\hat{p}} = \sqrt{\hat{p}\hat{q}\left(\frac{1}{n_1} + \frac{1}{n_2}\right)}$$

where $\hat{p}$ is the pooled estimate of the common value of $p_1$ and $p_2$.

**Table 9-3 Approximating the Sampling Distribution of $d\hat{p} = \hat{p}_1 - \hat{p}_2$ in Practice**

1. If the 4 outcomes, the 2 successes and the 2 failures, are all larger than 5, then the distribution of sample differences will be approximately normal.

2. The mean of the differences will equal the true population difference, $p_1 - p_2$, which by the assumption of the null hypothesis is zero. We write

   $$\mu_{d\hat{p}} = p_1 - p_2 = 0$$

3′. Since $p_1$ and $p_2$ are assumed to have the same value, we pool the samples to obtain $\hat{p}$ and $\hat{q}$.

   Recall that $\hat{p}$ is the pooled estimate of the common value of $p_1$ and $p_2$,

   $$\text{where } \hat{p} = \frac{S_1 + S_2}{n_1 + n_2} \quad \text{and} \quad \hat{q} = 1 - \hat{p}$$

   The formula to estimate the standard deviation becomes

   $$\hat{\sigma}_{d\hat{p}} = \sqrt{\hat{p}\hat{q}\left(\frac{1}{n_1} + \frac{1}{n_2}\right)}$$

Note that, whereas Table 9-1 describes the theoretical distribution of $d\hat{p}$, in practice we cannot know the correct values of $\sigma_{d\hat{p}}$. Instead, we use the approximation given in Table 9-3. The principal difference between the two tables is item number 3′ in Table 9-3.

## A Second Example

### ❑ EXAMPLE 9.2 This One Is a Killer

A researcher is comparing the safety records of two popular automobile models. She is interested in the percentage of accidents in which the driver is killed. She has heard that the Boomer model is more dangerous than its competitor, the Zoomer. Checking recent records, she finds that over the past several months in a large metropolitan area there were 423 accidents involving the Zoomer. In 34 of these the driver was killed. For the Boomer there were 580 accidents, and in 58 of these the driver was killed. Does this indicate, at the 0.01 significance level, that a driver who gets in an accident is more likely to be killed if he or she is driving a Boomer?

#### SOLUTION

This is a two-sample binomial hypothesis test for comparison of percentages.

Let population 1 be all drivers of Boomers who have an accident. Let population 2 be all drivers of Zoomers who have an accident.

$p_1 = P$(a driver of Boomer in an accident is killed).

$p_2 = P$(a driver of Zoomer in an accident is killed).

$H_0$: There is no difference in the percentages of drivers killed in an accident; $p_1 = p_2$.

$H_a$: A higher percentage of Boomer drivers are killed in an accident; $p_1 > p_2$ or $p_1 - p_2 > 0$ (a one-tail test on the right).

(If we had let population 1 be the Zoomer drivers, we would have a one-tail test on the left.)

$$n_1 = 580 \qquad \hat{p}_1 = 58/580 = 0.10 \qquad \hat{q}_1 = 0.90$$

$$n_2 = 423 \qquad \hat{p}_2 = 34/423 = 0.08 \qquad \hat{q}_2 = 0.92$$

Since the 4 outcomes (58 were killed, 532 lived, 34 were killed, and 389 lived) are all larger than 5, the sampling distribution of differences is approximately normal. The pooled estimate of $p$ is

$$\hat{p} = \frac{S_1 + S_2}{n_1 + n_2} = \frac{58 + 34}{580 + 423} = \frac{92}{1003} = 0.09$$

So $\hat{q} = 0.91$. Therefore,

$$\hat{\sigma}_{d\hat{p}} = \sqrt{\hat{p}\hat{q}\left(\frac{1}{n_1} + \frac{1}{n_2}\right)}$$

$$= \sqrt{0.09(0.91)\left(\frac{1}{580} + \frac{1}{423}\right)} = \sqrt{0.000335} \approx 0.018$$

This distribution is illustrated in Figure 9-3.

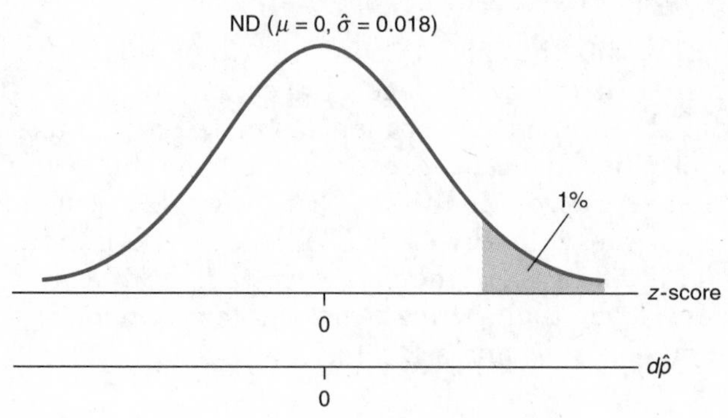


**Figure 9-3**

Using $z_c = 2.33$, we obtain

$$d\hat{p}_c = \mu + \hat{\sigma} z_c$$

$$= 0 + 0.018(2.33) = 0.04$$

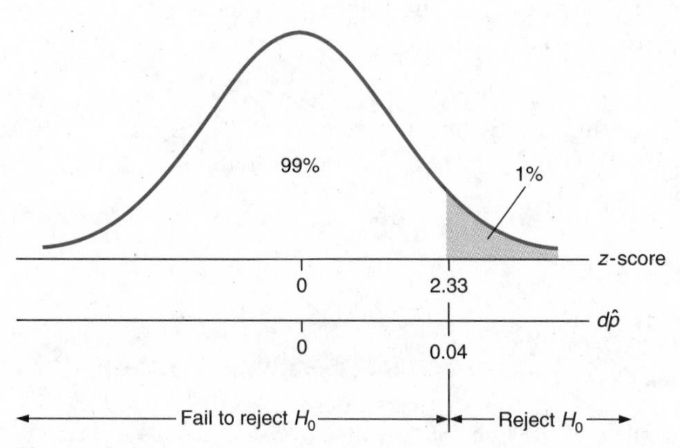


**Figure 9-4**

The critical value of $d\hat{p}$ is 0.04, so the decision rule is that if the sample difference $d\hat{p}$ is to the right of 0.04, we will reject the null hypothesis.

The observed difference is $\hat{p}_1 - \hat{p}_2 = 0.10 - 0.08 = 0.02$, which is not in the rejection region. We have failed to establish at $\alpha = 0.01$ that the fatality rate is higher in the Boomers. We report that the difference between the two fatality rates is *not statistically significant.* ❑

## STUDY AIDS

### Chapter Summary

In this chapter you have learned:

- To distinguish between sample statistics and population parameters by use of the hat (^) symbol to indicate sample statistics which estimate population parameters
- When to pool data from two samples assumed to be taken from the same population
- How to perform a two-sample binomial hypothesis test to compare percentages

### Vocabulary

You should be able to explain the meaning of each of these terms:

1. Two-sample hypothesis test
2. *p*-hat
3. Distribution of differences
4. Sampling distribution
5. Pooled estimate
6. Estimate of $\sigma_{d\hat{p}}$
7. Critical difference

### Symbols

You should be able to use and explain the meaning of each of these symbols:

1. $n_1, n_2$
2. $S_1, S_2$
3. $p_1, p_2, q_1, q_2$
4. $\hat{p}_1, \hat{p}_2, \hat{q}_1, \hat{q}_2$
5. $d\hat{p}, d\hat{p}_c$
6. $\mu_{d\hat{p}}$
7. $p, q, \hat{p}, \hat{q}$
8. $\sigma_{d\hat{p}}, \hat{\sigma}_{d\hat{p}}$

### Formulas

You should know when and how to use each of these formulas:

1. $\mu_{d\hat{p}} = p_1 - p_2$
2. $\hat{\sigma}_{d\hat{p}} = \sqrt{\hat{p}\hat{q}\left(\frac{1}{n_1} + \frac{1}{n_2}\right)}$
3. $\hat{p} = \frac{S_1 + S_2}{n_1 + n_2}$
4. $\hat{q} = 1 - \hat{p}$
5. Sample difference $d\hat{p} = \hat{p}_1 - \hat{p}_2$
6. Critical difference $d\hat{p}_c = \mu_{d\hat{p}} + z_c\hat{\sigma}_{d\hat{p}}$

## EXERCISES

9-1 Randy Semple, a conscientious statistics student, did a field project for Chapter 8. He read in the newspaper that 28% of the families in New York City own a dog. He tested whether 28% of the families in his community owned a dog. This, of course, was a one-sample hypothesis test. Now that he has studied Chapter 9, he realizes that the New York City figure must have been derived from a sample also, and he wonders if he should have done a two-sample test. Does he have enough data?

9-2 It is unlikely that 2 different types of coins have *exactly* the same probability of coming up heads. Winnie and Pooh wish to see if they can spot this difference for pennies and dimes by setting up an experiment at the 0.05 significance level. They get 100 newly minted pennies and 50 newly minted dimes, toss them 10 times each, and count the number of heads. For the pennies they have $n_1 = 1000$, and $S_1 = 490$ heads. For the dimes they have $n_2 = 500$, and $S_2 = 240$ heads.

(a) Define $p_1$ and $p_2$, and state the hypotheses.

(b) Is this a one-tail or a two-tail test?

(c) Find $\hat{p}_1$, $\hat{p}_2$, and $\hat{p}$.

(d) Find $\hat{\sigma}_{d\hat{p}}$.

(e) Find the decision rule for this experiment.

(f) Compute $d\hat{p} = \hat{p}_1 - \hat{p}_2$, and state the conclusion reached for this experiment.

9-3 Executive Airlines is analyzing its passenger service. One question is: Which of 2 flights carries a larger percentage of people on business trips? Over a period of 1 month they interview a random sample of passengers. On the first flight, 130 out of 200 were on business trips. On the second flight, 120 out of 200 were on business trips.

(a) Using $\alpha = 0.01$, find the decision rule for this test.

(b) Compute $d\hat{p}$, and state your conclusion.

9-4 One way of comparing 2 colleges is to look at the percentage of students in each who hold part-time jobs during the school year. A survey was taken at 2 colleges, with these results. At each college 100 students were interviewed. At Cardinal College 70% of those interviewed held jobs. At Bishop's 75% held jobs.

(a) Using $\alpha = 0.05$, find the decision rule for this test.

(b) Compute $d\hat{p}$, and state your conclusion.

9-5 In certain medical procedures a small plastic tube must be permanently attached to a patient's vein. This is a potential source of blood clots. A clinical trial could be done to see if putting such patients on a daily low dose of aspirin would reduce the occurrence of blood clots. In one such experiment the results of Dr. P. LaCebo were that 6 out of 19 patients on aspirin developed clots, while 18 out of 25 patients who were not given aspirin developed clots. Is this evidence at $\alpha = 0.01$ in favor of using the aspirin?

9-6 A teacher reads of a new approach to teaching a difficult idea. She has 2 classes of equal background, intelligence, and ability. She teaches 1 class of 20 students the traditional way, and after testing finds that 12 have grasped the topic. She teaches the second class of 25 students by the new technique, and 16 are found to have grasped the topic. Using $\alpha = 0.01$, is this evidence that the improvement is statistically significant?

9-7 To investigate the hypothesis that Roman Catholics and Orthodox Jews have different attitudes toward birth control, Foster gathered 2 random samples. Among 60 Catholics questioned, 42 opposed birth control, while among 60 Orthodox Jews, 29 opposed birth control. What is his conclusion if $\alpha = 0.05$?

9-8 The manager of Marty's Paint Mart wants to determine the effect advertising has on her sales. She has a paint which ordinarily sells for \$10.98. She decides to run an ad calling this paint a "special item" on sale for \$10.98, on Friday, Saturday, and Sunday only. To find out whether or not the advertisement is effective in increasing the percentage of sales of this item, she gathers 2 random samples from her customers. On the weekend prior to the ad, she found that 12 out of 100 randomly selected customers bought the paint. On the sale weekend, she found that 21 out of 110 randomly selected customers bought the paint. Decide at the 0.05 significance level if the ad was effective.

9-9 Peggy, a dietitian in a large university, suspects that the proportion of male college students who eat 3 meals a day is greater than the proportion of female college students who eat 3 meals a day. Two random samples are gathered. Of 500 men interviewed, 432 said they usually eat 3 meals a day, while of 500 women interviewed, 401 said they usually eat 3 meals a day. Determine if this is a statistically significant difference with $\alpha = 0.01$.

9-10 Oscar, a Ph.D. candidate, is doing a survey concerning the children's television program "Sesame Street." He wants to know if there is a difference between the proportion of suburban children who regularly watch the program and the proportion of inner-city children who regularly watch the program. In 2 random samples, he finds 24 out of 30 suburban children who watch it and 19 out of 25 inner-city children who watch it. Test with $\alpha = 0.05$.

9-11 Checking the records of deaths in a veterans' hospital, Luke found that of 50 nonsmokers, 6 had lung cancer, while of 60 smokers, 15 had lung cancer. Using a one-tail test, decide if this difference is statistically significant at the 0.05 level.

9-12 Dustin is attending his graduation party. Since the guests are primarily his parents' friends, he decides to while away the time by doing a two-sample binomial hypothesis test. He notices that there is a difference between men and women! He thinks that a greater proportion of men extinguish their matches by blowing them out. He finds that 7 of 13 men blew out their matches while only 6 of 14 women blew out their matches. Test if there is a difference at the 0.01 significance level. What are the populations?

9-13 Wally and Berta are arguing about which of them has more ESP. They split a pair of honest dice between themselves and alternately toss 1 die each. After Wally rolls his die and looks at the result, Berta attempts, without looking, to announce the correct result. Then Berta rolls her die and Wally tries to determine the result. After 60 rolls each, Wally was correct 20 of 60 times, while Berta was right 15 of 60 times.

(a) Does this indicate any difference in their ESP at the 0.05 significance level?

(b) Did Wally show any extraordinary ESP at the 0.01 significance level?

(c) Did Berta?

9-14 Prof. Laura Hardy is studying whether primitive man could have had webbed feet. In her work she examined the toes of 1000 schoolchildren. Of the boys examined, 45 of 500 had webbing between their second and third toes. Of the girls, 33 of 500 did. In some children, the webbing was present between all the toes, and Prof. Hardy ignored these cases, as this phenomenon is unknown among other primates. Do these figures indicate a difference between the percentage of boys and the percentage of girls with webbing between their second and third toes? Use $\alpha = 0.01$.

9-15 Recently, 1457 persons between 18 and 64 years of age were surveyed. Similarly, 2797 people over 65 were surveyed. A total of 50% of the younger group reported infirmities, yet only 23% of the senior citizens reported infirmities. Perform a one-tail hypothesis test on these data at the 0.01 significance level. What exactly are you testing?

9-16 An oil company has 2 methods of deciding where to drill for oil. Using method A, Dennis has been successful 10 of the last 30 tries. Using method B, Felicia has been successful 7 of the last 27 times. Does this indicate at $\alpha = 0.05$ that method A is superior?

5 9-17 Miss Simpson* runs a summer camp for the children of professors of statistics. Last year, she gathered the following data from the third graders:

**Last Year**

| | Boys | Girls |
|---|---|---|
| Like peanut butter and jelly | 250 | 80 |
| Do not like peanut butter and jelly | 350 | 120 |

Since $\hat{p}_1 - \hat{p}_2 = 250/600 - 80/200 = -0.017$, she thought that approximately 2% more third-grade girl campers liked peanut butter and jelly than did third-grade boy campers.

This year she decided to repeat her survey of third graders. Her results are as follows:

**This Year**

| | Boys | Girls |
|---|---|---|
| Like peanut butter and jelly | 20 | 50 |
| Do not like peanut butter and jelly | 130 | 200 |

This time $\hat{p}_1 - \hat{p}_2 = 20/150 - 50/250 = -0.067$, still indicating that a higher percentage of girls prefer peanut butter and jelly.

Sammy Smarts, one of her charges, suggests that she combine the data for both years, providing larger samples. Miss Simpson does so and gets the following:

**Combined Data for Both Years**

| | Boys | Girls |
|---|---|---|
| Like peanut butter and jelly | 270 | 130 |
| Do not like peanut butter and jelly | 480 | 320 |

This time $\hat{p}_1 - \hat{p}_2 = 270/750 - 130/450 = +0.071$, a greater percentage of boys(!) preferring peanut butter and jelly.

Wilma Wise, another camper, asks whether any of these results are *statistically significant.* We need to perform 3 hypothesis tests on the 3 sets of data to answer Wilma's query. Let the males in the class perform the first and third tests, while the females do the second and third. Use $\alpha = 0.05$.

Comment on your results.

*See Steven M. Day, "Simpson's Paradox and Major League Baseball's Hall of Fame," *The AMATYC Review,* vol. 15, no. 2, spring 1994, pp. 26–36.

9-18 In a recent National Mathematics Aptitude Examination, a sample of 500 adults (ages 26–35) and a sample of 500 teenagers (age 17) were given the following problem. A fruit punch is to be made with equal amounts of lemonade, limeade, orange juice, and ginger ale. You want to make 2 gallons of punch. How much ginger ale should you use? It was found that 30% of the teenagers got the correct answer and 36% of the adults got the correct answer. Does this support the idea that the older people are better at this type of question? Use $\alpha = 0.01$. (By the way, what is the correct answer?)

9-19 In a study of who was at fault in bicycle-automobile collisions, a survey revealed that cyclists 12 years old or younger were probably responsible for 92% of the accidents in which they were involved. The ratio dropped to 43% for cyclists 20–24 years old, and to 34% for those over 25.

(a) If the survey included 500 bicycle-auto accidents involving cyclists 12 or younger, 300 bicycle-automobile accidents involving 20- to 24-year-olds, and 400 accidents involving those over 25, test the hypothesis that the percentage of 20–24-year-olds who are responsible for collisions is greater than the percentage of over-25 cyclists who are responsible. Use the 0.05 significance level.

(b) In a certain city an intensive bike safety program was initiated in the elementary schools. After 1 year, the reports detailing responsibility for accidents of this type were summarized as follows:

| Age Group | Responsible for Accident |
|---|---|
| Under 12 | 240 of 300 |
| 20–24 | 40 of 100 |
| over 25 | 25 of 76 |

Do these results indicate that this kind of program reduces the responsibility for accidents in the under-12 age group? Use $\alpha = 0.05$.

9-20 A government-funded study of 2100 adults (ages 26–35) and 1700 teenagers (age 17), whose instruction was with the "new" math, showed that some consumers lose hundreds of dollars annually because they can't apply math to everyday purchases. Dr. Noah Progress, a member of the commission doing the survey, speculated that the nation has lost a generation of students who were taught how math works rather than how to use it. "They can't apply math to everyday problems," he said. Are his conclusions valid at the 0.01 significance level based on the evidence that only 40% of the 17-year-olds and 45% of the adults could calculate the lowest price per ounce of a box of rice?

9-21 A physician was studying the use of anticoagulants in treating acute heart attacks. She found that of 1104 patients who received anticoagulants, 8.3% died within 21 days of the attack. Of 1226 patients who did not receive this treatment, 27.3% died within 21 days. Show that this difference is significant at the 0.01 significance level.

9-22 In a random sample of the visiting records at the Sea View retirement home, George Stephens gathers the following data:

| | Patient Is Visited Regularly | Patient Is Not Visited Regularly | |
|---|---|---|---|
| Patient has grandchildren | 20 | 40 | 60 |
| Patient does not have grandchildren | 10 | 30 | 40 |
| | 30 | 70 | |

A typical conclusion often drawn is that since 20/30, or 2/3, of those who are regularly visited have grandchildren, someone who has grandchildren is more likely to be regularly visited. But this is false, because 20/60, or 1/3 (which is less than 1/2), of the grandparents receive visitors. We can consider 2 different experiments here:

(a) What is the probability that a person who is visited regularly is a grandparent? What is the probability that a person who is not visited regularly is a grandparent? Let population 1 be those patients who are visited regularly. Let population 2 be those who are not visited regularly. At $\alpha = 0.01$, do the data indicate that there is a different percentage of grandparents in population 1 than in population 2?

(b) What is the probability that a grandparent is visited regularly? What is the probability that a nongrandparent is visited regularly? Designate as population 1 the grandparents, and as population 2 the nongrandparents. At $\alpha = 0.01$, do the data indicate that there is a different percentage of patients who receive regular visits in population 1 than in population 2?

9-23 Donald Wahn was doing a survey on attitudes toward school bussing. He interviewed 400 students and found that 100 approved of school bussing. He interviewed 100 teachers, and 10 approved of school bussing. Show that this indicates a difference in attitude at the 0.01 significance level.

9-24 (a) Winny rolls a die. She considers rolling an odd number as winning, and an even number as losing. About what percentage of games should Winny win?

(b) Lucy draws a card from an ordinary 52-card deck. She considers drawing a heart as losing. About what percentage of games should Lucy lose?

(c) Winny rolls the die 20 times and records the percentage of times she wins as $\hat{p}_1$; Lucy draws 50 cards and records the percentage of times that she loses as $\hat{p}_2$. Drew subtracts $\hat{p}_1 - \hat{p}_2$ and records the differences as $d\hat{p}$. They repeat this procedure many times, obtaining a large distribution of $d\hat{p}$'s. Estimate the mean of this distribution.

(d) Normally, Drew draws a graph of the $d\hat{p}$'s. Explain why the graph that Drew drew should be approximately a normal curve.

(e) Since $p_1 \neq p_2$ the standard deviation of the $d\hat{p}$'s is found by a different formula which does not involve pooling:

$$\sigma_{d\hat{p}} = \sqrt{\frac{p_1 q_1}{n_1} + \frac{p_2 q_2}{n_2}} = \sqrt{\frac{0.50(0.50)}{20} + \frac{0.25(0.75)}{50}} = \sqrt{0.01625} = 0.13$$

Let $D$ be the number that separates the bottom 95% of Drew's outcomes from the top 5%. Find $D$.

9-25 Instead of testing a hypothesis that $p_1 = p_2$, we can test that $p_1$ is 20% more than $p_2$. $\mu_{d\hat{p}}$ is still $p_1 - p_2$, but we do *not* pool the experimental results (why?), and

$$\hat{\sigma}_{d\hat{p}} = \sqrt{\frac{\hat{p}_1\hat{q}_1}{n_1} + \frac{\hat{p}_2\hat{q}_2}{n_2}}$$

Using this idea, test the hypothesis that 20% more young females smoke than young males do. Use the data that of 200 young females interviewed, 120 smoked, and of 400 young males interviewed, 150 smoked. Let $\alpha = 0.05$.

## CLASS SURVEY QUESTION

Assuming that the breakdown of hair color in your class is representative of hair color in the whole school, perform the following experiment. Test whether the percentage of brown-haired females in the school is the same as the percentage of brown-haired males.

## FIELD PROJECTS

Review the general instructions for field projects given at the end of Chapter 8, and then select one of the following:

1. Perform an experiment similar to Exercise 9-2.
2. Perform an experiment to decide whether or not there is a significant difference between the percentage of males who are left-handed and the percentage of females who are left-handed.
3. Design and carry out a two-sample binomial test of your own choosing. (*Hints:* Differences between age groups, political groups, sexes, religions, races, etc.; on fashions; politics; preferences for food, literature, music, etc.)

### *Experimental Design and Research Methods*

Sir Ronald Fisher (1890–1962) was the most influential statistician of this century. His pioneering work was rooted in the fields of agriculture and genetics. In 1925 he published *Statistical Methods for Research Workers,* and in 1935, *Design of Experiments.*

## EXAMPLE OF A TWO-SAMPLE BINOMIAL FIELD PROJECT

***Part I, Proposal***

Is there any difference between the percentage of Washington College women who attend religious services regularly and the percentage of local non-college women who attend religious services regularly?

I will ask 30 women at Washington College the following questions:

1. I am doing a survey for my college statistics class. Would you answer two questions for me with a yes or no answer?
2. Are you a student at Washington College?
3. Do you attend religious services at least twice a month?

I will continue this survey until I get 30 women who answer yes to questions 1 and 2 and either yes or no to question 3.

I will ask 30 women at the Eilís' Garden shopping center the following questions.

1. I am doing a survey for my college statistics class. Would you answer two questions for me with a yes or no answer?
2. Did you attend college?
3. Do you attend religious services at least twice a month?

I will continue this survey until I get 30 women who answer yes to question 1, no to question 2, and either yes or no to question 3. I can then formulate the following hypotheses:

Let $p_1 = P$(a Washington College woman attends religious services regularly) and let $p_2 = P$(a non-college woman attends religious services regularly).

$H_a$: The percentage of Washington College women who attend religious services is different from the percentage of local non-college women; $p_1 \neq p_2$ or $p_1 - p_2 \neq 0$ (a two-tailed test).

$H_0$: The two percentages are the same; $p_1 = p_2$ or $p_1 - p_2 = 0$.

***Part II, Report***

On Tuesday, April 27, from 3:15 to 4:30 P.M. I questioned 47 women in the Washington College cafeteria. Of these, 4 would not answer question 1, 3 were not students, 8 would not answer question 3, and 2 answers were vague, neither yes nor no. Of the 30 who answered, 13 said yes. Therefore, $\hat{p}_1 = 13/30 = 0.43$ and $n_1 = 30$.

On Wednesday, April 28, from 1:20 to 3:05 P.M. I questioned 56 women at Roosevelt Field. Of these, 10 would not answer question 1, and 16 had attended college. Of the 30 who responded, 15 said yes. Therefore, $n_2 = 30$ and $\hat{p}_2 = 15/30 = 0.50$. Since the 4 sample outcomes ($n_1\hat{p}_1$, $n_1\hat{q}_1$, $n_2\hat{p}_2$, and $n_2\hat{q}_2$) equal 13, 17, 15, and 15, and are all larger than 5, we can use the normal distribution. $\alpha = 0.05$. This is a two-tail test. Therefore, $z_c = \pm 1.96$.

$$\hat{p} = \frac{13 + 15}{30 + 30} = \frac{28}{60} = 0.47$$

$$\hat{\sigma}_{d\hat{p}} = \sqrt{\hat{p}\hat{q}\left(\frac{1}{n_1} + \frac{1}{n_2}\right)} = \sqrt{0.47(0.53)\left(\frac{1}{30} + \frac{1}{30}\right)} = \sqrt{0.166} \approx 0.13$$

This distribution is represented in Figure 9-5.

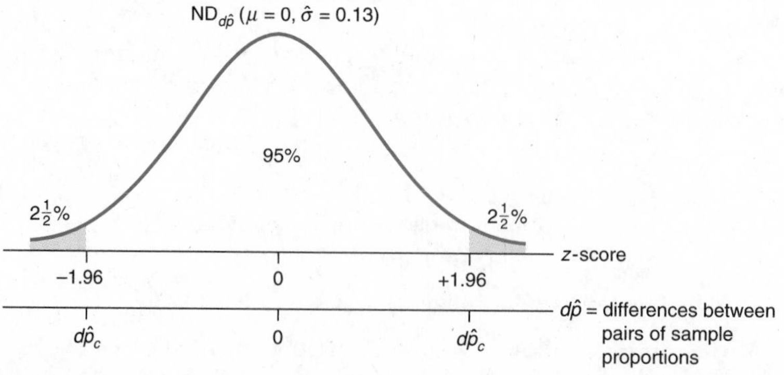


Figure 9-5

The critical difference is then

$$d\hat{p}_c = \mu_{d\hat{p}} + z_c \hat{\sigma}_{d\hat{p}}$$

$$= 0 + (\pm 1.96)(0.13) = \pm 0.25$$

My decision rule will be to reject $H_0$ if the sample difference is less than $-0.25$ or greater than 0.25.

The sample difference $d\hat{p} = \hat{p}_1 - \hat{p}_2 = 0.43 - 0.50 = -0.07$. Therefore, I have failed to reject the null hypothesis; that is, I have failed to prove that there is any significant difference (at the 0.05 significance level) between the attendance at religious services of Washington College women and local non-college women. I feel that it is unlikely that those who did not respond would have changed the outcome significantly.

# 10

# Hypothesis Testing with Sample Means: Large Samples

The average family has 2.7 children.

**CONTENTS**



**GOALS**

At the end of this chapter you will have learned:

- What statisticians consider a large sample when dealing with averages
- The central limit theorem concerning the sampling distribution of means taken from many equal-sized large samples
- How to perform both one- and two-sample hypothesis tests of means

# 10.1 Sample Means

A young man being offered a job as a secretary in a large company asked the personnel director what the average age of the secretaries in the company was. She replied that the company employed several hundred secretaries, and she did not know the correct answer. But, looking around the personnel office at the 38 secretaries there, she said that the average age in her office was about 20 and that the secretaries had been selected at random from the secretarial pool. The young man figured that the average for the whole company could not be much different from that, and so he agreed to work for the firm.

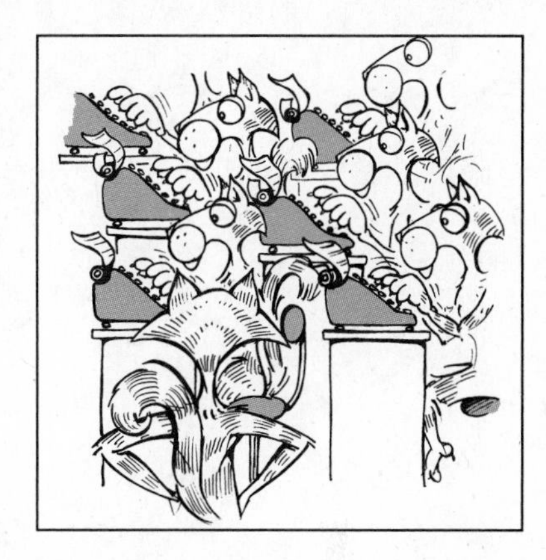

The young man has informally performed a type of hypothesis test which is very commonly done by statisticians. He came to a conclusion about a **population mean**, denoted by $\mu$, on the basis of what he knew about the mean of a sample taken from that population. Such a mean is referred to briefly as a **sample mean** and denoted $m$. In this chapter we will explain formally how to conduct hypothesis tests with sample means, but the basic ideas are exactly the same as those of the hypothesis-testing procedures you already know. Precisely the same steps are carried out. The only differences will be these:

**population mean**

**sample mean**

1. The hypotheses will be statements about $\mu$ instead of $p$.
2. The formulas for computing the mean and the standard deviation of the normal curve will be changed.

## 10.1-1 Theoretical Distribution of Sample Means

Imagine that the young man of our example goes from office to office and, for each office, separately computes the mean age of the secretaries. By the end of the day he would have a long list of sample means, and these averages would certainly vary. Such a list of averages is called a **distribution of sample means**.

**distribution of sample means**

Theoretically, you can imagine that the ages of all the secretaries in the company could be written down, one per slip of paper, and put into a giant drum from which they could be picked at random. Suppose the young man picked 38 slips at a time, computed the mean, wrote it down, and returned the slips to the drum to be remixed. You can see that he could do this hour after hour, getting a longer and longer distribution of sample means. A typical distribution of sample means could be listed as shown in Table 10-1.

**Table 10-1 Distribution of Sample Means**

| Sample Number (Each Sample Composed of 38 Ages) | Sample Mean $m$ (Average Age of the 38 Secretaries Sampled) |
|---|---|
| 1 | 26 |
| 2 | 22 |
| 3 | 19 |
| 4 | 21 |
| 5 | 29 |
| 6 | 20 |
| 7 | 20 |
| 8 | 19 |
| 9 | 18 |
| 10 | 26 |
| 11 | 21 |
| 12 | 20 |
| 13 | 20 |
| 14 | 19 |
| 15 | 22 |
| 16 | 21 |
| 17 | 20 |
| . | . |
| . | . |
| . | . |

The mean of the distribution of sample means is denoted by

$\mu_m$ or $\mu_{\bar{X}}$ (read: mu sub $m$ or mu sub $X$ bar)

The standard deviation of the distribution of sample means is denoted by

$\sigma_m$ or $\sigma_{\bar{X}}$ (read: sigma sub $m$ or sigma sub $X$ bar)

**central limit theorem**

Mathematicians have analyzed this type of distribution and have learned some useful facts about what would appear after many, many sample means have been computed. The situation is summarized by the **central limit theorem**.

### *Central, Main, Key, . . .*

In 1810 Pierre Simon Laplace (1749–1827), combining some of the ideas of Carl Friedrich Gauss (1777–1855) with his own work, developed the central limit theorem, (where "central" has the meaning of "main" or "key," not "middle").

## 10.1-2 Central Limit Theorem

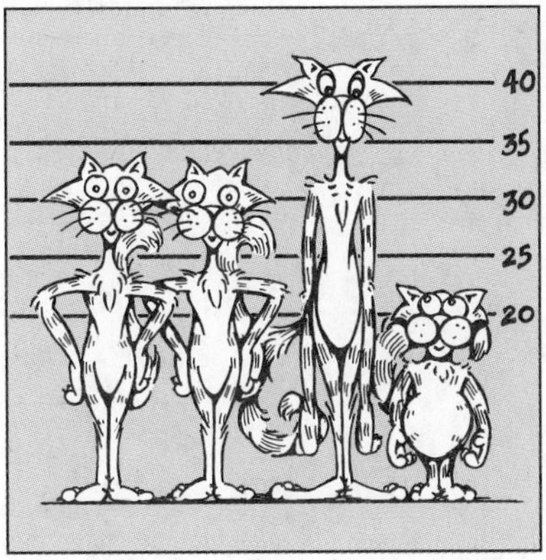

Under most circumstances the sampling distribution of the means of all possible **large, equal-sized** random samples taken from a population with mean $\mu$ and standard deviation $\sigma$ has, *in theory,* three characteristics:

1. The shape of the graph of the means is approximately normal.
2. The mean of the means $\mu_m$ is the same as the mean of the original population $\mu$ that is

$$\mu_m = \mu$$

Large versus small.

3. The standard deviation of the means is smaller than the standard deviation of the original population. How much smaller depends on the sample size

$$\sigma_m = \frac{\sigma}{\sqrt{n}}$$

How close the sampling distribution is to normal in any given application depends mainly on two factors:

1. The size $n$ of each of the many samples. In theory, the larger the sample size, the more closely the distribution is to normal in shape. For most applications, a sample in which $n$ is larger than 30 will lead to a distribution of means close enough to normal so that calculations based on the normal curve table will be reasonable. Informally, samples of more than 30 values are referred to as **large samples**.

   **large samples**

2. The "shape" of the original population. The closer the original population is to normal, the smaller the samples we can use and still expect our sample means to have approximately a normal distribution. In many statistical applications the populations are such that you can assume that the theorem holds. We will assume that there is no problem on this account.

This theorem is really quite useful because it gives the properties of distributions of sample means. This in turn allows a statistician to estimate how far the population mean is likely to be from the experimental sample mean. And since, in many situations, the main question concerns the population mean, the theorem allows you to "connect" your sample evidence to the question of interest.

For example, let us see how the theorem can help with the secretary problem. We can indicate in a sketch (see Figure 10-1) the relation between the distribution of the ages of the individual secretaries and the distribution of sample means as given by the central limit theorem.

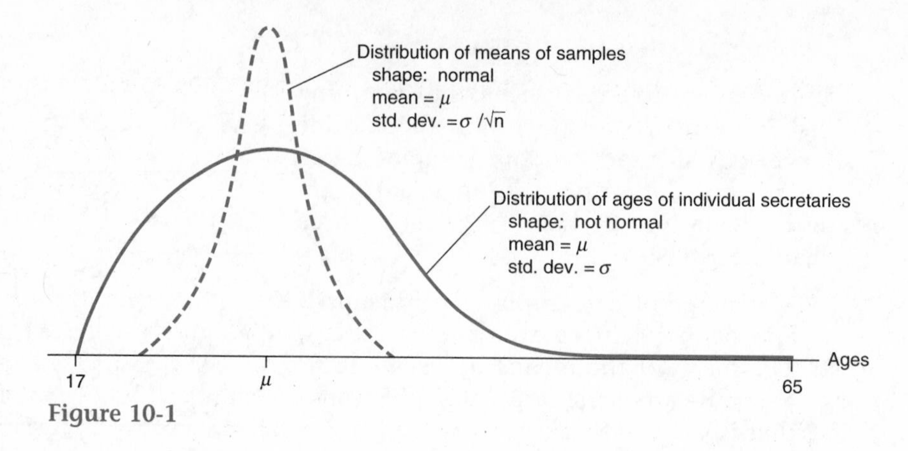


**Figure 10-1**

We would know, for example, that most of the sample means cluster near the population mean. More precisely, since the distribution of sample means is normal, we know that only 5% of the sample means are more than 1.96 standard deviations away from the population mean. Another way of looking at this is as follows. If we select one sample at random, the probability that the mean of this sample will be more than 1.96 standard deviations from the population mean is less than 0.05.

This information can be very helpful. For example, suppose the claim were made that the mean age of secretaries in the company was 20, but when the young man took his sample of 38 secretaries, he found that his sample mean was 47. He would think in this situation, "This sample mean is so far away from the supposed true mean that the probability of my having obtained this result is extremely small. Either a fantastically unlikely event has happened, or somebody lied to me about the population mean." He would make his job decision according to how he felt about these two alternatives. In formal hypothesis tests, when the sample mean is so far from the claimed population mean that the probability of its occurrence is less than some specified small number $\alpha$, statisticians do not decide that they have just witnessed a fantastically unlikely event. They reason that what they witnessed was likely and, on this basis, reject the claim about the population mean.

## ❑ EXAMPLE 10.1 Lts.' Wts.

Suppose we take as our population the weights of all U.S. Army lieutenants. Also, suppose it happens to be true that the mean weight $\mu$ equals 159 pounds, and the standard deviation $\sigma$ equals 24 pounds. Describe the theoretical distribution of sample means that you would get by taking many, many random samples of size 36.

SOLUTION

(a) The distribution of sample means would be approximately normal, since $n = 36$ is larger than 30.

(b) $\mu_m = \mu = 159$ pounds

(c) $\sigma_m = \dfrac{\sigma}{\sqrt{n}} = \dfrac{24}{\sqrt{36}} = \dfrac{24}{6} = 4$ pounds

Recall that the standard deviation measures the variability or spread of a distribution. In this example, the standard deviation of the population $\sigma$ reflects the variability among individual weights. These individual weights might range from 120 to 200. By contrast, the standard deviation of the sample means $\sigma_m$ reflects the variability among the *means* of samples of 36 weights. Even though it would not be unusual to find an *individual* lieutenant who weighed 120 pounds, it would be quite unlikely to obtain a random sample of 36 lieutenants with a *mean* of 120 pounds. That is because in each sample there will probably be both light and heavy people whose weights will tend to balance one another. These sample means vary much less than the individual weights, and hence $\sigma_m$ is smaller than $\sigma$. This distribution of sample means is represented in Figure 10-2.

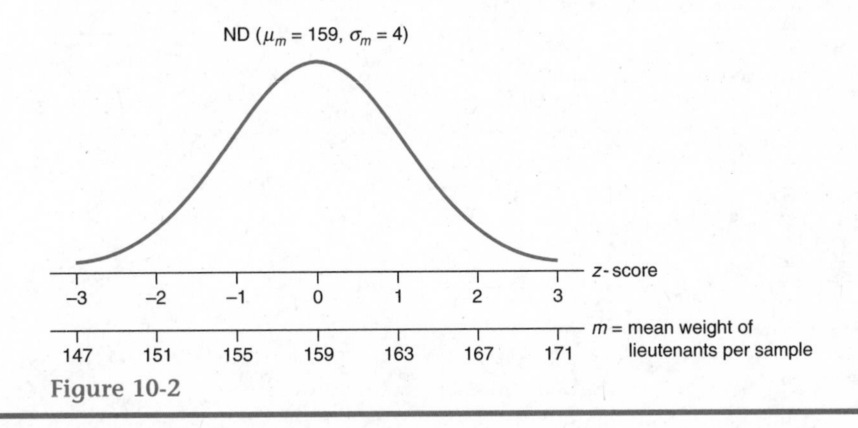


Figure 10-2

## 10.1-3 Estimating the Standard Deviation of a Distribution of Sample Means

In hypothesis tests based on the central limit theorem, it is necessary to know the standard deviation of the population in order to calculate the standard deviation of the distribution of sample means. Recall that

$$\sigma_m = \frac{\sigma}{\sqrt{n}}$$

Often, however, statisticians wish to use the theorem when they do not know $\sigma$. They must then do two things:

1. Find an estimate for $\sigma$. You will recall that we denote this by $s$.
2. Use $s$ to get an estimate for $\sigma_m$ denoted by $s_m$. That is,

$$s_m = \frac{s}{\sqrt{n}}$$

---

### ❑ EXAMPLE 10.2 Family Garbage, a One-Sample Hypothesis Test of a Mean

A newspaper states that a family in Alton, Rhode Island, on average, produces 5.2 pounds of organic garbage per day. A public health officer feels that the figure is probably incorrect. To test this, an experiment is set up to be analyzed at the 0.05 significance level. Forty families are chosen at random, and their organic garbage for 1 day is weighed. The data are shown in Table 10-2. Carry out the health officer's test of the newspaper's claim.

**Table 10-2 Results of Garbage-Weighing Experiment for 1 Random Sample of 40 Families**

| Family Number | Pounds of Garbage $X$ | Family Number | Pounds of Garbage $X$ |
|---|---|---|---|
| 1 | 2.6 | 21 | 4.3 |
| 2 | 4.8 | 22 | 4.2 |
| 3 | 5.0 | 23 | 4.1 |
| 4 | 7.3 | 24 | 4.0 |
| 5 | 2.2 | 25 | 3.6 |
| 6 | 3.4 | 26 | 3.8 |
| 7 | 4.6 | 27 | 7.0 |
| 8 | 5.8 | 28 | 6.2 |
| 9 | 5.0 | 29 | 5.5 |
| 10 | 4.0 | 30 | 4.3 |
| 11 | 3.1 | 31 | 4.2 |
| 12 | 2.2 | 32 | 3.2 |
| 13 | 5.1 | 33 | 2.7 |
| 14 | 4.7 | 34 | 4.0 |
| 15 | 4.8 | 35 | 4.0 |
| 16 | 3.0 | 36 | 3.2 |
| 17 | 7.3 | 37 | 4.1 |
| 18 | 7.1 | 38 | 4.0 |
| 19 | 6.2 | 39 | 4.2 |
| 20 | 6.0 | 40 | 5.5 |
| | | $n = 40$ | $\Sigma X = 180.3$ |

**SOLUTION**

1. The procedure he will use is a one-sample, large-sample hypothesis test of a mean.
2. The population being studied is families in Alton.
3. $H_a$: The claim made about the organic garbage produced by families in Alton is false; $\mu \neq 5.2$ pounds (a two-tail test).
4. $H_0$: The claim made about the organic garbage produced by families in Alton is true; $\mu = 5.2$ pounds.
5. $\alpha = 0.05$.
6. $n = 40 > 30$. Therefore the distribution of sample means will be approximately normal.
7. The data gathered was recorded in the table above.
8. Under the assumption of the null hypothesis, we have

$$\mu_m = \mu = 5.2 \qquad \text{and} \qquad \sigma_m = \frac{\sigma}{\sqrt{n}}$$

Since $\sigma$ is unknown, he must compute $s$. Recall the formula we used in Chapter 2 was $s = \sqrt{\dfrac{\Sigma(X^2) - \dfrac{(\Sigma X)^2}{n}}{n - 1}}$

From the data, we compute

$$n = 40$$

$$\Sigma X = 180.3$$

$$(\Sigma X)^2 = 32{,}508.09$$

$$\Sigma(X^2) = 883.65$$

$$s = \sqrt{\frac{\Sigma(X^2) - \dfrac{(\Sigma X)^2}{n}}{n - 1}}$$

$$= \sqrt{\frac{883.65 - \dfrac{32{,}508.09}{40}}{40 - 1}} = \sqrt{\frac{883.65 - 812.70}{39}}$$

$$= \sqrt{\frac{70.95}{39}} = \sqrt{1.82} \approx 1.35$$

We now use $s$, the estimate of $\sigma$, to calculate $s_m$, the estimate for $\sigma_m$. Therefore,

$$s_m = \frac{s}{\sqrt{n}} = \frac{1.35}{\sqrt{40}} = \frac{1.35}{6.32} = 0.21$$

The distribution of sample means is illustrated in Figure 10-3.

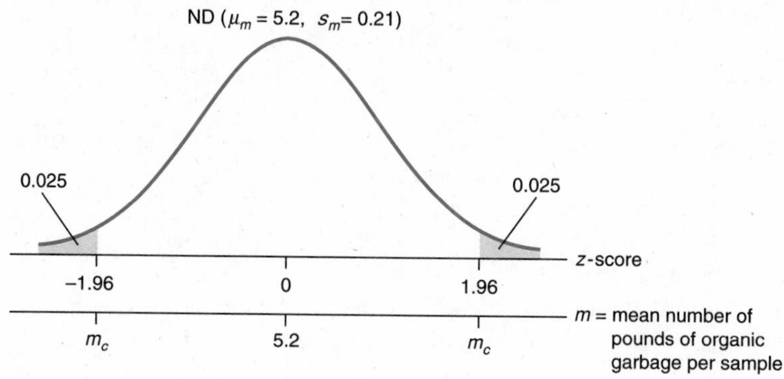


Figure 10-3

Using $z_c = \pm 1.96$, we find

$$\begin{aligned} m_c &= \mu_m + z_c s_m \\ &= 5.2 + (\pm 1.96)(0.21) \\ &= 5.2 \pm 0.41 = 4.8 \text{ and } 5.6 \end{aligned}$$

9a. *Decision rule:* A sample mean outside the range 4.8 to 5.6 will lead to rejection of the null hypothesis that the mean of the population is 5.2 pounds of garbage.

9b. *Outcome of experiment.* The sample mean is

$$m = \frac{\Sigma X}{n} = \frac{180.3}{40} = 4.5 \text{ pounds of garbage.}$$

9c. *Conclusion.* The sample outcome, $m = 4.5$, is outside the range 4.8 to 5.6. Based on this evidence, we reject the null hypothesis. The health officer was correct: The average amount of organic garbage in the population is not 5.2 pounds. Evidently, it is less. ❑

## Exercises

10-1 In the class survey you took in Chapter 1 you found the ages of this class. If you were to find the mean, the range, and the standard deviation of these ages (you may have done this in Chapter 2, but you can answer this question whether or not you know the numerical values), how would they compare with the mean, range, and standard deviation of the ages of the entire school? (*Hint:* Do you think that this class contains the oldest student on campus? The youngest?)

10-2 A random sample of one-family dwellings was taken in each of the following cities. The mean number of bathrooms per dwelling was calculated. The results are given below. Calculate the mean of these means.

| | | | |
|---|---|---|---|
| Bellaire | 3.5 | Belle Rive | 4.4 |
| Bellamy | 2.0 | Belle River | 1.6 |
| Bella Union | 4.1 | Belleriver Station | 3.4 |
| Bella Vista | 1.9 | Belle Rose | 2.8 |
| Bellbrook | 2.2 | Bellerose | 1.7 |
| Bell Buckle | 2.4 | Bellevalley | 2.6 |
| Bellburn | 3.6 | Belleview | 3.5 |
| Bell City | 2.6 | Belle View | 2.4 |
| Belle | 3.7 | Bellevue | 5.0 |
| Belle Center | 1.8 | Bellview | 1.8 |
| Belle Creek | 4.0 | Bellvue | 2.2 |
| Belledune | 1.7 | Belview | 2.0 |
| Bellefleure | 2.8 | Belvue | 1.9 |
| Bellflower | 2.1 | Bellmont | 2.0 |
| Bellfontaine | 3.8 | Bellmore | 3.5 |
| Bellefonte | 3.9 | Bellranch | 2.9 |
| Belle Glade | 3.0 | Belle Ville | 4.3 |
| Belle Haven | 3.1 | Belleville | 1.8 |
| Belle Isle | 1.6 | Bellville | 2.7 |
| Belle Mead | 1.5 | Bellwood | 2.5 |
| Belle Meade | 2.3 | Belwood | 1.9 |
| Bellemead | 3.2 | Belmont | 2.3 |
| Bellmead | 2.5 | Belton | 0.8 |
| Belleplains | 2.7 | Beltrami | 3.6 |
| Belle-Plaine | 3.3 | Belvedere | 4.2 |
| Belle Plains | 2.9 | Belvidere | 1.9 |

10-3 A commuter buys peanuts from a vending machine each evening on his way home from work. On the last 40 purchases he received the following numbers of peanuts per purchase: 12, 10, 0, 5, 15, 16, 20, 3, 12, 0, 12, 10, 9, 11, 8, 13, 15, 20, 18, 19, 20, 0, 14, 13, 15, 16, 15, 19, 11, 10, 10, 10, 3, 8, 2, 0, 0, 20, 12, and 12.

Considering these 40 purchases as 1 sample of size 40 from the population:

(a) What is the population being sampled?
(b) Compute $m$, the mean of the sample.
(c) Estimate $\mu$, the mean of the population.
(d) Compute $s$ to estimate the standard deviation of the population.
(e) Using the central limit theorem, describe the sampling distribution of sample means.

(f) Estimate $\mu_m$, the mean of the distribution of sample means.
(g) Find $s_m$, the estimate of the standard deviation of the distribution of sample means.

10-4 A random sample of the contributions of physicians to the United Fund were taken; 50 doctors were sampled. The results, in dollars, were as follows: 100, 95, 92, 92, 91, 90, 86, 85, 81, 80, 76, 76, 73, 73, 70, 70, 69, 69, 67, 66, 65, 61, 57, 52, 50, 49, 48, 47, 45, 39, 35, 35, 35, 35, 35, 30, 30, 30, 25, 25, 20, 20, 15, 15, 10, 10, 9, 5, 5, and 0. Answer parts (a) to (g) from the previous exercise.

10-5 A study was done in Europe recently to investigate the health hazards of working long hours in front of computer or word processor video displays. It found that it took an average of 2.6 hours before a certain symptom of eye strain developed. If a similar experiment in the United States using $n = 100$ showed $m = 2.8$ hours, with $s = 0.5$ hours, would this indicate that the American results are in conflict with the European results? $\alpha = 0.05$. Would you think that the difference between 2.8 and 2.6 hours is important in any practical way?

10-6 Brad Brandt bands brants. As a government ecologist, he has put leg bands on thousands of brants in order to study their migratory habits. He bands, on the average, 50 brants per week. This distribution of the number of brants banded is approximately normal with a standard deviation of 7 brants banded. His supervisor periodically checks up on him and the number of brants banded.

(a) What is the probability that the supervisor randomly picks a week in which Brad has banded less than 40 birds? Remember, the number of bandings is a whole number.
(b) If the supervisor randomly picks 36 weeks of the last 3 years' work, what is the probability that she gets an average of less than 45 birds banded per week? *Hint:* Remember, we treat averages as continuous data.

10-7 Ms. Kupp, the owner of Mae's cosmetic firm, commissioned a study of her wholesale customers' buying habits. Among the results, Mae learned that her customers bought, on the average, 50 tubes of Passion Flower lipstick per month. This distribution was approximately normally distributed with a standard deviation of 7 tubes of lipstick.
(a) What is the probability that a customer picked at random bought 40 or more tubes of Passion Flower? Remember, the number of tubes purchased is a whole number.
(b) If a random sample of 36 customers is taken, what is the probability that the sample average is more than 45 tubes of Passion Flower? Remember, averages are to be treated as continuous data.

10-8 A college admissions officer believes that this year's freshman class is superior in math aptitude to previous freshman classes. The mean score on the math aptitude test for previous classes was 470 with $\sigma = 120$.

(a) Assuming that $\sigma$ is the same for this year's class, if you test a random sample of 400 freshman at the 0.01 significance level, what is the critical aptitude test score to support the officer's belief?

(b) If the sample of this year's freshmen has a mean score $m = 490$, would this support the officer's belief?

10-9 A newspaper article states that college students at a large state university campus spend on the average $56 a year on illegal drugs. A student investigator wishes to test this hypothesis at the 0.05 significance level. He gets a random sample of 144 students and finds that $m = \$70$ and $s = \$54$ for this sample.

(a) What are the critical amounts of money for the test?

(b) Does the evidence support the claim of the newspaper article?

10-10 In an inspection of a chemical plant, the inspector, Foster Tuma, takes a specified quantity of a particular chemical, puts it on a microscope slide, and counts the number of microorganisms in it. The mean number of microorganisms in this quantity of chemical is supposed to be 1000. In an inspection 1 day, Foster finds for a random sample of 36 slides that $m = 1030$ and $s = 180$. If the mean truly is 1000, find the probability that Mr. Tuma would get a sample mean of 1030 or more.

10-11 A claim is published that in a certain area of high unemployment, $195 is the average amount spent on food per week by a family of four. A home economist wants to test this claim against the suspicion that the true average is lower than $195. She surveys a random sample of 36 families from the locality and finds $m = 193.20$ and $s = \$4.80$. Testing with $\alpha = 0.05$, what should be her conclusion?

10-12 The mean grade point average (GPA) of graduating college seniors who have been admitted to graduate school is 3.1, where an A is given 4 points. At Ivy University a random sample of 36 incoming graduate students yielded a GPA of 3.2 with a standard deviation of 0.24. Can we claim that the students going to Ivy have better grades than the national average, using the 0.01 significance level?

10-13 A normal distribution of diameters of marbles has a mean equal to 0.52 inches and a standard deviation equal to 0.04 inches. If we took a random sample of 50 marbles, what is the probability that our sample mean will lie between 0.53 and 0.55 inches?

10-14 Genito Sean DiIbn, an employee of Roberto's Rose Field, Inc., having been pricked once too often, is investigating the number of thorns per inch on the rosebushes he tends. Here are the data he gathered:

| | | | | | | | | | | | | |
|---|---|---|---|---|---|---|---|---|---|---|---|---|
| 9 | 12 | 7 | 7 | 8 | 9 | 11 | 10 | 6 | 9 | 13 | 7 | 8 |
| 12 | 2 | 8 | 10 | 7 | 13 | 6 | 5 | 8 | 10 | 6 | 13 | 6 |
| 11 | 6 | 9 | 0 | 7 | 4 | 9 | 8 | 7 | 7 | 9 | 3 | 6 |
| 9 | | | | | | | | | | | | |

Is it likely that these data come from a population with $\mu = 9$? Test with $\alpha = 0.05$.

10-15 Leda uses swan eggs to create mammoth omelets. Lately the eggs seem to be getting smaller. Her supplier insists that the average weight has remained at 4.1 ounces. Leda weighs the next 3 dozen eggs she receives and acquires these data:

| 3.9 | 4.2 | 3.7 | 3.7 | 3.8 | 3.9 | 4.1 | 4.0 | 3.6 | 3.9 | 4.3 |
|---|---|---|---|---|---|---|---|---|---|---|
| 3.7 | 3.8 | 4.1 | 3.2 | 3.8 | 4.0 | 3.7 | 4.3 | 3.6 | 3.5 | 3.8 |
| 4.0 | 3.6 | 4.3 | 3.6 | 4.2 | 3.6 | 3.9 | 3.0 | 3.7 | 3.4 | 3.9 |
| 3.8 | 3.7 | 3.7 | | | | | | | | |

Do these data support the supplier's contention or do they support Leda's suspicions? Test with $\alpha = 0.01$.

10-16 Professors Chiao and Gonzalez are trying to simulate the sampling distribution of the means of reading scores of fourth graders in the United States. From an article in a magazine they obtain the averages of many, many fourth grade classes and draw a graph of this distribution of sample means. It appears bell shaped. They average their data and find the mean of the sample means, and they use the formula

$$s_m = \sqrt{\frac{\Sigma m^2 - \frac{(\Sigma m)^2}{n}}{n - 1}}$$

to calculate the standard deviation. One of their students asks, "Why didn't you use the easy formula ($s_m = \frac{\sigma}{\sqrt{n}}$) from the central limit theorem to find the standard deviation instead of doing all those calculations?" How would you respond?

## 10.2 Two-Sample Tests of Means—Large Samples

Very often a statistician wishes to compare the means of two populations. One way to do this is to examine the **difference between the means of samples** taken from each of the populations.

### ❑ EXAMPLE 10.3 Sex Discrimination?

A spokeswoman for a women's group wishes to present evidence to support the claim that in their first year of employment, male scientists in industry are paid more than female scientists doing the same work. She gathers data from two random samples, as shown in the following table.

| | Male | Female |
|---|---|---|
| Sample size $n$ | 100 | 86 |
| Sample mean $m$ | \$37,400 | \$36,300 |
| Estimate of population standard deviation based on sample data $s$ | \$1200 | \$1000 |

It is clear that for the 186 scientists sampled the men are better paid on the average. The difference between the two sample means is \$37,400 − \$36,300 = \$1100. The question for the statistician is: Is this a statistically significant difference? Should we infer from it that the mean of the entire population of first-year male scientists is higher than the mean of the entire population of first-year female scientists? Or is it possible that the means of the two populations are equal, and this outcome happened just by chance—just because of the 186 people we happened to select?

When we are faced with the question "Is the difference between two observed sample means statistically significant?", then the situation has to be analyzed in a manner similar to the one used in Chapter 9, for testing the difference between two-sample proportions. In theory, we can think of the sampling of male and female scientists being repeated many, many times with pairs of samples of 100 men and 86 women. If we actually did this, our results might look like those listed in Table 10-3. Just as we used *dp* to represent the difference between two proportions, so now we use **dm** to represent the **d**ifference between the two **m**eans.

**Table 10-3**

| Results of | Mean Salary for Men $m_1$ | Mean Salary for Women $m_2$ | Differences of Means dm $= m_1 - m_2$ |
|---|---|---|---|
| First sampling | \$37,400 | \$36,300 | \$1100 |
| Second sampling | 37,200 | 36,700 | 500 |
| Third sampling | 37,100 | 37,300 | −200 |
| Fourth sampling | 37,104 | 37,102 | 2 |
| . | . | . | . |
| . | . | . | . |
| . | . | . | . |

When both samples are large—that is, both $n_1$ and $n_2$ are bigger than 30—then the distribution of numbers in the third column, the **differences of sample means**, is approximately normal, and the mean of these sample differences is equal to the true difference $\mu_1 - \mu_2$. Notice that if the men actually had the larger salaries, then the differences would tend to be positive, and so

**differences of sample means**

$\mu_{dm}$ would be greater than zero. If the women actually had the larger salaries, then the differences would tend to be negative, so that $\mu_{dm}$ would be less than zero. If there were really no differences between the male and female salaries, then some differences would be positive and some negative, and they would tend to cancel each other out, so that $\mu_{dm}$ would be zero. (Our interpretation of $\mu_{dm}$ being positive or negative depended upon which population was arbitrarily called population 1.) We summarize these ideas with the following theorem:

**Theorem about the Sampling Distribution of the Difference of the Means of Two Large Samples**

SUPPOSE

1. There are two populations, call them population 1 and population 2.
2. A large random sample of size $n_1$ is picked from population 1, and a separate and independent large random sample of size $n_2$ is picked from population 2.*
3. The mean of each sample is computed.
4. The difference between these two means is written down.
5. Steps 2, 3, and 4 are repeated many times (in theory—an infinite number of times), giving a long list of differences. This list of differences is called the **sampling distribution of the differences of means** (denoted by "dm").

**sampling distribution of differences of means**

CONCLUSION

The following statements are true about the distribution of sample means:

1. The distribution of differences is approximately normal.
2. The mean of the differences $\mu_{dm} = \mu_1 - \mu_2$. If we are assuming for our null hypothesis that $\mu_1 = \mu_2$, then $\mu_{dm} = 0$.
3. The standard deviation of the differences $\sigma_{dm}$ is given by the following formula:

$$\sigma_{dm} = \sqrt{\frac{\sigma_1^2}{n_1} + \frac{\sigma_2^2}{n_2}}$$

Usually the experimenter does not know $\sigma_1$ and $\sigma_2$, in which case one can estimate these values by $s_1$ and $s_2$. Using $s_1$ and $s_2$, we get an estimate for $\sigma_{dm}$:

$$s_{dm} = \sqrt{\frac{s_1^2}{n_1} + \frac{s_2^2}{n_2}}$$

*For good results, $n_1$ and $n_2$ should each be larger than 30. In general, the larger the samples, the closer the distribution of differences is to normal.

## Application of the Theorem

Let us apply the theorem to the problem of whether male scientists are paid more than female scientists during their first year of employment in industry. Recall the given information.

| | Male | Female |
|---|---|---|
| Sample size $n$ | 100 | 86 |
| Sample mean $m$ | \$37,400 | \$36,300 |
| Estimate of population standard deviation based on sample data $s$ | \$1200 | \$1000 |

### SOLUTION

Using the data in the table, we conduct a hypothesis test at the 0.05 significance level, calling the males population 1 and the females population 2.

1. The procedure we will use is a two-sample, large-sample hypothesis test of means.
2. The two populations being tested are the salaries of first-year male and female scientists in industry.
3. $H_a$: The mean salary for male scientists is larger than the mean salary for female scientists; $\mu_1 > \mu_2$ (a one-tail test).
4. $H_0$: The mean salaries are equal; $\mu_1 = \mu_2$.
5. $\alpha = 0.05$.
6. $n_1 = 100 > 30$ and $n_2 = 86 > 30$. Therefore the distribution of differences is approximately normal.
7. The data gathered were recorded in the table above.
8. Under the assumption of the null hypothesis, we have

$$\mu_{dm} = \mu_1 - \mu_2 = 0 \quad \text{and} \quad \sigma_{dm} = \sqrt{\frac{\sigma_1^2}{n_1} + \frac{\sigma_2^2}{n_2}}$$

Since $\sigma_1$ and $\sigma_2$ are unknown, we estimate them by $s_1$ and $s_2$. Thus, we estimate $\sigma_{dm}$ by $s_{dm}$:

$$\begin{aligned} s_{dm} &= \sqrt{\frac{s_1^2}{n_1} + \frac{s_2^2}{n_2}} \\ &= \sqrt{\frac{(1200)^2}{100} + \frac{(1000)^2}{86}} = \sqrt{\frac{1{,}440{,}000}{100} + \frac{1{,}000{,}000}{86}} \\ &= \sqrt{14{,}400 + 11{,}627.9} = \sqrt{26{,}027.9} = 161.33 \end{aligned}$$

This distribution is illustrated in Figure 10-4.

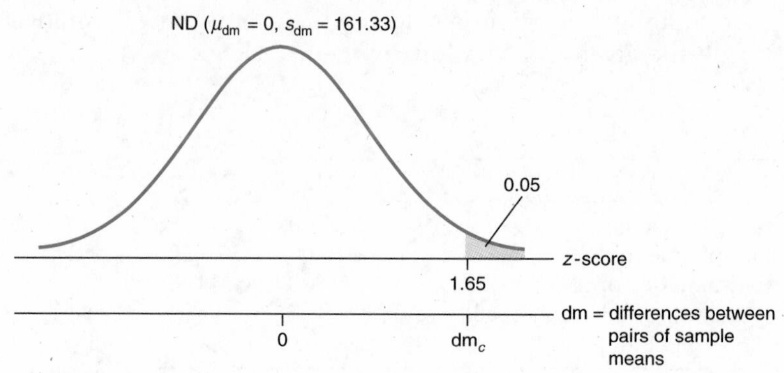


Figure 10-4

Using $z_c = 1.65$, we find

$$\begin{aligned} dm_c &= \mu_{dm} + z_c s_{dm} \\ &= 0 + 1.65(161.33) = \$266.19 \end{aligned}$$

9a. The decision rule is that we will reject the null hypothesis if the outcome is greater than \$266.19.

9b. The outcome in this study is the difference in mean salaries, \$37,400 − \$36,300 = \$1100.

9c. *Conclusion:* We reject the null hypothesis, and conclude that the difference we observed is statistically significant. It is probably true that the average salary of the female 1st-year scientist employed in industry is less than that of her male counterpart.

Under the assumption that the two populations have the same mean, we asked: "What is the probability that the difference between the mean of a randomly picked sample from population 1 and the mean of a randomly picked sample from population 2 will be as large as \$1100?" Our analysis shows that the probability of this happening is less than 0.05. We conclude that since we did get a difference of \$1100, then the two populations probably do not have the same mean, and the spokeswoman is most likely correct. ❑

## STUDY AIDS

### Chapter Summary

In this chapter you have learned:

- When dealing with averages, samples of more than 30 numbers are considered to be large samples
- How to describe the sampling distributions of the means of equal-sized samples according to the central limit theorem

- How to perform one- and two-sample hypothesis tests using means taken from large samples.

**Vocabulary**

You should be able to explain the meaning of each of these terms:

1. Population mean
2. Sample mean
3. Distribution of sample means
4. Central limit theorem
5. Large samples
6. Differences of sample means
7. Sampling distribution of the differences of sample means

**Symbols**

You should understand the meaning of the following symbols:

1. $\mu$
2. $\sigma$
3. $s_1, s_2$
4. $m$
5. $\mu_m$
6. $\sigma_m$
7. $s_m$
8. $\mu_1, \mu_2$
9. $\mu_{\text{dm}}$
10. $m_1, m_2$
11. dm
12. $\sigma_{\text{dm}}$
13. $s_{\text{dm}}$
14. $n, n_1, n_2$

**Formulas**

You should know when and how to use the following formulas:

*For one-sample tests:*

1. $\mu_m = \mu$

2. $s = \sqrt{\dfrac{\Sigma(X^2) - \dfrac{(\Sigma X)^2}{n}}{n - 1}}$

3. $s_m = \dfrac{s}{\sqrt{n}}$

4. $m_c = \mu_m + z_c s_m$

5. Experimental outcome, $m = \dfrac{\Sigma X}{n}$

*For two-sample tests:*

6. $\mu_{\text{dm}} = \mu_1 - \mu_2$ (If $H_0$ states that $\mu_1 = \mu_2$, then $\mu_1 - \mu_2 = 0$)

7. $s_{\text{dm}} = \sqrt{\dfrac{s_1^2}{n_1} + \dfrac{s_2^2}{n_2}}$

8. $\text{dm}_c = \mu_{\text{dm}} + z_c s_{\text{dm}}$

9. Experimental outcome, $\text{dm} = m_1 - m_2$

## EXERCISES

10-17 A design engineer, Wilbur Orville, wants to compare 2 mechanisms for use in pilot ejection seats. He makes 50 of each and subjects them to a stress test. He records the amount of stress (in pounds) that causes each to malfunction. Here are the results. Should this evidence be conclusive in favor of model 1? Let $\alpha = 0.01$.

| Model 1 | Model 2 |
|---|---|
| $m$ = 600 pounds | $m$ = 550 pounds |
| $s$ = 75 pounds | $s$ = 75 pounds |

10-18 A teacher used 2 different teaching methods in 2 similar statistics classes of 35 students each. Then each class took the same exam. In one class we get $m = 82$ and $s = 4$. In the other class we have $m = 77$ and $s = 7$. Test to see if we have evidence that one method is better than the other at the 0.05 significance level.

10-19 In an unusual experiment, Professor Stever had some students take an exam while hanging upside down and had another group of students take the exam while lying on the floor. The results were as follows. Hanging group: $m = 52$, $s = 10$, $n = 36$. Lying group, $m = 60$, $s = 7$, $n = 36$. Does this indicate a statistically significant difference in performance? Use $\alpha = 0.01$.

10-20 Have the high school averages of a college's entering freshman class gone up if 1 year the mean high school average of 80 freshmen picked at random is 82.5 with $s = 2.5$, while the next year the mean high school average of 84 freshmen picked at random is 83.1 with $s = 2.6$? Use $\alpha = 0.05$.

10-21 It is suspected for most people the older they get, the poorer their hearing becomes. A hearing test was given to a group of 40 boys (age 10) and a group of 40 men (age 50). A high score on the test means that the person could hear high-pitched sounds. The mean score for the boys was 200 with $s = 20$. The mean score for the men was 170 with $s = 20$. Show that this is statistically significant at the 0.05 significance level.

10-22 In a comparison of buying habits, the following data were obtained from 2 samples, each consisting of 64 nuns: 20 years ago nuns bought an average of 2.3 habits per year with $s = 0.3$. Today the average is 0.4 habits per year with $s = 0.2$. Using $\alpha = 0.01$, does this indicate a change in the buying habits of nuns buying habits?

10-23 A school psychologist in California administered a standardized aptitude test in arithmetic to a group of 75 randomly picked sixth-grade students who had come to California from Vietnam the previous year. She gave the same test to 75 randomly picked sixth-graders who had attended California elementary schools from first grade. The mean score for the Vietnamese students was 150 with $s = 25$. The California students had a mean score of 100 with $s = 40$. Show that the Vietnamese scores are higher at the 0.01 significance level.

10-24 A group of 40 left-handed people were asked to pick up 10 pennies quickly with their right hands. Then a group of 80 right-handed people were asked to pick up 10 pennies quickly with their left hands. The length of time each person took was recorded. The following information was gathered:

$$n_1 = 40, \quad m_1 = 2.8 \text{ seconds}, \quad s_1 = 1.0 \text{ second}$$

$$n_2 = 80, \quad m_2 = 3.2 \text{ seconds}, \quad s_2 = 2.0 \text{ second}$$

Is this difference statistically significant at $\alpha = 0.05$?

10-25 In a carefully controlled experiment, Etherea raised 35 sunflower plants by reciting a tender poem by Kahlil Gibran to each plant whenever she fed and watered it. She also raised 35 other sunflower plants without talking to them at all. After 1 month the results were as follows. Of the plants she talked to, the mean growth was 10.1 inches with $s = 1$ inch. For the others, the mean growth was 9.8 inches with $s = 1$ inch. Does this indicate at $\alpha = 0.05$ that reciting the poem is associated with superior growth?

10-26 Professor Signo Diferens wants to see if there is any significant difference between the average grades of students who hand in their test papers early and those who hand them in later. In a recent test of 80 students, the first 40 papers had an average grade of 83 with $s = 10$. The last 40 papers had an average grade of 78 with $s = 6$. Does this information indicate a statistically significant difference? Use $\alpha = 0.05$.

10-27 Two authors' books were sampled for the number of letters per word. Data were as follows:

| Doctor Suisse | | | | | | Beatrix (Bunny) Potts | | | | | |
|---|---|---|---|---|---|---|---|---|---|---|---|
| 6 | 1 | 4 | 2 | 3 | 10 | 7 | 2 | 5 | 3 | 4 | 2 |
| 6 | 5 | 9 | 5 | 2 | 5 | 7 | 6 | 1 | 6 | 3 | 6 |
| 6 | 4 | 6 | 5 | 4 | 5 | 7 | 5 | 7 | 6 | 5 | 6 |
| 6 | 5 | 5 | 6 | 9 | 6 | 7 | 6 | 6 | 7 | 1 | 7 |
| 11 | 6 | 4 | 8 | 9 | 7 | 2 | 7 | 5 | 9 | 1 | 8 |
| 5 | 6 | 6 | 4 | 7 | 9 | 6 | 7 | 7 | 5 | 8 | 1 |
| 11 | 12 | 4 | 1 | 2 | 1 | 2 | 3 | 5 | 2 | 3 | 2 |
| 6 | 6 | 7 | 8 | 5 | 5 | 7 | 7 | 8 | 9 | 6 | 6 |
| 4 | 3 | | | | | 5 | 4 | | | | |

Test to discover if there is a statistically significant difference in average word length between these 2 authors. Use the 0.01 significance level.

10-28 Two samples of purchases recently charged at Pat's Peacocks and Pharmaceuticals produced the following data. Perform a hypothesis test at the 0.05 significance level to see if the average purchases are the same for each credit card.

| Wisa | | | Faster Kard | | |
|---|---|---|---|---|---|
| $140.19 | 31.61 | 510.20 | $859.80 | 68.38 | 489.79 |
| 29.22 | 562.73 | 121.03 | 70.77 | 437.26 | 878.96 |
| 243.38 | 22.45 | 732.40 | 756.61 | 77.54 | 267.59 |
| 88.41 | 37.87 | 1348.05 | 11.58 | 62.12 | 951.94 |
| 83.57 | 363.39 | 953.66 | 16.42 | 636.60 | 46.33 |
| 22.64 | 89.90 | 268.78 | 77.35 | 10.09 | 731.21 |
| 647.74 | 66.28 | 274.13 | 352.25 | 33.71 | 725.86 |
| 515.33 | 635.16 | 1187.25 | 484.66 | 364.83 | 1002.74 |
| 13.43 | 78.54 | 495.66 | 86.56 | 21.45 | 504.33 |
| 66.10 | 737.52 | 636.00 | 33.89 | 262.47 | 363.99 |
| 48.12 | 99.66 | 731.70 | 51.87 | 00.33 | 268.29 |
| 173.04 | 1.07 | 1099.09 | 826.95 | 8.92 | 1000.90 |

## CLASS SURVEY QUESTION

If you have data for more than 30 students, do a hypothesis test to see whether or not the average of the last digit of all Social Security numbers is 4.5 (which is the average of 0, 1, 2, . . . , 9). What about the average of the fifth digit?

## FIELD PROJECTS

Select one of the following as a special project. After your proposal has been approved by your instructor, complete the project.

1. To test whether or not people can *randomly* pick a number from 1 to 10, ask 100 people to pick a number from 1 to 10 inclusive. The null hypothesis is that the mean of the 100 replies should be 5.5 (the mean of 0, 1, 2, 3, . . . , 10). Test at the 10% significance level.
2. Test whether male and female students on your campus carry, on average, the same amount of coin change with them.
3. Perform a large-sample hypothesis test of your own choosing. Do either a one- or a two-sample test. Recall that for a one-sample test, you must have a legitimate justification for the mean given by the null hypothesis. This often comes from a newspaper, magazine, or textbook. Some projects which students have performed include the following:

   *One-Sample Tests*

   (a) A newspaper claimed that people averaged a certain amount of money spent on entertainment each week. Test to see if the average in your neighborhood is different.

   (b) A report claimed that people averaged a certain amount of time watching TV each week. Is the average at your place of employment different?

   *Two-Sample Tests*

   (a) Is there a difference between the average number of cigarettes smoked per day by waiters and nurses in your locality?

   (b) Test the difference between the average test grades for two different teachers.

   (c) Test the difference between the average age of teaching faculty and the average age of the rest of the college staff.

## CLASS EXPERIMENTS

1. Test to see if there is a statistically significant difference between the pulse rates of the males and the females. What are the populations?
2. Test to see if there is a statistically significant difference between the number of paid hours of work in the previous 7 days for the males and females.
3. (Variability of sample means) Compute $\sigma$, the standard deviation of the population, $X$: 1, 2, 3, 4. Now calculate $m$ for every possible sample of size 2. There are 16 such samples: {1,1}, {1,2}, {1,3}, {1,4}, {2,1}, {2,2}, etc. Finally, compute $\sigma_m$, the standard deviation of the 16 means, and confirm that $\sigma_m = \frac{\sigma}{\sqrt{n}}$, where $n = 2$, the size of each sample.

# 11

# Hypothesis Testing with Sample Means: Small Samples

William S. Gossett, a British statistician, worked in a Guinness brewery where he took small samples. In 1908, under the pseudonym "A. Student," he wrote a paper detailing his work. He referred to the resulting curves as *t-curves*. As a result, they are often referred to as **Student's *t*-curves**.

**CONTENTS**



**GOALS**

At the end of this chapter you will have learned:

- The concept of degrees of freedom
- How to perform one- and two-sample tests of sample means when one or both samples are small, and the conditions under which such tests are valid
- How to perform a matched-pairs test on the differences between scores in two samples

## 11.1 Student's *t*-Distribution

Ideally, in any statistical study the largest possible samples should be used. Larger samples are more likely to be representative of the population, and statistics based on larger samples are likely to be closer to the population parameters. However, in many important and valuable studies large samples are simply not available. For example, data on new medical treatments may first arise in studies with only a few patients. Similarly, in the manufacture of expensive products, quality-control procedures may destroy the product (like a crash test of a new car), and it does not make economic sense to destroy a large number of them. Thus, some studies necessarily involve small samples.

In this book we discuss small studies separately, because the sampling distributions of some important statistics are not normal when the statistics come from small samples, and a new curve is needed to replace the familiar normal curve.

In our earlier discussion of the central limit theorem (Chapter 10), we said that the sampling distribution of means of *large* samples of size $n$ from any population is approximately normal in shape, with

$$\mu_m = \mu \qquad \text{and} \qquad \sigma_m = \frac{\sigma}{\sqrt{n}}$$

In any given hypothesis test, of course, we construct a particular normal curve, usually taking for $\mu$ a value given in our null hypothesis, and taking $s$ from the sample data as our estimate of $\sigma$. This procedure can be shown to produce good results both in theory and in practice. That is, when we set $\alpha = 0.05$, for example, in many repeated tests, we actually do reject a true hypothesis about 5% of the time.

**small samples**

However, about 90 years ago the statistician W. S. Gossett showed that if you use this same procedure when you have **small samples**, Type I errors will be made more than 5% of the time. Basically, this happens because, in repeated experiments with small samples, the values of $s$ tend to be quite variable, so that if you use

$$m_c = \mu + z_c s$$

**Student's *t*-scores**

to establish the critical value for the rejection region, you tend to run more of a chance of a Type I error than $z_c$ would indicate. To solve this problem, what Gossett did, from our point of view, was to come up with different sets of critical scores—**Student's *t*-scores**—to be used in place of the critical $z$-scores, depending on how big the sample actually is. These scores are called **critical values of *t*.** They are printed in Table D-5 for two-tail tests and in Table D-6 for one-tail tests. Gossett showed that even these $t$-scores are not always reliable. They are reliable, however, in the case where the original population from which we are taking our sample is near normal to begin with.

### Criminal Fingers

Gossett investigated the sampling distribution of the variance $s^2$ and showed it has a chi-square distribution and that there is no correlation between the variance and the mean. From this he deduced the probability distribution which he called $t$. (Chi-square distributions and correlation are discussed in Chapters 13 and 14.)

In his investigation he checked 750 small samples (with $n = 4$) from published data on the height and middle-finger length of 3000 criminals. He did this by shuffling 3000 pieces of cardboard containing the data!

For example, using $t$-scores to analyze small samples of weights of loaves of bread made by Ackamee bakery would probably be all right, since the distribution of these weights is probably close to normal. But an experiment based on small samples from a nonnormal variable should not be analyzed using $t$-scores. For example, an experiment to test a claim about the mean annual income of faculty at a certain university should *not* be analyzed with $t$-scores if we can only sample, say, 20 faculty members, because incomes are known to be usually distributed in a nonnormal pattern.

In summary, the procedure we have been using with the normal curve and $z$-scores is correct for large-sample testing from any population. Student's $t$-scores are correct for small-sample testing from normal populations. There is no simple general approach for small samples from nonnormal populations. Certain tests, called "nonparametric tests," may be useful in such cases. See Chapter 16 for a brief introduction to this topic.

When you draw the curve for a particular experiment using a small sample, you will be using a particular set of critical $t$-scores corresponding to the sample size. A curve which corresponds to a set of $t$-scores is called a ***t*-curve**. (See Figure 11-1.) There are many $t$-curves, and for each sample size there is

***t*-curve**

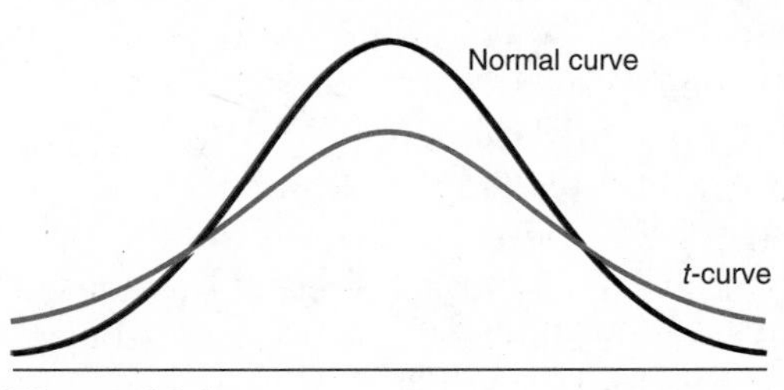

**Figure 11-1**

a slightly different curve. The $t$-curves are similar to the normal curve in that they are symmetrical and bell-shaped. For your rough sketches, there is no need to distinguish between the shapes. When drawn precisely, however, it can be seen that the $t$-curves are somewhat flatter than the normal curve, and that the smaller the sample size, the flatter is the curve. Conversely, the larger the sample size, the more closely the $t$-curve resembles the normal curve. Once the sample size is larger than 30, the $t$-curve and the normal curve are practically the same. Many researchers, therefore, use $t$-curves for samples of size 30 or less and the normal curve for samples larger than 30; although it is always correct to use $t$-curves.

In small-sample testing, the critical $t$-scores are numerically larger than the critical $z$-scores from large-sample testing, and the smaller the sample, the larger the critical $t$-values must be. You will see that this is the case by looking at the $t$-tables (Tables D-5 and D-6). This should make sense, because in using only a small sample we should require more conclusive evidence to reject a null hypothesis.

A mean from a small sample should really be quite different from that predicted by the null hypothesis if we are going to use it to reject the null hypothesis. By using larger critical $t$-values we are, in practice, requiring stronger evidence to reject a null hypothesis.

### *Babbage*

Father of the modern computer, Charles Babbage (1791–1871) was among those who early on recommended a mathematical approach to economics. He played a key role in establishing the London Statistical Society.

He also founded the statistical section of the British Association, which was subdivided into sections, each concerned with a particular branch of science. Charles Dickens mocked the sometimes comical activities of what he referred to as the "British Ass," calling their activities "Ditchwateristics."

By the way, Babbage hated organ-grinders and tried to have them outlawed.

*Source:* Taken in part from Anthony Hyman, *Charles Babbage, Pioneer of the Computer,* Princeton, NJ: Princeton University Press, 1982.

## Degrees of Freedom and Using the *t*-Table

To use the $t$-table, you must find the entry that corresponds to your particular sample size. You would therefore expect the table to have a column labeled $n$, for sample size. However, it turns out that this same table can be used in other problems (for example, problems involving two different samples of different sizes) where that label would not make sense. It is usual, instead, to have a column labeled **degrees of freedom**. It is not obvious why this name should be used. Technically, it is used because the $t$-curve is related to another curve we will study later (called the *chi-square curve*) for which the phrase "degrees of freedom" makes more intuitive sense. For our current purposes it is sufficient to know that in a hypothesis test about a population mean where we are working with a single small sample of size $n$, the correct numerical value of the degrees of freedom for $t$ is $n - 1$.

**degrees of freedom**

### Examples of Degrees of Freedom

The phrase "degrees of freedom" comes up often in statistical work. Here are some contexts in which it may appear.

---

### ❑ EXAMPLE 11.1 Degrees of Freedom in a Triangle

Suppose that Carl arbitrarily selects two angles of a triangle as 40° and 60°. Furthermore, he informs Carla that the third angle is 80°. He was free to pick different values for the first 2 angles, but once they were selected, he was not free in his choice of the third angle, because the 3 angles must add up to 180°. Thus we say that in the selection of the values for the 3 angles of a triangle there are only 2 degrees of freedom. ❑

### ❑ EXAMPLE 11.2 Degrees of Freedom for a Given Average

Consider a teacher who wishes to pick 5 numbers for an example in which she desires that the mean of the numbers be 10. She is free to pick any 4 numbers she wishes, but the fifth number is determined because the sum must be 50. We say that in selecting 5 numbers whose mean is 10, there are only 4 degrees of freedom. ❑

### ❑ EXAMPLE 11.3 Degrees of Freedom in a Table

Suppose some statisticians wanted to pick 200 persons so that 100 were male and 100 were female, including 50 Democrats and 150 Republicans. They would need 4 groups—namely, male Democrats, female Democrats, male Republicans, and female Republicans. However, for the sizes of these 4 groups, there is only 1 degree of freedom.

Table 11-1

| | Democrats | Republicans | |
|---|---|---|---|
| Male | 20 | | 100 |
| Female | | | 100 |
| | 50 | 150 | 200 |

As you can see from Table 11-1, if they arbitrarily select 20 male Democrats, they must pick 30 female Democrats, 80 male Republicans, and 70 female Republicans. ❑

### ❑ EXAMPLE 11.4 Degrees of Freedom for the Standard Deviation

In computing the standard deviation of 10 numbers, we use the deviations from the mean. These deviations must always add up to zero. Therefore, when we compute the standard deviation of 10 numbers, there are only 9 degrees of freedom. In general, when we are trying to estimate the standard deviation of a population by using a sample of size $n$, there are $n - 1$ degrees of freedom. ❑

### Illustration of a t-Test

The following example uses data from a small sample to test a claim about a population mean.

---

❑ **EXAMPLE 11.5 Eagle Scouts**

A claim is made that Eagle Scouts in Plainfield County have a mean knot-tying score of 117. Doubting that this is true, we take a random sample of 16 Eagle Scouts. The mean score of the sample is not 117 knots, but 118.2. The estimate $s$ of the standard deviation of the population was calculated to be 1.6. Test at the 0.05 significance level, assuming that the knot-tying abilities of all Eagle Scouts in Plainfield are in fact distributed in a pattern not too different from normal.

**SOLUTION**

Since 16 is less than 30, we will do a one-sample, small-sample $t$-test of a mean.

*Population:* All the Eagle Scouts in Plainfield

*Sample:* 16 Eagle Scouts from Plainfield

$H_0$: The claim is correct; $\mu = 117$.

$H_a$: The claim is wrong; $\mu \neq 117$. (two-tail test)

$$m = 118.2$$

$$s = 1.6$$

$$n = 16 \leq 30 \qquad \text{(Therefore, we will use the } t\text{-distribution.)}$$

$$\alpha = 0.05$$

The number of degrees of freedom is given by $n - 1 = 16 - 1 = 15$

$$\mu_m = \mu = 117$$

$$s_m = \frac{s}{\sqrt{n}} = \frac{1.6}{\sqrt{16}} = \frac{1.6}{4} = 0.4$$

Looking up 0.025 and 0.975 at 15 degrees of freedom in Table D-5, we find the critical $t$-scores, $t_c = \pm 2.13$. Converting these to raw scores, we get

$$m_c = \mu_m + t_c s_m$$

$$= 117 + (\pm 2.13)(0.4) = 117 \pm 0.85 = 116.15 \text{ and } 117.85$$

We use the notation $tD(\mu_m = \quad , s_m = \quad)$ to indicate a *t*-distribution as shown in Figure 11-2.

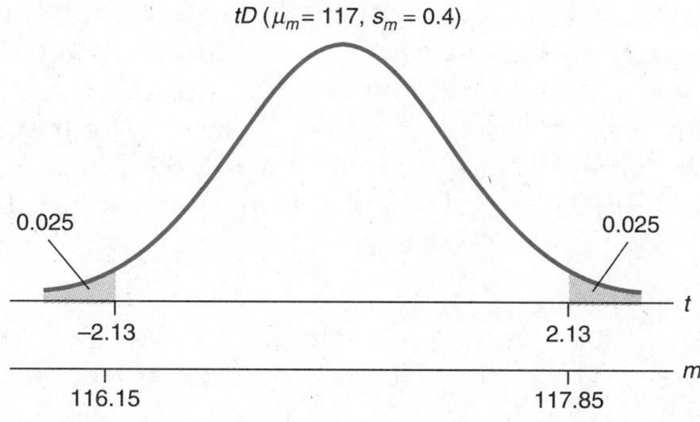


**Figure 11-2**

Therefore, our decision rule is to reject the null hypothesis if we get a sample mean less than 116.15 or greater than 117.85. Since our outcome was 118.2, we have good evidence that the mean knot-tying score of Eagle Scouts in Plainfield is not 117 knots; it's higher.

You will notice that the only difference between this hypothesis test and one with $n$ larger than 30 is that we used $t_c = \pm 2.13$ rather than $z_c = \pm 1.96$. ❑

---

As in the case of *z*-scores, some people prefer to convert raw scores to *t*-scores. They would write $t = \dfrac{m - \mu}{s_m} = \dfrac{118.2 - 117}{0.4} = 3.00.$

Since this is greater than the critical *t*-value of 2.13, they would also reject the null hypothesis.

### Exercises

11-1 Which one of the following statements is true?
(a) Every normal curve is bell-shaped.
(b) Every bell-shaped curve is a normal curve.

11-2 If you did not know any better, and you performed a *z*-test instead of a *t*-test on data from a small sample, would you increase or decrease $\alpha$? What about $\beta$?

11-3 It is claimed that a large corporation discriminates against its women employees in its promotion practices. Over many years, the mean time before first promotion for its male employees has been 3 years. A random sample of 20 females who had worked there many years showed a mean time of 3.8 years before first promotion, with $s = 1.2$ years. Using $\alpha = 0.05$, see if the data support the claim. Assume that these data are from a normal distribution.*

11-4 A species of peccary formerly thought to be extinct was recently discovered living in Paraguay. The basis for this claim is that several different skull measurements are very near to the measurements of the ancient fossils. For example, a sample of 13 Paraguayan peccaries had a mean width across the canines of 60.8 millimeters with $s = 1.8$ millimeters. The "extinct" peccary has a mean width of 60.0 millimeters across the canines. Assuming that these measurements are from a normal distribution, show at $\alpha = 0.05$ that you cannot reject the hypothesis that both species have the same average width across the canines.

11-5 The average IQ of Martians is quoted to be 260. Thinking that this figure is too large, an earthling tested the crew of the first Martian spaceship to land in Bayonne, New Jersey. The average IQ of the 8 crew members was 250 with $s = 8$. Test at $\alpha = 0.01$, assuming that the distribution of Martian IQs is near normal.

11-6 The secretary of an association of professional landscape gardeners claims that the average cost of services to customers is $90 per month. Feeling that this figure is too low, we question a random sample of 10 customers. Our sample yields a mean cost of $125 with $s = \$20$. Test at the 0.05 significance level. Assume that such costs are normally distributed.

11-7 A claim is made that the mean height of police officers is 5 feet 11 inches. To test whether this claim is true or not, a random sample of 25 police officers was gathered. This sample had a mean height of 5 feet 10 inches with $s = 3$ inches. Test at the 0.05 significance level. Assume that heights are normally distributed.

11-8 If a population is *known* to be normally distributed, and the value of $\sigma$ is also *known* (rather than estimated), then the $z$-scores may be used to test hypotheses regardless of the size of $n$. Repeat the previous exercise with $\sigma = 3$ instead of $s = 3$.

11-9 The mean score on a standardized psychology test is supposed to be 50. Believing that a group of psychologists will score higher, we test a random sample of 11 psychologists. Their mean score is 45 with $s = 3$. Test at $\alpha = 0.01$. Assume that such scores are distributed near normally.

11-10 Does the evidence support the idea that the average lecture consists of 3000 words if a random sample of the lectures of 16 professors had a mean of 3472 words with $s = 500$ words? Use $\alpha = 0.01$. Assume that lecture lengths are approximately normally distributed.

---

*There exist tests—beyond the scope of this book—by which one can judge whether or not it is reasonable to assume that a distribution is approximately normal.

11-11 An astronomer is testing the claim that the mean brightness of a certain star is now more than 30 units. She is able to get 6 readings on the star during her experiment as shown in the table. Using $\alpha = 0.01$, does this indicate a new mean brightness of more than 30?

| Reading | Brightness |
|---|---|
| 1 | 30 |
| 2 | 30 |
| 3 | 29 |
| 4 | 31 |
| 5 | 32 |
| 6 | 33 |

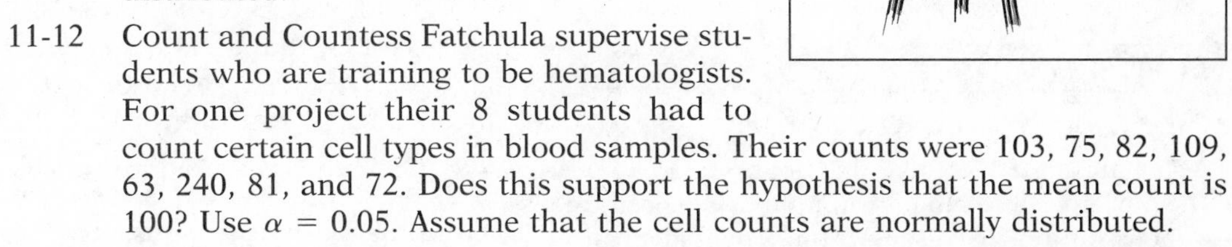

Assume that such readings are normally distributed.

11-12 Count and Countess Fatchula supervise students who are training to be hematologists. For one project their 8 students had to count certain cell types in blood samples. Their counts were 103, 75, 82, 109, 63, 240, 81, and 72. Does this support the hypothesis that the mean count is 100? Use $\alpha = 0.05$. Assume that the cell counts are normally distributed.

11-13 Skippy, the skeptic, doubts a statistics teacher's claim that 95% of the means of samples of size 9 will have a *t*-score less than 1.86. A computer program is available that picks a number randomly from a normal distribution with mean equal to zero. Skippy programs the computer to pick 9 of these random numbers and to then compute $m$, $s$, and $s_m$. Finally, using

$$t = \frac{m - \mu}{s_m}$$

the computer finds the *t*-score for the sample mean.

Skippy then has the computer repeat this experiment 500 times; 468 times the *t*-score is less than 1.86! Therefore, she claims that the teacher is wrong. Test using $\alpha = 0.05$.

## 11.2 Two-Sample *t*-Tests

The problem of testing the difference between two means when one or both samples are small (less than 30) is simplest when we can make two additional assumptions, and we shall restrict our attention to such cases.

1. Both populations are approximately normal.
2. The variances of the two populations are approximately equal.*

When these assumptions are reasonable, it follows that the distribution of the differences of sample means can be related to the $t$-distribution. To be able to use the $t$-table, we will need to know the appropriate degrees of freedom. To calculate the critical difference, we will need a formula for $s_{\text{dm}}$, the standard deviation of the sample differences. We give both of these results:

$$\text{Degrees of freedom} = (n_1 - 1) + (n_2 - 1) = n_1 + n_2 - 2$$

$$s_{\text{dm}} = \sqrt{\frac{s_p^2}{n_1} + \frac{s_p^2}{n_2}} = \sqrt{s_p^2\left(\frac{1}{n_1} + \frac{1}{n_2}\right)}$$

where $s_p$, the *pooled* estimate of the variance, is found by

$$s_p^2 = \frac{(n_1 - 1)(s_1^2) + (n_2 - 1)(s_2^2)}{n_1 + n_2 - 2}$$

**Weighted Mean**

Roger Cotes (1682–1716) was one of the first statisticians to recognize that not all data had the same importance. In his work published posthumously in 1722 he discusses the weighted mean.

Notice that $s_p$ is a *weighted* average of the two variance estimates. According to assumption 2, both sample variances are estimating the same true variance. So it makes sense to combine both estimates. The weighting simply assigns more weight to the estimate from the bigger sample. If both samples happen to be the same size, then

$$s_p^2 = \frac{s_1^2 + s_2^2}{2}$$

***Note*** If the number of degrees of freedom given by $n_1 + n_2 - 2$ is not in our table, we will approximate it by the nearest value in the table. If the number of degrees of freedom is greater than 500, we use the last line of the table. You will notice that these entries are the critical $z$-scores.

### ❑ EXAMPLE 11.6 College Bored Scores

A claim is made that Hahvahd business students have higher Bored Scores than Hahvahd medical students. In an attempt to establish this claim at the 0.01 significance level, we select a random sample from each population and

*In Chapter 15 we give a test for this second assumption.

administer the Snooze Boredom Test to each person. (Assume that boredom scores are normally distributed in both populations.) In a sample of 15 business students, the mean Bored score was 120 with $s_1 = 10$, while in a sample of 18 medical students the mean was 110 with $s_2 = 5$.

### SOLUTION

A two-sample, small-sample test of means.

*Population 1:* All Hahvahd business students in the last 5 years

*Population 2:* All Hahvahd medical students in the last 5 years

*Sample 1:* 15 business students

*Sample 2:* 18 medical students

$H_a$: The mean Bored score of business students is higher than the mean Bored score of medical students at Hahvahd; $\mu_{dm} = \mu_1 - \mu_2 > 0$. (Therefore, we have a one-tail test on the right.)

$H_0$: The mean Bored score for both groups of students is the same; $\mu_{dm} = \mu_1 - \mu_2 = 0$.

*Sample data:*

$$n_1 = 15 \qquad n_2 = 18$$

$$m_1 = 120 \qquad m_2 = 110$$

$$s_1 = 10 \qquad s_2 = 8$$

$$\text{Degrees of freedom} = 15 + 18 - 2 = 31 \qquad \alpha = 0.01$$

Since the number of degrees of freedom, 31, is not in our table, we use the nearest value, namely, degrees of freedom = 30. Looking up the critical $t$-score, we find $t_c = +2.46$.

We find the pooled value of $s^2$:

$$s_p^2 = \frac{(n_1 - 1)s_1^2 + (n_2 - 1)s_2^2}{n_1 + n_2 - 2}$$

$$s_p^2 = \frac{14(100) + 17(64)}{15 + 18 - 2} = \frac{2488}{31} = 80.26$$

Then we use $s_p^2$ to get $s_{dm}$:

$$s_{dm} = \sqrt{s_p^2\left(\frac{1}{n_1} + \frac{1}{n_2}\right)} = \sqrt{80.26\left(\frac{1}{15} + \frac{1}{18}\right)} = \sqrt{9.81} = 3.13$$

We have $tD(\mu_{dm} = 0, s_{dm} = 3.13)$, so that

$$\begin{aligned} dm_c &= \mu_{dm} + t_c s_{dm} \\ &= 0 + 2.46(3.13) = 7.70 \end{aligned}$$

Therefore, our decision rule is to reject the null hypothesis if we get a difference greater than 7.70. The experimental outcome, dm = 120 − 110 = 10, is greater than 7.70. Therefore, we reject the null hypothesis at the 0.01 significance level and claim that for the past 5 years Hahvahd business students had, on average, higher Bored scores than Hahvahd medical students. ❑

## Exercises

*In these exercises, assume that the populations being sampled are approximately normal and that their variances are approximately equal.*

11-14 Goodbenefits Manufacturing Company replaced the Ratari game machine in the employees' lunchroom because of the number of wrist injuries. Dr. Leech, the company doctor, notes that with the new Satari game machine there has been an average of 38 wrist injuries per day over the past 14 working days, with a standard deviation of 4 wrist injuries. Checking back in his previous records, he discovers that with the old Ratari machine there was an average of 40 injuries over the last 10 days they had the machine. The standard deviation was also 4. If the difference is statistically significant ($\alpha = 0.05$), Dr. Leech thinks he can write an article, which will be published in the local medical journal. Is it?

11-15 To test whether or not there is a difference between the mean grade point averages of male math majors and female math majors at State U, 2 random samples were gathered. For the 35 males, we find that $m_1 = 3.1$ with $s_1 = 0.2$. For the 20 females, $m_2 = 3.2$ with $s_2 = 0.15$. Test at $\alpha = 0.01$.

11-16 To find out whether fluoride treatment is effective in reducing the occurrence of cavities, a test was conducted in Dentine County. A total of 15 children who had fluoride treatment were compared with 15 who did not. Of those who used fluoride, the mean number of cavities $m_1$ was 0.8 with $s_1 = 0.7$. For those who did not use fluoride, $m_2$ was 1.4 with $s_2 = 0.9$. Test at the 0.05 significance level.

11-17 Exemptions from final exams may improve the grades of students. To test this idea, 40 students of similar background and ability in French I were picked at random and placed in 2 classes of French II, 20 students in each class. In one class, the students were told that if they averaged 90 or more on all classwork during the term, they would be exempt from the final exam. In the second class, the students were told that no exemptions would be given. The average mark on classwork in the first class was 83.3 with $s_1 = 9.4$. The average in the second class was 80.7 with $s_2 = 11.1$. Test using $\alpha = 0.05$.

11-18 An inventor states that his new method for packing eggs is safer than the old method. The director of shipping at Ova-Dairy, Inc., has 2 shipments of eggs packed, 1 by the former method and 1 by this new method. After the eggs are delivered, a sample of 10 gross from each shipment was inspected, and the number of cracked eggs in each gross was counted. In the first sample, the average number of cracked eggs per gross $m_1$ was 7.2 with $s_1 = 3.2$. In the second sample, $m_2$ was only 4.1 with $s_2 = 3.0$. Test at the 0.01 significance level. (*Hint:* $n_1 = 10$ and $n_2 = 10$.)

11-19 To see whether there is a statistically significant difference between the age of owners of convertibles and the age of owners of sedans, a Madison Avenue advertising agency gathered 2 random samples in Mobile, Alabama. It was found that 15 convertible owners had an average age of 29.2 years with $s_1 = 5$, and 35 sedan owners had an average age of 24.8 years with $s_2 = 8$. Test at the 0.05 significance level.

11-20 A study was conducted recently of the temperatures of spark plugs when cars were driven at 50 miles per hour and at 55 miles per hour. The study was done on new police cars in Security, Kansas. It found that for 22 spark plugs tested at 50 miles per hour, the temperatures in degrees F were as follows: 900°, 920°, 860°, 890°, 910°, 820°, 950°, 880°, 930°, 870°, 900°, 850°, 910°, 830°, 930°, 950°, 840°, 860°, 900°, 900°, 960°, and 890°. The 16 spark plugs tested at 55 miles per hour had temperatures of 1000°, 930°, 900°, 1100°, 950°, 1070°, 890°, 1050°, 910°, 1090°, 870°, 1130°, 1000°, 1110°, 1050°, and 950°. Do these data suggest that the average temperature at 55 miles per hour is different from that at 50 miles per hour? Use the 0.01 significance level.

11-21 Amy and Josh go outside in the evenings to collect toads in Harvey Cedars, New Jersey. In June they caught 50 toads with a mean weight of 1.1 ounces and with $s = 0.05$ ounce. In August they caught 25 toads with a mean weight of 2.1 ounces and $s = 0.07$ ounce. Does this indicate at the 0.01 significance level that August toads are heavier than June toads?

11-22 People were asked how long after a meal they became hungry. A random sample of 10 patrons of How Long's restaurant yielded a mean of 1.3 hours and a standard deviation of 0.5 hour. A sample of 15 patrons of Sum Fun's restaurant had a mean of 1.7 hours and a standard deviation of 0.2 hour. Is this difference statistically significant at $\alpha = 0.05$?

11-23 Is there a statistically significant difference between the bowling averages of male nurses and those of businessmen, if last year's league play in Bowling Green, Ohio, produced the following data: 10 male nurses averaged 180 with $s = 10$, and 15 businessmen averaged 170 with $s = 10$. Use $\alpha = 0.05$.

11-24 Casey is batting against a baseball-pitching machine. The machine pitches either fast or very fast. The distances of all of Casey's fly balls are recorded. Test at the 0.05 significance level to see if the very fast pitches tend to be hit farther. The data in feet were as follows:

| | | | | | | |
|---|---|---|---|---|---|---|
| Fast | 250 | 300 | 275 | 280 | 350 | 350 |
| Very fast | 300 | 290 | 400 | 325 | 330 | 400 |

11-25 Are Krispy toasters quicker than No-Burn toasters, at the 0.01 significance level, if the times it took to toast a slice of bread were as follows:

No-Burn 60, 55, 70, 65, 80 seconds
Krispy 70, 65, 65, 60, 60 seconds

11-26 The 2 formulas for $s_p^2$ given in this chapter are

$$s_p^2 = \frac{(n_1 - 1)s_1^2 + (n_2 - 1)s_2^2}{n_1 + n_2 - 2}$$

and

$$s_p^2 = \frac{s_1^2 + s_2^2}{2}$$

How different are they? Calculate $s_p$ by both formulas, using various values of $s_1$, $s_2$, $n_1$, and $n_2$, such as the following:

| $s_1$ | $s_2$ | $n_1$ | $n_2$ |
|---|---|---|---|
| 4 | 6 | 100 | 101 |
| 50 | 60 | 70 | 80 |
| 10 | 10 | 50 | 500 |
| . | | | |
| . | | | |
| . | | | |

If you have access to a computer, you can easily compare many values.

## 11.3 A *t*-Test for Paired Differences, Matched-Pair Designs

### *Ramanujan*

The esoteric theories of the Indian genius Ramanujan (1887–1920) have been used to solve an important problem in statistical mechanics, explaining such things as how liquid helium disperses through a crystal lattice of carbon.

—Robert Kanigel
*The Man Who Knew Infinity,*
New York: C. Scribner's, 1991.

In Exercise 11-17 we looked at an experiment designed to see if exemptions from final exams improve grades. The two-sample *t*-test showed "no significant difference." For that experiment we randomly placed 40 students in two classes of 20 students each. We say that the two samples were chosen independently of one another because there is no particular connection between the students in one class and those in the other. There are other ways of deciding which students go into which class. For example, suppose we first rank the students from 1 to 40 in terms of their past grades in French I. We then pair them off as shown in Table 11-2.

**Table 11-2**

| Student Numbers |
|---|
| 1 and 2 |
| 3 and 4 |
| 5 and 6 |
| 7 and 8 |
| 9 and 10 |
| 11 and 12 |
| 13 and 14 |
| 15 and 16 |
| 17 and 18 |
| 19 and 20 |
| 21 and 22 |
| 23 and 24 |
| 25 and 26 |
| 27 and 28 |
| 29 and 30 |
| 31 and 32 |
| 33 and 34 |
| 35 and 36 |
| 37 and 38 |
| 39 and 40 |

This puts two students of similar ability in each pair. One student has an odd number, and one has an even number. Now we decide which student in each pair goes into the first class and which one goes into the second class. We toss a coin 20 times, once for each pair. We let heads mean that the odd-numbered student goes into the first class. Tails means that the odd-numbered student goes into the second class. If our 20 tosses result in

HTTTT HTTHT THTHT HHHTH

then our two classes would be as shown in Table 11-3.

**Table 11-3**

| Pair Number | Student in First Class | Student in Second Class |
|---|---|---|
| 1 | 1 | 2 |
| 2 | 4 | 3 |
| 3 | 6 | 5 |
| 4 | 8 | 7 |
| 5 | 10 | 9 |
| 6 | 11 | 12 |
| 7 | 14 | 13 |
| 8 | 16 | 15 |
| 9 | 17 | 18 |
| 10 | 20 | 19 |
| 11 | 22 | 21 |
| 12 | 23 | 24 |
| 13 | 26 | 25 |
| 14 | 27 | 28 |
| 15 | 30 | 29 |
| 16 | 31 | 32 |
| 17 | 33 | 34 |
| 18 | 35 | 36 |
| 19 | 38 | 37 |
| 20 | 39 | 40 |

The experiment has been substantially changed because each student in one class has been *paired off* with a student of similar ability in the second class. Statisticians say that this is a new experimental design. For this type of design, it is better not to analyze these data as a two-sample $t$-test on the difference of the sample means, because that method ignores the fact that the pairs are matched. In the case of matched pairs, the ordinary two-sample $t$-test is a weaker approach because it is not as likely to pick up a small but real difference between the two populations. The two-sample $t$-test would not notice, for example, the situation where every student in the first class did better than his or her "partner" in the second class, if the first class's students were only a little better than their partners. For a somewhat less obvious illustration, let us suppose that the grades for classwork were paired as shown in Table 11-4.

We would then have a sample of 20 differences which we can consider to be a sample of size 20 from the population of all such differences. We denote these differences by the letter $d$. As usual, we denote the mean of a sample (of $d$'s) by the symbol $m$ (or $m_d$). Assuming that the population of all $d$'s is approximately normal, then our experiment reduces to a one-

**Table 11-4**

| Pair Number | Grade in French II for Student in First Class (Exemptions Available) | Grade in French II for Student in Second Class (No Exemptions Allowed) | Difference |
|---|---|---|---|
| 1 | 100 | 98 | +2 |
| 2 | 96 | 100 | −4 |
| 3 | 97 | 95 | +2 |
| 4 | 92 | 90 | +2 |
| 5 | 91 | 91 | 0 |
| 6 | 93 | 88 | +5 |
| 7 | 79 | 80 | −1 |
| 8 | 79 | 83 | −4 |
| 9 | 81 | 71 | +10 |
| 10 | 86 | 85 | +1 |
| 11 | 90 | 88 | +2 |
| 12 | 80 | 77 | +3 |
| 13 | 80 | 74 | +6 |
| 14 | 82 | 78 | +4 |
| 15 | 82 | 80 | +2 |
| 16 | 75 | 71 | +4 |
| 17 | 71 | 60 | +11 |
| 18 | 73 | 69 | +4 |
| 19 | 65 | 72 | −7 |
| 20 | 75 | 65 | +10 |

sample hypothesis test about sample means, where our raw score is now $d$ instead of $X$.

## ❑ EXAMPLE 11.7 Matched Pairs

Test the hypothesis that the exemptions make a difference using the data in Table 11-2.

### SOLUTION

We perform a one-sample test of means (of differences) using a small sample.

*Population:* All differences in grades for matched-pair students of French II when taught with and without the hope of an exemption.

*Sample:* The differences for the 20 students shown in Table 11-4.

$H_a$: Exemptions tend to raise grades, $\mu > 0$. (One-tail test)
$H_0$: Exemptions make no difference; $\mu = 0$.

$n = 20$ (Therefore, we have a $t$-test with 19 degrees of freedom.)

The data consist of the 20 differences shown in Table 11-4. We need to compute $s$ and $s_m$. From Table 11-4 we create Table 11-5 in order to compute $m$, $s$, and $s_m$.

**Table 11-5**

| Pair Number | Difference $d$ | $d^2$ |
|---|---|---|
| 1 | +2 | 4 |
| 2 | −4 | 16 |
| 3 | +2 | 4 |
| 4 | +2 | 4 |
| 5 | 0 | 0 |
| 6 | +5 | 25 |
| 7 | −1 | 1 |
| 8 | −4 | 16 |
| 9 | +10 | 100 |
| 10 | +1 | 1 |
| 11 | +2 | 4 |
| 12 | +3 | 9 |
| 13 | +6 | 36 |
| 14 | +4 | 16 |
| 15 | +2 | 4 |
| 16 | +4 | 16 |
| 17 | +11 | 121 |
| 18 | +4 | 16 |
| 19 | −7 | 49 |
| 20 | +10 | 100 |
| | 52 | 542 |

$$m = \frac{52}{20} = 2.6$$

$$s = \sqrt{\frac{542 - \frac{52^2}{20}}{19}} = 4.63$$

$$s_m = \frac{4.63}{\sqrt{20}} = 1.03$$

Therefore, we have $tD(\mu_m = 0, s_m = 1.03)$, with 19 degrees of freedom.

Taking $\alpha = 0.05$, we get $t_c = +1.73$, thus

$$m_c = \mu_m + t_c s_m$$
$$= 0 + 1.73(1.03) = 1.8$$

Our decision rule is then to reject the null hypothesis if the outcome is greater than 1.8. Since our outcome, $m = 2.6$, is greater than 1.8, we reject the null hypothesis. Exemptions apparently raise grades. ❑

---

**matched pair**
**paired differences**

This type of experiment is referred to as a **matched-pair** experiment, and the *d*'s are called the **paired differences**. This is the usual type of experiment found when we do "before-and-after" comparisons or when we compare siblings. If the subjects in an experiment can be "matched up" in some reasonable way, this type of one-sample test has greater power than the ordinary two-sample *t*-test. With it we are more likely to detect a difference if there is one.

***Note*** A paired-differences test can also be performed with more than 30 pairs, in which case we would perform a large-sample test of the sample mean.

## STUDY AIDS

### Chapter Summary

In this chapter you have learned:

- How to use *t*-curves and *t*-scores
- The concept of degrees of freedom
- How to do hypothesis tests using means from small samples
- The conditions under which such tests are valid
- How to do a paired-differences test—a before-and-after-type test

### Vocabulary

You should be able to explain the meaning of each of these terms:

1. Small samples
2. Student's *t*-scores
3. *t*-curve
4. Degrees of freedom
5. Matched pair
6. Paired differences

### Symbols

You should understand the meaning of each of these symbols:

1. $t$
2. $t_c$
3. $d$
4. $tD(\mu_m = 3, s_m = 1)$

**Formulas**

You should know when and how to use each of these formulas:

***For One-Sample Tests***

1. Degrees of freedom $= n - 1$
2. $\mu_m = \mu$
3. $s_m = \dfrac{s}{\sqrt{n}}$
4. $m_c = \mu_m + t_c s_m$
5. Experimental outcome: $m = \dfrac{\Sigma X}{n}$

***For Two-Sample Tests***

6. Degrees of freedom $= n_1 + n_2 - 2$
7. $\mu_{dm} = \mu_1 - \mu_2$. If $H_0$ states that $\mu_1 = \mu_2$, then $\mu_1 - \mu_2 = 0$.
8. $s_{dm} = \sqrt{s_p^2\left(\dfrac{1}{n_1} + \dfrac{1}{n_2}\right)}$ where
9. $s_p^2 = \dfrac{(n_1 - 1)(s_1^2) + (n_2 - 1)(s_2^2)}{n_1 + n_2 - 2}$
10. $\text{dm}_c = \mu_{dm} + t_c s_{dm}$
11. Experimental outcome: $d_m = m_1 - m_2$

## EXERCISES

11-27 The same data were used in two hypothesis tests in this chapter. In Exercises 11-17, using a $t$-test on the difference between 2 sample means, we failed to show that granting exemptions affected grades. However, in Example 11.7, using a $t$-test for paired differences, we proved that exemptions did affect the grades. Explain how the same data can lead to different conclusions.

11-28 In a study of the effects of cigarette smoking on blood conditions, 11 people had their blood sampled before and after they smoked a single cigarette. Each blood sample then had a chemical added to it which causes blood clots. A numerical value was given to describe the amount of clotting. Here are the results.

| Person | Difference $d$ in Clotting Measure Before and After Smoking |
|---|---|
| 1 | 2 |
| 2 | 4 |
| 3 | 10 |
| 4 | 12 |
| 5 | 16 |
| 6 | 15 |
| 7 | 4 |
| 8 | 27 |
| 9 | 9 |
| 10 | −1 |
| 11 | 15 |

You will find that $m = 10.3$ and $s = 8.0$. Using $\alpha = 0.01$, show that this is a statistically significant effect of smoking.*

11-29 (a) In order to test which brand of gasoline gives better mileage, the Acme Cab Company tried Axon gas in 10 cabs and Flug in 10 similar cabs. The cabs using Axon averaged 16.3 miles per gallon with $s_1 = 4.2$ miles per gallon. The cabs using Flug averaged 16.9 miles per gallon with $s_2 = 4.0$ miles per gallon. Test at the 0.01 significance level for any difference between the 2 brands.

(b) If Axon were used in 10 cabs for 1 week and then Flug were used in the same 10 cabs, we would have a paired-difference test. If $m = -0.6$ with $s = 0.39$, test at $\alpha = 0.01$.

11-30 (a) Assume that $X$ represents the number of children that 7 randomly selected first-born males produced, and $Y$ represents the number of children that 7 randomly selected first-born females produced.

| $X$ | 3 | 2 | 4 | 1 | 2 | 0 | 5 |
|---|---|---|---|---|---|---|---|
| $Y$ | 4 | 0 | 6 | 1 | 0 | 3 | 2 |

Test to see if $\mu_x > \mu_y$ at $\alpha = 0.05$.

(b) Now assume that $X$ represents the number of children of the male sibling and $Y$ represents the number of children of the female sibling in a random sample of pairs consisting of a brother and a sister. Test $\mu_x > \mu_y$ again at $\alpha = 0.05$.

11-31 Data were collected on 10 adults who enrolled in the Looze-A-Pound weight-reduction program. Their weights were recorded before and after the program. For the 10 pairs of differences, $m$ was 5.8 pounds lost with $s = 5$. Using $\alpha = 0.05$, is this evidence that the program works?

11-32 Dr. Quack has invented a new way to teach reading. A total of 40 sixth-graders were paired by reading ability and randomly divided into 2 sections. In 1 section, 20 were taught by Dr. Quack and, in the other section, 20 were taught in the ordinary manner. At the end of the course, the differences between the reading abilities of the pairs of students were computed, letting $d$ equal the Quack score minus the ordinary score. The average difference was $-3.6$ with $s = 4.0$. Is Dr. Quack's method significantly different? Use $\alpha = 0.05$.

11-33 An analysis of tennis at the Nassau Country Club included a study by Commissioner P. Recka of certain errors. He studied the play of 10 professionals, recording how often they missed on a backhand return and how often they missed on a forehand return, to see if there was any significant difference. Letting $d$ equal the backhand errors minus the forehand errors, he found for these 10 players (over the course of 100 errors) a mean value of $d = 10$ with $s = 6$. Does this indicate, at $\alpha = 0.05$, that in general more errors are committed on backhand shots?

11-34 The difference between the length of the left arm and the length of the right arm ($d$ equals left length minus right length) was measured on 200 right-handed marines. If

*Problem based on discussion in Glantz, *Primer of Biostatistics*, McGraw-Hill Book Company, New York, 1981.

$m = -0.05$ inches with $s = 0.015$ inches, test to see if marines have right arms that are longer than left arms. Use $\alpha = 0.05$.

11-35 A total of 10 students took a statistics test, and then they attended Dr. Fleece's Quickie Course in Imperative Statistics. After having completed the course, they took a test equivalent to the first one. Here are their grades.

| Student | Before Course | After Course |
|---|---|---|
| Re | 15 | 18 |
| Jeanne | 10 | 14 |
| Barbe | 13 | 15 |
| Tommy | 19 | 18 |
| Joe | 20 | 18 |
| Anne | 14 | 16 |
| Mary Frances | 3 | 17 |
| John | 18 | 17 |
| Paul | 16 | 14 |
| Tim | 16 | 15 |

Does the doctor's course improve grades? Use $\alpha = 0.05$.

11-36 A sample of fully grown cats were fed a diet of Fatkat brand food. The results of the diet are shown in the following table.

| Cat | Weight Before | Weight After |
|---|---|---|
| Tabby | 12 | 7 |
| Maurice | 7 | 7 |
| Samantha | 5 | 7 |
| Snow White | 7 | 6 |
| Toby | 3 | 9 |
| Licorice | 10 | 9 |
| Felix | 3 | 3 |
| Krazy | 4 | 5 |
| Apples | 6 | 7 |
| Puccini | 8 | 8 |
| Tiger | 7 | 8 |
| StatKat | 9 | 9 |

Test at the 0.05 significance level to see if Fatkat affects the cats' wieghts.

11-37 A music therapist, Alex Rosewell, thought that 13 minutes of a combination of Jewish and Irish traditional fiddle music would put a teenager in a good mood. He subjected 9 teenagers to the treatment, evaluating their moods with the Dowfeld Mood-Indicator Test. Analyze his data at $\alpha = 0.05$, and decide if the evidence supports his theory.

**Dowfeld Mood-Indicator Scores**

| Teenager | Before | After |
|---|---|---|
| 1 | 20 | 24 |
| 2 | 12 | 15 |
| 3 | 20 | 23 |
| 4 | 21 | 25 |
| 5 | 25 | 25 |
| 6 | 20 | 21 |
| 7 | 19 | 22 |
| 8 | 21 | 19 |
| 9 | 19 | 24 |

## CLASS SURVEY QUESTION

Consider this class as a random sample of your school. Are your class data strong enough to show that the average height of the males in this school is greater than the average height of the females in this school?

## FIELD PROJECT

Do a one-sample or a two-sample field project similar to any of those in Chapter 10, but using small samples, or do a project using a paired-difference design.

## CLASS EXPERIMENT

Do some experiment similar to those suggested in Chapter 10, but using small samples.

# Sample Test for Chapters 8, 9, 10, and 11

1. A statistical hypothesis is a statement about a __________ (parameter, statistic).
2. What is a Type I error? What does it have to do with "significance level"?
3. Why don't we conduct hypothesis tests using $\alpha = 0$?
4. We are testing to see if two populations have the same mean. We wish to use small samples and to be able to use a standard $t$-test. What two assumptions must be reasonable about the two populations?

   *For Problems 5 to 9:*
   (a) State a null and an alternative hypothesis.
   (b) State a decision rule.
   (c) Give a conclusion in good English mentioning the specific variables in the problems.
5. A carnival spinning wheel is one-quarter red. This means that the wheel should stop on red about one-fourth of the time. A student wonders if the wheel is honest. What would your decision be if the wheel came up red 12 out of 60 trials? $\alpha = 0.05$. Is this a one- or a two-tail test?
6. A study of a random sample of 100 parking receipts at a large airport indicated that the average time a car stayed in the "short-term" lot was 2.6 hours, with $s = 45$ minutes. Does this evidence imply, at $\alpha = 0.05$, that the true mean parking time is more than 2 hours and 15 minutes? (*Hint:* Be careful of units. You might want to change all values to minutes, or all to hours).
7. We wish to compare two new models of solar-powered cooking units. The test is to see how long each kind takes to boil 1 quart of water (which is at 50°F to begin with). Five trials are done on each type of unit. The results are as shown.

| | Time to Boil Water, Minutes |
|---|---|
| Sunny Boil | 4.7, 5.1, 5.5, 5.7, 4.0 |
| Sol-Heato | 4.2, 4.2, 4.7, 4.5, 4.4 |

   Is this conclusive evidence that one is superior to the other? Use $\alpha = 0.05$.
8. In a study for her Ph.D. thesis, Dr. Payne subjected 100 volunteers to the following experiment. The 100 people were split randomly into two groups of 50. In group 1 they were shown a film about heroic soldiers who endured suffering and emerged victorious. In group 2 they were shown a similar film,

but in it the soldiers finally lost the battle. Then each volunteer was asked to endure a very uncomfortable noise as long as he or she could. Do these results indicate that the films had any effect on the volunteers' behavior? Would you use a one-tail or a two-tail test? Use $\alpha = 0.05$.

| Group | Minutes of Noise Endured | |
|---|---|---|
| | $m$ | $s$ |
| 1 | 4.4 | 1 |
| 2 | 3.6 | 1 |

9. Ten sets of 8-year-old identical male twins were used in a study of the effect of teaching techniques. One twin in each pair was taught some vocabulary words by computer tutorial without teacher assistance. The other was taught by the teacher without machine aids. At the end they were all given the same test. The results are shown below.

| Names | Scores with Computer | Scores with Teacher |
|---|---|---|
| Abe and Babe | 116 | 123 |
| Bob and Rob | 112 | 170 |
| Clark and Mark | 100 | 140 |
| Don and Ron | 186 | 108 |
| Ed and Fred | 173 | 163 |
| Frank and Hank | 198 | 153 |
| Gary and Larry | 178 | 119 |
| Harry and Barry | 140 | 140 |
| Jake and Blake | 173 | 171 |
| Kenneth and Percival | 159 | 181 |

(a) Why were twins used?
(b) What assumptions were made about their vocabulary skills at the start of the experiment?
(c) At $\alpha = 0.05$, what does the study indicate? Do the analysis two ways: as a paired $t$-test and as an unpaired $t$-test.

10. Your Aunt Tilly is not mathematically inclined. Upon hearing that you were studying statistics in college, she wrote to you asking, Just what is a hypothesis test? Answer her as clearly as you can.

# 12

# Confidence Intervals

**CONTENTS**



**GOALS**

At the end of this chapter you will have learned:

- How to use sample statistics to estimate population parameters with point estimates, interval estimates, and confidence intervals
- How to interpret the percentage in a confidence interval

# 12.1 Confidence Intervals for Proportions in Binomial Experiments

An advertising agency wants to know the percentage of families in Nassau County that own a personal computer (PC). It would be very difficult and time-consuming to find the exact answer because that would mean checking every family in the county. So the agency selects a random sample of 500 families in the county instead. Suppose they find that 340 of the families own a PC. This is 340/500, or 68%. They would conclude that about 68% of the families in the county own a PC. The single best estimate for the true answer is 68%. They write $\hat{p} = 0.68$.

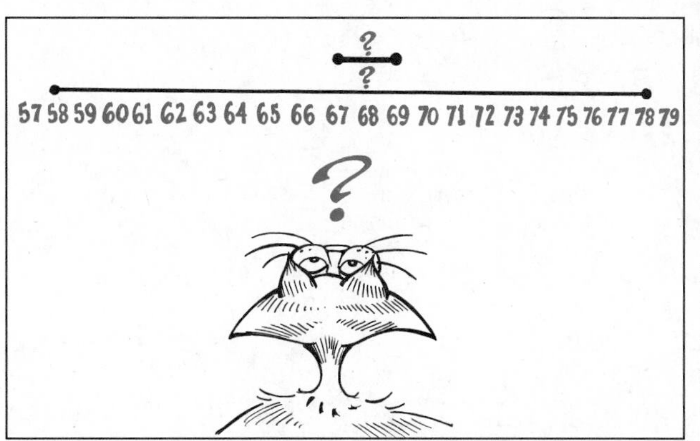


In most statistical applications it is not enough to say that the correct value of $p$ is *"about"* 68%. After all, that is a rather vague statement. Just what do we mean by "about" 68%? Do we mean that we believe the correct value is some value between 58% and 78%? Or do we mean that we believe the correct value is between 67% and 69%?

When we give a range of values that we think includes the true value of some population parameter, this range of values is called an **interval estimate**. Such an interval is usually assigned a probability, and then it is called a **confidence interval.** The higher the probability assigned, the more confident we are that the interval does, in fact, include the true value. Let us develop this idea further in the example below.

**interval estimate**

**confidence interval**

**point estimate**

*Definitions:* A **point estimate** is *one* value that is believed to be near a parameter from some population of interest.

An **interval estimate** is a *range* of values that is believed to contain a parameter.

A **confidence interval** is an interval estimate *and an associated probability* that the parameter is within the given interval.

## ❑ EXAMPLE 12.1 Families with PCs

**95% confidence interval**

Using a random sample of 500 families, compute an interval estimate for the percentage of families in Nassau County that own a PC. We want this interval to have a probability of 0.95 of containing the true percentage of families. In short, find the **95% confidence interval** for the percentage of families that own a PC.

### SOLUTION

Imagine that we *repeatedly* pick 500 families at random from Nassau County and record the number that own a PC. We would get a table like Table 12-1.

**Table 12-1**

| Sample Number | $n$ | Number with PC | Percentage $\hat{p}$ with PC |
|---|---|---|---|
| 1 | 500 | 340 | 68.0 |
| 2 | 500 | 345 | 69.0 |
| 3 | 500 | 304 | 60.8 |
| ⋮ | ⋮ | ⋮ | ⋮ |
| 300 | 500 | 375 | 75.0 |
| ⋮ | ⋮ | ⋮ | ⋮ |
| 1000 | 500 | 320 | 64.0 |
| ⋮ | ⋮ | ⋮ | ⋮ |

## A New Formula for $\hat{\sigma}$

We let $p = P$(a family owns a PC), the true percentage of families that own a PC. If $np > 5$ and $nq > 5$, then it follows that, in the long run, this distribution of percentages $\hat{p}$ would be approximately normal with mean equal to $p$ and standard deviation equal to $\sqrt{\frac{pq}{n}}$, where $n = 500$ families. Note that because the distribution is normal (Figure 12-1), the sample values for $\hat{p}$ will tend to cluster near the true value $p$.

In reality, we do the experiment only once. Suppose that when we check the 500 families, we find that 68% have PCs. On the basis of this we want to compute a confidence interval for $p$ (which we assume is near 68%). We have picked one of the many possible random samples that could have been selected from the county. We are faced with two possibilities: either we have picked a rare sample with a value of $\hat{p}$ far from $p$, or we have picked a more common sample with a value of $\hat{p}$ near $p$. Statisticians will always assume that one of the more common ones has been chosen. For instance, if we are computing the 95% confidence interval, we will assume that we have picked a sample with a value of $\hat{p}$ which occurs in the most-common 95% of the cases. (If we are computing the 99% confidence interval, we will assume that we have one of the most-common 99% of the cases.)

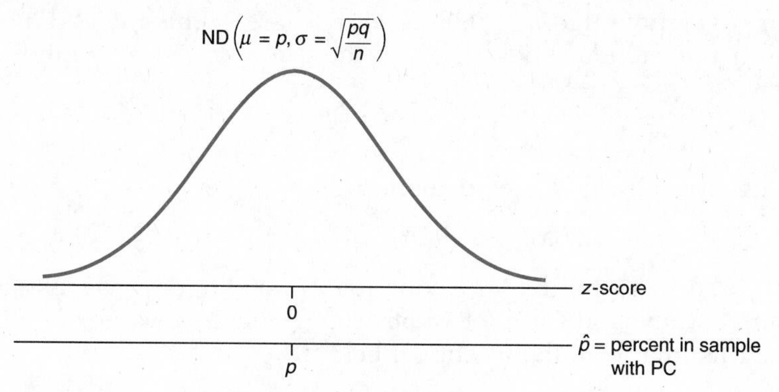


Figure 12-1 Sampling distribution of $\hat{p}$.

Let us follow through the reasoning which leads to the 95% confidence interval. You can find any other confidence interval by similar reasoning. Recall that our sample yielded a point estimate of $\hat{p} = 0.68$. Therefore we assume that $\hat{p} = 0.68$ is within the middle of 95% of the normal distribution based on the true value of $p$ (Figure 12-2).

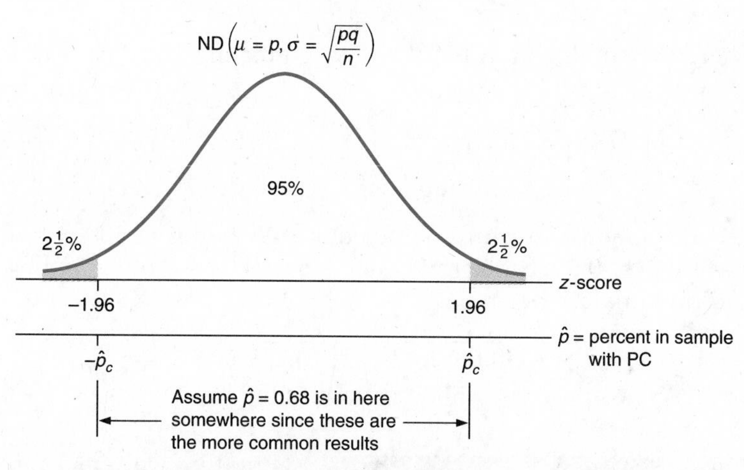


Figure 12-2 The middle 95% of the $\hat{p}$s.

## *Moral Certainty*

Jakob Bernoulli I (1713) introduced the idea of "moral certainty." For him, an event was morally certain (as opposed to absolutely certain) if its probability was no less than 99.9%. Bernoulli formulated his Law of Large Numbers to find the sample size necessary to achieve moral certainty in binomial experiments. This is the first known formal treatment of statistical error. In an illustration Bernoulli shows that to estimate the proportion of white pebbles in a mix of white and black pebbles and be morally certain that he was within 2% of the correct proportion, he needed a random sample of size 25,550. Such huge samples discouraged practical use of this theory. Later work by DeMoivre, which took advantage of the normal curve, showed that such large samples were not necessary.

If we assume that $\hat{p} = 0.68$ is somewhere in this interval, then $\hat{p} = 0.68$ is within 1.96 standard deviations of the true value of $p$. Symbolically, this is written as

$$p - 1.96\sigma < 0.68 < p + 1.96\sigma$$

Algebraically, this is equivalent to

$$0.68 - 1.96\sigma < p < 0.68 + 1.96\sigma$$

Therefore, to get the 95% confidence interval for $p$, we just add $1.96\sigma$ to the sample value and subtract $1.96\sigma$ from the sample value of $\hat{p}$. We can summarize this in a formula, as shown below.

*95% Confidence Interval for* p:

$$p \text{ lies between } \hat{p} - 1.96\sigma \text{ and } \hat{p} + 1.96\sigma$$

In a real application, of course, we do not know the exact value of $\sigma$ because it depends on the unknown $p$. So we use

$$\hat{\sigma} = \sqrt{\frac{\hat{p}\hat{q}}{n}}$$

Therefore the formula for the 95% confidence interval that we actually use is

*95% Confidence Interval for* p:

$$p \text{ lies between } \hat{p} - 1.96\hat{\sigma} \text{ and } \hat{p} + 1.96\hat{\sigma}$$

Let us apply this formula to our data. We had $n = 500$, $\hat{p} = 0.68$. Therefore, $\hat{q} = 1 - \hat{p} = 0.32$. Since $n\hat{p} = 340 > 5$ and $n\hat{q} = 160 > 5$, the distribution of $\hat{p}$ is approximately normal.

**margin of error**

$$\hat{\sigma} = \sqrt{\frac{\hat{p}\hat{q}}{n}} = \sqrt{\frac{0.68(0.32)}{500}} = \sqrt{\frac{0.2175}{500}} = \sqrt{0.0004} = 0.02$$

Finally, we calculate $1.96\hat{\sigma} = 1.96(0.02) = 0.04$. Therefore, we estimate that $p$ is between $0.68 - 0.04$ and $0.68 + 0.04$, or that $p$ is between 0.64 and 0.72. In symbols, $0.64 < p < 0.72$. In many media reports the value of the expression $1.96\hat{\sigma}$ is referred to as the **margin of error** or the error term for the estimate. In this example the margin of error is 4%.

*Conclusion:* On the basis of our sample data, the 95% confidence interval for the percentage of families that

### Caution!

Be careful when interpreting reports. Suppose the news mentions a survey showing that 55% of voters approve of the President's plan for health care with a margin of error of ±3%. You should realize that this is a popular way of reporting 95% confidence intervals. In particular, the report does not mean the researchers are *absolutely certain* that the true approval percentage is between 52% and 58%, but rather that there is only a 5% chance that this interval missed the true percentage.

own PCs is 64% to 72%. For this problem, we have answered our opening question, "What do we mean by 'about' 68%?" In this problem, "about" 68% means "between 64% and 72%." ❑

***Note*** The interval we arrived at in the previous example is a 95% confidence interval because the estimation procedure we used to get it has probability equal to 0.95 of producing an interval which contains the true percentage. In the procedure outlined in Table 12-2, we would find that about 95% of the confidence intervals we calculate do indeed "capture" the true parameter. When people informally report that they are "95% sure" a parameter lies in some interval, they really mean that the *procedure* will capture the parameter 95% of the time.

Table 12-2

| Sample Number | $n$ | Number with PC | Percentage $\hat{p}$ with PC | 95% Confidence Interval |
|---|---|---|---|---|
| 1 | 500 | 340 | 68.0 | 0.64 to 0.72 |
| 2 | 500 | 345 | 69.0 | 0.65 to 0.73 |
| 3 | 500 | 304 | 60.8 | 0.57 to 0.65 |
| . | . | . | . | . |
| . | . | . | . | . |
| . | . | . | . | . |
| 300 | 500 | 375 | 75.0 | 0.71 to 0.79 |
| . | . | . | . | . |
| . | . | . | . | . |
| . | . | . | . | . |
| 1000 | 500 | 320 | 64.0 | 0.60 to 0.68 |
| . | . | . | . | . |
| . | . | . | . | . |
| . | . | . | . | . |

In summary, the formulas for confidence intervals for $p$ are:

*95% confidence interval:* $p$ lies between $\hat{p} - 1.96\hat{\sigma}$ and $\hat{p} + 1.96\hat{\sigma}$

*99% confidence interval:* $p$ lies between $\hat{p} - 2.58\hat{\sigma}$ and $\hat{p} + 2.58\hat{\sigma}$

In general, $p$ lies between $\hat{p} - z_c\hat{\sigma}$ and $\hat{p} + z_c\hat{\sigma}$ where $z_c$ corresponds to the particular confidence interval we seek. Observe that the value of $z$ is the *same value* that you would use in a corresponding two-tail hypothesis test.

### A Quick Estimate

A conservative estimate for the margin of error can be found by taking $p = q = 0.5$ and $z_c = 2$. This leads to a margin of error equal to $1/\sqrt{n}$ and a 95% confidence interval of $\hat{p} \pm 1/\sqrt{n}$.

### Determining the Sample Size Needed for a Desired Margin of Error

As requested by the Dean of Furniture, you are estimating the percentage of left-handed students at State University. What size sample is needed to ensure that the margin of error of a 95% confidence interval is less than, say, 3%? That is, the quantity $1.96\sigma$, which we add to and subtract from our sample value, should be less than 3%. Algebraically, we write $1.96\sigma < 0.03$. Now since

$$\sigma = \sqrt{\frac{pq}{n}}$$

we have

$$1.96\sqrt{\frac{pq}{n}} < 0.03$$

Solving this for $n$, we obtain

$$n > \left(\frac{1.96}{0.03}\right)^2 pq$$

We don't know $p$ and $q$, but it is true that no matter what they are, their product $pq$ cannot be bigger than 0.25. (Try it! Remember $p$ plus $q$ must equal 1.) So if we replace $pq$ by 0.25, we get a value of $n$ that meets our requirements:

$$n > (1.96/0.03)^2(0.25) = 1067.1$$

**Margin of Error and Sample Size**

Note that the margin of error depends on the value of $n$ and that larger values of $n$ result in smaller margins of error.

Any sample bigger than this—that is, $n > 1068$—will give us our required accuracy.

In general, the formula for $n$ needed to achieve a desired accuracy is

$$n > \left(\frac{z_c}{a}\right)^2 pq$$

where $a$ is the desired accuracy, or margin of error.

---

❑ **EXAMPLE 12.2 Divorced Parents**

You want to know the percentage of students whose biological parents are divorced. You want to be within 4% of the correct value on a 95% confidence interval. What size sample do you need?

**SOLUTION**

The desired accuracy is $a = 0.04$.

$$n > \left(\frac{z_c}{a}\right)^2 pq = \left(\frac{1.96}{0.04}\right)^2 (0.25) = 600.25$$

Any random sample of size 601 or more is sufficient. ❑

---

### Exercises

12-1 When speaking about a 95% confidence interval, why is it not correct to say that $\alpha = 0.05$?

12-2 Explain the 2.58 in the formula for the 99% confidence interval for $p$.

12-3 Find a formula for the 90% confidence interval for $p$.

12-4 In a medical screening test for a certain disease, some proportion of the patients will be **false positives.** That is, they will show a reaction to the test, but on further examination will be found not to have the disease. When doctors design a screening test, they need an estimate of the proportion of positives that will be false in continued use of the test. They must perform a pilot study to get such an estimate. In 1 such pilot study 10 out of 100 positives were false positives. Give a 95% confidence interval for the proportion of positives that will be false.

12-5 Smee, the manager of a hardware store, read that about 53% of dog owners buy dog houses. Wondering what the true value of this percentage is, Smee sees that the sample size used in the research was $n = 100$. Find the 99% confidence interval for this percentage.

12-6 A newspaper reports that of 1064 soldiers in the 173rd Airborne Brigade, 16% said they used marijuana "about every day" or "more often." Find the 95% confidence interval for this estimate.

12-7 Private Walter Parte, one of General Custer's privates, took a random sample of 50 men in his regiment. Of them, 30 felt that they had the enemy outnumbered. Find the 95% confidence interval for the proportion of men who felt this way.

12-8 A random sample of 90 prisoners in a large state prison revealed that 20% were college graduates. Find the 99% confidence interval in the population sampled for the percentage of prisoners who are college graduates.

12-9 Professor Kwizee's tests this year have contained 100 true-false questions, and so far this term 80 of them have been true. Presuming that he will continue in this vein, find the 90% confidence interval for the proportion of "trues" on the next test.

12-10 If the early returns on election night are considered a random sample, and Mayor Fogbottom is leading her opponent with 52% of the vote, based on the first 400 votes counted, should she be 99% confident of victory?

12-11 At a certain beach in New Jersey there is a local rule which says that only town residents may use the beach. One Sunday, 80 people (at random) on the beach are checked. It turns out that 68 are residents.

(a) What is the 95% confidence interval for the percentage of nonresidents using the beach?

(b) Based on your answer to part (a), if there are 700 people on the beach, about how many are nonresidents?

12-12 Sanford R. Brochure, the noted mail-advertising magnate, wants to verify the percentage of the population of Upperdownunder that discard the mail he sends them without even opening it. How many residents should be sampled if he wants to be 95% sure that he is within about 2% of the correct value of $p$?

12-13 Not every binomial experiment is as simple as a heads-tails or yes-no problem. An 1800-step computer program simulates the joint use of sonar by a navy destroyer and a helicopter to locate an enemy submarine. At the end of each run of this program, either the submarine has been spotted or it has escaped.

(a) When the program was run 100 times, the submarine was located 32 times. Estimate $p = P$(a submarine is located). Find the 95% confidence interval for your estimation.

(b) Repeat part (a) for the data $n = 500$ and the submarine located 139 times.

(c) Repeat part (a) for the data $n = 1000$ and the submarine located 297 times.

(d) We know that computer time costs money. On the basis of your answers to parts (b) and (c) alone, was it worthwhile running the program 1000 times, or would 500 times have been just as useful?

(e) Show that for $n = 38{,}000$(!) we can estimate $p$ to within $\pm\frac{1}{2}\%$.

12-14 Milly Meter, a theater manager, wants to know what percentage of the elementary-school children in West Islip have seen *Snow White.*

(a) What size samples does she need to be 99% sure that she is within 3% of the correct value?

(b) What size sample does she need to be 95% sure that she is within 1% of the correct value?

12-15 Coolo Keeno watches a lot of TV. She wishes to estimate the percentage of scenes that last less than 5 seconds. How many scenes must she time if she wishes to be 99% sure that she is within 5% of the correct value?

## 12.2 Confidence Intervals for Means Based on Large Samples

To estimate the mean of a population we typically use the mean of a random sample. For example, the average time it takes for an anesthetic to take effect could be estimated by trying it on a random sample of 100 patients and finding the mean for the sample. If we found the mean of the sample to be, say, $m = 4.6$ minutes, we could use this point estimate as our best guess at the mean of the population. We would expect, if we put the anesthetic into general use, that the mean time to take effect would be *about* 4.6 minutes. To give meaning to the word "about," we can find the 95% confidence interval for the mean. To do this when the sample size $n$ is greater than 30, we use the central limit theorem (Chapter 10). We reason that the sample of patients we studied is an ordinary sample, one of the usual 95% of the samples with $m$ near the true mean, and not one of the very rare 5% of the samples. Thus, we can be reasonably sure that the true mean is not more than 1.96 standard deviations away from our sample mean. This is illustrated in Figure 12-3.

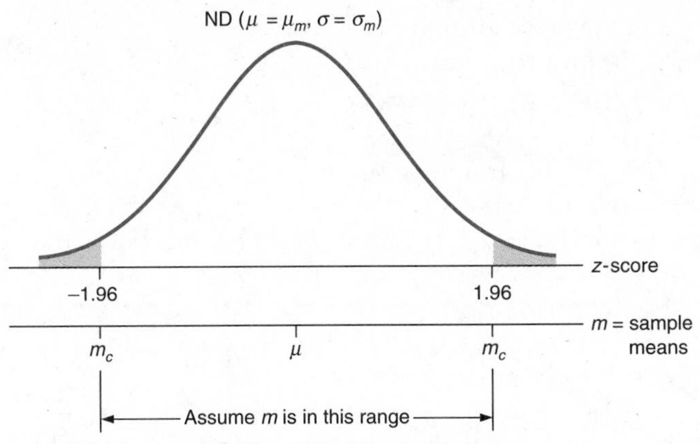


**Figure 12-3** Sampling distribution of $m$.

Following the reasoning that we used in the binomial experiment at the beginning of this chapter, we are led to similar formulas for the confidence intervals of population means.

The formulas for confidence intervals for $\mu$ are

*95% confidence interval:* $\mu$ lies between $m - 1.96s_m$ and $m + 1.96s_m$
*99% confidence interval:* $\mu$ lies between $m - 2.58s_m$ and $m + 2.58s_m$

In general, for large samples, $\mu$ lies between $m - z_c s_m$ and $m + z_c s_m$, where $z_c$ is the appropriate critical $z$-score.

---

### ❑ EXAMPLE 12.3 Anesthetic Times

Suppose that our sample of 100 patients led us to compute $m = 4.6$ minutes with $s = 1.1$ minutes. What is the 95% confidence interval for the average time it takes for the anesthetic to take effect?

**SOLUTION**

$$s_m = \frac{s}{\sqrt{n}} = \frac{1.1}{\sqrt{100}} = \frac{1.1}{10} = 0.11 \qquad \text{and} \qquad n = 100 > 30$$

Hence the error term is $1.96s_m = 1.96(0.11) = 0.22$. Therefore, the 95% confidence interval for $\mu$ is between $4.6 - 0.22$ and $4.6 + 0.22$, or $\mu$ is between 4.4 and 4.8 minutes. ❑

---

### ❑ EXAMPLE 12.4 The Toothpick Factory

Find the 99% confidence interval for the average age of workers in the A-Maize toothpick factory if a random sample of only 100 workers there had a mean age of 35.2 with $s = 10.3$.

SOLUTION

$$s_m = \frac{s}{\sqrt{n}} = \frac{10.3}{\sqrt{100}} = \frac{10.3}{10} = 1.03$$

Since $n = 100 > 30$, we use a $z$-score.

Therefore, we find $2.58s_m = 2.58(1.03) = 2.7$; hence, the confidence interval for $\mu$ is from $35.2 - 2.7$ to $35.2 + 2.7$, or $\mu$ lies between 32.5 and 37.9. ❑

---

### Exercises

12-16 A rule of thumb used by some people to obtain a quick confidence interval is to simply go 2 standard deviations in each direction from the mean. What percent confidence interval does this rule yield for data based on large samples?

12-17 Find the point estimate and the 95% confidence interval for the mean weight of a Venusian if a random sample of 36 Venusians weighed 16, 22, 31, 28, 15, 20, 20, 21, 22, 35, 28, 27, 25, 24, 20, 18, 19, 31, 17, 18, 20, 15, 25, 24, 27, 18, 20, 31, 29, 23, 20, 20, 19, 31, 30, and 20 *glymphs* (the unit for Venusian weight).

12-18 Find the point estimate and the 99% confidence interval for the mean number of persons riding in a subway car between midnight and 6 A.M. if a random sample of 40 subway cars had the following number of persons: 0, 0, 1, 3, 7, 13, 15, 20, 20, 20, 23, 25, 29, 30, 35, 35, 36, 36, 36, 36, 37, 40, 41, 41, 41, 43, 44, 47, 50, 56, 59, 60, 60, 61, 63, 69, 70, 71, 89, and 103.

12-19 Josh manages a radio station. *Each week* he reports the number of minutes that the station broadcast a song from the "top 10" list. Last year the mean of these 52 weekly reports was 3523 minutes with $s = 162$ minutes. Next year the station will keep the same programming format. Find a 95% confidence interval for next year's mean time per week given to the "top 10" list.

12-20 Find the 95% confidence interval for the mean salary of teachers in Constick County if a random sample of 100 teachers had a mean salary of $29,000 with $s = \$1000$.

12-21 Find the 99% confidence interval for the mean number of hours of TV watched by children per day if a random sample of 49 children had a mean of 4.6 hours with $s = 2.9$ hours.

12-22 Find the 95% confidence interval for the mean number of pieces of junk mail received per week in the United States if a random sample of 1000 homes had a mean of 23.1 pieces with an estimated standard deviation of 4.1 pieces.

12-23 Find the 98% confidence interval for the mean number of minutes of commercials per hour if a random sample of 48 hours of TV yielded a mean of 15.2 minutes per hour and $s = 1.5$ minutes per hour.

12-24 Find the 95% confidence interval for the average time spent waiting for a prescription to be filled at Rex's Rx Center if a random sample of 64 customers showed an average waiting time of 12.3 minutes with a standard deviation of 5.0 minutes.

12-25 A sample of 400 students turned in statistics projects to Professor Rick Williams. The average sample size was 110.3 and the standard deviation was 12.2. Find the 99% confidence interval for the average sample size of student projects.

12-26 The commissioner of Little League surfing in Spearfish, South Dakota, is planning the annual picnic. He knows from past years that the average number of hot dogs consumed is 6.5 per person, with 95% confidence interval from 6.1 to 6.9 hot dogs. He expects all 507 Little Leaguers to show up for the picnic.
(a) Using the average 6.5, how many hot dogs should he order?
(b) If he wants to be sure that he does not run out of hot dogs as he did last year (before the riot), how many should he order?
(c) If he orders as in part (b), using the high end of the interval, and it turns out that the mean number of hot dogs this year is near the low end of the interval, how many hot dogs will be left over?
(d) If he does get stuck as in part (c), and he has to eat 2 hot dogs a day until the leftovers are gone, how long will it take him to finish the hot dogs?

## 12.3 Confidence Intervals for Means Based on Small Samples

In this section dealing with small-samples, we will follow the same procedures used above to find confidence intervals, except that we will use critical $t$-scores instead of critical $z$-scores. The reasoning behind this is given in the chapter on small sample tests (Chapter 11).

***Note:*** Recall that when using $t$-scores we must have approximately normal distributions for the original data.

Thus, the formula for a confidence interval for a mean becomes

$$\mu \text{ is between } m - t_c s_m \text{ and } m + t_c s_m$$

### ❑ EXAMPLE 12.5 Average Gas Mileage

A lawyer must drive from a northern suburb to Chicago's Loop every day to get to work. In order to decide whether he should take the train instead, he computes his gas mileage each day for 15 days. He finds that he has a mean mileage of 12.2 miles per gallon, with an estimated standard deviation of 2.1

miles per gallon. Find the 95% confidence interval for his mileage, assuming that gas mileage is normally distributed.

**SOLUTION**

$$s = 2.1 \quad \text{and} \quad n = 15.$$

Therefore,

$$s_m = s/\sqrt{n} = 2.1/\sqrt{15} = 0.54$$

Since $n$ is not greater than 30, we use a $t$-score with degrees of freedom equal to $n - 1 = 14$.

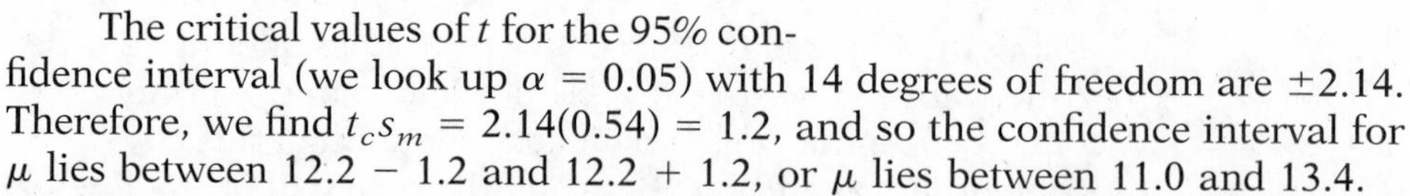

The critical values of $t$ for the 95% confidence interval (we look up $\alpha = 0.05$) with 14 degrees of freedom are $\pm 2.14$. Therefore, we find $t_c s_m = 2.14(0.54) = 1.2$, and so the confidence interval for $\mu$ lies between $12.2 - 1.2$ and $12.2 + 1.2$, or $\mu$ lies between 11.0 and 13.4.

Thus, the lawyer is 95% confident that the true gas mileage lies between 11.0 and 13.4 miles per gallon. He can use this information to decide whether it is cheaper to drive or to take the train. ❑

## Exercises

*In these exercises, assume that the populations being sampled are approximately normal.*

12-27 If you mistakenly used a $z$-score instead of a $t$-score for a small-sample confidence interval, would the resulting interval be too wide or too narrow?

12-28 In an overcrowded computer lab, students must wait to get at a computer terminal. John De Newman recorded the times he waited for his last 10 visits as 15, 14, 12, 15, 15, 16, 14, 15, 16, and 14 minutes. Estimate his average waiting time, and find the 99% confidence interval for your estimate.

12-29 Naomi makes and sells earrings. Over the past year the mean *monthly* income from the business was \$814 with $s = \$148$. She expects next year's sales to be similar. Find a 95% confidence interval for next year's mean *monthly* income.

12-30 Watch Out! Inc., a consumer organization, wants to estimate the life of Extra-Strong lightbulbs. With 25 randomly purchased bulbs, they find the mean life to be 998 hours, with a standard deviation $s$ of 30 hours. Find the 95% confidence interval for the mean life of the bulbs.

12-31 A doctor desires to estimate the mean age of women having hysterectomies in New York City municipal hospitals. She finds that the mean age in a random sample of 19 patients is 39.7 years, with $s$ equal to 3.2 years. Find the 99% confidence interval for the mean age of women having hysterectomies in these hospitals.

12-32 The Gypsy Taxi Cab Company of Brooklyn, New York, desires to know the mean life of tires on its cabs. A random sample of 28 tires from the cabs had a mean life of 17,821 miles, with a standard deviation of 1206 miles. Find the 95% confidence interval for the life of the tires.

12-33 An efficiency expert for Kleen-Wipe, a chalkboard eraser company, gathers a random sample of 16 workers' hourly production. He finds that the mean number of erasers produced per worker per hour is 123.1, with $s$ equal to 8.0. Find the 90% confidence interval for the mean number of erasers produced by a worker in 1 hour.

12-34 Peggy Babbcock of Fredericksburg, Virginia, won the rapid-speaking contest 3 times, rapidly using the phrase "red leather, yellow leather." Her rate of speaking was 7.3, 7.1, and 7.2 syllables per second. Find the 95% confidence interval for her average rate of speaking at contests. (If you can say her name 3 times rapidly, you will get second prize.)

# 12.4 Two-Sample Confidence Intervals for Differences

***Abbé Mersenne***

The Mersenne Academy, founded by the friar minor Abbé Marin Mersenne (1588–1648) was a center for the development of the science of probability. Mersenne carried on an extensive correspondence with Descartes, Fermat, and Galileo. Every week Descartes, Desargues, and Pascal gathered along with many others to discuss mathematics and science. These meetings were the forerunners of the Royal Society of London and the Académie des Sciences of Paris.

Just as we have estimated the value of a binomial probability $p$ or a population mean $\mu$ by taking a sample from one population, so too we can estimate the difference between two probabilities, $p_1 - p_2$, or the difference between two means, $\mu_1 - \mu_2$, by taking separate and independent samples from each of two populations.

## 12.4-1 Differences between Two Proportions

❑ **EXAMPLE 12.6 Blonds vs. Redheads**

Suppose we are interested in the difference between the proportion of blonds at Men's Institute of Tennis who wear contact lenses and the proportion of redheads there who wear contact lenses. If $p_1 = P$(a blond at MIT wears contact lenses) and $p_2 = P$(a redhead at MIT wears contact lenses), then we wish to estimate the value of the parameter $p_1 - p_2$. We could gather two random samples of students at MIT, one a sample of blonds and the other a sample of redheads. If 90 of 300 blonds wore contact lenses, while 24 of 200 redheads wore contact lenses, we would estimate $p_1$ and $p_2$ by

$$\hat{p}_1 = 90/300 = 0.30$$
$$\text{and} \quad \hat{p}_2 = 24/200 = 0.12$$

Then our point estimate of $p_1 - p_2$ would be $\hat{p}_1 - \hat{p}_2 = 0.30 - 0.12 = 0.18$. Therefore, about 18% more blonds than redheads wear contact lenses. To calculate the 99% confidence interval for this estimate, we note that $n_1\hat{p}_1$, $n_1\hat{q}_1$, $n_2\hat{p}_2$, and $n_2\hat{q}_2$ are equal to the 4 outcomes 90, 210, 24, and 176. Since these outcomes are all greater than 5, the distribution of all possible estimates would be approximately normal, and the middle 99% will be between $z = -2.58$ and $z = 2.58$. If we let $d\hat{p}$ symbolize $\hat{p}_1 - \hat{p}_2$, then

$$\mu_{d\hat{p}} = p_1 - p_2 \quad \text{and} \quad \sigma_{d\hat{p}} = \sqrt{\frac{\hat{p}_1\hat{q}_1}{n_1} + \frac{\hat{p}_2\hat{q}_2}{n_2}}$$

**Note** The question is, "How *different* are $p_1$ and $p_2$?" We are *not* assuming that $p_1$ equals $p_2$, and hence, we do not pool our sample results as we did in Chapter 9. That's why the formula is different.

The confidence interval is given by

$$p_1 - p_2 \text{ is between } d\hat{p} - z_c\hat{\sigma}_{d\hat{p}} \text{ and } d\hat{p} + z_c\hat{\sigma}_{d\hat{p}}$$

Computing $\hat{\sigma}_{d\hat{p}}$, we have

$$\hat{\sigma}_{d\hat{p}} = \sqrt{\frac{0.30(0.70)}{300} + \frac{0.12(0.88)}{200}} = 0.035$$

We find $z_c\hat{\sigma}_{d\hat{p}} = 2.58(0.035) = 0.09$. Thus, the 99% confidence interval for $p_1 - p_2$ is from $0.18 - 0.09$ to $0.18 + 0.09$, or $p_1 - p_2$ lies between 0.09 and 0.27.

The difference is that anywhere from 9 to 27% more blondes at MIT wear contact lenses. We have conclusive evidence that a larger percentage of blondes wear contact lenses, but we cannot pin down the difference any more than to say it is somewhere between 9 and 27%. If we wish to have a narrower confidence interval with this size sample, then we will have to run more than a 1% risk of error. Correspondingly, if we wish to estimate the difference between the percentage of blondes and the percentage

### *Confidence Intervals vs. Hypothesis Testing*

*Note:* There is a similarity between 95% confidence intervals and two-tail hypothesis tests. For example, when a 95% confidence interval for $p_1 - p_2$ includes the number 0, this is roughly equivalent to failing to reject the null hypothesis $p_1 = p_2$ with $\alpha = 0.5$. Conversely, if the interval does not contain 0, this is analogous to rejecting the null hypothesis. Some statisticians prefer the confidence-interval approach over hypothesis testing.

of redheads more closely and not increase the risk of error, we will have to get larger samples. ❑

## 12.4-2 Differences between Two Means

Depending upon whether or not both $n_1$ and $n_2$ are larger than 30, the confidence interval for $\mu_1 - \mu_2$ is given by

$$\mu_1 - \mu_2 \text{ is between dm} - z_c s_{\text{dm}} \text{ and dm} + z_c s_{\text{dm}}, \text{ or}$$

$$\mu_1 - \mu_2 \text{ is between dm} - t_c s_{\text{dm}} \text{ and dm} + t_c s_{\text{dm}}$$

### ❑ EXAMPLE 12.7 Window Sizes Using Large Samples

Find the 95% confidence interval for the difference between the average size of a window in a one-family house built over 10 years ago and the average window size in a newer house if 2 random samples gave the following data:

400 windows in older houses, $m_1 = 15.6$ square feet, $s_1 = 20.2$ square feet
500 windows in new houses, $m_2 = 19.0$ square feet, $s_2 = 24.8$ square feet

SOLUTION

We first calculate dm $= m_1 - m_2 = 15.6 - 19.0 = -3.4$. For a point estimate we can say that windows in new houses are about 3.4 square feet larger than windows in older houses. Since $n_1 = 400$ and $n_2 = 500$ are both larger than 30, the sampling distribution of differences is approximately normal and we can use $z_c = \pm 1.96$, and

$$s_{\text{dm}} = \sqrt{\frac{s_1^2}{n_1} + \frac{s_2^2}{n_2}} = \sqrt{\frac{(20.2)^2}{400} + \frac{(24.8)^2}{500}} = 1.50$$

We find $z_c s_{\text{dm}} = 1.96(1.50) = 2.9$. Hence, the confidence interval for $\mu_1 - \mu_2$ is from $-3.4 - 2.9$ to $-3.4 + 2.9$, or $\mu_1 - \mu_2$ lies between $-6.3$ and $-0.5$.

We are 95% sure that the average size of windows in newer houses is at least 0.5 but at most 6.3 square feet larger than the average size of windows in houses over 10 years old. ❑

### ❑ EXAMPLE 12.8 Space Travel Using Small Samples

Corresponding to our work in Chapter 11 on small-sample hypothesis tests, we will present an approach for the case in which it is reasonable to assume that:

1. Both populations are *normal.*
2. $\sigma_1$ and $\sigma_2$ are approximately *equal.*

This allows us to use the $t$-distribution in the analysis. Under these assumptions, we use the pooled version of $s_{dm}$. (You should refer to a more extensive text for an approach to use when the second assumption is not reasonable.)

(a) In a pilot study of the time it takes to fly from Earth to the planet Hangon, a sample of 10 human pilots averaged 8.9 velans with $s = 1.1$. Forty Hangon pilots averaged 7.1 velans with $s = 1.2$. Estimate the difference between the average times of these two populations.
(b) Find a 95% confidence interval for your estimate.

SOLUTIONS

(a) $m_1 - m_2 = 8.9 - 7.1 = 1.8$. Our best point estimate is that the average time of Earth pilots is 1.8 velans more than the average time of the Hangon pilots.

(b)
$$s_p^2 = \frac{(n_1 - 1)s_1^2 + (n_2 - 1)s_2^2}{n_1 + n_2 - 2}$$

$$= \frac{9(1.1)^2 + 39(1.2)^2}{48} = \frac{67.05}{48} \approx 1.40$$

$$s_{dm} = \sqrt{s_p^2\left(\frac{1}{n_1} + \frac{1}{n_2}\right)} = \sqrt{1.40\left(\frac{1}{10} + \frac{1}{40}\right)} \approx 0.42$$

Since $n_1 = 10$ and is smaller than 30, we have a $t$-distribution with $n_1 + n_2 - 2 = 10 + 40 - 2 = 48$ degrees of freedom. Looking up the critical $t$-scores for the middle 95% of the outcomes in Table D-5 (see Appendix D), we use 50 degrees of freedom, the entry nearest to 48, and find $t_c = \pm 2.01$. We find $t_c s_{dm} = 2.01(0.42) = 0.84$. Thus, the confidence interval $\mu_1 - \mu_2$ is from $1.8 - 0.84$ to $1.8 + 0.84$, or $\mu_1 - \mu_2$ lies between 1.0 and 2.6.

We are 95% sure that human pilots average at least 1.0 velans more than Hangon pilots do for the trip from Earth to Hangon. ❑

## STUDY AIDS

**Chapter Summary**

In this chapter you have learned:

- The difference between a point estimate, an interval estimate, and a confidence interval
- How to calculate confidence intervals for binomial estimates and sample mean estimates for both one- and two-sample experiments

- How to correctly interpret the 95% probability in a 95% confidence interval in terms of repeating the *procedure*.

**Vocabulary**

You should be able to explain the meaning of each of these terms:

1. Interval estimate
2. Confidence interval
3. Point estimate
4. The percentage named in a confidence interval (95% confidence interval)
5. Margin of error

**Formulas**

You should know when and how to use each of these formulas:

***For Distribution of Sample Proportions***

1. $\mu = p$
2. $\sigma = \sqrt{\dfrac{pq}{n}}$
3. $\hat{\sigma} = \sqrt{\dfrac{\hat{p}\hat{q}}{n}}$
4. $p$ lies between $\hat{p} - z_c\hat{\sigma}$ and $\hat{p} + z_c\hat{\sigma}$
5. $n > \left(\dfrac{z_c}{a}\right)^2 pq$

***For Distribution of Sample Means***

6. $s_m = \dfrac{s}{\sqrt{n}}$
7. $\mu$ lies between $m - z_c s_m$ and $m + z_c s_m$ (large samples)
8. $\mu$ lies between $m - t_c s_m$ and $m + t_c s_m$ (small samples)

***For Differences between Two Proportions***

9. $\hat{\sigma}_{d\hat{p}} = \sqrt{\dfrac{\hat{p}_1\hat{q}_1}{n_1} + \dfrac{\hat{p}_2\hat{q}_2}{n_2}}$
10. $p_1 - p_2$ lies between $d\hat{p} - z_c\hat{\sigma}_{d\hat{p}}$ and $d\hat{p} + z_c\hat{\sigma}_{d\hat{p}}$

***For Differences between Two Means***

11. $s_{\text{dm}} = \sqrt{\dfrac{s_1^2}{n_1} + \dfrac{s_2^2}{n_2}}$ (large samples)
12. $s_p^2 = \dfrac{(n_1 - 1)s_1^2 + (n_2 - 1)s_2^2}{n_1 + n_2 - 2}$ (small samples)
13. $s_{\text{dm}} = \sqrt{s_p^2\left(\dfrac{1}{n_1} + \dfrac{1}{n_2}\right)}$ (small samples)
14. $\mu_1 - \mu_2$ lies between $\text{dm} - z_c s_{\text{dm}}$ and $\text{dm} + z_c s_{\text{dm}}$ (large samples)
15. $\mu_1 - \mu_2$ lies between $\text{dm} - t_c s_{\text{dm}}$ and $\text{dm} + t_c s_{\text{dm}}$ (small samples)

## EXERCISES

12-35 In the upcoming election for president of the Student Government Association at Taxes A&M, a random survey of agricultural students found that 160 students intended to vote for "Bubba" Knox, while 340 were going to vote for Betty Jean Sue Faulkner. A similar survey of the mining students found that 220 would vote for Bubba, while 360 preferred Betty Jean Sue.

(a) Calculate the 95% confidence interval for the difference in the percentages of students from each group voting for Bubba.

(b) Calculate the 95% confidence interval for the difference in the percentages of students from each group voting for Betty Jean Sue.

(c) Is there an easier way to answer part (b)?

12-36 Find a 99% confidence interval for the difference between the percentage of Jewish students who joined Hillel and the percentage of Catholic students who joined Newman at Luther University, if a random sample of 500 persons showed that 20 of 80 Jewish students belonged to Hillel, while 24 of 120 Catholic students belonged to Newman.

12-37 A sample of 500 high school students and 500 adults in Reading, Pennsylvania, showed that 350 students and 250 adults owned library cards. Find the point estimate and the 95% confidence interval for the difference between the percentage of each group that owns a library card.

12-38 Suppose that data from 2 samples at Ocean County Country Club are used to compute a 95% confidence interval for the difference between the means of the 2 populations. These same data are also used to perform a hypothesis test at the analogous 0.05 significance level. If the confidence interval goes from a negative value to a positive value, would you generally expect to reject the null hypothesis or not to reject it?

12-39 Georgia and Greg, Florida newspaper reporters, investigated prices of blood glucose monitors at 10 randomly selected pharmacies in Riche County and 10 randomly selected pharmacies in Poore County. Here are the results. Assuming that prices are normally distributed, find a 95% confidence interval for the difference between the mean price of a monitor in the 2 counties.

| Riche | Poore |
|---|---|
| 59 | 89 |
| 79 | 109 |
| 70 | 100 |
| 55 | 85 |
| 60 | 90 |
| 58 | 88 |
| 49 | 79 |
| 50 | 80 |
| 50 | 80 |
| 50 | 80 |

12-40 Dr. Showvan spent years gathering the IQ data below.

| | Male | Female |
|---|---|---|
| $n$ | 10,000 | 10,000 |
| $m$ | 100.3 | 99.9 |
| $s$ | 10.8 | 11.1 |

He then performed a test on the difference between the average IQ of American males and the average IQ of American females. He calculated

$$\text{dm}_c = 0 \pm 1.96\sqrt{\frac{10.8^2}{10{,}000} + \frac{11.1^2}{10{,}000}} = \pm 0.30$$

$$\text{dm} = 100.3 - 99.9 = 0.4$$

Hence he concluded that the average IQ of American males is higher than the average IQ of American females with $\alpha = 0.05$. He claims that his results are "statistically significant."

(a) Show that the 99% confidence interval for this difference is $0.0 < \text{dm} < 0.8$.

(b) Discuss the difference between "statistical significance" and "importance."

3 12-41 A teacher desires to estimate the difference in reading levels between children from two-parent homes and children from one-parent homes in his community. Using 2 random samples of fourth-grade youngsters, he finds that 19 children from two-parent homes had a mean reading level $m_1$ of 5.1 with standard deviation $s_1 = 1.4$ and that 13 children from one-parent homes had a mean reading level $m_2$ of 3.8 with standard deviation $s_2 = 2.1$. Find a 99% confidence interval for the difference in the reading levels.

12-42 The Gypsy Taxi Cab Company of Brooklyn, New York, is still checking tires. They now want to know if cab drivers under 25 are harder on tires than are older drivers. Of their 76 cabs, 32 are driven exclusively by the younger drivers and the remaining 44 by the older drivers. The younger drivers average 17,482 miles for a set of tires with a standard deviation of 1320 miles, while the older drivers average 17,728 miles with a standard deviation of 981 miles. Find a 99% confidence interval for the estimate of the true difference in mileage.

12-43 *Gotcha*, the consumer magazine, is testing the lives of 2 kinds of flashlight batteries. Britelite claims to give more life in normal use but is more costly than ordinary batteries. They randomly purchase 50 Britelite batteries and 50 ordinary batteries. The Britelite batteries are found to have a mean life of 17.5 months of normal use with a standard deviation of 1.1 months. The ordinary batteries have a mean life of 14.7 months with a standard deviation of 1.3 months. Find a 95% confidence interval for the time difference in the lives of the batteries.

12-44 Albert Sechsauer, an efficiency expert, claims that an additional afternoon break in a factory will result in more production on an assembly line. In the prior month, with 21 working days, the line produced a mean of 72.3 items per day with a standard deviation of 3.4 items. In the 22 working days of the first month with the break, the line produced a mean of 70.6 items with a standard deviation of 2.1 items.

(a) Find a 99% confidence interval for the estimate of the true difference between the 2 production averages.

(b) Without the break, in 21 days the line had a mean of 3.1 defective items per day with a standard deviation of 0.43 item. In the 22 days after the break was instituted, the line had a mean of 2.4 defective items with a standard deviation of 0.53 item. Find a 99% confidence interval for the estimate of the difference between the number of defects.

\ 12-45 Two terrific Irish traditional musicians often attend the Sunday afternoon informal sessions in their home towns. They are always teased, though, about showing up late.

Over the last 52 weeks, Billy was late every time except the first Sunday of every month. Liz was on time just for the 8 Sundays in March and April. Calculate the 95% confidence interval for the difference in the percentage of sessions for which the 2 people were late.

## CLASS SURVEY QUESTION

Find the point estimate and a confidence interval for the percentage of students in your school that are left-handed.

## FIELD PROJECTS

1. Estimate the mean of a population, the difference between the means of two populations, the proportions of a binomial population, or the difference between two proportions. Find a confidence interval for your estimate. Some examples which other students have tried on campus include estimates of:
   (a) Average age of faculty
   (b) Difference between average age of male and female faculty
   (c) Average amount of coin change carried by students
   (d) Percentage of students who use public transportation to get to campus
2. Obtain a newspaper report which contains an estimated percentage, the sample size, and a margin of error. Use the methods of this chapter to calculate the margin of error yourself. Is your result close to the one in the report? They may not agree exactly because the reported study may have used a complex sampling design and not just a simple random sample.

# 13

# Chi Square

In 1900 Karl Pearson (1857–1936) invented the chi-square statistic to demonstrate that the roulette wheel at Monte Carlo was not fair.

**CONTENTS**



**GOALS**

At the end of this chapter you will have learned:

- How to test two variables to discover whether they are statistically independent
- How to test whether or not a random variable conforms to a predetermined model

## 13.1 Chi-Square Tests of Independence

In Chapter 9 we saw how we could compare two proportions. For example, we could compare men and women in some community on their attitudes toward building a nuclear power plant nearby. If we should find only a small difference between the percentage of men and the percentage of women who are in favor of building the plant, we would say that the difference is not statistically significant. Another way to express this finding is to say that men and women respond similarly to the question, or that a person's sex *has nothing to do* with his or her attitude on building the plant. In the technical language of statistics, we can say that the two variables—(1) *sex* and (2) *opinion on building the nuclear plant*—are **independent**. In this chapter we shall look further at the idea of statistical independence.

independent

### ❑ EXAMPLE 13.1 Pros and Cons of a Nuclear Power Plant

Suppose we took a random sample of 200 people in this community and found that of 50 men and 150 women sampled, 60 people favored building the nuclear power plant, 100 were against it, and 40 had no opinion. We can summarize these results in Table 13-1, called a **contingency table.**

contingency table

Table 13-1

| | Men | Women | Row Totals |
|---|---|---|---|
| In favor of building plant | | | 60 |
| Against building plant | | | 100 |
| No opinion | | | 40 |
| Column totals | 50 | 150 | 200 = $n$, or sample size |

If sex and opinions on building the nuclear plant are independent, the same *percentage* of men and women should be in favor of it (that is, men are no more likely than women to be for building it). Notice that 60 of the 200 people interviewed, or 30%, were in favor of building the plant. Therefore, if sex and opinion on nuclear power plants are independent, we expect about

30% of the males *and* 30% of the females to be in favor. Similarly, we expect100/200, or 50% of the males and 50% of the females, to be against it. Also, we expect 40/200, or 20% of males and 20% of the females, to have no opinion. Thus, the expected results under the null hypothesis of independence should be as listed in Table 13-2.

**expected results**

**Table 13-2 Expected Results if Sex and Opinion Are Independent**

| | Men | Women | Row Totals |
|---|---|---|---|
| In favor of building plant | $\frac{60}{200}$ of 50 | $\frac{60}{200}$ of 150 | 60 |
| Against building plant | $\frac{100}{200}$ of 50 | $\frac{100}{200}$ of 150 | 100 |
| No opinion | $\frac{40}{200}$ of 50 | $\frac{40}{200}$ of 150 | 40 |
| Column totals | 50 | 150 | 200 = $n$, or sample size |

This leads to Table 13-3.

**Table 13-3 Expected Results**

| | Men | Women | Row Totals |
|---|---|---|---|
| In favor of building plant | 1 15 | 2 45 | 60 |
| Against building plant | 3 25 | 4 75 | 100 |
| No opinion | 5 10 | 6 30 | 40 |
| Column totals | 50 | 150 | 200 = $n$, or sample size |

**cells**

For convenience, we have numbered all **cells**, or boxes, of the contingency table from left to right.

A simple way to obtain these expected results is to multiply the column total times the row total for each cell in Table 13-2 and divide by the sample size. For example, in cell 1 (men in favor) we have column total = 50, row total = 60, and sample size = 200; therefore, the expected result is

$$\frac{50(60)}{200} = 15$$

$$\text{In cell 2 (women in favor), we have} \quad \frac{150(60)}{200} = 45$$

$$\text{In cell 3 (men against), we have} \quad \frac{50(100)}{200} = 25$$

$$\text{In cell 4 (women against), we have} \quad \frac{150(100)}{200} = 75$$

$$\text{In cell 5 (men with no opinion), we have} \quad \frac{50(40)}{200} = 10$$

$$\text{In cell 6 (women with no opinion), we have} \quad \frac{150(40)}{200} = 30$$

In general, to find the expected result for any cell of the table we may use the following formula:

$$\text{Expected result} = \frac{\text{(column total)(row total)}}{\text{sample size}}$$

**observed results**

In Table 13-4, we give the **observed results** actually obtained from our random sample of 200 people.

**Table 13-4 Observed Results**

| | Men | Women | Row Totals |
|---|---|---|---|
| In favor of building plant | (1) 17 | (2) 43 | 60 |
| Against building plant | (3) 22 | (4) 78 | 100 |
| No opinion | (5) 11 | (6) 29 | 40 |
| Column totals | 50 | 150 | 200 = $n$, or sample size |

Let us now examine the difference between the observed results and the expected results (Table 13-5).

**Table 13-5**

| Cell | Category | Observed Result $O$ | Expected Result $E$ | Difference $O - E$ |
|---|---|---|---|---|
| 1 | Men in favor | 17 | 15 | 2 |
| 2 | Women in favor | 43 | 45 | −2 |
| 3 | Men against | 22 | 25 | −3 |
| 4 | Women against | 78 | 75 | 3 |
| 5 | Men, no opinion | 11 | 10 | 1 |
| 6 | Women, no opinion | 29 | 30 | −1 |

Is this group of differences (2, −2, −3, 3, 1, and −1) large or small? If large, we will claim that opinions on building nuclear power plants *are* related to sex. If small, we will be unable to make this claim. We are looking for one number which will indicate overall whether or not the differences are large. Note two problems. The first problem is that the sum of these differences is zero, and this sum will always be zero. For this reason we cannot use the mean of these differences. As you recall, we encountered this problem before when we studied the standard deviation. As we did at that time, we will now work with the squares of the differences. They are shown in Table 13-6.

**Table 13-6**

| Cell | Observed Result $O$ | Expected Result $E$ | Difference $O - E$ | Difference Squared $(O - E)^2$ |
|---|---|---|---|---|
| 1 | 17 | 15 | 2 | 4 |
| 2 | 43 | 45 | −2 | 4 |
| 3 | 22 | 25 | −3 | 9 |
| 4 | 78 | 75 | 3 | 9 |
| 5 | 11 | 10 | 1 | 1 |
| 6 | 29 | 30 | −1 | 1 |

The second problem is this. The squared difference of 4 in cell 1 is the same as the squared difference in cell 2. However, in cell 1 the expected result was 15, while in cell 2 the expected result was 45. A squared difference of 4 is more significant when you expected 15 responses than when you expected 45 responses; 4/15 (or 0.27) is 3 times greater than 4/45 (or 0.09). To take relative size into account, we divide each squared difference by the expected result for its cell. The sum of these numbers is called the statistic $X^2$, which is the sample estimate of a parameter called **chi-square**, $\chi^2$.

**chi square**

***Note*** As is common, we use the English $X^2$ for the statistic and the Greek $\chi^2$ for the parameter. In symbols, the formula for this statistic is

$$X^2 = \sum \frac{(O - E)^2}{E}$$

We now compute $X^2$ for the data in Table 13-4. The results are shown in Table 13-7, where we see that

$$X^2 = \sum \frac{(O - E)^2}{E} = 0.97$$

Table 13-7

| Cell | $O$ | $E$ | $O - E$ | $(O - E)^2$ | Ratio of Differences Squared to Expected Result $(O - E)^2/E$ |
|---|---|---|---|---|---|
| 1 | 17 | 15 | 2 | 4 | 4/15 = 0.27 |
| 2 | 43 | 45 | −2 | 4 | 4/45 = 0.09 |
| 3 | 22 | 25 | −3 | 9 | 9/25 = 0.36 |
| 4 | 78 | 75 | 3 | 9 | 9/75 = 0.12 |
| 5 | 11 | 10 | 1 | 1 | 1/10 = 0.10 |
| 6 | 29 | 30 | −1 | 1 | 1/30 = 0.03 |
| | | | Sum = 0 | | Sum = 0.97 |

If the expected results should turn out to be equal to or very close to the observed results, then the differences would be zero or near zero, and so the value of $X^2$ would be near zero. This result should occur if there is no relationship between the variables.

**tests of independence**

On the other hand, if $X^2$ is far away from zero, there is a high probability that the variables are not independent, that there is some statistical relationship between them.

In these **chi-square tests of independence** the null hypothesis will be that the **variables are independent;** that is,

$H_0$: Opinion on building nuclear power plants and one's sex are independent; $\chi^2 = 0$.

Our motivated, or alternative, hypothesis is that the **variables are dependent;** that is,

$H_a$: Opinions on building nuclear power plants and one's sex are not independent; $\chi^2 > 0$.

### Helpful Hint

In a chi-square test of independence, the null hypothesis has the form $H$: (*variable 1*) and (*variable 2*) are independent; $\chi^2 = 0$.

These tests of independence are always one-tail tests because we wish to see if the $X^2$ statistic is significantly *greater* than 0. In this problem we must now decide whether 0.97, the value of $X^2$, is significantly larger than zero. As we did in previous chapters with the normal and $t$-distributions, we compare a number based on sample data with a critical value. Table D-7 (see Appendix

D) lists critical values of the statistic $X^2$ for different significance levels and degrees of freedom for one-tail tests.

The theoretical distribution from which these values are taken is called the **chi-square distribution.** In technical language, we say that the sampling distribution of $X^2$ is "chi-square." This is similar to earlier statements which said that the sampling distribution of the statistic $m$ is normal or that the sampling distribution of the statistic $S$ is binomial.

**chi-square distribution**

Table D-7 is called a **chi-square table**, and problems such as this power plant one are referred to as "chi-square tests of independence." We can assume that it is reasonable for us to use these critical values if the sample size is large enough. This will usually be the case if each expected value is greater than 5. Occasionally it may be necessary to combine categories to achieve expected values which are all greater than 5.

**chi-square table**

You will also note that the entries for the critical chi-square values in Table D-7 are listed according to degree of freedom, as well as significance level. For a chi-square test with a 3 by 2 table, as in this example, there are 2 degrees of freedom. (The way to determine the number of degrees of freedom will be discussed in the next section.) Thus, to perform this chi-square test at the 0.05 significance level, we find the critical value for 2 degrees of freedom and $\alpha = 0.05$ to be $X_c^2 = 5.99$.

Since the computed value, 0.97, is smaller than $X_c^2$, the differences we found between the observed and expected results are not large enough to reject the hypothesis of independence at the 0.05 significance level. Thus, we have been unable to show a statistical relationship between opinion on building nuclear power plants and sex. Differences between men and women on this topic are not statistically significant. ❑

## 13.2 Degrees of Freedom in a Contingency Table

In Example 13.1 we said that a 3 by 2 table has 2 degrees of freedom. Let us illustrate this concept. In the previous discussion you were told that of the 200 people, 50 were men and 150 were women; it was also stated that 60 of them favor building a nuclear power plant, 100 oppose it, and the remaining 40 have no opinion. Now suppose we know only one cell of Table 13-4, the observed results. For example, cell 1 indicates that 17 men favor building a nuclear power plant. From this information you could figure out what the entry would have to be in cell 2 in order to have the correct total of 60 (Table 13-8).

**Table 13-8**

| | Men | Women | Row Totals |
|---|---|---|---|
| In favor of building plant | 1 17 | 2 | 60 |
| Against building plant | 3 | 4 | 100 |
| No opinion | 5 | 6 | 40 |
| Column totals | 50 | 150 | 200 = $n$ |

Clearly, you can see that the entry in cell 2 must be 60 − 17 = 43. However, cells 3 to 6 are not yet determined. If we had one of these cells given, such as 22 people in cell 3, then all the remaining cells could be determined (Table 13-9).

**Table 13-9**

| | Men | Women | Row Totals |
|---|---|---|---|
| In favor of building plant | 1 17 | 2 | 60 |
| Against building plant | 3 22 | 4 | 100 |
| No opinion | 5 | 6 | 40 |
| Column totals | 50 | 150 | 200 = $n$ |

Now we solve for the remaining cells as follows:

Cell 2: 60 − 17 = 43

Cell 4: 100 − 22 = 78

Cell 5: 50 − (17 + 22) = 50 − 39 = 11

Cell 6: 40 − 11 = 29

Thus you would obtain the results shown in Table 13-10.

**Table 13-10**

| | Men | Women | Row Totals |
|---|---|---|---|
| In favor of building plant | 1 17 | 2 43 | 60 |
| Against building plant | 3 22 | 4 78 | 100 |
| No opinion | 5 11 | 6 29 | 40 |
| Column totals | 50 | 150 | 200 = $n$ |

We say that a 3 by 2 table has 2 degrees of freedom because with 2 cells given, all the other cells can be determined by knowing the totals.

Suppose we had a 4 by 3 table with the totals given, as shown in Table 13-11.

Table 13-11

| | | | | Row Totals |
|---|---|---|---|---|
| | 1 | 2 | 3 | 200 |
| | 4 | 5 | 6 | 200 |
| | 7 | 8 | 9 | 100 |
| | 10 | 11 | 12 | 500 |
| Column totals | 400 | 200 | 400 | 1000 = sample size $n$ |

How many cells would you need to know before you could fill them all in? You can see that if we pick 6 numbers, leaving the last row and the last column blank, we can determine the remaining numbers (see Table 13-12).

Table 13-12

| | | | | Row Totals |
|---|---|---|---|---|
| | 1 50 | 2 60 | 3 ■ | 200 |
| | 4 70 | 5 50 | 6 ■ | 200 |
| | 7 30 | 8 30 | 9 ■ | 100 |
| | 10 ■ | 11 ■ | 12 ■ | 500 |
| Column totals | 400 | 200 | 400 | 1000 = sample size $n$ |

Thus, we get

$$
\begin{aligned}
\text{Cell } 3&: \quad 200 - (50 + 60) &= 90 \\
\text{Cell } 6&: \quad 200 - (70 + 50) &= 80 \\
\text{Cell } 9&: \quad 100 - (30 + 30) &= 40 \\
\text{Cell } 10&: \quad 400 - (50 + 70 + 30) &= 250 \\
\text{Cell } 11&: \quad 200 - (60 + 50 + 30) &= 60 \\
\text{Cell } 12&: \quad 500 - (250 + 60) &= 190
\end{aligned}
$$

so that the complete table is as shown in Table 13-13.

Table 13-13

| | | | | Row Totals |
|---|---|---|---|---|
| | 1   50 | 2   60 | 3   90 | 200 |
| | 4   70 | 5   50 | 6   80 | 200 |
| | 7   30 | 8   30 | 9   40 | 100 |
| | 10   250 | 11   60 | 12   190 | 500 |
| Column totals | 400 | 200 | 400 | 1000 = sample size $n$ |

Therefore, a 4 by 3 table has 6 degrees of freedom. In general, if a table has $R$ rows and $C$ columns, then it has $(R - 1)(C - 1)$ degrees of freedom. For example,

| Table Size | $(R - 1)(C - 1)$ | Degrees of Freedom |
|---|---|---|
| 5 by 2 | $(5 - 1)(2 - 1) = 4(1)$ | 4 |
| 4 by 3 | $(4 - 1)(3 - 1) = 3(2)$ | 6 |
| 6 by 7 | $(6 - 1)(7 - 1) = 5(6)$ | 30 |

In general, the formula for degrees of freedom is

**degrees of freedom**

$$\text{Degrees of freedom} = (R - 1)(C - 1)$$

Recall Example 11.3, where we showed that a 2 by 2 contingency table had only 1 degree of freedom. This agrees with our new formula: $(2 - 1)(2 - 1) = (1)(1) = 1$

## 13.3 A Shorter Way to Compute $X^2$

The previous discussion about computing $X^2$ led to the formula

$$X^2 = \sum \frac{(O - E)^2}{E}$$

As we have seen before, mathematicians often find an equivalent formula which is easier to use. A more convenient formula for $X^2$ is

$$X^2 = \sum \frac{O^2}{E} - n$$

We illustrate this by redoing the calculation of $X^2$ for Example 13.1. The calculations are shown in Table 13-14.

Table 13-14 Calculating $X^2$

| Cell | $O$ | $E$ | $O^2$ | $O^2/E$ |
|---|---|---|---|---|
| 1 | 17 | 15 | 289 | 19.27 |
| 2 | 43 | 45 | 1849 | 41.09 |
| 3 | 22 | 25 | 484 | 19.36 |
| 4 | 78 | 75 | 6084 | 81.12 |
| 5 | 11 | 10 | 121 | 12.10 |
| 6 | 29 | 30 | 841 | 28.03 |
| | $n = \Sigma O = 200$ | | | $\Sigma(O^2/E) = 200.97$ |

Therefore,

$$X^2 = \sum \frac{O^2}{E} - n$$

$$= 200.97 - 200 = 0.97$$

This is the same value as that computed with the previous formula.

## ❑ EXAMPLE 13.2 Student Residence

A random sample of 200 students on a university campus led to the following results about where students lived (Table 13-15).

Table 13-15 Observed Results

| Major Field | Live with Parents | Live on Campus | Other Arrangements | Row Totals |
|---|---|---|---|---|
| Business | 30 | 6 | 14 | 50 |
| Education | 7 | 17 | 6 | 30 |
| Fine arts | 9 | 23 | 43 | 75 |
| Social science | 6 | 17 | 22 | 45 |
| Column totals | 52 | 63 | 85 | 200 = sample size $n$ |

To determine if there is a relationship between major field and place of residence, test the null hypothesis that residence is *independent* of major field. Use the 0.01 significance level.

SOLUTION

*Procedure:* A chi-square test of independence

*Population:* All students at the campus in question

*Sample:* 200 students selected at random

$H_a$: Residence and major field are dependent; $\chi^2 > 0$.

$H_0$: Residence and major field are independent; $\chi^2 = 0$.

*Significance level:* $\alpha = 0.01$

*Degrees of freedom:* The contingency table is composed of 4 rows and 3 columns, hence degrees of freedom $= (R - 1)(C - 1) = (4 - 1)(3 - 1) = 3(2) = 6$

Since there are 6 degrees of freedom, we must use the formula for $E$, the expected values, in *at least* 6 of the cells; the remainder of the values of $E$ can be found by subtraction.

### Helpful Hint

Remember that a chi-square test of independence is always a one-tail test.

$$\text{Expected result} = \frac{(\text{column total})(\text{row total})}{\text{sample size}}$$

$$\text{Cell 1:} \quad E = \frac{52(50)}{200} = 13.000$$

$$\text{Cell 2:} \quad E = \frac{63(50)}{200} = 15.750$$

$$\text{Cell 4:} \quad E = \frac{52(30)}{200} = 7.800$$

$$\text{Cell 5:} \quad E = \frac{63(30)}{200} = 9.450$$

$$\text{Cell 7:} \quad E = \frac{52(75)}{200} = 19.500$$

$$\text{Cell 8:} \quad E = \frac{63(75)}{200} = 23.625$$

These expected results are shown in Table 13-16.

**Table 13-16 Expected Results**

| Major Field | Live with Parents | Live on Campus | Other Arrangements | Row Totals |
|---|---|---|---|---|
| Business | 1 13.000 | 2 15.750 | 3 | 50 |
| Education | 4 7.800 | 5 9.450 | 6 | 30 |
| Fine arts | 7 19.500 | 8 23.625 | 9 | 75 |
| Social science | 10 | 11 | 12 | 45 |
| Column totals | 52 | 63 | 85 | 200 = sample size |

***Note*** When using a calculator, it may sometimes be easier and/or quicker to continue to find the rest of the expected values by the formula. Either method is correct.

We can find the rest of the table by subtraction:

Cell 3: $E = 50 - (13.000 + 15.750) = 21.250$

Cell 6: $E = 30 - (7.800 + 9.450) = 12.750$

Cell 9: $E = 75 - (19.500 + 23.625) = 31.875$

Cell 10: $E = 52 - (13.000 + 7.800 + 19.500) = 11.700$

Cell 11: $E = 63 - (15.750 + 9.450 + 23.625) = 14.175$

Cell 12: $E = 45 - (11.700 + 14.175) = 19.125$

You can check the above calculations by adding the cells in the last column under "other arrangements" to see if you get the column total of 85. This column acts as a check, since we did not use it in our calculations.

In Table 13-17 we calculate the sample statistic, $X^2$. The first few significant digits of $O^2/E$ are shown as they appear on a calculator.

**Table 13-17 Calculating $X^2$**

| Cell | $O$ | $E$ | $O^2$ | $O^2/E$ |
|---|---|---|---|---|
| 1 | 30 | 13.000 | 900 | 69.230··· |
| 2 | 6 | 15.750 | 36 | 2.285··· |
| 3 | 14 | 21.250 | 196 | 9.223··· |
| 4 | 7 | 7.800 | 49 | 6.282··· |
| 5 | 17 | 9.450 | 289 | 30.582··· |
| 6 | 6 | 12.750 | 36 | 2.823··· |
| 7 | 9 | 19.500 | 81 | 4.153··· |
| 8 | 23 | 23.625 | 529 | 22.391··· |
| 9 | 43 | 31.875 | 1849 | 58.007··· |
| 10 | 6 | 11.700 | 36 | 3.076··· |
| 11 | 17 | 14.175 | 289 | 20.388··· |
| 12 | 22 | 19.125 | 484 | 25.307··· |
| | $\sum O = 200$ | All $E$s > 5 | | $\sum(O^2/E) = 253.752\cdots$ |

Rounding off, we have $\Sigma(O^2/E) = 253.75$. Your value may vary slightly if you round off at each step. We rounded the final answer to two decimal places because Table D-7 (Appendix D) expresses the critical values of $X^2$ correct to two decimal places.

In Table 13-17, notice that we indicated that all the expected values are greater than 5 under the column of expected values ($E$). Thus, the sample is large enough for us to use the critical values of $X^2$ in Table D-7.

**Helpful Hint**

If you want $X^2$ correct to two decimal places, work with *at least* one more, that is, with three decimal places, until the final calculation is performed, and then round off. If you *leave all intermediate results in the calculator memory,* there is *no need* to round off until you are finished.

Since the sample size is 200, we take $n = 200$ and calculate

$$X^2 = \sum \frac{O^2}{E} - n = 253.75 - 200 = 53.75$$

*Critical value:* Since there are 6 degrees of freedom and we are using the 0.01 significance level, the critical value $X_c^2$ from Table D-7 is 16.81.

Our usual three-step ending to a hypothesis test is as follows:

1. *Decision rule:* We will reject the null hypothesis of independence if the sample outcome is greater than the critical value $X_c^2$, 16.81.
2. *Outcome:* $X^2 = 53.75$
3. *Conclusion:*
   (a) *Technical language:* Since the value of $X^2$ from the experiment, 53.75, is larger than 16.81, we reject the null hypothesis of independence and claim that there is some relationship between major field and place of residence.
   (b) *Ordinary English:* Compare the $O$ and $E$ values above. Note that in cells 1 and 5 we observe about twice what we expect, while in cells 2, 6 and 7 we observe less than $\frac{1}{2}$ of the expected. From the former we see that more business majors live with parents, and more education majors live on campus; from the latter we infer that fewer business majors live on campus, fewer education majors have other arrangements, and fewer fine arts majors live with parents than would occur if residence and major were independent. This sheds light on the nature of the dependence. ❑

**Helpful Hint**

Notice that we use the Greek symbol for chi, $\chi$, when referring to the population parameter in the two *hypotheses;* however we use the English letter $X$ when referring to the sample outcome in the *decision rule* and the *conclusion.*

## 13.4 2 by 2 Contingency Tables

A 2 by 2 chi-square analysis may also be done as a two-sample binomial hypothesis test of the equality of two proportions (see Chapter 9). In fact, there is an exact algebraic equivalence between the two approaches. The value of $X^2$ in the 2 by 2 test is equal to the square of the value of $z$ you get in the $dp$ test from the formula

$$z = \frac{\hat{p}_1 - \hat{p}_2}{\sigma_{d\hat{p}}}$$

You can notice also that the critical values for $X^2$ for 1 degree of freedom are in fact the squares of the critical values of $z$. For example, in a two-tail test with $\alpha = 0.05$, $z_c = 1.96$, while $X_c^2$ for the 2 by 2 test at $\alpha = 0.05$ is 3.84, and $1.96^2 = 3.84$.

The 2 by 2 chi-square test is one of the most commonly used tests in applied statistics. This is because in so many experiments the most basic questions involve splitting variables into just two categories, such as "success, failure," "male, female," "old, young."

Furthermore, it can be shown by algebra that the value of $X^2$ in the 2 by 2 case can be easily calculated directly from the observed values without first figuring out the expected values. Here is a sample 2 by 2 table of observed values and the simplified formula.

| | | |
|---|---|---|
| $a$ | $b$ | $g$ |
| $c$ | $d$ | $h$ |
| $e$ | $f$ | $n$ |

The formula for $X^2$ is

$$X^2 = \frac{(ad - bc)^2 n}{efgh}$$

## EXAMPLE 13.3 The Side Effects of Medicines

Two medicines are being compared regarding a particular side effect; 60 similar patients are split randomly into two groups, one on each drug. The results are presented in Table 13-18.

**Table 13-18 Observed Values**

| | Side Effects | No Side Effects | Row Totals |
|---|---|---|---|
| Kure/All | 10 | 20 | 30 |
| Quik-Fix | 5 | 25 | 30 |
| Column totals | 15 | 45 | $60 = n$ |

As a two-sample binomial test we would test the hypothesis that the probability of side effects with Kure/All equals that with Quik-Fix. As a chi-square test we test the equivalent hypothesis that the two variables drug type and side effects are independent—that is, which drug a patient takes makes no difference as far as the likelihood of side effects is concerned.

**SOLUTION**

*Procedure:* A chi-square test of independence

*Populations:* All patients who would be given these drugs

$H_a$: Drug types and side effects are dependent; $\chi^2 > 0$.

$H_0$: Drug types and side effects are independent; $\chi^2 = 0$.

*Critical value:* With $\alpha = 0.05$ and 1 degree of freedom, $X_c^2 = 3.84$.

*Decision rule:* We will reject the null hypothesis if the outcome is greater than 3.84.

*Outcome:* Using the formula above

$$X^2 = \frac{[10(25) - 5(20)]^2 60}{15(45)(30)(30)} = \frac{150^2(60)}{607{,}500} = 2.22$$

*Conclusion:* We fail to reject the null hypothesis. We do not have enough evidence to show that drug types and likelihood of side effects are dependent; they may well be independent. Any difference in the rate of side effects is not statistically significant. ❑

## 13.5 Goodness-of-Fit Test (Contingency Tables with Only One Row)

All the examples so far have had tables with at least two rows. We can do chi-square tests on data in contingency tables with only one row (or, equivalently, only one column). The expected values in the experiment will come from the null hypothesis, and the formula for the degrees of freedom will be simpler.

goodness-of-fit

Our examples will be the simplest of the type of tests called **goodness-of-fit** tests. Such tests are used to determine whether sample observations fall into categories in the way they "should" according to some ideal model. When they come out as expected, we say that the data fit the model. The chi-square statistic helps us to decide whether the fit of the data to the model is good.

### ❑ EXAMPLE 13.4 A Carnival Wheel of Fortune

A carnival wheel of fortune for a July 4th fair is divided into 5 equal areas colored red, blue, red, white, and blue. The wheel is spun 50 times, and the results are 25 red, 18 blue, and 7 white. Should you decide that the wheel is biased at the 0.05 significance level?

## SOLUTION

*Procedure:* A chi-square test of goodness-of-fit

*Population:* All spins of this carnival wheel

*Sample:* 50 spins of this carnival wheel

$H_a$: The wheel is biased; $\chi^2 > 0$.

$H_0$: The wheel is fair; $\chi^2 = 0$.

We always obtain the expected values from the null hypothesis. However, the method used when we had more than one row and more than one column doesn't apply when we have only one row or only one column. In this particular problem, because the null hypothesis says the wheel is fair, we expect red = 2/5(50) = 20, blue = 2/5(50) = 20, and white = 1/5(50) = 10.

**Expected Results**

| Color | Red | Blue | White |
|---|---|---|---|
| Frequency | 20 | 20 | 10 |

**Observed Results**

| Color | Red | Blue | White |
|---|---|---|---|
| Frequency | 25 | 18 | 7 |

As usual, we now combine these results into one table (Table 13-19) and calculate $X^2$.

Table 13-19

| Cell | $O$ | $E$ | $O^2$ | $O^2/E$ |
|---|---|---|---|---|
| 1 | 25 | 20 | 625 | 31.250 |
| 2 | 18 | 20 | 324 | 16.200 |
| 3 | 7 | 10 | 49 | 4.900 |
| | $n = \sum O = 50$ | All $E$s > 5 | $\sum O^2/E =$ | 52.350 |
| | | | $-n =$ | $-50$ |
| | | | $X^2 =$ | 2.350 |

### Degrees of Freedom for a Contingency Table with One Row

If we wish to fill in the 3 numbers in a 3 by 1 contingency table so that their total is 50, it should be clear that we can arbitrarily pick any 2 of them. Therefore, in our wheel of fortune problem we have 2 degrees of freedom.

***Note*** For the simplest problems where the data are displayed in a $C$ by 1 contingency table, the data should be analyzed using a chi-square test with $C - 1$ degree of freedom. You can use this rule throughout the text.

*Critical value:* With $\alpha = 0.05$ and degrees of freedom $= 2$, we have $X_c^2 = 5.99$

*Decision rule:* We will reject the null hypothesis if the outcome $X^2$ is greater than 5.99.

*Outcome:* $X^2 = 2.35$

*Conclusion:* We fail to reject the null hypothesis. The wheel may be honest. We do not have enough evidence at the 0.05 significance level to prove that it is biased. ❑

---

## STUDY AIDS

### Chapter Summary

In this chapter you have learned:

- How to use two chi-square tests:

  A test for independence when both the number of rows and the number of columns exceed 1

  A goodness-of-fit test where there is only one row
- How to find degrees of freedom in each of these two different tests
- How to find expected values in each of these two different tests

### Guidelines for a Chi-Square Test

1. Each observation in the sample falls into one and only one cell of the table.
2. The sample size is large enough so that the expected value of every cell of the table is larger than 5.
3. Compute $X^2$.
4. The outcome $X^2$ as calculated above is compared to the critical value $X_c^2$ from Table D-7 (Appendix D). Which value $X_c^2$ we choose depends on the significance level and the degrees of freedom.
5. If $X^2$ is larger than the critical value $X_c^2$, then reject the null hypothesis.

**Vocabulary**
You should be able to explain the meaning of each of these terms:

1. Independent
2. Contingency table
3. Expected results
4. Cell
5. Observed results
6. Chi square
7. Tests of independence of variables
8. Chi-square distribution
9. Chi-square table
10. Statistical relationship between variables
11. Degrees of freedom
12. Test of goodness-of-fit

**Symbols**
You should understand the meaning of each of these symbols:

1. $O$ 2. $E$ 3. $\chi^2, X^2, X_c^2$ 4. $R$ 5. $C$

**Formulas**
You should know when and how to use each of these formulas:

1. $$\text{Expected result} = \frac{(\text{column total})(\text{row total})}{\text{sample size}}$$

2. $$X^2 = \sum \frac{O^2}{E} - n \quad \text{or} \quad X^2 = \sum \frac{(O - E)^2}{E}$$

3. $$X^2 = \frac{(ad - bc)^2 n}{efgh} \quad \text{for 2 by 2 tables}$$

***For Degrees of Freedom***

4a. For tests of independence with both $R > 1$ and $C > 1$,

$$\text{degrees of freedom} = (R - 1)(C - 1)$$

4b. For goodness-of-fit tests with either $R = 1$ (or $C = 1$),

$$\text{degrees of freedom} = C - 1 \text{ (or } R - 1)$$

## EXERCISES

13-1 (a) Friends of yours are planning a chi-square field project on people's attitudes toward the construction of a new state highway. They show you this contingency table which they intend to use. What mistake have they made?

| | For | Against | Undecided | Row Totals |
|---|---|---|---|---|
| Age 30 and under | | | | 150 |
| Over age 30 | | | | 150 |
| Column totals | 100 | 100 | 100 | $n = 300$ |

(b) After they corrected their error (on your advice), their results were as follows.

| | For | Against | Undecided | Row Totals |
|---|---|---|---|---|
| Age 30 and under | 100 | 40 | 10 | 150 |
| Over age 30 | 50 | 60 | 40 | 150 |
| Column totals | 150 | 100 | 50 | $n$ = 300 |

With these data they computed $X^2 = 38.67$. State their null hypothesis and their conclusions. Use $\alpha = 0.05$.

(c) Other friends tried a similar project. Their data were:

| | For | Against | Undecided | Row Totals |
|---|---|---|---|---|
| Age 30 and under | 15 | 13 | 2 | 30 |
| Over age 30 | 9 | 9 | 2 | 20 |
| Column totals | 24 | 22 | 4 | $n$ = 50 |

Show that not all *E*s are greater than 5. What can these friends do in order to complete the chi-square analysis?

13-2 For each of the following contingency tables, find the degrees of freedom and the appropriate critical value of $X^2$ at the indicated significance level.

(a) 8 by 2, $\alpha = 0.05$

(b) 7 by 1, $\alpha = 0.01$

(c) 3 by 4, $\alpha = 0.05$

(d) 5 by 5, $\alpha = 0.01$

13-3 Students in Uptudate College's English Comp courses may use a word processor for their assignments if they wish. The following contingency table reports their grades last semester. Analyze these data using the methods of this chapter.

| Grades | Always Used Word Processor | Sometimes Used Word Processor | Never Used Word Processor |
|---|---|---|---|
| A or B | 38 | 20 | 2 |
| C or D | 20 | 18 | 22 |
| F | 3 | 21 | 36 |

(a) Name the procedure you are using.

(b) What is the population being sampled?

(c) State the alternative and the null hypotheses.

(d) Find the degrees of freedom.

(e) Find the critical value of $X^2$, using $\alpha = 0.05$.

(f) Find the expected results.

(g) Calculate $X^2$.

(h) State the conclusion.

(i) Does this show that using a word processor improves grades?

13-4 The Maid of White Wheat Tavern in Agoura, California, is offering a special on a half dozen of their rum drinks. Damon, the barkeep, wants to see if there is any connection between a patron's dress and the drink ordered. He collected the following data last week and gave it to his friend Pythias, who is a student of statistics, to analyze. At the 0.05 significance level, what should Pythias conclude?

| Drink Ordered | T-shirts, Shorts, Jeans, etc. | Sports Shirts, Skirts, Dress Pants, etc. |
|---|---|---|
| Strawberry daiquiri | 10 | 21 |
| Piña colada | 10 | 12 |
| Strawberry colada | 9 | 15 |
| Mai Tai | 14 | 11 |
| Chi Chi* | 7 | 20 |
| Long Island Iced Tea | 25 | 6 |

13-5 A mathematics teacher, Professor Drew Thomas, wanted to know if there is a relationship between grades in Calculus 1 and the ability to pass Calculus 2. A random sample of students who completed Calculus 2 was taken, with the following results.

| Grade in Calculus 1 | Passed Calculus 2 | Failed Calculus 2 | Row Totals |
|---|---|---|---|
| A | 14 | 1 | 15 |
| B | 18 | 4 | 22 |
| C | 26 | 12 | 38 |
| D | 5 | 15 | 20 |
| Column totals | 63 | 32 | 95 = sample size $n$ |

What would you claim about the relationship of grades in Calculus 1 and passing Calculus 2? Use $\alpha = 0.05$.

*Also known as "Chi-Square," a drink favored by students of statistics.

13-6 A sample of Franciscans were questioned as to their smoking habits. Based upon the data below, are sex and smoking independent in this community? Use $\alpha = 0.05$.

| | |
|---|---|
| Priests and Brothers, smokers | 100 |
| Priests and Brothers, nonsmokers | 250 |
| Nuns, smokers | 50 |
| Nuns, nonsmokers | 100 |

13-7 Because of a great number of applicants, Ms. Lawrah, Dean of Admissions at Private University, must establish a new standard for admission. It was suggested that she turn down anyone whose College Board scores averaged less than 600, but some of the admissions committee believe that there is no significant difference between previously admitted students who averaged 600 or more and those who averaged under 600 in their potential to graduate. To determine who was right, they took a random sample of records of students who entered the university 5 years ago.

| | Graduated | Withdrew Voluntarily | Flunked Out | Row Totals |
|---|---|---|---|---|
| Scored 600 or more | 48 | 7 | 5 | 60 |
| Scored under 600 | 76 | 13 | 11 | 100 |
| Column totals | 124 | 20 | 16 | 160 = sample size $n$ |

Do the data present enough evidence to establish that 600 is a good cutoff point for admission? Use $\alpha = 0.05$.

13-8 In a congressional district in Los Angeles a random sample of voters were asked which of the 3 candidates they voted for. The results were as follows.

| Candidates | White | Black | Mexican-American | Row Totals |
|---|---|---|---|---|
| Blanca | 2 | 4 | 19 | 25 |
| Whyte | 3 | 28 | 4 | 35 |
| Weiss | 25 | 13 | 2 | 40 |
| Column totals | 30 | 45 | 25 | 100 = sample size $n$ |

Do the data present sufficient evidence to claim that voting habits depend on ethnic background? Test at the 0.01 significance level.

13-9 Eddy Torre notices that various authors prefer different symbols for footnotes and for cross-referencing. Wondering if there is any connection between the type of manuscript and the symbols, he gathers the following data for the 3 most popular[§] symbols used: the asterisk, the octothorpe, and the dagger.

| | *Symbol Used* | | |
|---|---|---|---|
| Type of Manuscript | * | # | † |
| Historical | 74 | 42 | 51 |
| Physical sciences | 73 | 29 | 93 |
| Social sciences | 102 | 68 | 78 |
| Computing | 41 | 111 | 28 |

Perform a chi-square hypothesis test on these data allowing for a 5% probability of a Type I error.

13-10 A psychologist was studying patterns of selfishness in various family groups. As part of the study a test designed to measure selfishness was given to a random sample of 600 people who said that they were not going to have any more children. Here are the results.

| | *Score on Selfishness Exam* | | | |
|---|---|---|---|---|
| Number of Children Now | Low | Medium | High | Row Totals |
| None | 30 | 40 | 50 | 120 |
| 1 | 40 | 50 | 60 | 150 |
| 2 | 50 | 50 | 40 | 140 |
| 3 | 60 | 30 | 20 | 110 |
| More than 3 | 40 | 30 | 10 | 80 |
| Column totals | | | | 600 |

At $\alpha = 0.05$, does this indicate a relationship between score on selfishness exam and number of children?

[§]Excluding numerals

13-11 In horse races at Upsand Downs racetrack, is the order of winning independent of the pole positions at the start of the race? Use $\alpha = 0.01$. The results of 100 horse races were as follows.

| Pole Position | Win | Place | Show |
|---|---|---|---|
| 1 | 30 | 19 | 11 |
| 2 | 16 | 12 | 16 |
| 3 | 20 | 18 | 16 |
| 4 | 8 | 17 | 11 |
| 5 | 9 | 9 | 16 |
| 6 | 6 | 12 | 9 |
| 7 | 6 | 6 | 16 |
| 8 | 5 | 7 | 5 |

13-12 Jason Jason, M.A., M.A., Ph.D, Ph.D, brought 32 children on a nature study hike. Of them, 16 had older brothers or sisters and 16 did not. He noted that of the first 16 to ask duplicate questions, 10 had older brothers or sisters. Should Jason Jason accept this as evidence at the 0.05 significance level that there is a connection between having older siblings and asking duplicate questions? Should Jason Jason accept this as evidence at the 0.05 significance level that there is a connection between having older siblings and asking duplicate questions?

13-13 Judy and Pat are arguing about competitiveness between girl and boy skiers. Judy claims that boys are more competitive, while Pat holds there is really no difference. Using the Trerice-Murphy competitiveness ranking scale, they observe and rank many young skiers over 1 winter. Do the results shown support Pat, or do they support Judy? Use $\alpha = 0.05$.

*Trerice-Murphy Ranks of Competitiveness*

| | High | Low |
|---|---|---|
| Girls | 30 | 20 |
| Boys | 22 | 18 |

13-14 Ms. Edwards and Mr. Kelvin teach kindergarten. They each have 20 students. Their approaches to zippers, overshoes, etc., at dismissal time are very different. They observe that by the time half the children are able to dress themselves, 11 are from Ms. Edwards's class. Does this evidence indicate any difference between the 2 approaches at $\alpha = 0.05$?

13-15 Tom performed the following field project. He walked directly toward a person who was approaching him in the middle of the quadrangle on campus. He recorded the sex of each person and whether they turned to the right or to the left. (He disregarded 2 collisions, 1 fight, and 1 person whose sex he could not determine right away.) His data were as follows.

| | Turned to Left | Turned to Right |
|---|---|---|
| Male | 57 | 43 |
| Female | 43 | 57 |

Using $\alpha = 0.05$, finish Tom's project. (Save your work for use in Exercise 13-35.)

13-16 In Exercise 9-22 we posed 2 questions for the data:

| | Patient Is Visited Regularly | Patient Is Not Visited Regularly |
|---|---|---|
| Patient has grandchildren | 20 | 40 |
| Patient does not have grandchildren | 10 | 30 |

Now we ask a third question: Are having grandchildren and being visited regularly independent? With $\alpha$ still 0.01, as it was in Exercise 9-22, we find $X^2 = 79$ and $X_c^2$ for 1 degree of freedom is 6.635. Since 0.79 is less than 6.635, we fail to reject.

(a) In Exercise 9-22(a) we found $\mu = 0$, $\hat{\sigma} = 0.107$, and $d\hat{p} = 0.0952$. Show that the $z$-score for $d\hat{p} = 0.0952$ is 0.89.

(b) In Exercise 9-22(b) we found $\mu = 0$, $\hat{\sigma} = 0.0935$, and $d\hat{p} = 0.083$. Find the $z$-score for $d\hat{p} = 0.083$.

(c) We found $X^2 = 0.79$ above. Find $X = \sqrt{0.79}$.

(d) $X_c^2 = 6.635$. Find $X_c$.

(e) Compare your answer in part (c) with the $z$-score for $z_c$ in parts (a) and (b). Also, compare your answer in part (d) with $z_c$ in a 2-tail test with $\alpha = 0.01$.

(f) What can you conclude about the 3 questions that we asked regarding these data?

13-17 In Example 13.3, a student used the short formula to calculate $X^2 = 2.22$. Show that the previous, longer method using $X^2 = \sum \frac{O^2}{E} - n$ gives the same result.

13-18 In a certain game a coin is tossed twice. Mary is tossing the coin, and John is getting suspicious about the results. To test for fairness, he decides to do a 3 by 1 chi-square test. He incorrectly uses the following expected values for the next 120 games.

| No Heads | 1 Head | 2 Heads | |
|---|---|---|---|
| 40 | 40 | 40 | $n = 120$ |

What are the correct expected values?

13-19 Dr. Al Timiter theorized that 40% of pilots drink coffee and cola, and 10% drink only coffee, 20% drink only cola, and the rest abstain from caffeine. In a recent pilot study, you observed the following data.

| Both | Only Coffee | Only Cola | Neither |
|---|---|---|---|
| 106 | 10 | 50 | 10 |

Should you reject his theory, using $\alpha = 0.05$?

13-20 In Example 5.2 we discussed Marty's spinning a spinner 4 times. Marty repeated this game 810 times and observed the following data.

| | *Number of Wins in Every 4 Spins* | | | | |
|---|---|---|---|---|---|
| | 4 | 3 | 2 | 1 | 0 |
| Number of times it happened | 10 | 100 | 200 | 300 | 200 |
| Number of times it was expected | | | | | |

(a) Complete the chart above.

(b) Using $\alpha = 0.01$, should he consider the spinner biased?

13-21 If $P(\text{a boy}) = P(\text{a girl}) = 0.5$, then our knowledge of binomials leads us to expect that in families with 2 children we would have about 25% with 2 girls, 50% with 1 girl and 1 boy, and 25% with 2 boys. Is this true in Irtusk, if a random sample of 1000 families show 100 with 2 boys, 600 with 1 boy and 1 girl, and 300 with 2 girls? Use $\alpha = 0.05$.

13-22 Using $\alpha = 0.01$, are 2 monopoly dice fair if the last 360 tosses produced the following results?

| | *Total on Each Toss* | | | | | | | | | | |
|---|---|---|---|---|---|---|---|---|---|---|---|
| | 2 | 3 | 4 | 5 | 6 | 7 | 8 | 9 | 10 | 11 | 12 |
| Number of times total occurred | 4 | 16 | 24 | 46 | 55 | 70 | 57 | 43 | 25 | 15 | 5 |

13-23 If a particular species of bug is distributed *randomly* in a wooded area, then probability theory can be used to predict how many bugs should be found under rocks picked up at random in the area. For 100 rocks, the predictions are as follows.

| | *Number of Bugs Under Rock* | | | | | |
|---|---|---|---|---|---|---|
| | 0 | 1 | 2 | 3 | 4 | 5 |
| Expected number | 38 | 35 | 17 | 6 | 3 | 1 |

An ecology student went to such a wooded area to count bugs under 100 rocks and found the following.

| | *Number of Bugs Under Rock* | | | | | |
|---|---|---|---|---|---|---|
| | 0 | 1 | 2 | 3 | 4 | 5 |
| Actual count | 33 | 3 | 15 | 33 | 15 | 1 |

Interpret these results by a chi-square test.

13-24 In 1975 the Bicentennial Committee surveyed federal employees, raising 2 issues. Without being identified as such, a portion of the Declaration of Independence was given to the government employees to read, and they were then asked the questions below. The data collected are shown below; note that a large number of people refused even to respond to the first question.

*Question 1:* Would you sign such a declaration?

| | Employees of the Pentagon | Employees of the Congress | Other |
|---|---|---|---|
| Would sign | 41 | 94 | 30 |
| Would not sign | 159 | 106 | 55 |

*Question 2:* Do you recognize this quotation?

| | Recognized Quote | Failed to Recognize Quote |
|---|---|---|
| Would sign | 121 | 615 |
| Would not sign | 130 | 1434 |

Analyze these 2 sets of data at $\alpha = 0.01$, and interpret your results.

13-25 Pat asked a random sample of students and former students of Lax University if they would pose in the nude for the centerfold of a magazine. Of 100 current students 30 said that they would, but of 200 former students only 40 would pose nude.

(a) Find the 95% confidence interval for the difference between the proportion of students who would pose in the nude and the proportion of former students who would not.

(b) Do a two-sample binomial hypothesis test on these data with $\alpha = 0.05$. What is the null hypothesis?

(c) Do a chi-square hypothesis test on these data with $\alpha = 0.05$. What is the null hypothesis?

13-26 Little Joe created a random-number table by tossing 2 dice. He counted the number of times each number appeared. He let a toss of a 10 correspond to 0, a toss of an 11 correspond to a 1, and he disregarded all 12s. His dice tosses were

4, 6, 7, 11, 8, 9, 10, 5, 8, 7, 7, 4, 6, 9, 4, 12, 7, . . .

giving him the random digits

4, 6, 7, 1, 8, 9, 0, 5, 8, 7, 7, 4, 6, 9, 4, 7, . . .

He continued this until he had a total of 100 outcomes. Counting the number of times each digit appeared, he found the following.

| | Digit | | | | | | | | | |
|---|---|---|---|---|---|---|---|---|---|---|
| | 0 | 1 | 2 | 3 | 4 | 5 | 6 | 7 | 8 | 9 |
| Number of appearances | 10 | 8 | 4 | 7 | 9 | 11 | 13 | 16 | 12 | 10 |

(a) Test the hypothesis that all the digits are *equally likely* to appear. Use $\alpha = 0.05$.

(b) Probability theory of randomly tossed dice leads to the following expected values (which are *not* equally likely).

| | Digit | | | | | | | | | |
|---|---|---|---|---|---|---|---|---|---|---|
| | 0 | 1 | 2 | 3 | 4 | 5 | 6 | 7 | 8 | 9 |
| Number of appearances | 8.6 | 5.7 | 2.9 | 5.7 | 8.6 | 11.4 | 14.3 | 17.1 | 14.3 | 11.4 |

Test that the dice behave according to this theory at the 0.01 significance level.

5 13-27 (a) A newspaper report listed the total number of homicides each month in Egress Junction, Nebraska, over a 10-year period. Use these data to test the hypothesis that homicides in all months are equally likely. Use $\alpha = 0.01$.

| Month | Number of Homicides |
|---|---|
| January | 834 |
| February | 744 |
| March | 789 |
| April | 829 |
| May | 867 |
| June | 823 |
| July | 1024 |
| August | 985 |
| September | 875 |
| October | 973 |
| November | 869 |
| December | 1042 |

$\Sigma O = \frac{10654}{12} = 887.83$

(b) Test the hypothesis that the percentage of murders that occur in December is greater than the percentage occurring in July.

(c) True or false? For this period of time the average number of murders per day in February is more than the average number per day in March. (Assume 3 leap years in this 10-year period.)

13-28 A group of 50 people each toss a penny 7 times.

(a) Assuming the coin is fair and using Pascal's triangle, figure out what percentage of the people should get 0 heads, 1 head, 2 head, . . . , 7 heads.

(b) About how many people should be in each of these categories?

(c) The observed results in 1 group are listed below. Is this evidence, at the 0.05 significance level, that the coin is not fair? (You may have to combine some categories if all the expected values are not more than 5.)

| Number of Heads | Number of People |
|---|---|
| 0 | 0 |
| 1 | 0 |
| 2 | 3 |
| 3 | 7 |
| 4 | 12 |
| 5 | 13 |
| 6 | 11 |
| 7 | 4 |
| | $n = 50$ |

4 13-29 In Chapter 11 we often made the simplifying assumption that a population was normal. This exercise demonstrates how we can use chi-square to do a goodness-of-fit test for normality. A quality-control engineer for an engineering firm, which makes rocket engines, tests a sample of 260 relay switches taken from a production line. Supposedly they operate at a mean speed of 50 microseconds with a standard deviation of 10 microseconds.

(a) If the population is normal with a mean of 50 and a standard deviation of 10, and $n = 260$, how many values would you expect to fall into each of the following categories?

(1) Less than 30 ($z < -2$)

(2) From 30 to 40 ($-2 \leq z < -1$)

(3) From 40 to 50 ($-1 \leq z < 0$)

(4) From 50 to 60 ($0 \leq z < 1$)

(5) From 60 to 70 ($1 \leq z < 2$)

(6) At least 70 ($2 \leq z$)

(b) The data for the sample of 260 relay speeds were as follows.

| Speed, Microseconds | Number of Relay Switches |
|---|---|
| Less than 30 | 15 |
| From 30 to 40 | 30 |
| From 40 to 50 | 85 |
| From 50 to 60 | 80 |
| From 60 to 70 | 40 |
| At least 70 | 10 |

Use the expected data found in part (a) to perform a 6 by 1 chi-square test with $\alpha = 0.05$. Is it reasonable to assume that these speeds are from a normal population?

13-30 At the annual meeting of the Association of Master Chefs of Bayonne, 1000 chefs were asked to pour out 1.5 ounces of virgin olive oil without measuring. The results were grouped as follows.

| | *Ounces* | | | | | | |
|---|---|---|---|---|---|---|---|
| | 0.4 to 0.7 | 0.7 to 1.0 | 1.0 to 1.3 | 1.3 to 1.6 | 1.6 to 1.9 | 1.9 to 2.2 | 2.2 to 2.5 |
| Number of chefs | 80 | 100 | 0 | 400 | 300 | 100 | 20 |

(a) The distribution of all chefs' estimates of 1.5 ounces of virgin olive oil was claimed to be a normal distribution, with $\mu = 1.6$ ounces and $\sigma = 0.3$ ounce. Assuming this claim to be true, how many of the 1000 chefs would you expect in each cell of the contingency table shown above?

(b) Do a chi-square goodness-of-fit test on this claim. Use $\alpha = 0.05$.

13-31 In a medical study of chronic pain, 3 groups of patients were given different pills. In group 1, the pills were just sugar; the patients in group 2 got aspirin; the patients in group 3 got a new experimental medicine. Because the patients were randomly assigned to their groups, neither the patients nor the evaluating doctors knew which patients got which pills until the experiment was over. Such an experiment is call **double-blind.** Do a chi-square test at $\alpha = 0.05$, and interpret your results.

| | Got Relief | Did Not Get Relief | Row Totals |
|---|---|---|---|
| Sugar | 70 | 30 | 100 |
| Aspirin | 80 | 20 | 100 |
| Experimental medicine | 85 | 15 | 100 |
| Column totals | 235 | 65 | 300 = sample size |

13-32 The Duodecimal Society of America tested a sample of 300 Americans about their beliefs concerning counting and measurement. The experiment went as follows. First, the people in the sample were given a short lecture clarifying 3 ideas.

(1) How we naturally measure many things in 12s (e.g., inches, dozens, and grosses; months and hours; ounces in the Troy pound used by jewelers and druggists; the new telephone dialing systems), and yet we count with 10 symbols (0, 1, 2, 3, 4, 5, 6, 7, 8, and 9).
(2) How we could change our measurements to some artificial system based on 10 similar to 1 of the many decimal metric systems used in various parts of the globe today.
(3) How we could learn to count in dozens using 12 symbols (0, 1, 2, 3, 4, 5, 6, 7, 8, 9, *, and #) and be able to retain our natural measurements.

After the lecture they were asked to pick 1 of these 3 responses as best expressing their beliefs:

A The dozenal system of counting and measuring is easier, and hence the world will 1 day adopt it.

B The dozenal system of counting and measuring is easier, but it will never be adopted by many people.

C The dozenal system of counting and measuring is not any easier.

Their responses were gathered in the following contingency table.

| Profession | Choice A | Choice B | Choice C | Undecided |
|---|---|---|---|---|
| Involves both measuring and arithmetic | 45 | 35 | 10 | 10 |
| Involves arithmetic but not measurement | 35 | 35 | 20 | 10 |
| Involves neither measurement nor arithmetic | 30 | 30 | 30 | 10 |

Perform a hypothesis test on the data with $\alpha = 0.05$.

13-33 Repeat some of the exercises in Chapter 9, using the methods of this chapter.

13-34 Some statisticians argue for using what is called the **Yates correction factor** in the 2 by 2 chi-square problem. The use of the correction factor makes the value of $X^2$ smaller. The purpose is to ensure that the probability of a Type I error is not more than it is stated to be. If you wish to use the Yates correction factor in order to be on the conservative side, the simplified formula becomes

$$X^2 = \frac{\left(|ad - bc| - \frac{n}{2}\right)^2 n}{efgh}$$

where $|ad - bc|$ is the absolute value of $ad - bc$.

Redo the hypothesis test of Example 13.3 using this correction factor. Do you expect $X^2$ to be greater than or less than the previously calculated 2.22?

\ 13-35 In Exercise 13-15 Tom did a 2 by 2 project without using the correction factor given in the previous exercise.

(a) Tim, too lazy to gather his own data, copied Tom's information. To make his project appear different, Tim used the correction factor. Finish Tim's project.

(b) Chet, aware of what Tom and Tim were doing, solved the problem as a binomial two-sample problem, taking as his null hypothesis that the same percentage of males and females turned left. Finish Chet's project.

(c) Compare the results in the Tom-Tim-Chet problem.

## CLASS SURVEY QUESTIONS

1. It has been noted in many populations that sex and smoking are not independent. Test to see whether this is true on your campus.
2. Are hair coloring and eye coloring independent at your school? Do a chi-square hypothesis test using your sample data.

## FIELD PROJECTS

Design and perform a chi-square test of your choosing. Some projects that students have done on campus include investigating the relationship between

1. Sex of student and student's major
2. Class size and number of students who drop the course
3. Students' grades and how close they sit to the teacher
4. Faculty rank and amount of time faculty member spends with students
5. Religion and membership in religious clubs on campus

# Sample Test for Chapters 12 and 13

1. Explain why a 100% confidence interval is useless.
2. A random sample of 200 people chosen from the large membership list of the American Therapist Association included 30 females.
   (a) What is the 95% confidence interval for the percentage of females in the association?
   (b) Under what conditions would you expect your answer in part (a) to estimate the percentage of females in the *profession*?
3. A 95% confidence interval is desired for the percentage of persons who perform perfectly on a sample test. What size sample should be used if we want to be confident that our estimate is no more than 3% off the true value?
4. One day, New York City officials monitored the 8 A.M. to 4 P.M. shift in the emergency medical services department. This department receives emergency phone calls for ambulances. They reported that for 88 calls the average response time was 28.8 minutes.
   (a) If we wish to compute a confidence interval based on these data, what statistical population would it refer to—that is, of what population do you think 28.8 is the sample mean? Give some conditions under which this would be reasonable. Why might it be unreasonable?
   (b) If we assume that $\sigma$ is about 15 minutes, what is the 95% confidence interval?
5. We wish to estimate the difference between the percentage of male and female sales managers who earn more than $40,000 per year in a large industry. A random sample of 50 male and 50 female managers is chosen, and their salaries are recorded. It is found that 36 of the males earn more than $40,000 while 18 of the females do. Find the 95% confidence interval for the difference in percentages earning more than $40,000 per year. What are some of the reasons often given to explain why such an imbalance might exist?
6. Dr. Noah Tall observed the frequencies shown below in a random sample of father-son pairs in a large city. Interpret the results using a chi-square test. Use $\alpha = 0.05$.

| | | *Father* | |
|---|---|---|---|
| *Son* | Shorter Than 5 Feet, 6 Inches | 5 Feet, 6 Inches to 6 Feet | Taller Than 6 Feet |
| Shorter Than 5 Feet, 6 Inches | 50 | 400 | 10 |
| 5 Feet, 6 Inches to 6 Feet | 150 | 2000 | 200 |
| Taller Than 6 Feet | 5 | 300 | 60 |

7. A random sample of students at Wayup High School was asked 2 questions. (a) Do you own more rock or more rap recorded music? (b) Are you for or against Proposition 31? Do the results indicate that answers to these 2 questions are statistically independent? Use $\alpha = 0.05$.

| | *Proposition 31* | |
|---|---|---|
| Listen to | For | Against |
| More rock | 20 | 60 |
| More rap | 30 | 90 |

8. A coin is suspected of bias. It is tossed 3 times, and the number of heads is counted. This is repeated 40 times with the following results.

| Number of Heads in 3 Tosses | Number of Times This Occurred |
|---|---|
| 3 | 11 |
| 2 | 16 |
| 1 | 9 |
| 0 | 4 |

(a) Find the expected number of times for each of the 4 possible outcomes if the coin were fair.
(b) Analyze the data with a chi-square test using $\alpha = 0.05$.

# 14

# Correlation and Prediction

**CONTENTS**



**GOALS**

At the end of this chapter you will have learned:

- The meaning of correlation between two variables, and how to measure it with a coefficient of correlation
- How to draw a scattergram and a best-fitting line
- How to predict the value of a variable from the formula for the best-fitting line
- How to describe the effect one variable has upon the value of another

## 14.1 Correlation Coefficients

correlation coefficient

It frequently happens that statisticians want to describe with a single number a relationship between two sets of scores. A number that measures a relationship between two sets of scores is called a **correlation coefficient**. There are several correlation coefficients for measuring various types of relationships between different kinds of measurements. In this text we will illustrate the basic concepts of correlation by discussing only the Pearson correlation coefficient, which is one of the more widely used correlation coefficients.

The statistic is named for its inventor, Karl Pearson, one of the founders of modern statistics. It is denoted by $r$ and is used to measure what is called the **linear relationship** between two sets of measurements. The corresponding parameter is denoted by the Greek letter $\rho$ (rho).

To explain how $r$ works and what is meant by a linear relationship, we will look at a few overly simplified examples. It is unlikely that a real application of the correlation coefficient would be made with so few pieces of data.

Imagine that six students are given a battery of tests by a vocational guidance counselor with the results shown in Table 14-1.

### *Karl Pearson*

In his three-volume work, *The Life, Letters and Labours of Francis Galton,* Karl Pearson (1857–1936) gives Galton due credit for the concepts of correlation. However it was Pearson's own work that was much more important in the development of correlation as it is used today. Between 1895 and 1898 Pearson worked out a formula for estimating the correlation coefficient and assessing the accuracy of the estimate.

### *Greek Rho*

Francis Ysidro Edgeworth (1845–1926), a link between Galton and Pearson, introduced $\rho$ as the symbol for the coefficient of correlation parameter.

Table 14-1

| Student | Interest in Retailing | Interest in Theater | Math Aptitude | Language Aptitude |
|---|---|---|---|---|
| Pat | 51 | 30 | 525 | 550 |
| Sue | 55 | 60 | 515 | 535 |
| Inez | 58 | 90 | 510 | 535 |
| Arnie | 63 | 50 | 495 | 520 |
| Gene | 85 | 30 | 430 | 455 |
| Bob | 95 | 90 | 400 | 420 |

scattergram

The counselor might want to see if there are any correlations among these sets of marks. For example, it looks as if the people who do well in math also do well in language skills. Let us draw a graph called a **scattergram** to investigate this relationship. To draw this scattergram, we first draw a vertical axis and a horizontal axis, one for the math scores and the other for the language scores. In the scattergram shown in Figure 14-1 we have put the math scores on the horizontal axis, but that is not important. We could have put them on the vertical axis. Notice how the math axis is labeled. The reported math scores ranged from 400 to 525, and so the math axis must be labeled in a way that all those scores can be recorded. We therefore chose to make a series of equally spaced marks labeled from 400 to 525 counting by 25 at a time. Similarly, the vertical axis was labeled from 420 to 550 to cover the range of the language scores.

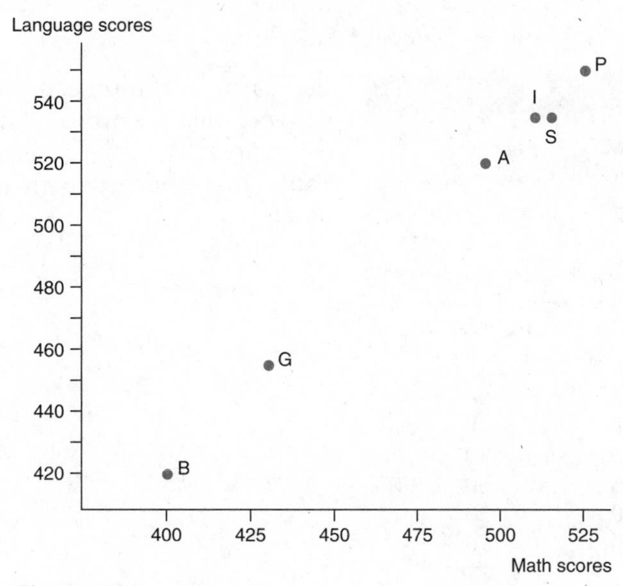


**Figure 14-1**

After both axes are drawn and labeled, we use one dot for each person. The dot is placed so that it is directly over the person's math score, and directly to the right of the person's language score. For example, the dot for Pat's scores is over the 525 on the math axis, and to the right to 550 on the language axis. In Figure 14-1 we have put each person's initial next to his or her dot to help you read the graph, but the initials are usually left off. For convenience, we repeat the math scores and language scores in Table 14-2.

**Table 14-2**

| Student | Math Aptitude | Language Aptitude |
|---|---|---|
| Pat | 525 | 550 |
| Sue | 515 | 535 |
| Inez | 510 | 535 |
| Arnie | 495 | 520 |
| Gene | 430 | 455 |
| Bob | 400 | 420 |

You will notice three things about the scattergram:

1. There is one point for each pair of scores, six points in all.
2. The points are arranged approximately in a straight line. When this happens, we say that there is good *line*ar correlation between the two variables (in this case, math skill and language skill).
3. The higher numbers in the math column of the table correspond to the higher numbers in the language column. This causes the line to slope up to the right. This is called **positive correlation**.

**positive correlation**

## Math Scores vs. Theater-Interest Scores

> ***Sir Francis Galton***
>
> In 1888 Sir Francis Galton (1822–1911) described an "index of co-relation" centered on the median rather than the mean.

Let us now compare the relationship between math scores and theater-interest scores. We repeat these scores in Table 14-3 and graph them in Figure 14-2.

Table 14-3

| Student | Math Aptitude | Interest in Theater |
|---|---|---|
| Pat | 525 | 30 |
| Sue | 515 | 60 |
| Inez | 510 | 90 |
| Arnie | 495 | 50 |
| Gene | 430 | 30 |
| Bob | 400 | 90 |

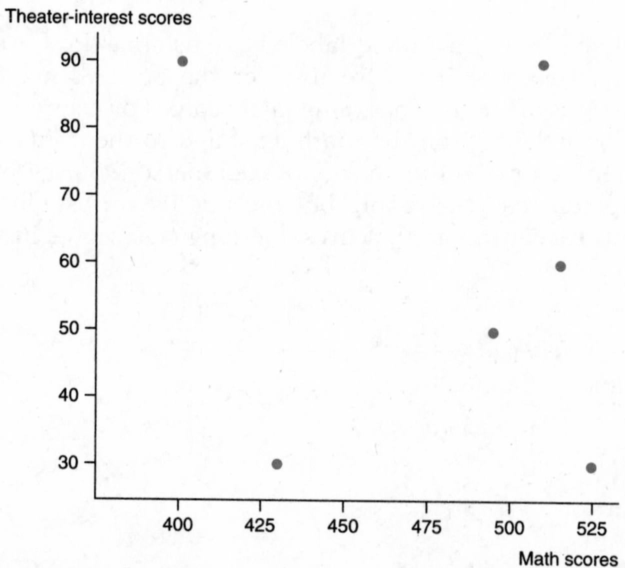


Figure 14-2

In Figure 14-2 you notice that there is no special tendency for the points to appear in a straight line. We say that there is *littie* linear correlation or even **zero linear correlation**,* between the math scores and the theater-interest scores. Also note that it is not necessary for both variables to be scored on the same scale, since the correlation coefficient describes the *pattern* of the scores, not the actual values.

zero correlation

## Math Scores vs. Retailing-Interest Scores

In Table 14-4 and Figure 14-3 we compare math scores to retailing-interest scores.

Table 14-4

| Student | Math Aptitude | Interest in Retailing |
|---|---|---|
| Pat | 525 | 51 |
| Sue | 515 | 55 |
| Inez | 510 | 58 |
| Arnie | 495 | 63 |
| Gene | 430 | 85 |
| Bob | 400 | 95 |

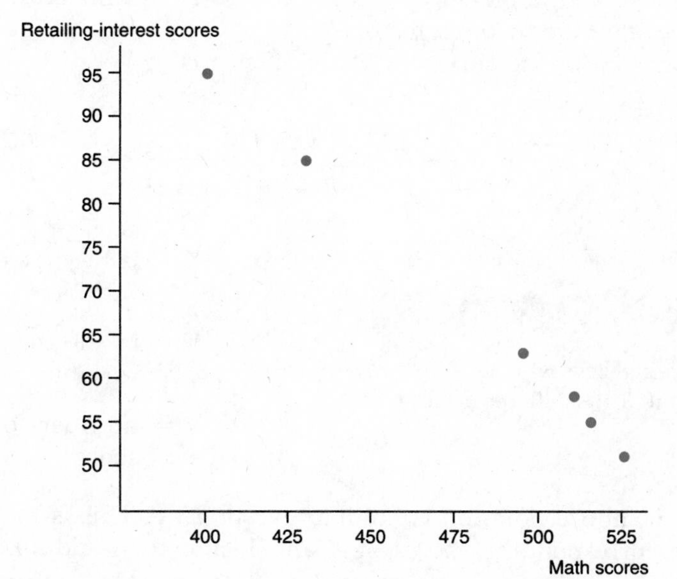


Figure 14-3

*There may, of course, be a nonlinear relationship between the variables.

In Figure 14-3 there is a tendency for the points to lie in a line that slopes down to the right. This is called **negative correlation**. This happens because the higher scores in the column for math correspond to the lower scores in the column for retailing interest.

**negative correlation**

### Computation of *r*

Pearson defined the coefficient of correlation $r$ so that it has a minimum possible value of $-1$ and a maximum possible value of $+1$. When the sample points lie *exactly* in a line sloping *down to the right,* we say there is *perfect negative correlation:* $r = -1$. When the sample points lie *exactly* in a line sloping *up to the right,* we say there is *perfect positive correlation:* $r = +1$. When there is *no tendency* for the points to lie in a straight line, we say there is *no correlation:* $r = 0$. If $r$ is near $+1$ or $-1$, we say that we have *high correlation.* If $r$ is near zero, we say that we have *low correlation.*

We would expect that, for Figure 14-1, the value of $r$ is near 1; for Figure 14-2, $r$ is near 0; and, for Figure 14-3, $r$ is near $-1$. We will show that this is true after we state the formula for $r$.

The formula for the correlation coefficient $r$ is

$$r = \frac{n\Sigma XY - (\Sigma X)(\Sigma Y)}{\sqrt{n\Sigma X^2 - (\Sigma X)^2}\ \sqrt{n\Sigma Y^2 - (\Sigma Y)^2}}$$

where $X$ = label for one of the variables

$Y$ = label for the other variable

$n$ = number of *pairs* of scores

**An Easier Way**

Some calculators have special keys which compute $r$ automatically after the data have been entered.

All of the above notation is familiar except $\Sigma XY$. This is found by multiplying the corresponding values of $X$ and $Y$, and then adding all of these products.

### Illustration of *r* Computations

For the correlation depicted in Figure 14-1, we can tabulate the data as shown in Table 14-5.

Table 14-5

| Math Aptitude $X$ | Language Aptitude $Y$ | $X^2$ | $Y^2$ | $XY$ |
|---|---|---|---|---|
| 525 | 550 | 275,625 | 302,500 | 288,750 |
| 515 | 535 | 265,225 | 286,225 | 275,525 |
| 510 | 535 | 260,100 | 286,225 | 272,850 |
| 495 | 520 | 245,025 | 270,400 | 257,400 |
| 430 | 455 | 184,900 | 207,025 | 195,650 |
| 400 | 420 | 160,000 | 176,400 | 168,000 |
| $\Sigma X = 2875$ | $\Sigma Y = 3015$ | $\Sigma X^2 = 1,390,875$ | $\Sigma Y^2 = 1,528,775$ | $\Sigma XY = 1,458,175$ |

Solving for $r$,

$$r = \frac{n\Sigma XY - (\Sigma X)(\Sigma Y)}{\sqrt{n\Sigma X^2 - (\Sigma X)^2}\sqrt{n\Sigma Y^2 - (\Sigma Y)^2}}$$

$$= \frac{6(1,458,175) - 2875(3015)}{\sqrt{6(1,390,875) - (2875)^2}\sqrt{6(1,528,775) - (3015)^2}}$$

$$= \frac{80,925}{\sqrt{79,625}\sqrt{82,425}} = 0.999$$

which, as we expected, is near +1.

Similarly, for the correlation shown in Figure 14-2, we find that $\Sigma X = 2875$; $\Sigma Y = 350$; $\Sigma X^2 = 1,390,875$; $\Sigma Y^2 = 24,100$; and $\Sigma XY = 166,200$. Thus,

$$r = \frac{n\Sigma XY - (\Sigma X)(\Sigma Y)}{\sqrt{n\Sigma X^2 - (\Sigma X)^2}\sqrt{n\Sigma Y^2 - (\Sigma Y)^2}}$$

$$= \frac{6(166,200) - 2875(350)}{\sqrt{6(1,390,875) - (2875)^2}\sqrt{6(24,100) - (350)^2}}$$

$$= \frac{-9050}{\sqrt{79,625}\sqrt{22,100}} = -0.22$$

which, as we predicted, is close to zero.

For the correlation shown in Figure 14-3, we get $\Sigma X = 2875$; $\Sigma Y = 407$; $\Sigma X^2 = 1,390,875$; $\Sigma Y^2 = 29,209$; and $\Sigma XY = 190,415$. Thus,

$$r = \frac{n\Sigma XY - (\Sigma X)(\Sigma Y)}{\sqrt{n\Sigma X^2 - (\Sigma X)^2}\sqrt{n\Sigma Y^2 - (\Sigma Y)^2}}$$

$$= \frac{6(190,415) - 2875(407)}{\sqrt{6(1,390,875) - (2875)^2}\sqrt{6(29,209) - (407)^2}}$$

$$= \frac{-27,635}{\sqrt{79,625}\sqrt{9605}} = -0.999$$

and again, this is close to our expected value, −1.

## Exercises

14-1 Imagine that you are a school principal and that you have calculated the coefficient of correlation between last year's grades in Señor Oldways' class in Spanish 2 and this year's grades in Señorita Moderna's class in Spanish 3. What would it mean if $r$ were close to $-1$? Close to $+1$? Close to 0?

14-2 Students taking Professor Hardrocque's course in geology compared their grades on the final exam with their final course grades. They calculated $r = 0.12$. What does this indicate?

14-3 Professor Hadit, frustrated by students' run-on sentences, announces, "There is no correlation between the length of your sentences and your grades." The next day, Bradley tells Eilís, who was absent, that their professor said, "The longer your sentences are, the worse your grade will be." Are these statements equivalent? Why or why not?

14-4 (a) Draw a scattergram for the information in the following table.

| | $X$ | $Y$ | $X^2$ | $Y^2$ | $XY$ |
|---|---|---|---|---|---|
| | 1 | 1 | | | |
| | 2 | 1 | | | |
| | 3 | 2 | | | |
| | 4 | 3 | | | |
| Totals | | | | | |

(b) Do you think that $r$ is positive, negative, or 0?
(c) Complete the table.
Find:#

| | | |
|---|---|---|
| (d) $n$ | (h) $\Sigma X^2$ | (l) $n\Sigma XY - (\Sigma X)(\Sigma Y)$ |
| (e) $\Sigma X$ | (i) $\Sigma Y^2$ | (m) $\sqrt{n\Sigma X^2 - (\Sigma X)^2}\,\sqrt{n\Sigma Y^2 - (\Sigma Y)^2}$ |
| (f) $\Sigma Y$ | (j) $(\Sigma X)^2$ | |
| (g) $\Sigma XY$ | (k) $(\Sigma Y)^2$ | (n) $r$ |

14-5 (a) Draw a scattergram for the information in the following table.
(b) Do you think that $r$ is near $+0.8$, $+0.3$, 0, $-0.3$, or $-0.8$?
(c) Complete parts (c) to (n) of the previous exercise.

| | $X$ | $Y$ | $X^2$ | $Y^2$ | $XY$ |
|---|---|---|---|---|---|
| | 10 | 1 | | | |
| | 5 | 9 | | | |
| | 3 | 3 | | | |
| | 8 | 9 | | | |
| | 0 | 0 | | | |
| Totals | | | | | |

14-6 Follow the instructions of the previous exercise, using these data:

| | $X$ | $Y$ | $X^2$ | $Y^2$ | $XY$ |
|---|---|---|---|---|---|
| | 2 | −1 | | | |
| | 2 | 3 | | | |
| | 0 | −2 | | | |
| | 3 | 4 | | | |
| | 4 | 4 | | | |
| | 1 | 0 | | | |
| Totals | | | | | |

14-7 Below is a table giving the lengths of some pieces of lumber in feet and in inches.

| Length, in Inches | Length, in Feet |
|---|---|
| 12 | 1 |
| 36 | 3 |
| 60 | 5 |
| 48 | 4 |
| 24 | 2 |
| 72 | 6 |

(a) Draw a scattergram for the data.
(b) Intuitively guess whether $r$ is near 1, −1, or 0.
(c) Compute $r$.

14-8 Here is a table showing what some people's ages will be in April 2008, and the years of their births. Draw a scattergram, guess at the value of $r$, and then compute $r$.

| Age in April 2008 | Year of Birth |
|---|---|
| 21 | 1987 |
| 22 | 1986 |
| 23 | 1984 |
| 24 | 1984 |
| 26 | 1981 |

14-9 For a group of mothers and their firstborn children, what kind of correlation (negative, positive, or 0) would you expect between the mothers' ages and the childrens' ages. Explain your answer.

*Save your work and your results from exercises 14-10 to 14-14. They will be used in exercises later in the chapter.*

14-10 A random sample of men were stopped in a shopping center and asked their shoe sizes and the number of ties that they owned. Would you expect there to be any correlation between the 2 variables? Here are the data. Draw a scattergram and compute *r*.

| | Shoe Size | Number of Ties Owned |
|---|---|---|
| Sam | 7½ | 10 |
| Tom | 9 | 17 |
| Alf | 9 | 17 |
| Tim | 11 | 4 |
| Ike | 8½ | 10 |
| Sol | 8 | 1 |
| Ted | 13 | 6 |
| Ira | 10 | 9 |
| Cal | 10 | 11 |
| Sid | 10 | 10 |

14-11 Draw a scattergram and compute *r* for these grades of 6 students in Preparatory English and in the first semester of Survey of English Literature:

| Prep. English | English Lit. |
|---|---|
| 50 | 45 |
| 57 | 57 |
| 68 | 60 |
| 75 | 75 |
| 80 | 84 |
| 89 | 93 |

14-12 An experiment was done to see if there was any correlation between the volume of water in a fish tank and the average length to which goldfish grew when they were hatched and raised in that tank. Here are the results. Draw a scattergram and compute *r*.

| Fish Tank Volume, in Gallons | Average Length of Fish, in Inches |
|---|---|
| 0.5 | 1.8 |
| 1 | 2.1 |
| 2 | 2.2 |
| 4 | 2.9 |
| 5 | 3.3 |

14-13 An experiment was performed where an object was dropped through a certain liquid. The distance it traveled was recorded every second for 6 seconds. Here are the results.

| Time, in Seconds | Distance Object Traveled through Liquid, in Feet |
|---|---|
| 0 | 0 |
| 1 | 1 |
| 2 | 4 |
| 3 | 9 |
| 4 | 16 |
| 5 | 25 |
| 6 | 36 |

The experimenter recognized that there was not a true linear relationship. However, it seemed that the relationship was nearly linear for the time period from 3 to 6 seconds. Check the experimenter's intuition by drawing a scattergram and computing $r$:

(a) For all the data
(b) For just the data from 3 to 6 seconds

14-14 Here are some stopping distances for a certain vehicle going at various speeds. Compute $r$ for the data. Describe this relationship in an ordinary English-language sentence.

| Speed, in Miles per Hour | Stopping Distance, in Feet |
|---|---|
| 30 | 90 |
| 40 | 150 |
| 50 | 240 |
| 60 | 370 |
| 70 | 530 |

14-15 For the weather data below, sketch a scattergram and compute $r$.

| Day | Daily High, °F | Daily Low, °F |
|---|---|---|
| Tuesday | 70 | 50 |
| Wednesday | 72 | 50 |
| Thursday | 66 | 48 |
| Friday | 73 | 51 |
| Saturday | 67 | 49 |

14-16 An investor investigating a possible correlation between 2 stocks noticed what might be a pattern connecting their prices. Sketch a scattergram and compute $r$ for the data to see if there seems to be a linear relationship between their prices.

| Date | Selling Price of BQT Stock | Selling Price of CRV Stock |
|---|---|---|
| January 1 | $47 | $22 |
| February 1 | $40 | $24 |
| March 1 | $30 | $26 |
| April 1 | $15 | $30 |

14-17 The data in the following table come from a sample of 20 employees at Squid Ink, Inc.

(a) Draw a scatterplot for all the data in order to display the relationship between number of years worked and salary. Does there seem to be a positive correlation between the variables? Compute $r$ and $r^2$.

(b) Mark the dots for females with an *F*, and mark those for males with an *M*. What impression does the scattergram make now? Find the correlation between years worked and salary for just the female employees. Repeat the procedure for the males.

| Years Employed | Salary, in $1000s | Sex |
|---|---|---|
| 1 | 25 | M |
| 1 | 33 | F |
| 2 | 17 | M |
| 2 | 30 | F |
| 3 | 29 | M |
| 3 | 33 | F |
| 4 | 30 | M |
| 4 | 24 | M |
| 5 | 33 | F |
| 5 | 32 | F |
| 6 | 43 | F |
| 6 | 32 | M |
| 7 | 21 | M |
| 7 | 36 | F |
| 8 | 26 | M |
| 8 | 41 | F |
| 9 | 31 | M |
| 9 | 43 | F |
| 10 | 43 | F |
| 10 | 33 | M |

*Source:* This exercise is based on one created by Prof. L. Roethel of Nassau Community College.

14-18 Here is a set of data showing the relationship between salt intake and blood pressure in two different populations. Population N is made up of Dr. Payne's "normal" patients. Population S comprises Dr. Hirtz' patients who suffer from Salswitch disease.

(a) Draw a scatterplot for all the data in order to display the relationship between salt intake and blood pressure. Is there any apparent correlation? Compute $r$ and $r^2$.

(b) Mark the dots for population N with an *N,* and mark those for population S with a *S.* What impression does the scattergram make now? Find the correlation between salt intake and blood pressure just for population N. Repeat for population S.

| Units of Salt Intake Daily | Blood Pressure | Population |
|---|---|---|
| 100 | 81 | N |
| 100 | 107 | S |
| 110 | 84 | N |
| 110 | 102 | S |
| 120 | 94 | S |
| 120 | 88 | N |
| 130 | 108 | N |
| 130 | 97 | S |
| 140 | 88 | S |
| 140 | 114 | N |
| 150 | 89 | S |
| 150 | 111 | N |
| 160 | 75 | S |
| 160 | 117 | N |
| 170 | 132 | N |
| 170 | 83 | S |
| 180 | 71 | S |
| 180 | 142 | N |
| 190 | 80 | S |
| 190 | 155 | N |

14-19 State 2 variables for which there would probably be:

(a) Negative correlation

(b) Zero correlation

(c) Positive correlation

For each example, explain your choice.

## 14.2 Testing the Significance of $r$

Suppose that we examined an entire population and computed $\rho$, the correlation coefficient for two variables. If this coefficient equaled zero, we would say that there is no correlation between these two variables in this population. Consequently, when we examine a random sample taken from a population, then a *sample* value of $r$ near zero is interpreted as reflecting little or no linear correlation between the variables *in the population.* A sample value of $r$ far from zero (near $+1$ or $-1$) indicates that there *is some* correlation in the population. The statistician must decide when a sample value of $r$ is far enough from zero to be statistically significant—that is, when it is sufficiently far from zero to reflect positive or negative correlation in the population.

This test for significance would have to be carried out in different ways, depending on what the distributions of the two variables in the population were. The simplest case results when it can be assumed that **both variables are normally distributed**. We will assume this normality throughout this section of the text.

In testing a sample value of $r$ for statistical significance, we take as the null hypothesis "There is no correlation between the two variables in question." Using English $r$ for the sample correlation and Greek $\rho$ (rho) for the population correlation, we have:

### *Unusual Research*

Francis Galton, cousin of Charles Darwin, was independently wealthy and thus able to devote time to his favorite subject—measurement. He once proposed and started a statistical investigation of the efficacy of praying. He tried to construct a "beauty map" of the British Isles by classifying people as attractive, indifferent, or repellent. He claimed that for beauty London ranked highest and Aberdeen ranked lowest. He humorously suggested measuring boredom by counting the number of times people fidgeted during a sermon.

$H_0$: There is no correlation in the population; $\rho = 0$.

and *one* of these three alternatives:

$H_a$: There is some correlation in the population; $\rho \neq 0$ (a two-tail test).
$H_a$: There is some positive correlation in the population; $\rho > 0$ (a one-tail test).
$H_a$: There is some negative correlation in the population; $\rho < 0$ (a one-tail test).

We will be able to reject the null hypothesis when our sample value of $r$ is farther from zero than some critical value. Tables D-9 and D-10 (see Appendix D) contain lists of critical values of $r$ arranged according to the total number of pairs of scores in the sample $n$, the significance level of the test $\alpha$, and the number of tails in the test. Values of $r$ are given in this table without signs. You must determine whether the critical values of $r$ are positive, negative, or both, from the alternative hypothesis.

### *Critical Values*

Karl Pearson was the first to derive the critical values of $r$ by describing its sampling distribution.

### ❑ EXAMPLE 14.1 Verbal Aptitude and Income

In a study of academic success, a random sample of 30 public school children in Plaintalk, New Hampshire, was taken. For each child this survey recorded verbal aptitude (as determined by a commonly accepted test) and the annual income of the child's family. The 30 pairs of scores were used to compute $r$. It was found that $r = 0.46$. Does this support the hypothesis that there is positive correlation between verbal aptitude and family income in the children of this community? Test at the 0.05 significance level.

#### SOLUTION

We will perform a hypothesis test on the significance of $r$.

*Population:* All public school children in Plaintalk, New Hampshire

*Sample:* 30 public school children from Plaintalk

$H_0$: There is no correlation between verbal aptitude and family income in the population of Plaintalk; $\rho = 0$.

$H_a$: There is some positive correlation between verbal aptitude and family income in the population of Plaintalk; $\rho > 0$ (a one-tail test).

*Critical value:* For $\alpha = 0.05$ and $n = 30$, we get $r_c = +0.31$.

*Decision rule:* We will reject the null hypothesis if $r > +0.31$.

*The outcome:* $r = 0.46$.

*Conclusion:* We reject the null hypothesis and conclude that the coefficient of correlation in the population is greater than zero. There is *some* positive linear correlation in this population between family income and verbal aptitude. Higher incomes are associated with higher verbal aptitude. ❑

**Note** The existence of some nonzero correlation in a population does not mean that the correlation is *high*, merely that it is not zero. Unless the correlation is far enough from zero, it may be of very little practical value.

> **Caution!**
>
> The fact that the correlation exists does not by itself prove anything about the *reasons* for the correlation. Researchers must decide independently (1) what causes the correlation and (2) whether or not the correlation and its causes are of any practical concern.

The significance test may verify the existence of the correlation in the given population; it does not establish cause and effect. For example, it would not be hard to show that there is a positive correlation between the average salary of elementary school teachers in Montpelier, Vermont, and the number of pizzas sold in Montpelier over the past 10 years. Can you think of any reasons why this is so?

## ❑ EXAMPLE 14.2 Eating and Age at Death

A research team wondered if there was any correlation between the amount of food a freely feeding laboratory rat ate when it was 50 days old and the age at which it died. They planned to experiment with 30 rats who would be left alone to eat as much as they chose. What are their two hypotheses? What values of $r$ would they have to find to establish a correlation between diet and longevity if they use $\alpha = 0.05$?

### SOLUTION

We will perform a hypothesis test on the statistical significance of $r$.

*Population:* All laboratory rats of the type under consideration.
*Sample:* 30 such rats.

$H_0$: For this type of rat there is no correlation between the amount of food rats eat at age 50 days and the age at which they die.

$H_a$: For this type of rat, there is some correlation between the amount of food rats eat at age 50 days and the age at which they die. $\rho \neq 0$ (a two-tail test).

*Critical value:* For $\alpha = 0.05$ and $n = 30$, we get $r_c = \pm 0.36$.

*Decision rule:* We will reject the null hypothesis if $r$ is either less than $-0.36$ or greater than $+0.36$.

If they find that their sample value of $r$ is more than $+0.36$ or less than $-0.36$, this will be evidence in favor of a correlation between food consumption and life span. A significant negative value of $r$ would indicate an association of eating a little with living a long time. ❑

### Exercises

*In the following exercises, assume that both populations are normal so that the Pearson correlation is appropriate.*

14-20 A hypothesis test shows that there is high positive correlation between grades at the Police Academy and success "on the street" as a police office. Yet Sergeant Experience notes that sometimes a candidate who did well at the academy fails on the beat. Similarly, he notes that a few officers do well on the job who did not do well at the academy. Just what information is conveyed by high positive correlation anyhow?

*Exercises 14-21 to 14-26*

*You have previously computed* r *for these exercises (in Exercises 14-10 through 14-14). Now we will state what a researcher was testing for. You should then perform a hypothesis test on the significance of* r *with* $\alpha = 0.01$.

14-21 (Data from Exercise 14-10).
The researcher was testing for nonzero correlation.

| | Shoe Size | Number of Ties Owned |
|---|---|---|
| Sam | 7½ | 10 |
| Tom | 9 | 17 |
| Alf | 9 | 17 |
| Tim | 11 | 4 |
| Ike | 8½ | 10 |
| Sol | 8 | 1 |
| Ted | 13 | 6 |
| Ira | 10 | 9 |
| Cal | 10 | 11 |
| Sid | 10 | 10 |

14-22 (Data from Exercise 14-11.)
Test for positive correlation.

| Prep. English | English Lit. |
|---|---|
| 50 | 45 |
| 57 | 57 |
| 68 | 60 |
| 75 | 75 |
| 80 | 84 |
| 89 | 93 |

14-23 (Data from Exercise 14-12.)
The experimenter was testing for nonzero correlation.

| Fish Tank Volume, in Gallons | Average Length of Fish, in Inches |
|---|---|
| 0.5 | 1.8 |
| 1 | 2.1 |
| 2 | 2.2 |
| 4 | 2.9 |
| 5 | 3.3 |

14-24 [Data from Exercise 14-13(a).]

Test for positive correlation.

| Time, in Seconds | Distance Object Traveled through Liquid, in Feet |
|---|---|
| 0 | 0 |
| 1 | 1 |
| 2 | 4 |
| 3 | 9 |
| 4 | 16 |
| 5 | 25 |
| 6 | 36 |

14-25 [Data from Exercise 14-13(b).]

Test for positive correlation.

| Miles per Hour | Distance Object Traveled through Liquid, in Feet |
|---|---|
| 3 | 9 |
| 4 | 16 |
| 5 | 25 |
| 6 | 36 |

14-26 (Data from Exercise 14-14.)

Test for positive correlation.

| Speed, in Miles per Hour | Stopping Distance, in Feet |
|---|---|
| 30 | 90 |
| 40 | 150 |
| 50 | 240 |
| 60 | 370 |
| 70 | 530 |

14-27 Suppose you thought that there was some correlation between the length of a male college student's hair and his political beliefs. Imagine that some clever psychology professor has designed a test of political belief. When a person takes this test, the score can run anywhere from 0 (extreme right-wing beliefs) to 200 (extreme left-wing beliefs). You get a random sample of 25 male students. You score each student for the 2 variables. These are the results.

| Student Number | Hair Length | Test Score |
|---|---|---|
| 1 | 0.5 | 50 |
| 2 | 2.0 | 140 |
| 3 | 1.0 | 60 |
| 4 | 2.5 | 80 |
| 5 | 3.0 | 115 |
| 6 | 1.5 | 75 |
| 7 | 4.5 | 170 |
| 8 | 3.5 | 120 |
| 9 | 2.5 | 95 |
| 10 | 3.0 | 120 |
| 11 | 1.0 | 85 |
| 12 | 4.0 | 160 |
| 13 | 2.0 | 100 |
| 14 | 2.5 | 100 |
| 15 | 4.5 | 165 |
| 16 | 1.5 | 90 |
| 17 | 3.0 | 105 |
| 18 | 2.5 | 105 |
| 19 | 2.0 | 85 |
| 20 | 3.5 | 140 |
| 21 | 5.0 | 180 |
| 22 | 2.5 | 130 |
| 23 | 4.0 | 150 |
| 24 | 3.0 | 100 |
| 25 | 2.0 | 80 |

Test for positive correlation at the 0.05 significance level.

14-28 An educational testing laboratory is developing a new test to measure computer-programming aptitude. They wish to develop 2 different forms of the test. Theoretically, a person should get the same score, no matter which form of the test he or she takes. To determine whether or not both forms give about the same results, they are administered to 30 people. The results are as follows.

| Candidate | Form A | Form B |
|---|---|---|
| 1 | 99 | 80 |
| 2 | 97 | 95 |
| 3 | 97 | 87 |
| 4 | 90 | 88 |
| 5 | 89 | 83 |
| 6 | 83 | 90 |
| 7 | 80 | 85 |
| 8 | 80 | 78 |
| 9 | 75 | 40 |
| 10 | 70 | 76 |
| 11 | 69 | 70 |
| 12 | 69 | 71 |
| 13 | 68 | 70 |
| 14 | 68 | 72 |
| 15 | 68 | 68 |
| 16 | 67 | 63 |
| 17 | 67 | 60 |
| 18 | 65 | 64 |
| 19 | 65 | 81 |
| 20 | 65 | 65 |
| 21 | 63 | 60 |
| 22 | 62 | 61 |
| 23 | 61 | 59 |
| 24 | 60 | 50 |
| 25 | 59 | 58 |
| 26 | 50 | 40 |
| 27 | 43 | 51 |
| 28 | 40 | 70 |
| 29 | 20 | 19 |
| 30 | 3 | 0 |

Test for positive correlation at the 0.01 significance level. Interpret your results.

14-29 The following test scores were collected at random from Ms. Betty's School for Young People.

| Student | Reading | Spelling | Math | Music |
|---|---|---|---|---|
| Sam | 20 | 7 | 100 | 10 |
| Samantha | 15 | 7 | 70 | . . . |
| Toni | 25 | 10 | 60 | 3 |
| Anthony | 35 | 8 | 90 | 9 |
| Salvatore | 30 | 9 | . . . | 20 |
| Sally | 50 | 8 | 80 | 15 |
| Pat | 40 | 10 | 80 | 5 |

Perform a hypothesis test at the 0.05 significance level for each of the following:

(a) Is there positive correlation between the reading scores and the spelling scores?
(b) Is there negative correlation between the spelling scores and the math scores?
(c) Is there nonzero correlation between the spelling scores and the music scores?

## 14.3 Prediction Based on Linear Correlation

If it has been determined that there is sufficiently high* linear correlation between two variables, we can try to represent the correspondence by an ideal line—a line that best represents the linear correspondence. We can then write the formula which determines this line, and use this formula to predict, for instance, which value of the $Y$ variable corresponds ideally to any given value of the $X$ variable.

For example, let us suppose that at State U grades in English 1 and English 2 have a high positive correlation. Suppose we have found a formula which predicts grades in English 2 from grades in English 1. Given that a grade in English 1 was 85, the formula predicts a grade of 81 in English 2. Clearly, if 10 students had grades of 85 in English 1, we do not expect all 10 to get an 81 in English 2. In fact, maybe none of them will actually get an 81. The predicting formula really says that our best estimate of their *mean* grade in English 2 will be 81. On the other hand, if we do want to predict one student's grade, the best point estimate we can make will be this mean, 81. In this section we will show how to get the formula for the line which is used to get the best point estimates. The important topic of evaluating the reliability of these estimates will not be discussed, but, roughly speaking, the closer the points in the scattergram approximate a straight line, the more reliable the estimate will be. (See Exercise 14-42.)

### ❑ EXAMPLE 14.3 Predicting College Success

For several years, admission to a certain college has been on an open-enrollment basis. Because of recent enrollment growth, Dr. Dean, the dean of admissions, now finds it necessary to turn some applicants down. Dean Dean wants to turn down those applicants who would probably flunk out anyway. A random sample of former students' records are pulled from the files. It is noticed that there is strong positive correlation between student scores on a

*What is considered to be a high correlation varies with the field of application.

certain aptitude test and their college grade point average (GPA) at the time they leave the college (either graduating or withdrawing). The scattergram for the data is shown in Figure 14-4.

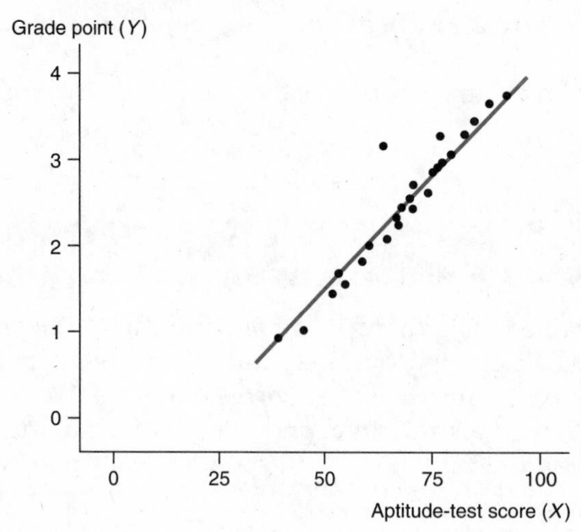


**Figure 14-4**

To the scattergram has been added the line that best represents a linear correspondence between the two variables. This line shows, for any given value of *X*, what the corresponding value of *Y* is according to this ideal line. For example, if someone applies for admission who has scored 75 on the aptitude test, this line predicts a GPA of about 2.9. This may be seen by finding 75 on the horizontal axis *X*, then looking straight up until you hit the predicting line, and then looking directly to the left to see where you hit the vertical axis *Y*. Follow the dashed arrow on the diagram in Figure 14-5.

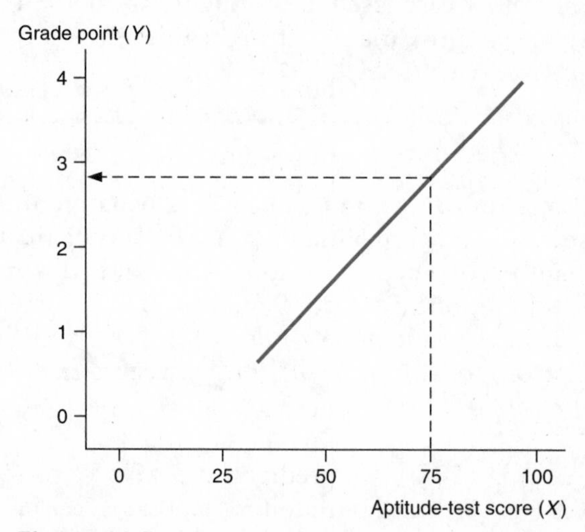


**Figure 14-5**

Similarly if a student scores 62 on the aptitude test, we get a predicted value on the GPA of about 2.1. A student with an aptitude-test score of 50 would have a predicted GPA of about 1.4. ❑

---

*Note* The interpretation of the scattergram says *nothing* about the *reasons* for the correlation, the nature of the test questions, or the intelligence of the students. It merely acknowledges that a pattern exists and that as long as the population from which their applicants come remains the same, and as long as the college's curriculum and grading system remain the same, then it is likely that the predictions are reasonable.

**best-fitting line**

**line of regression**

In this prediction problem, the statistician, after having decided that there is a strong correlation, must then produce the formula for making the predictions. This formula is called the formula of the **best-fitting line** or **line of regression**.

The formula for the best-fitting line is

$$Y_{\text{predicted}} = m_y + b(X - m_x)$$
$$= \bar{Y} + b(X - \bar{X})$$

where $m_x$, $m_y$, $\bar{X}$, and $\bar{Y}$ are sample means, $n$ is the number of pairs of scores, and where

$$b = \frac{n\Sigma XY - (\Sigma X)(\Sigma Y)}{n\Sigma X^2 - (\Sigma X)^2}$$

### *Helpful Hint*

You should choose the labels for your variable so that you assign $Y$ to the variable whose value you want to predict, and $X$ to the variable whose value you know.

You can always compute the value of $Y_{\text{predicted}}$, no matter how the points of the scattergram are arranged, but its accuracy in making predictions declines as $r$ gets closer to zero. An important topic, but one that we omit in this elementary text, is a test for the reliability of the predicted values of $Y$.

### *The Best-Fitting Line*

The line of regression, or the best-fitting line, is based on the "least squares" method. This was first published in 1805 by Adrien-Marie Legendre (1752–1833), although Gauss had apparently developed the same idea previously.

The method is called "least squares" because it *minimizes* the sum of the squares of the distances of the data points from the regression line.

Current use of regression lines is rooted in the work of G. U. Yule (1871–1951), who was a young associate of Karl Pearson. His 1911 book, *Introduction to the Theory of Statistics*, was influential for more than 50 years.

---

### ❑ EXAMPLE 14.4 Grade Point Averages

Let us apply the formula to the following set of data (Table 14-6) to try to predict the GPA corresponding to aptitude test scores of 60, 70, and 80 in a sample of 25 pairs of scores.

**Table 14-6 Sample Data**

| *Aptitude Test* $X$ | *GPA* $Y$ | $XY$ | $X^2$ |
|---|---|---|---|
| 40 | 0.88 | 34.20 | 1600 |
| 45 | 1.02 | 45.90 | 2025 |
| 53 | 1.56 | 82.68 | 2809 |
| 54 | 1.75 | 94.50 | 2916 |
| 55 | 1.63 | 89.65 | 3025 |
| 60 | 1.90 | 114.00 | 3600 |
| 62 | 2.07 | 128.34 | 3844 |
| 65 | 3.21 | 208.65 | 4225 |
| 66 | 2.12 | 139.92 | 4356 |
| 68 | 2.38 | 161.84 | 4624 |
| 68 | 2.40 | 163.20 | 4624 |
| 69 | 2.52 | 173.88 | 4761 |
| 70 | 2.63 | 184.10 | 4900 |
| 72 | 2.52 | 181.44 | 5184 |
| 72 | 2.79 | 200.88 | 5184 |
| 75 | 2.72 | 204.00 | 5625 |
| 75 | 2.90 | 217.50 | 5625 |
| 76 | 2.95 | 224.20 | 5776 |
| 77 | 3.01 | 231.77 | 5929 |
| 77 | 3.35 | 257.95 | 5929 |
| 80 | 3.10 | 248.00 | 6400 |
| 84 | 3.41 | 286.44 | 7056 |
| 86 | 3.52 | 302.72 | 7396 |
| 89 | 3.75 | 333.75 | 7921 |
| 94 | 3.82 | 359.08 | 8836 |
| $\Sigma X = 1732$ | $\Sigma Y = 63.91$ | $\Sigma XY = 4669.59$ | $\Sigma X^2 = 124{,}170$ |

From the table, we compute:

$$n = 25$$

$$m_X = \Sigma X/n = 1732/25 = 69.28$$

$$m_Y = \Sigma Y/n = 63.91/25 = 2.56$$

$$b = \frac{n\Sigma XY - (\Sigma X)(\Sigma Y)}{n\Sigma X^2 - (\Sigma X)^2} = \frac{25(4669.59) - 1732(63.91)}{25(124{,}170) - (1732)^2} = 0.06$$

$$Y_{\text{predicted}} = m_y + b(X - m_x)$$

$$= 2.56 + 0.06(X - 69.28)$$

For $X = 60$, we get

$$Y_{\text{predicted}} = 2.56 + 0.06(60 - 69.28) = 2.0$$

For $X = 70$, we get

$$Y_{\text{predicted}} = 2.56 + 0.06(70 - 69.28) = 2.6$$

For $X = 80$, we get

$$Y_{\text{predicted}} = 2.56 + 0.06(80 - 69.28) = 3.2$$

Therefore, for aptitude-test scores of 60, 70, and 80, we have predicted average final GPAs, of 2.0, 2.6, and 3.2, respectively. ❑

## 14.3-1 Interpretation of $b$*

In the formula for the best-fitting line, the value of $b$ tells you *how much you can expect* Y *to change when you change* X *by 1 unit.*

### ❑ EXAMPLE 14.5 Piston Pressures

The results of a test where the weight placed on a piston and the pressure it exerted on a fluid were measured are given below.

| Weight $X$, in pounds | 100 | 200 | 300 |
|---|---|---|---|
| Pressure $Y$, in pounds per square inch | 17 | 18 | 19 |

We can calculate $\bar{X} = 200$, $\bar{Y} = 18$, and $b = 0.01$ for these data. When $X$ is 200, the predicted value of $Y$ is 18. What will the predicted value of $Y$ be if $X$ is 230?

**SOLUTION**

Recall that $b$ represents the change in $Y$ per *one* unit change in $X$. Since $b = 0.01$ and the change in $X$ is $230 - 200 = 30$, the expected change in $Y$ is $(b)(30) = 0.01(30) = 0.3$. So $Y_{\text{predicted}} = 18 + 0.3 = 18.3$ pounds per square inch. ❑

*Geometrically, $b$ can be interpreted as the slope of the best-fitting line.

## 14.4 The Coefficient of Determination

Over the past few years, Dr. Giuseppe O'Reilly has found a high correlation, $r = 0.9$, between the grades of his students in Hebrew 3 and their grades in Hebrew 4. His regression formula came out to be $Y = 1.2(X - 80.5) + 77.6$, where $Y$ equals the Hebrew 4 grade and $X$ equals the Hebrew 3 grade of a particular student. If he lets $X = 90$, he computes a grade of 89. As we stated previously, this does not mean that all students who receive a grade of 90 in Hebrew 3 will definitely get a grade of 89 in Hebrew 4, but rather that we expect their average grade in Hebrew 4 to be close to 89. The grades in Hebrew 4 earned by all the students who got a 90 in Hebrew 3 will have some variation, and this variation can be attributed to many things: increased or decreased interest in the language, response to parental pressures, changed study habits, successes or failures in social life, as well as previous knowledge of the material such as measured by grades in Hebrew 3.

In general, it is true in any regression study that the actual observed $Y$ values will vary somewhat from the predicted $Y$ values. In a study where the actual $Y$ values come out very close to the predicted ones, we say the regression has a lot of **strength**. Furthermore, it turns out that the statistic $r^2$ can be used to measure this strength. Mathematically, it can be shown that

**strength**

$$r^2 = \frac{\text{variance of predicted } Y \text{ values}}{\text{variance of actual } Y \text{ values}}$$

In words, we say that $r^2$ measures the proportion of variance of the actual $Y$'s that can be "explained by" the linear relationship between $X$ and $Y$. The statistic $r^2$ is called the **coefficient of determination**.

**coefficient of determination**

In our example $r = 0.9$, thus $r^2 = 0.81 = 81$ percent. Therefore, we say that 81 percent of the *variance* of the grades in Hebrew 4 is due to the **linear correlation** between the grades in Hebrew 3 and the grades in Hebrew 4. In an informal sense we can say that the closer $r^2$ is to 1, the more strongly the values of $Y$ *depend* on the values of $X$, that is, the more likely it is that $Y$ will turn out to be close to what it is predicted to be. In the extreme case of perfect correlation ($\rho^2 = 1$), $Y$ is completely determined by $X$. There is no other source of variation that will cause $Y$ to vary from its predicted value. At the other extreme, that of no correlation ($\rho = 0$), we can say that $X$ and $Y$ are statistically independent (assuming, as we said earlier, that $X$ and $Y$ are normally distributed).

**linear correlation**

***Note*** If we measure the strength of regression by the coefficient of determination, then, for example, a regression with $r = 0.8$ has *more than twice* the strength as one with $r = 0.4$. If $r_1 = 0.8$, then $r_1^2 = 0.64 = 64$ percent, while if $r_2 = 0.4$, then $r_2^2 = 0.16 = 16$ percent, and the first regression is about *4 times* as strong as the second.

## STUDY AIDS

### Chapter Summary

In this chapter you have learned:

- The meaning of a coefficient of correlation and how to calculate it
- How to draw a scattergram and a best-fitting line
- How to use the formula for a best-fitting line to predict results
- How to use $r^2$ to describe the strength of regression

### Vocabulary

You should be able to explain the meaning of each of these terms:

1. Correlation coefficient
2. Scattergram
3. Positive correlation
4. Zero correlation
5. Negative correlation
6. Best-fitting line
7. Line of regression
8. Strength of regression
9. Coefficient of determination
10. Linear correlation

### Symbols

You should understand the meaning of each of these symbols:

1. $r$ 2. $\rho$ 3. $b$ 4. $Y_{\text{predicted}}$ 5. $r^2$

### Formulas

You should know when and how to use each of these formulas:

1. $$r = \frac{n\Sigma XY - (\Sigma X)(\Sigma Y)}{\sqrt{n\Sigma X^2 - (\Sigma X)^2}\sqrt{n\Sigma Y^2 - (\Sigma Y)^2}}$$

2. $$b = \frac{n\Sigma XY - (\Sigma X)(\Sigma Y)}{n\Sigma X^2 - (\Sigma X)^2}$$

3. $Y_{\text{predicted}} = m_Y + b(X - m_X)$ or $Y_{\text{predicted}} = \bar{Y} + b(X - \bar{X})$

## EXERCISES

*In each of the following exercises, assume that the correlation is high enough to allow for reasonable prediction.*

14-30 The graph on the following page appeared in a medical report on hypertension.* The 2 variables are average daily salt intake (in grams) and percentage of people suffering hypertension (high blood pressure). The 5 dots represent 5 different communities.

---

***Hypertension Update*, vol. 1, Health Learning Systems, Bloomfield, N.J., 1979.

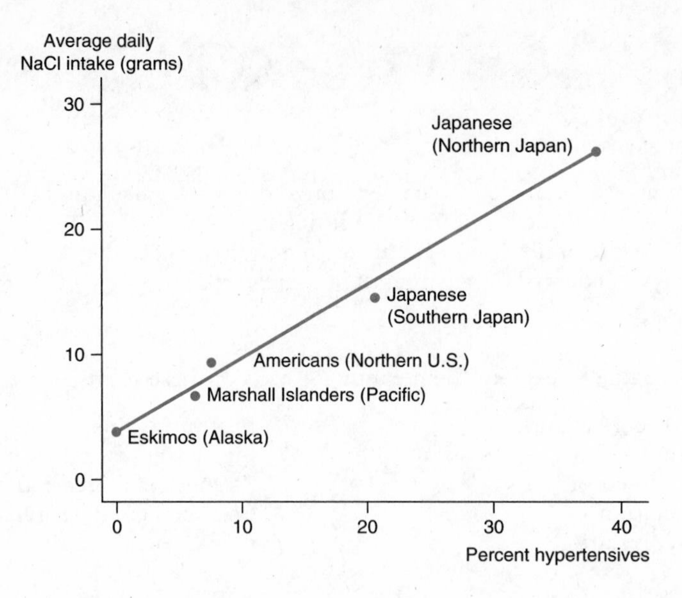


(a) Does there seem to be a positive correlation between the 2 variables?

(b) Give at least 2 different possible explanations for the pattern of results.

(c) If you assume the relationship would hold up for other locations, about what percentage of the population would suffer hypertension if the average daily salt intake were 20 grams of salt?

14-31 A regression formula based on the high correlation between grades in Math 001, a preparatory mathematics course, and subsequent grades in Math 112, Introductory College Math, was found to be $Y_{\text{predicted}} = 1.5X - 40$.

(a) Using the formula for a student with a grade of 82 in Math 001, find the predicted grade for Math 112.

(b) It was noted that, of all 47 students who received an 82 in Math 001, not 1 scored the predicted 83 in Math 112. Comment on this fact.

14-32 The following data have been collected from a random sample of 10 families in a community.

| Total Income, in $1000 | Percentage Spent on Food |
|---|---|
| 13.2 | 42 |
| 14.0 | 37 |
| 15.0 | 33 |
| 16.0 | 32 |
| 17.0 | 30 |
| 17.0 | 27 |
| 18.2 | 25 |
| 20.0 | 23 |
| 21.0 | 22 |
| 22.0 | 20 |

(a) Compute $r$.

(b) Compute $b$.

(c) Write the formula for predicting the percentage spent on food when you know the income.

(d) Predict what percentage a family with an income of $15,000 spends on food.

14-33 Professor Gonzales has a theory that grades on his final exams are related to where the students have been sitting all term. He takes a random sample of 15 students and records their grades and the distance between their seats and his desk.

| Distance, in feet | 3 | 4 | 5 | 6 | 7 | 8 | 8 | 8 | 9 | 10 | 12 | 12 | 12 | 14 | 15 |
|---|---|---|---|---|---|---|---|---|---|---|---|---|---|---|---|
| Grade | 85 | 98 | 93 | 83 | 78 | 70 | 83 | 65 | 80 | 63 | 60 | 57 | 71 | 40 | 90 |

(a) Compute $r$.

(b) Compute $b$.

(c) Write the formula for predicting final grades from seat position.

(d) Predict the average grade for people who sit 11 feet from Professor Gonzales.

14-34 The secretary of Gamma Gamma Gamma fraternity has done some research on the grading practices of Professor Neumann. By examining a random sample of term papers from Professor Neumann's classes, he has assembled the following data: mean weight of papers = 8.3 ounces, mean grade = 76.6, $r = 0.84$, and $b = 5.8$. Predict the mean grade for students who turn in papers weighing 6 ounces.

14-35 A representative of the fishing industry in Arid, Arizona has gathered the following monthly data relating the amount of a certain pollutant in the fishing waters and the amount of fish caught: mean amount of pollutant = 36 units per sample of water, mean amount of fish caught = 50 barrels, $r = 0.7$, and $b = -1.4$. Predict the amount of fish caught if the amount of pollutant is 48 units per sample of water.

14-36 Bill Board, the director of an ad agency, found that for 1 of his products there was a negative correlation between the expense of a campaign ($X$) and the revenue it generated ($Y$)! It was found that $b = -2.5$, $m_X = 1.2$ million dollars, and $m_Y = 30$ million dollars. What revenue does his regression predict for an ad campaign which costs 2.2 million dollars?

14-37 Three researchers were attempting to devise a convenient method for measuring evaporation of water from soils and crops. They discovered a relation between evaporation and what is called "net thermal radiation." Here are their data collected from 1 field on 8 different days scattered throughout the growing season, and rearranged in order:

| Day | $Y$ = Units of Evaporation in Calories per Square Centimeter | $X$ = Units of Net Thermal Radiation in Calories per Square Centimeter |
|---|---|---|
| 1 | 17.0 | −87 |
| 2 | 17.5 | −86 |
| 3 | 19.0 | −84 |
| 4 | 21.0 | −86 |
| 5 | 55.0 | −62 |
| 6 | 70.0 | −55 |
| 7 | 83.0 | −45 |
| 8 | 90.0 | −41 |

(a) Draw a scattergram for these data.

(b) Compute $r$.

(c) Find the best-fitting line.

(d) Predict net thermal radiation if evaporation is 50.0.

14-38 If the coefficient of correlation between chemistry grades and math grades is 0.7, approximately what percentage of the variation in students' chemistry grades is due to the relationship between their chemistry grades and their math grades?

14-39 If the coefficient of correlation between GPAs and annual income is found to be 0.8, about what percentage of the variation in annual income can be attributed to other factors, such as ingenuity, personal initiative, or marrying the boss's daughter?

14-40 Miss Krobar found that in her third-grade class the weights of the children have a correlation with both their ages and their heights. For the weights and ages, she found $r_A = 0.7$. For the weights and heights, she found $r_B = 0.8$.

(a) What percentage of the variation in weights can be attributed to the relationship between weight and age?

(b) What percentage of the variation in weights can be attributed to the relationship between weight and height?

(c) Discuss the paradox implied by your answers to parts (a) and (b).

14-41 The coefficient of correlation between typing speed and typing errors was found to be 0.4. The variation in typing errors due to inattention is 4 times as great as the variation due to speed. Find the coefficient of correlation between typing errors and inattention.

14-42 These 2 scattergrams produce the same formula for the best-fitting line. Which one will give more-reliable estimates? Why?

## CLASS SURVEY QUESTION

Would you expect to find positive, negative, or no correlation between the heights of males in this class and the heights of their fathers? Compute the coefficient of correlation, and test to see whether or not this is a significant correlation.

## FIELD PROJECTS

Compute $r$ for two variables of your choosing. Test the hypothesis that the coefficient of correlation is zero. If it is not zero, you may wish to include the formula for prediction and some pertinent predictions.

Some projects of this type which students have done include investigating correlations between:

1. Speed they drive and time it takes them to get to campus
2. Number of days left in the month and number of parking tickets placed on cars at a certain location
3. Length of time a customer spends in a small gift shop and total amount of purchase

# 15

# Tests Involving Variance

> Every good gift comes from God in Whom there is no variance.
>
> —The Bible

**CONTENTS**



**GOALS**

At the end of this chapter you will have learned:

- How to test whether the variance of a population is equal to a given value.
- How to test whether or not two populations have equal variances.
- How to perform two-tail *F*-tests by changing them into one-tail tests and using $\frac{1}{2}\alpha$.
- How to test whether two *or more* population *means* are equal by examining two *variance* estimates calculated by different methods.

# Introduction

Many important statistical hypotheses are concerned with variability. Two common measures of variability used in hypothesis tests are the *standard deviation* and the square of the standard deviation, the *variance.* You have already worked quite a bit with the standard deviation. One of its advantages is that it measures variability in whatever units are used for measuring the variable. For example, we might say that the average distance from the floor to the doorknobs in a certain apartment house is 36 *inches,* with a standard deviation of $\frac{1}{4}$ *inch.* In contrast to this, if you use the variance to measure variability, the variance is not in the original units of the problem and so is not as easy to interpret intuitively. In the example just stated, the variance is $(\frac{1}{4})^2$, or $\frac{1}{16}$, but the unit is "inches squared," which makes no intuitive sense. In this chapter, however, we will see that certain hypotheses should be tested by looking at sample variances rather than sample standard deviations, because statistics based on the sample variances can be meaningfully compared with certain well-established tables of critical values. Note that where we use the letter $s$ to represent standard deviations calculated from samples, we will use $s^2$ to represent variances calculated from samples. Our basic formulas are

$$s = \sqrt{\frac{\Sigma X^2 - \frac{(\Sigma X)^2}{n}}{n - 1}} \quad \text{and} \quad s^2 = \frac{\Sigma X^2 - \frac{(\Sigma X)^2}{n}}{n - 1}$$

In this chapter we are going to illustrate three types of problems.

1. The test of a claim about the variance of a population. For example, a claim might say that the heights in inches of 6-year-old children in the Los Angeles public school system have a variance equal to 4. We would have $H_0$: The claim that the variance is 4 is true; $\sigma^2 = 4$. This is equivalent to claiming that the standard deviation is 2 inches. This type of claim will be tested by computing a statistic and comparing it with a critical value from a chi-square table. (This is a new application of the chi-square table.) We will call these tests **one-sample tests of variance** (via chi-square).

**one-sample tests of variance**

2. A test to compare the variances of two populations. An example might be to test the claim that in sea-farming lobsters, one diet produces more erratically sized lobsters than another diet. The null hypothesis will be $H_0$: The variances are the same; $\sigma_1^2 = \sigma_2^2$. The computations will involve computing the variances of two samples, one from each population, then using these to compute a statistic which can be compared to critical values in a new table, called the $F$-table. We will call these tests **two-sample comparisons of variance** (via the $F$-test).

**two-sample comparisons of variance**

**comparison of the means of several populations**

**analysis of variance**

3. We can use the $F$-table and sample variances to analyze the hypothesis that the *means* of several populations are equal. This type of test is an extension of the two-sample mean tests already developed in Chapter 11. We will call these tests **comparison of the means of several populations**, or **analysis of variance** (ANOVA).

## 15.1 One-Sample Tests of Variance

The Deep Dark Device Department of the Kynda Klever Kamera Company manufactures darkroom timers—"the kind you wind," as their ads proclaim. Sample Sam, the quality-control man, always tests every timer they make to see if the assembly process is working properly. Sam does the test by setting each timer to the "30 seconds" mark, and then timing it electronically to see how long it actually takes to ring. He repeats this 11 times for each timer. He knows from past experience that if he would test a properly working timer thousands of times and make a frequency chart and histogram of his results, he would see an approximately normal distribution of the times with a mean of 30 seconds. (See Figure 15-1.)

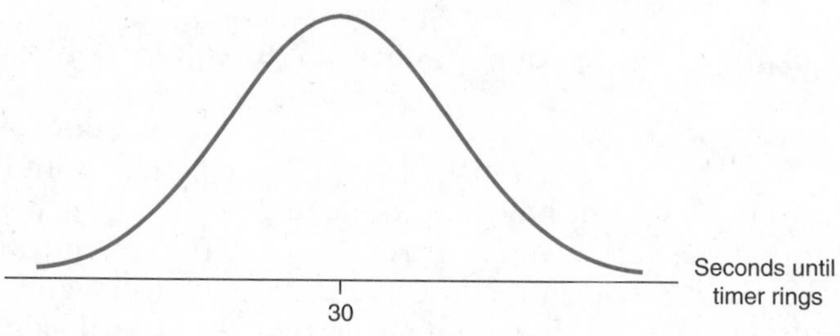


Figure 15-1

Sam's concern is, Does this distribution have a comparatively small variance? This means that this particular timer behaves consistently and is therefore more reliable in the darkroom. For example, if for two particular timers his tests resulted in the two graphs shown in Figure 15-2, this would indicate that timer A is more reliable than timer B.

Sam has decided to reject any timer that appears to have a standard deviation of more than $\frac{1}{2}$ second or, equivalently, a variance of more than $(\frac{1}{2})^2$ or $\frac{1}{4}$. Of course, Sam cannot afford to test each timer thousands of times.

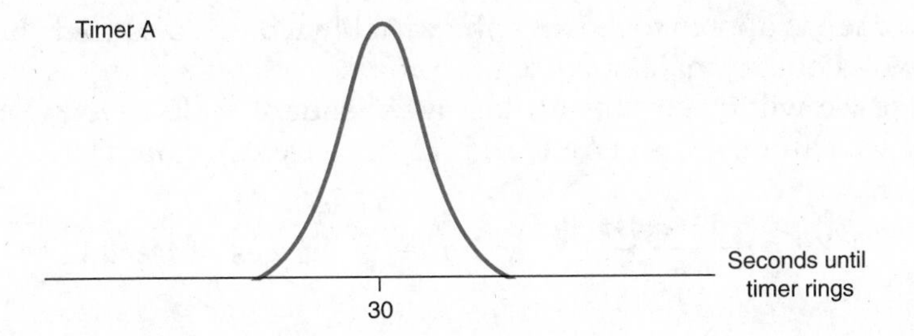


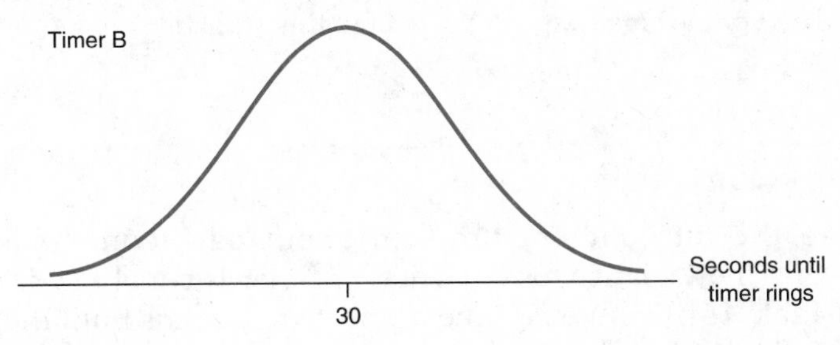


**Figure 15-2**

Recall that he tests each timer only 11 times, and he must judge on the basis of these 11 trials what might happen if he were to continue thousands of times. He therefore is in a hypothesis-testing situation. His motivated hypothesis, the one that will cause him to reject a timer, is $H_a$: The variance is more than $\frac{1}{4}$; $\sigma^2 > \frac{1}{4}$. You might guess the general idea of how this test is actually carried out. The first step is to get an estimate of $\sigma^2$ from the variability among the 11 test readings. This would be $s^2$. The second step is to see if $s^2$ is significantly larger than $\frac{1}{4}$.

It turns out that if you try to compare the sample value of $s^2$ with the theoretical value $\sigma^2$ by subtracting $\sigma^2$ from $s^2$ (as we essentially did with means in Chapters 10 and 11), there is no convenient set of critical values to use as a decision rule. However, it has been shown that if you divide $s^2$ by $\sigma^2/(n-1)$, you can use the table of critical values for a chi-square distribution with $n - 1$ degrees of freedom to get your decision rule. In short, the test statistic is

$$\frac{s^2}{\sigma^2/(n-1)}$$

or, what is the same,

$$\frac{(n-1)s^2}{\sigma^2}$$

This approach is appropriate when the variable whose variance is being tested is approximately normal.

Since we will check this against a chi-square table at $n - 1$ degrees of freedom, we can call this statistic $X^2$. That is, we compute

$$X^2 = \frac{(n-1)s^2}{\sigma^2} \quad \text{with } n - 1 \text{ degrees of freedom}$$

### Helpful Hint

In the one-sample tests of variance that follow, it is helpful to not use $X$ for the name of the variable since we will use $X^2$ for this statistic.

## ❑ EXAMPLE 15.1 Darkroom Timers

Referring to the introductory problem about darkroom timers, test the hypothesis that the variance for a particular timer is equal to $\frac{1}{4}$ if the data are as given in Table 15-1, where $Y$ = the number of seconds until the timer rang. Use the 0.05 significance level.

Table 15-1 Actual Times for 30-Second Setting, in Seconds

| Test number | 1 | 2 | 3 | 4 | 5 | 6 | 7 | 8 | 9 | 10 | 11 |
|---|---|---|---|---|---|---|---|---|---|---|---|
| $Y$ | 29.2 | 29.5 | 30.2 | 30.0 | 30.0 | 30.3 | 28.6 | 30.0 | 29.8 | 29.7 | 30.2 |

SOLUTION

*Procedure:* This is a one-sample test of variance.

*Population:* All the times obtainable by setting this timer to 30 seconds. We are assuming that the distribution of $Y$ is near normal.

*Sample:* The 11 times obtained from setting this timer to 30 seconds.

$H_0$: The timer is acceptable; its variance equals $\frac{1}{4}$; $\sigma^2 = \frac{1}{4}$.

$H_a$: The timer is not acceptable; its variance is more than $\frac{1}{4}$; $\sigma^2 > \frac{1}{4}$ (a one-tail test on the right).

*Degrees of freedom:* We have $n = 11$, and therefore $n - 1 = 10$ degrees of freedom.

*Critical value:* The critical value of $X^2$ for 10 degrees of freedom with $\alpha = 0.05$ is $X_c^2 = 18.31$.

*Decision rule:* We will reject $H_0$ if $X^2$ is greater than 18.31. This is illustrated in Figure 15-3.

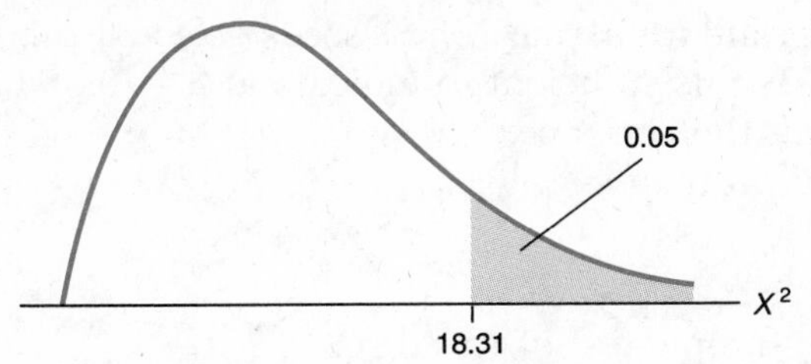


Figure 15-3 Chi-square distribution with one tail on the right.

*Outcome:* Using the data in Table 15-1, we compute $s^2$ and $X^2$ as follows.

| Test Number | $Y$ | $Y^2$ |
|---|---|---|
| 1 | 29.2 | 852.64 |
| 2 | 29.5 | 870.25 |
| 3 | 30.2 | 912.04 |
| 4 | 30.0 | 900.00 |
| 5 | 30.0 | 900.00 |
| 6 | 30.3 | 918.09 |
| 7 | 28.6 | 817.96 |
| 8 | 30.0 | 900.00 |
| 9 | 29.8 | 888.04 |
| 10 | 29.7 | 882.09 |
| 11 | 30.2 | 912.04 |
| Totals | $\Sigma Y = 327.5$ | $\Sigma Y^2 = 9753.15$ |

$$s^2 = \frac{\Sigma Y^2 - \frac{(\Sigma Y)^2}{n}}{n-1} = \frac{9753.15 - \frac{327.5^2}{11}}{10} = 0.2581818\ldots$$

$$X^2 = \frac{(n-1)s^2}{\sigma^2} = \frac{10(0.25818\ldots)}{\frac{1}{4}} = 10.33$$

*Conclusion:* Since 10.33 is not greater than 18.31, Sam would fail to reject the null hypothesis. This timer passes his inspection. ❑

❑ **EXAMPLE 15.2 Bean Leaf Growth**

A seed-packaging company has a type of bean seed which on the average grows its first leaves 15 days after being planted. The variance for the number of days until first leaves is 5. A new way of selecting the seeds is supposed to reduce this variance. A random sample of 61 seeds selected the new way is found to have $s^2 = 3.2$. Is this sufficient to indicate at $\alpha = 0.05$ that the new selection method reduces the variance?

## SOLUTION

*Procedure:* This is a one-sample test of variance.

*Population:* The number of days until first leaves appear on all the bean plants grown from seeds selected by the new method.

*Assumption:* The variable "number of days until first leaves" is close to being normally distributed.

*Sample:* The number of days for these 61 seeds to show first leaves.

$H_0$: The variability is unchanged with the new method. The variance equals 5; $\sigma^2 = 5$.

$H_a$: The new method reduces variability. The variance is less than 5; $\sigma^2 < 5$ (a one-tail test on the left).

*Critical value:* Since the degrees of freedom $= n - 1 = 61 - 1 = 60$ and $\alpha = 0.05$, we find in Table D-7 (see Appendix D) that $X_c^2 = 43.19$.* This situation is illustrated in Figure 15-4.

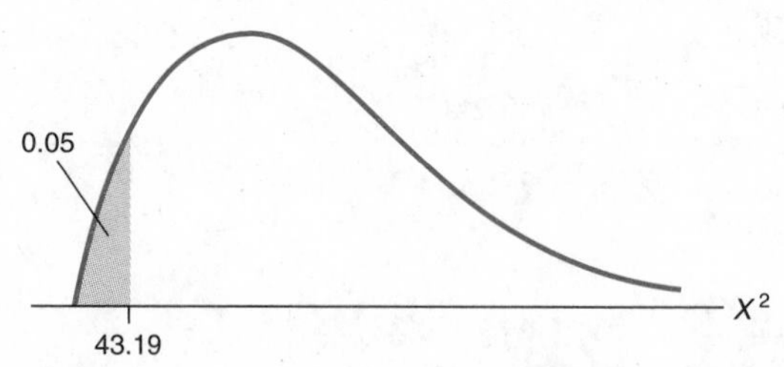


**Figure 15-4** Chi-square distribution with one tail on the left.

*Decision rule:* If $X^2$ turns out to be less than 43.19, we will reject $H_0$, and we will have evidence that the variance is no longer equal to 5, it has been reduced.

*Outcome:* $X^2 = \dfrac{(n-1)s^2}{\sigma^2} = \dfrac{60(3.2)}{5} = 38.4$

*This is the first time we have used the chi-square table for a critical value *on the left*. Be careful in reading the table.

*Conclusion:* Since the outcome, 38.4, is less than 43.19, we reject the null hypothesis. We have evidence that the new selection procedure does reduce the variance. ❑

**Note** If we have a two-tail test with an alternative hypothesis that the variance is *not equal* to some given variance, then we need to find two critical values of $X^2$. We will reject the null hypothesis if the value of $X^2$ is not between these two critical values.

## 15.2 Two-Sample Comparisons of Variance (via the *F*-Test)

Sometimes it is important to know if two populations have the same variance. For example, in a $t$-test which compares the means of two samples, we assumed that the variances of both populations involved were approximately equal. This is because the $t$-test really answers the question, "Could these two samples be from the same normal population?" When we reject the null hypothesis, we are saying "no" to this question. There are two reasons for two normal populations to be unequal—they could have different means or they could have different variances. If we can assume that the variances are equal, then the $t$-test becomes solely a test about the means of the populations. This is the way that we have been using it in our two-sample tests of means. All of this implies that if you perform a two-sample $t$-test for means and you end up rejecting the null hypothesis, then you should also check to rule out the possibility that your results are not caused by unequal variances in the two populations.

Consider some other cases in which you should check for unequal variance in two populations. Suppose you teach Advanced Knot Tying at the local Institute for Marine and Matrimonial Sciences. You think of two ways to teach slipknot tying, but you suspect that method B might produce more erratic results than method A. You try method B on one group of randomly selected students and method A on another. Then you give both groups the same exam. Suppose you get the same mean results in both samples. This means that the two methods are equally effective on the average. But suppose the variance in group B is significantly larger than the variance in group A. This would indicate that there is a comparatively wide gap between the best and worst students in group B. In short, method B is better for the good students and worse for the bad ones. You may decide not to use method B if you want to keep the class "together." Let's give an example of this type.

### 15.2-1 One-Tail Tests for Comparison of Variance

variance 1 > variance 2

In these one-tail tests we always name the population that we expect to have the larger variance population 1. Hence, the motivated hypothesis is always $\sigma_1^2 > \sigma_2^2$. The test statistic will be the ratio

$$F = \frac{s_1^2}{s_2^2}$$

where we put the sample variance that we expect to be larger on top. We therefore expect this fraction to be greater than 1. The question is, Is it significantly greater than 1? To decide this, we find a critical value $F_c$ in Table D-11, D-12, D-13, or D-14 (Appendix D).

**Note** Since the test statistic is always positive, the critical value is also always positive.

---

❑ **EXAMPLE 15.3 A Knotty Problem**

**F *Is for Fisher***

The letter *F* is used to honor Sir Ronald Fisher, who pioneered much of the work discussed in this chapter. The *F* was introduced by the statistician George Snedecor in the 1930s.

We have two methods of teaching slipknot tying. We want to see if method B produces exam scores with a greater variance than does method A. The scores on the slipknot final for method B are 75, 80, 90, 100, 110, 120, and 125. For method A they are 90, 95, 100, 100, 105, 105, and 110. We assume that both methods produce exam scores that are normally distributed.

**SOLUTION**

*Procedure:* This is a two-sample comparison of variance (via the *F*-test).

*Populations:* All people taught knot tying by these two methods. Because we are testing to see if the variance of the scores of the people taught by method B is greater, we name those people population 1.

*Samples:* The scores for those persons who took the slipknot final.

$H_0$: The methods are equivalent. The variances are equal; $\sigma_1^2 = \sigma_2^2$.

$H_a$: Method B has more erratic results. The first variance is greater than the second; $\sigma_1^2 > \sigma_2^2$ (a one-tail test on the right).

*Degrees of freedom:* In comparing two variances we separately find the number of degrees of freedom for the numerator and the denominator:

$$\text{Degrees of freedom (numerator)} = n_1 - 1 = 7 - 1 = 6$$

$$\text{Degrees of freedom (denominator)} = n_2 - 1 = 9 - 1 = 8$$

*Critical value:* Using $\alpha = 0.05$, we look at the table entry corresponding to this **pair** of degrees of freedom, and we find that $F_c = 3.58$. This is illustrated in Figure 15-5.

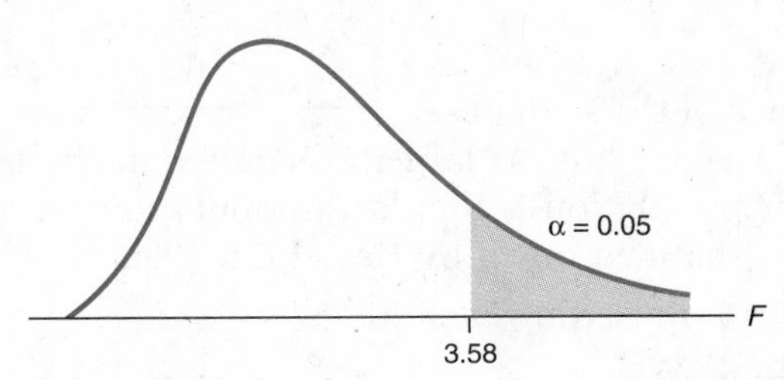


**Figure 15-5**

*Decision rule:* We will reject the null hypothesis if the outcome is greater than 3.58.

*Outcome:* From the given data, we calculate

$$n_1 = 7,\ m_1 = 100,\ s_1^2 = 375$$

$$n_2 = 9,\ m_2 = 100,\ s_2^2 = 37.5$$

The test statistic is

$$F = \frac{s_1^2}{s_2^2} = \frac{375}{37.5} = 10$$

*Conclusion:* We see that the outcome is well beyond the critical value. This indicates that there is a significant difference in variability between the results in group 1 and the results in group 2. The method used to teach knot tying in group 1 produces more-variable results than the method used to teach group 2. ❑

---

## ❑ EXAMPLE 15.4 Prison Sentencing

In a district court in Big City, the average sentence given for larceny by two judges is about the same. An investigator for a prison reform group has reason to believe that Justice Herthaykum, however, is much more erratic in sentencing the convicted than is Justice Tharthaygoe. The investigator checks random samples of 50 sentences given for this crime by each judge. The sentences given by Justice Herthaykum had $m = 18.2$ months with $s = 6.1$ months. The sentences given by Justice Tharthaygoe had $m = 18.5$ months with $s = 3.4$ months. Is this a statistically significant difference in variability? Use $\alpha = 0.05$.

**SOLUTION**

*Procedure:* This is a two-sample comparison of variance (via the $F$-test).

*Populations:* Because the investigator suspects that the variance of the sentences imposed by Justice Herthaykum is larger, we name all the larceny sentences he imposes as population 1. Similarly, population 2 consists of all the larceny sentences handed down by Her Honor, Tharthaygoe.

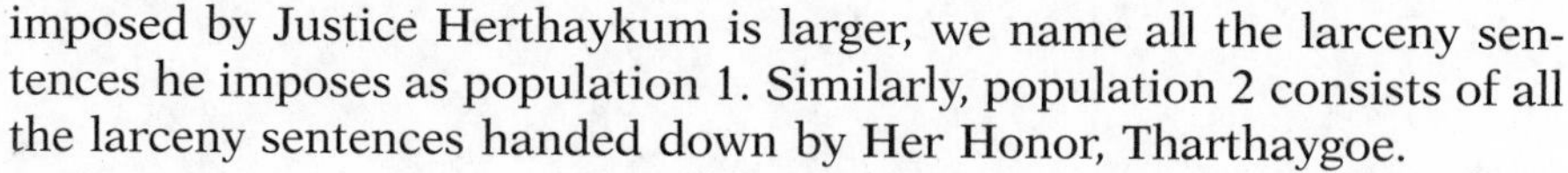

*Samples:* The 50 sentences imposed by each judge.

$H_0$: The variances are equal; $\sigma_1^2 = \sigma_2^2$.

$H_a$: The first variance is greater than the second; $\sigma_1^2 > \sigma_2^2$ (a one-tail test on the right).

*Critical value:* For $n_1 = 50$, we have $n - 1 = 49$ degrees of freedom. The same is true for $n_2$. We find from the tables that $F_c \approx 1.60$ (using the closest entry in the table). This is shown in Figure 15-6.

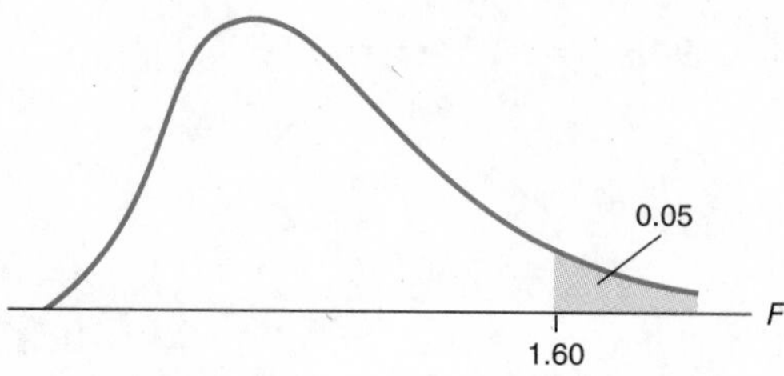


**Figure 15-6**

*Decision rule:* We will reject the null hypothesis if the outcome is greater than 1.60.

*Outcome:* The test statistic is

$$F = \frac{s_1^2}{s_2^2} = \frac{(6.1)^2}{(3.4)^2} = 3.22$$

*Conclusion:* Since this is larger than 1.60, we reject the null hypothesis. We have good evidence that Justice Herthaykum is more erratic than Justice Tharthaygoe. ❑

## 15.2-2 Two-Tail Tests for Comparison of Variance

In both of the last two examples, the alternative hypothesis $\sigma_1^2 > \sigma_2^2$ led to a one-tail test because we were motivated by the suspicion that one variance was larger than the other. In some problems, however, the alternative hypothesis is $\sigma_1^2 \neq \sigma_2^2$. We do not know ahead of time which variance will be larger. At the 0.05 significance level, for example, we would have a two-tail test, as illustrated in Figure 15-7, and either a very low $F$ or a very high $F$ will cause us to reject $H_0$.

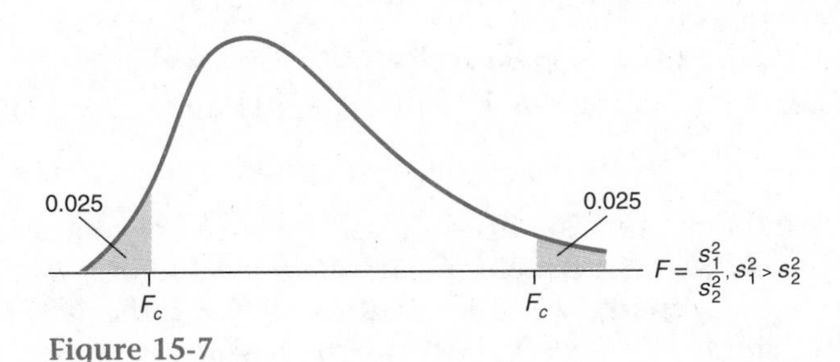


**Figure 15-7**

It is convenient to always put the larger sample variance on the top of fraction $F$ and, hence, do all tests on the right, maintaining only *one-half* of $\alpha$ (in this case 0.025). We have supplied separate $F$-tables for these two-tail tests (Tables D-12 and D-14 in Appendix D).

---

### ❑ EXAMPLE 15.5 Resort Temperatures

Is the variability in temperature the same in the two resort cities of Harvey-Cedars and Claremont? On the first day of every month the highest temperature of the day is recorded at Harvey-Cedars and Claremont. This is done for 1 year. The data collected showed that Harvey-Cedars had a mean high temperature of 70.2°F, with $s = 10.8$, and that Claremont had a mean high temperature of 74.1°F, with $s = 15.3$. Test the hypothesis that the high temperatures at Harvey-Cedars and Claremont are equally variable. Use $\alpha = 0.05$.

**SOLUTION**

*Procedure:* This is a two-sample comparison of variance hypothesis test (via the $F$-test).

*Populations:* The populations are all the daily high temperatures at Harvey-Cedars and all the daily high temperatures at Claremont.

Because $s = 15.3$ at Claremont and $s = 10.8$ at Harvey-Cedars, we choose the larger one for $s_1$, so that the daily high temperatures at Claremont become population 1.

*Samples:* The samples are the 12 monthly temperatures recorded in each city.

$H_0$: There is no difference in the variability of the temperatures in these two cities; $\sigma_1^2 = \sigma_2^2$.

$H_a$: There is a difference in the variability of the temperatures in these two cities; $\sigma_1^2 \neq \sigma_2^2$ (a two-tail test).

*Critical value:* The degrees of freedom for $n_1 = 12$ and $n_2 = 12$ will be

$$\text{Degrees of freedom (numerator)} = n_1 - 1 = 11, \text{ and}$$

$$\text{Degrees of freedom (denominator)} = n_2 - 1 = 11$$

Thus, at $\alpha = 0.05$ on a two-tail test we get $F_c = 3.48$.

*Decision rule:* We will reject the null hypothesis if the value of $F$ is greater than 3.48.

**Note** Even though this is a two-tail test, we have only one critical value because we deliberately put the larger value of $s$ in the numerator. (Essentially, we are using $\alpha = 0.025$ rather than 0.05. See the figures in Tables D-11 through D-14.)

*Outcome:* The test statistic is

$$F = \frac{s_1^2}{s_2^2} = \frac{(15.3)^2}{(10.8)^2} = 2.01$$

*Conclusion:* Since the $F$-statistic is smaller than the critical value of $F$, we fail to reject the null hypothesis of equal variances. The variability of high temperatures in the two cities may be the same; the difference in variability is not statistically significant. ❑

## 15.3 Using Sample Variances to Make Inferences about Means—An Introduction to the Analysis of Variance

ANOVA

In this section we want to show how analyzing sample variances can help answer questions about population means. This approach to comparing means is called analysis of variance. **An**alysis **o**f **va**riance is often abbreviated as **ANOVA**. Let us consider the kind of problem we usually solve by using a two-sample $t$-test for comparing means. In this kind of problem we assume that the two populations are both approximately normal and that they have the same variance. Then we test the null hypothesis that says they have equal means.

If they do have equal means, then the two populations are essentially the same for statistical purposes: they are both normal, and they have the same mean and the same variance. We will estimate the numerical value of this common variance. We will do this in two distinct ways. If the null hypothesis is true (the means are equal), then, according to statistical theory, these two distinct estimates of the variance should be "close" to one another as determined by an $F$-ratio test. We will take an obvious case for illustrative purposes.

*ANOVA*

Pierre-Simon Laplace (1749–1827) discussed the ideas of what we now refer to as "analysis of variance," or ANOVA, in 1827, a half century before they were fully developed.

## ❑ EXAMPLE 15.6 Brand D Brandy and Brand E Brandy

Consider two brands of brandy, brand D brandy and brand E brandy. Suppose as part of a quality-control procedure we report a measurement of sugar content. There are 5 test batches of each brand. The following data show that the sugar content of the two brandies are very different from each other.

| Sugar Content of Brand D Brandy, in Units of Sugar | Sugar Content of Brand E Brandy, in Units of Sugar |
|---|---|
| 8, 7, 8, 7, 8 | 13, 12.5, 12, 13, 12 |

From these data, we compute that $m_D = 7.6$, with $s_D^2 = 0.30$, and $m_E = 12.5$, with $s_2^E = 0.25$. Clearly, brand E brandy has a much higher sugar content than brand D brandy. The average amount of sugar in brand E brandy is much more than the average amount of sugar in brand D brandy. We want to show how this difference in means can also be seen indirectly by looking at variances. Recall that we are assuming that the two populations have the same variance. We are going to estimate this variance by two methods and then compare the two estimates.

**SOLUTION**

*Procedure:* Analysis-of-variance (ANOVA) hypothesis test of means

*Populations:* The sugar content in brand D brandy and in brand E brandy.

*Samples:* 5 test batches of each brandy.

$H_0$: The means are equal; $\mu_D = \mu_E$.

$H_a$: The means are not equal; $\mu_D \neq \mu_E$. (*Note:* This will always become a one-tail $F$-test of the variances.)

*Part 1: Estimation of Population Variance Based on Variability* **within** *Each Sample (Method 1)* We have two distinct samples of the same size from populations with the same variance. Recall that $s^2$ from a sample is a good estimate for the variance of the population. Therefore, we have two good estimates of the variance—namely, $s_D^2 = 0.30$ and $s_E^2 = 0.25$—and assuming that $\sigma_1^2 = \sigma_1^2$, it is reasonable to use the average of $s_D^2$ and $s_E^2$ as a good estimate of the value of the population variance. Therefore, we estimate the value of the variance of the population by

$$\frac{s_D^2 + s_E^2}{2} = \frac{0.30 + 0.25}{2} = 0.275$$

**mean square** ***within*** **groups**

You will remember from our original definition of variance that the variance is the mean of a set of squared deviations and is often called a *mean square.* In most textbooks the variance we just estimated is called a **mean square *within* groups** because each value of $s^2$ was based on the variability *within* the individual sample from which it was calculated. The term $s_D^2$ measures variability *within* sample D, and $s_E^2$ measures variability *within* sample E.

We will write $s_W^2$ for this mean square within groups, where the letter $W$ indicates "*w*ithin":

$$s_W^2 = \frac{s_D^2 + s_E^2}{2}$$

In our case, $s_W^2 = 0.275$.

In general, $s_W^2$ is

$$s_W^2 = \frac{\Sigma s^2}{N}$$

where $N$ is the number of samples we are averaging.

We have seen that $s_W^2$ is a good estimate of the variance of the two populations. Furthermore, it is true that $s_W^2$ is a good estimate of this variance *whether or not the populations also have equal means.*

*Part 2: Estimation of Population Variance Based on the Variability* **among** *the Sample Means (Method 2)* There is another way to estimate the variance of populations which have equal variance, but this way *only gives an accurate estimate if the populations also have equal means.* When the populations have equal means, then this method gives approximately the same value as $s_W^2$. But when the populations have different means, then this second method will give a value significantly larger than $s_W^2$. (This is the reason we have a one-tail test.)

If the two normal populations have the same variance and the same mean, then for statistical purposes they are considered to be one large population. The two samples can be considered to be two samples of size 5 from the same population. When we have several samples of the same size from a population, we can apply what we know about the relationship between the variance of a population and the variance of the means of samples taken from that population. Recall from the central limit theorem given in Chapter 10 that

$$\sigma_m = \frac{\sigma}{\sqrt{n}}$$

This is equivalent to

$$\sigma^2 = n\sigma_m^2$$

where $n$ is the size of each sample.

In ordinary words, this last formula says that you can find the variance of a population by multiplying $n$ by $\sigma_m^2$, the variance of the sample means. We can estimate $\sigma_m^2$ by calculating $s_m^2$ for as many sample means as we do have.

In our example we have taken two samples, so we have two sample means, $m_D = 7.6$, and $m_E = 12.5$. We compute $s_m^2$ from these two values:

| $m$ | $m^2$ |
|---|---|
| 7.6 | 57.76 |
| 12.5 | 156.25 |
| $\Sigma m = 20.1$ | $\Sigma m^2 = 214.01$ |

$$s_m^2 = \frac{\Sigma m^2 - \dfrac{(\Sigma m)^2}{N}}{N - 1} = \frac{214.01 - \dfrac{(20.1)^2}{2}}{1} = 12.00$$

***Note*** Capital $N$ is the number of equal size samples; lowercase $n$ is the size of each of these samples.

Now that we have $s_m^2$, we can apply the above idea that $\sigma^2 = n\sigma_m^2$ and estimate $\sigma^2$ by $s^2 = ns_m^2$, obtaining $5(12.00) = 60$.

**mean square *among* groups**

We call this second estimate of $\sigma^2$ the **mean square *among* groups** because it measures the variability among all the sample means. We write $s_A^2 = 60$, where the letter $A$ indicates "*a*mong." The formula then is

$$s_A^2 = ns_m^2$$

Thus, $s_A^2$ is our second estimate of the population variance. We now need to compare $s_A^2$ and $s_W^2$, our two estimates of the population variance.

*Part 3: Comparison of the Two Estimates, $s_A^2$ and $s_W^2$*

*Critical value:* We know that when the means are equal, then $\sigma_A^2 = \sigma_W^2$; but when the means are not equal, then $\sigma_A^2$ is larger than $\sigma_W^2$. Therefore, $s_A^2$ is the numerator and $s_W^2$ is the denominator for the $F$-statistic. To use the $F$-table we need to know the degrees of freedom for the numerator and denominator. The degrees of freedom in the numerator is $N - 1 = 2 - 1 = 1$, because $\sigma_A^2$ was computed from a list of two means. The degrees of freedom in the denominator is 8, because to obtain $\sigma_W^2$ we pooled two values of $\sigma^2$, each with degrees of freedom = 4. In general, we use degrees of freedom = $(n_1 - 1) + (n_2 - 1)$. Referring to Table D-11 for these two values of degrees of freedom at $\alpha = 0.05$, we get $F_c = 5.32$.

*Decision rule:* We will reject the null hypothesis if the outcome is greater than 5.32.

**F-ratio test**

*Outcome:* The test statistic is the **F-ratio**

$$F = \frac{\text{mean square } among \text{ groups}}{\text{mean square } within \text{ groups}} = \frac{s_A^2}{s_W^2} = \frac{60}{0.275} = 218.2$$

### Helpful Hint

*A precedes W alphabetically, and $s_A^2$ precedes (is on top of) $s_W^2$ in the F-ratio.*

*Conclusion:* You can see that the outcome, $F = 218.2$, is much larger than the critical value, which is what we expected, since the means of the two samples are not at all alike. A large value of $F$ indicates that there is great variability among the sample means—the means of the samples are not alike in value; that is, a large value of $F$ indicates that the samples probably came from populations with different means. We have evidence that the two brandies differ in mean sugar content. ❑

## ❑ EXAMPLE 15.7 The Brandy Repeated

We look at another brandy problem, but this time the sample values show that the two brands are practically the same in sugar content. Once again we assume both variables are near normally distributed with equal variances.

| Sugar Content of Brand G Brandy, in Units of Sugar | Sugar Content of Brand Z Brandy, in Units of Sugar |
|---|---|
| 8, 7, 8, 7, 8 | 7, 8, 8, 9, 9 |

We now have $m_G = 7.6$ with $s_G^2 = 0.30$ and $m_Z = 8.2$ with $s_Z^2 = 0.70$, and $N = 2$, since we have 2 samples with $n_1 = 5$ and $n_2 = 5$. (Recall the difference between capital $N$ and lowercase $n$.)

**SOLUTION**

*Procedure:* Analysis-of-variance (ANOVA) hypothesis test

*Populations:* The sugar content in all samples of brand G and brand Z brandies.

*Samples:* 5 test batches of each brandy.

$H_0$: The means are equal, thus the variance determined by $\sigma_A^2$ equals the variance determined by $\sigma_W^2$.

$H_a$: The means are not equal, thus the variance determined by $\sigma_A^2$ is larger than the variance determined by $\sigma_W^2$. (This is always a one-tail test of the variances.)

*Part 1: Mean Square* **within** *Groups (Method 1)*

$$s_W^2 = \frac{\Sigma s^2}{N} = \frac{s_G^2 + s_Z^2}{2} = \frac{0.30 + 0.70}{2} = 0.50$$

*Part 2: Mean Square* **among** *Groups (Method 2)*

| $m$ | $m^2$ |
|---|---|
| 7.6 | 57.76 |
| 8.2 | 67.24 |
| $\Sigma m = 15.8$ | $\Sigma m^2 = 125.00$ |

$$s_m^2 = \frac{\Sigma m^2 - \dfrac{(\Sigma m)^2}{N}}{N - 1} = \frac{125.00 - \dfrac{(15.8)^2}{2}}{1} = 0.18$$

$$s_A^2 = ns_m^2 = 5(0.18) = 0.90$$

*Part 3: Comparison of Estimates*

*Critical value:* Degrees of freedom (numerator) $= N - 1 = 2 - 1 = 1$, and

Degrees of freedom (denominator) $= (n_1 - 1) + (n_2 - 1)$

$= 4 + 4 = 8$

For $\alpha = 0.05$, $F_c = 5.32$

*Decision rule:* We will reject the null hypothesis if the outcome is greater than 5.32.

*Outcome:* $F = \dfrac{s_A^2}{s_W^2} = \dfrac{0.90}{0.50} = 1.8$

*Conclusion:* You see that this time $F = 1.8$ is smaller than $F_c$. Hence we fail to reject the null hypothesis. The sugar content of these two brandies may be the same. This is because $s_A^2$, the mean square among groups, decreased from 60 to 0.80, making the value in the top of the $F$ fraction smaller. Thus, the numerator of $F$ decreases when the sample means become more alike, and increases when they become more variable. If you find that the sample means are variable enough to cause $F$ to be larger than $F_c$, this is a good indication that the samples came from populations with different means. Therefore, if you get $F$ larger than $F_c$, you can reject a null hypothesis which says that the population means are equal. ❑

This method of comparing means by looking at variances is the simplest version of analysis of variance. One advantage it has over $t$-tests is that it is not limited to comparing only two samples. See the following example for a test asking, Are 3 means equal?

### ❑ EXAMPLE 15.8 The Rain in Spain

Señor L. Aquador is a professional rainmaker. He runs the Seedy Cloud Company. He has learned of three different ways to drop chemicals into the clouds over the plains in Spain. He wants to know if there is any difference in the results for these three methods. Using ANOVA, he can test the null hypothesis:

$H_0$: The average rainfall produced by each method is the same;

$$\mu_1 = \mu_2 = \mu_3.$$

A large value of $F$ will cause Señor Aquador to reject this hypothesis—he will have evidence that not all the methods are equal. (Be careful! This does not mean that they are all different, only that *at least one* of the means is different from the rest.) Here are the results after having tried each method 6 times. He records the rainfall in millimeters.

| Method 1 | Method 2 | Method 3 |
|---|---|---|
| 12 | 10 | 8 |
| 12 | 11 | 10 |
| 13 | 10 | 11 |
| 13 | 12 | 12 |
| 12 | 12 | 10 |
| 13 | 10 | 10 |

From these data he calculates that $m_1 = 12.5$ with $s_1^2 = 0.30$, $m_2 = 10.83$ with $s_2^2 = 0.97$, and $m_3 = 10.17$ with $s_3^2 = 1.77$. Using $\alpha = 0.05$, he tests the hypothesis that all the methods produce the same average rainfall.

**SOLUTION**

*Procedure:* Assuming that all three populations are normal with the same variance, we use ANOVA to perform an $F$-test for equal means.

*Population:* All the rainfall produced by the three methods.

*Samples:* The results of the 6 trials of each method.

$H_0$: The average rainfall produced by each method is the same;

$$\mu_1 = \mu_2 = \mu_3.$$

$H_a$: Not all the $\mu$'s are equal. At least two of the means are different from each other; $\mu_1 \neq \mu_2$, *or* $\mu_2 \neq \mu_3$, *or* $\mu_1 \neq \mu_3$ (always a one-tail test).

*Part 1: Mean Square within Groups (Method 1)*

Since we are considering 3 populations, $N = 3$.

$$s_W^2 = \frac{\Sigma s^2}{N} = \frac{0.30 + 0.97 + 1.77}{3} = 1.01$$

*Part 2: Mean Square among Groups (Method 2)*

| $m$ | $m^2$ |
|---|---|
| 12.50 | 156.25 |
| 10.83 | 117.29 |
| 10.17 | 103.43 |
| $\Sigma m = 33.50$ | $\Sigma m^2 = 376.97$ |

$$s_m^2 = \frac{\Sigma m^2 - \dfrac{(\Sigma m)^2}{N}}{N - 1} = \frac{376.97 - \dfrac{(33.50)^2}{3}}{2} = 1.44$$

Since each sample consists of 6 trials, $n = 6$, and

$$s_A^2 = ns_m^2 = 6(1.44) = 8.64$$

*Part 3: Comparison of Estimates*

*Critical value:*

$$\text{Degrees of freedom (numerator)} = N - 1 = 3 - 1 = 2, \text{ and}$$

$$\text{Degrees of freedom (denominator)} = (n_1 - 1) + (n_2 - 1) + (n_3 - 1) = 5 + 5 + 5 = 15$$

For $\alpha = 0.05$, $F_c = 3.68$

*Decision rule:* Reject the null hypothesis if the outcome exceeds 3.68.

*Outcome:* $F = \dfrac{s_A^2}{s_W^2} = \dfrac{8.64}{1.01} = 8.55$

*Conclusion:* He rejects the null hypothesis. The methods of producing rain do not all have the same average rainfall. (See Exercise 15-23.) ❑

### Some Comments

In our discussion of analysis of variance we have illustrated only the case where all samples are the same size and where we assume that there is only one factor responsible for any differences we observe. In the rainmaking example, we assume the differences in rainfall are due to the type of seeding and not to other factors such as differing weather conditions when each method was tried. The type of approach we have illustrated is often called **one-way analysis of variance**. But procedures have been developed for taking into account unequal sample sizes and multiple factors at work. These more advanced features of ANOVA are important to researchers and are described in more-comprehensive texts.

#### *An Early ANOVA*

In 1835 Adolphe Quetelet (1796–1874) used a crude ANOVA concept as he studied characteristics of the "average" person using such variables as the conviction rate of persons accused of a crime versus age, sex, and literacy.

## STUDY AIDS

### Chapter Summary

In this chapter you have learned:

- How to test whether the variance of a population is equal to a given value. This is a chi-square test which can be a one-tail test on either side or a two-tail test. (This is a new idea; previous chi-square tests were all one-tail on the right.)
- How to test whether two populations have equal variances. This is a one-tail $F$-test.
- How to perform two-tail $F$-tests by changing them into one-tail tests with $\frac{1}{2}\alpha$.
- How to test whether two or more population means are equal by analysis of variance.

### Vocabulary

You should be able to explain the meaning of each of these terms:

1. One-sample tests of variance
2. Two-sample comparisons of variance
3. Comparison of the means of several populations
4. Analysis of variance (ANOVA)
5. Mean square *within* groups
6. Mean square *among* groups
7. $F$-ratio test

**Symbols**

You should understand the meaning of each of these symbols:

1. $F_c$
2. $s_W^2$
3. $s_A^2$
4. $n$
5. $N$

**Formulas**

You should know when and how to use each of these formulas:

1. $s^2 = \dfrac{\Sigma Y^2 - \dfrac{(\Sigma Y)^2}{n}}{n - 1}$

2. $X^2 = \dfrac{(n - 1)s^2}{\sigma^2}$, with $n - 1$ degrees of freedom

3. $F = \dfrac{s_1^2}{s_2^2}$, $s_1^2 > s_2^2$ (separately calculating the degrees of freedom in the numerator and the denominator)

4. $s_W^2 = \dfrac{\Sigma s^2}{N}$ where uppercase $N$ is the *number* of populations sampled

5. $s_m^2 = \dfrac{\Sigma m^2 - \dfrac{(\Sigma m)^2}{N}}{N - 1}$ where uppercase $N$ is the *number* of populations sampled

6. $s_A^2 = ns_m^2$, where lowercase $n$ is the *common sample size*

7. $F = \dfrac{\text{mean square } among \text{ groups}}{\text{mean square } within \text{ groups}} = \dfrac{s_A^2}{s_W^2}$

## EXERCISES

*In the exercises that follow, be careful to distinguish between variance and standard deviation.*

15-1 In Chapter 11 we made 2 assumptions about the populations when we used the two-sample $t$-test. Now we can use an $F$-test to test 1 of these assumptions. What is the assumption?

15-2 Using $\alpha = 0.05$:

(a) Test the hypothesis that the weights of 6 boxes of Soggy Morning Cereal come from a distribution with variance equal to 0.30, if the weights in grams are 453.2, 453.0, 452.8, 452.9, 453.2, and 452.7.

(b) Assume that the weights are distributed normally, and test the hypothesis that the mean weight is 453.1 grams.

15-3 Every day for the month of July, Donna Shore marks the high-tide point on a certain beach. She puts a stick in the sand where the water comes farthest onto the beach directly in front of a permanent flagpole. Then she measures the distance from the flagpole to the stick.

(a) At $\alpha = 0.05$, does she have good evidence that the standard deviation of high-tide marks on her beach is more than 5 feet if she obtained an average of $m = 150$ feet with $s = 8$ feet?

(b) Assuming that the distances are distributed normally, test the hypothesis that the mean distance is 125 feet. Use $\alpha = 0.05$.

15-4 Lazar Beame, a pediatrician in a large maternity hospital, tells his son to record the number of hours, in the first 24 hours of a baby's life, that a baby sleeps. Here are the results that son Beame recorded for 10 babies: 18.6, 16.2, 22.4, 17.9, 18.9, 20.1, 17.0, 18.8, 19.0, and 21.4. Assume such measurements are distributed normally, and test the following hypotheses. Use $\alpha = 0.01$.

(a) Variance equals 4.

(b) $\mu = 18$ hours.

15-5 A series of measurements were made by the noted anthropologist, Dr. Sanford Q. Krotche.

(a) He measured the length of a certain bone in the limbs of 10 adult male skeletons that he dug up. Are Sandy's measurements consistent with his hypothesis that the standard deviation of these bone lengths is 10 millimeters? Test at $\alpha = 0.05$ if the lengths of bone, in millimeters, are 150, 160, 152, 157, 159, 148, 152, 153, 156, and 138.

(b) Sandy then found that the length of a similar bone from an unknown primate was 162 millimeters. If you assume that the lengths of these bones are normally distributed with standard deviation 10 millimeters and mean given by $m$ from part (a), what is the probability of finding a bone as large as 162 millimeters (that is, a bone which is 162 millimeters or longer)?

15-6 Reread Exercise 11-4 concerning peccaries. With $\alpha = 0.05$, test the claim that the standard deviation is less than 2 millimeters.

15-7 Reread Exercise 11-5 dealing with Martians. Is it reasonable at the 0.05 significance level to assume that $\sigma$ is less than 10?

15-8 Reread Exercise 11-6 regarding landscape gardeners. Is it reasonable to assume at $\alpha = 0.05$ that $\sigma$ is more than $15?

15-9 Reread Exercise 11-7 about the height of police officers. It was also claimed that the police officers are very similar to one another in height, with a standard deviation of not more than 1 inch. Is this supported by the data given? Use $\alpha = 0.05$.

15-10 Reread Exercise 11-9 about psychologists. Using $\alpha = 0.01$, test the hypothesis that the variance of the psychologists' scores is the same as the variance for the general public. This variance is 64.

15-11 For these following exercises from Chapter 11, check the assumption that the variances of the 2 populations sampled are equal. Use $\alpha = 0.05$.

(a) 11-15, GPAs

(b) 11-16, fluoride treatment

(c) 11-17, exemptions

(d) 11-18, egg packaging

15-12 Two typists, Martin and Marvin, type on the average about 60 words per minute. They were each tested 10 times, and their scores are shown below. Using $\alpha = 0.05$, test the hypothesis that their test scores are equally variable.

| Martin | Marvin |
|---|---|
| 50 | 52 |
| 55 | 57 |
| 60 | 50 |
| 65 | 64 |
| 62 | 65 |
| 50 | 66 |
| 61 | 56 |
| 60 | 61 |
| 62 | 67 |
| 65 | 66 |

15-13 A chemical was applied to the roots of 20 corn plants and 20 oat plants. The concentration of the chemical in the plant stem was measured 24 hours later. Using $\alpha = 0.01$, test the hypothesis that the chemical acts more erratically in the oat plants than it does in the corn plants. It was found that the mean concentration of chemical in corn plants was 440 units with $s = 40$ units, and the mean concentration of chemical in oat plants was 620 units with $s = 100$ units.

15-14 Doctors Robin and Jay Byrde were studying nectar production in the trumpet creeper. They discovered that the flower produces 2 distinct types of nectar, 1 that attracts ants, and 1 that attracts hummingbirds. In the course of their study they had to measure the heights of the petioles and corollas of the flowers. Robin measured 120 petioles and Jay measured 200 corollas. Are they justified in claiming that the 2 are equally variable in height if $s = 20$ units for the petioles and $s = 25$ units for the corollas? Use $\alpha = 0.05$.

15-15 Three scientists were investigating the possibility of developing a pill to cure jet lag. Working with laboratory rats, they found that one drug, theophylline, caused about an 18-hour forward shift in the rats' body rhythm. Another drug, phenobarbital, caused about a 12-hour backward shift. Each drug was given to 10 rats after the rats had been adjusted to a precise schedule. Using $\alpha = 0.01$, test the hypothesis that these pills are equally variable in their effects if the results shows that the theophylline had a mean average shift of $+18.2$ hours with a variability of 1.3 hours, and the phenobarbital had a mean average shift of $-12.1$ hours with a variability of 2.4 hours.

15-16 As part of a study on national income distribution, a student recorded the median family income, in dollars, in 1990 for 10 counties picked at random in the states of California and Washington. The results are shown. Test the hypothesis that median family income is a more variable quantity in California counties than in Washington counties. Use $\alpha = 0.05$.

| California | Washington |
|---|---|
| $45,037 | $26,652 |
| 39,823 | 29,631 |
| 26,563 | 32,842 |
| 27,407 | 30,555 |
| 37,841 | 29,128 |
| 53,670 | 29,907 |
| 31,842 | 44,555 |
| 37,694 | 27,124 |
| 41,289 | 23,147 |
| 32,923 | 25,244 |

15-17 In recent years there has been much medical interest in the effect of extended exercise on women's menstrual cycles. One study compared 3 groups of women who were similar except for the amount and intensity of regular running they did. They were grouped into runners, joggers, and a control group consisting of nonrunners. Each woman contributed data for 1 year, including the number of menstrual periods which she had. The results were as follows.*

| | Runners | Joggers | Control |
|---|---|---|---|
| Mean number of menses $m$ | 9.1 | 10.1 | 11.5 |
| Standard deviation | 2.4 | 2.1 | 1.3 |
| $n$ | 26 | 26 | 26 |

Is this evidence strong enough at $\alpha = 0.01$ to indicate that running affects the frequency of menstruation?

15-18 An ad agency designed 3 different sample displays for a product. To compare their effectiveness, they were randomly tried in 15 similar retail stores: 5 stores received a humorous display, 5 received a display featuring a family, and 5 received a display depicting a scientist. The displays were left up in each store for 1 week and the number of sales of the product in each store was recorded. Use ANOVA to analyze the results shown below. Decide at $\alpha = 0.05$ whether there is a statistically significant difference in mean sales according to which display was used.

*Data are based on a study by Dale et al., "Menstrual Dysfunction in Distance Runners," as mentioned by S. Glantz in *Primer of Biostatistics*, McGraw-Hill Book Company, New York, 1981.

**Number of products sold**

| Display 1 | Display 2 | Display 3 |
|---|---|---|
| 47 | 41 | 46 |
| 47 | 45 | 46 |
| 52 | 47 | 54 |
| 50 | 40 | 49 |
| 49 | 47 | 46 |
| $m$ = 49.0 | $m$ = 44.0 | $m$ = 48.2 |
| $s$ = 2.12 | $s$ = 3.32 | $s$ = 3.49 |

15-19 A study was done to see whether migrating birds are sensitive to magnetic fields. A large circular cage was built around a radio transmitter and antenna system. A few at a time, 51 birds were put into the cage and left alone when the transmitter was off. After 5 minutes in the cage, the direction each bird was facing was recorded with reference to a compass. It was found that the mean direction was 151° with a standard deviation of 65°. Then the transmitter was turned on, creating an unnatural magnetic field. The birds were replaced in the cage and the measurements taken again. This time the mean direction was 189° with $s = 80°$. Is this difference in mean direction significant at $\alpha = 0.05$?

(a) Do this test as a two-sample test of means.

(b) Do this test using ANOVA techniques.

15-20 Some scientists were studying the process by which a type of clam egg usually managed to be fertilized by only 1 sperm when in the presence of many sperm. They placed an equal suspension of clam eggs in each of 15 test tubes containing seawater. Then they prepared 2 concentrations of a clam sperm solution, which they called the low concentration and the high concentration.

To each tube, they added the low concentration of clam sperm. Then 5 seconds later they added the high concentration to tubes 1, 2, 3, 4, and 5, and 5 seconds after this they added the high concentration to tubes 6, 7, 8, 9, and 10. Then 5 seconds later they added the high concentration to tubes 11, 12, 13, 14, and 15. For each tube, they recorded the number of eggs that were fertilized by more than 1 sperm. The results are shown in the following table. Use the techniques of this chapter to show that there is less than a 1% probability that the differences among these three groups are just due to chance. Can you suggest an explanation of what is happening?

| High Concentration Added in First 5 Seconds | High Concentration Added in Next 5 Seconds | High Concentration Added in Last 5 Seconds |
|---|---|---|
| 800 | 700 | 40 |
| 820 | 704 | 32 |
| 798 | 685 | 31 |
| 808 | 680 | 39 |
| 790 | 720 | 46 |

15-21 The Persian biologist Dr. Ismar Kaz thought that certain strains of laboratory mice have genetically determined preferences for chocolate. Shown below are her results for 4 strains of mice. The experiment is conducted by having 2 containers, 1 for regular Mouse-Yummies and 1 for chocolate-flavored Mouse-Yummies, in the cage, and then measuring how much of each the different strains of mice eat each day. All values of $m$ and $s$ are recorded in milligrams per gram. For example, when we say $m_1 = 0.13$ milligrams per gram, this means that for the mice of strain 1, the average consumption is 0.13 milligrams of chocolate per day for each gram of body weight.

| Strain 1 | Strain 2 | Strain 3 | Strain 4 |
|---|---|---|---|
| $n_1 = 7$ | $n_2 = 7$ | $n_3 = 7$ | $n_4 = 7$ |
| $m_1 = 0.13$ | $m_2 = 0.11$ | $m_3 = 0.09$ | $m_4 = 0.07$ |
| $s_1 = 0.02$ | $s_2 = 0.02$ | $s_3 = 0.01$ | $s_4 = 0.01$ |

Is there a significant difference in the mean amounts of chocolate consumed at the 0.05 significance level?

15-22 In a wine-growing experiment conducted by the Julius Caesar Vineyards, Inc., 3 different, but similar, grape varieties were used to make a certain type of wine. The acidity of the finished wine was measured. Using $\alpha = 0.05$, test the hypothesis that there is no significant difference in acidity. The results for each type of grape are as follows.

| Type I | Type II | Type III |
|---|---|---|
| $m_I = 322$ units | $m_{II} = 350$ units | $m_{III} = 352$ units |
| $s_I = 20$ units | $s_{II} = 23$ units | $s_{III} = 18$ units |
| $n_I = 15$ samples | $n_{II} = 15$ samples | $n_{III} = 15$ samples |

15-23 In Example 15.8 we rejected the null hypothesis and concluded that either $\mu_1 \neq \mu_2$, or $\mu_2 \neq \mu_3$, or $\mu_1 \neq \mu_3$.

(a) Explain how we could test in order to determine which of these 3 inequalities are true and which are false.

(b) Do you think that $\mu_2 \neq \mu_3$? Perform a test to determine if $\mu_2 \neq \mu_3$. Use the 0.05 significance level.

## CLASS SURVEY QUESTION

Test to see whether the variance in the heights of the males on your campus is equal to the variance in the heights of the females.

## FIELD PROJECTS

Outline a project such as one of those indicated below. After your method has been approved by your instructor, perform the project you selected.

1. Gather two samples of similar data from two populations and test to see whether the variances in these populations are the same or not. For example, compare variability in points scored per game for two basketball players.
2. Gather three or more samples and use an ANOVA to test for equal population means. For instance, compare the mean hours of sleep per night for young, middle-aged, and old people.
3. If you performed the field project in Chapter 1, finding ages of students on your campus, pair off with another student who also collected a random sample of ages.
   (a) Compute the sample mean and standard deviation for each sample, and then use chi-square to test whether or not the data came from a normal population. (The degrees of freedom are 2 less than $C-1$ because $m$ and $s$ were calculated from sample data.)
   (b) Test to see if the variances of the samples confirm that both samples came from the same population by doing an $F$-test for comparing the variances.
   (c) Test to see if the means of the samples confirm that both samples came from the same population. Use a $t$-test or analysis of variance.

# 16

# Nonparametric Tests

Does a fair coin *alternate* heads and tails?

**CONTENTS**



**GOALS**

At the end of this chapter you will have learned:

- That nonparametric tests may be used when data do not meet all the requirements for *t*-tests
- Three specific nonparametric tests: a sign test, a runs tests, and a test for ranked data

## Introduction

In much of this book we have stressed that in order to carry out hypothesis tests we needed to make certain specific assumptions about the types of distributions from which we were sampling. For example, to do $t$-tests we needed to assume that the populations involved were approximately normal. An important part of statistics deals with tests for which we do not need to make such specific assumptions. These tests are called "**nonparametric**" tests, or **distribution-free** tests. There is not a sharp distinction between parametric and nonparametric tests, but most frequently the main idea is that in a nonparametric hypothesis test the statistician is not making some assumption about normality in the original data.

nonparametric

distribution-free

Some students ask, "What happens when I use a parametric test even though the data do not really meet the proper requirements?" Currently, much research goes on to see how well various statistical tests work when the populations do not meet the theoretical requirements. For example, in our introduction to the analysis of variance (ANOVA) in Chapter 15, we assumed that the populations were approximately normal. Therefore, ANOVA is a parametric test. But research has indicated that many times its results are still reliable even if the populations are not too close to normal. The idea of measuring how well a statistical test works when we violate some of its theoretical assumptions is associated with the concept of robustness. A test is **robust** if it still works well when the theoretical assumptions are violated. If you don't know that a given procedure is robust, then you may be better off using a nonparametric test specifically designed for your data.

robust

There are many nonparametric tests in common use, and we have picked just three simple cases as illustrations to convey the idea of how such tests work. The tests we will discuss are

1. A sign test for matched pairs of data
2. A runs test, useful for testing whether certain patterns of experimental outcomes are random
3. The Mann-Whitney $U$-test for comparing ranks in two samples

## 16.1 The Sign Test for Matched Pairs of Data

In this section we show that the data from matched-pairs experiments can lead to a familiar binomial hypothesis test. This approach is useful when the data are from populations which are not normally distributed.

## ❑ EXAMPLE 16.1 Maintaining Reading Comprehension

Patricia McZirk, the dean of a large school of theology, is thinking of instituting a course designed to maintain reading comprehension. She picks 13 of the students at random to try this course. They completed the course and were tested on the day they finished it. They were tested again 1 month later to see if they maintained the level they had achieved. Dean McZirk expects a drop in their scores. But, if the reading course works, there will be no statistically significant difference between scores at the end of the course and scores 1 month later. The results were as shown in Table 16-1.

**Table 16-1**

| Student | $X$<br>Score at<br>Course Ending | $Y$<br>Score after<br>1 Month |
|---|---|---|
| Abraham | 50 | 52 |
| Balaam | 48 | 51 |
| Cain | 46 | 46 |
| David | 50 | 49 |
| Esther | 62 | 50 |
| Felix | 80 | 70 |
| Gideon | 23 | 21 |
| Hosea | 30 | 33 |
| Israel | 45 | 46 |
| Job | 53 | 53 |
| Keturah | 49 | 48 |
| Laban | 51 | 48 |
| Moses | 46 | 48 |

Dean McZirk has no reason to believe that scores on this kind of exam are normally distributed, so a $t$-test on the paired differences is not appropriate. Analyze the data using a nonparametric test for matched pairs.

### SOLUTION

We can reason as follows. The null hypothesis assumes that there is no reason to expect $X$ to be greater than $Y$, that there is only random variation between the two scores for each student.

Therefore, if we ignore any pair of data in which the two test scores are the same, it is just as likely for $X$ to be greater than $Y$ as it is for $X$ to be less than $Y$. If we replace each pair of values in which $X$ is greater than $Y$ by a plus sign and each pair of values in which $X$ is less than $Y$ by a minus sign, the null hypothesis leads us to expect about the same number of each type of sign (see Table 16-2). A plus sign means the student did worse on the retest than on the first test. A minus sign means the student did better on the retest.

Table 16-2

| Student | Sign | Student | Sign |
|---|---|---|---|
| Abraham | − | Hosea | − |
| Balaam | − | Israel | − |
| Cain | 0 | Job | 0 |
| David | + | Keturah | + |
| Esther | + | Laban | + |
| Felix | + | Moses | − |
| Gideon | + | | |

You will see that this way of looking at the data allows us to phrase Dean McZirk's question as a binomial hypothesis test about plus and minus signs. We will say that the reading course works if the number of plus and minus signs is about the same, and that it doesn't work if there are a lot more pluses than minuses. Because we are no longer paying attention to the size of the differences, this is a less informative test than a *t*-test. The analysis we make on the pattern of plus and minus signs is called a **sign test.**

**sign test**

In a binomial hypothesis test there must be two outcomes: success and failure. In this example we will let success mean "a + appears" and failure "a − appears." Thus, if we ignore the results for Job and Cain because they each have a zero, we can set up a binomial hypothesis test as follows:

We do the sign test as a one-sample binomial hypothesis test.

*Population:* All the students at McZirk's school.

*Sample:* The 13 students who participated in the reading course.

$$\text{Let } p = P(\text{a person scores higher on the first test}) = P(X > Y) = P(\text{a sign is } +).$$

$H_0$: Original reading scores and scores 1 month later are the same. A plus sign is just as likely as a minus sign; $p = 0.50$.

$H_a$: The scores 1 month later are lower. There will be more plus than minus signs; $p > 0.50$ (a one-tail test).

Since we have ignored the results for 2 people in the sample, we have $n = 13 - 2 = 11$.

$$\mu = np = 11(0.50) = 5.5$$

$$\sigma = \sqrt{npq} = \sqrt{11(0.5)(0.5)} = 1.66$$

Since both $np$ and $nq = 5.5 > 5$, we can use the normal approximation to the binomial distribution for the number of successes: $ND_S(\mu_S = 5.5, \sigma_S = 1.66)$.

For $\alpha = 0.01$, the critical value is $z_c = +2.33$.

Therefore,

$$S_c = \mu_S + z_c\sigma_S = 5.5 + 2.33(1.66) = 9.4$$

Our decision rule is, We will reject $H_0$ if we get more than 9.4 plus signs. Since the outcome is 6 plus signs, we fail to reject $H_0$. We have failed to show that reading skills have decreased after 1 month. Thus, the course may help students to maintain reading comprehension. ❑

## ❑ EXAMPLE 16.2 Anxiety and Grades

As part of a study about the effect of anxiety on test performance, a class at Why Worry Tech was given a test under careful supervision. Although cheating was not specifically mentioned, it was almost certain that nobody could or did cheat. But the next day the class was *told* that because some of them cheated, an equivalent test would be given with several proctors in the room. The marks for both tests were as shown in Table 16-3.

Table 16-3

| Student | $X$ (1st Day) | $Y$ (2nd Day) | Student | $X$ (1st Day) | $Y$ (2nd Day) |
|---|---|---|---|---|---|
| Ellington | 70 | 66 | Mancini | 84 | 84 |
| Lombardo | 43 | 39 | Alpert | 78 | 70 |
| Prima | 91 | 85 | Severinsen | 92 | 69 |
| Dorsey | 89 | 92 | Welk | 83 | 84 |
| Basie | 73 | 72 | Floren | 75 | 74 |
| James | 64 | 63 | Cugat | 73 | 72 |
| Kostelanetz | 51 | 40 | Duchin | 89 | 83 |
| Bach | 83 | 88 | | | |

Do these data support the idea that the anxiety caused by the announcement lowered the test results? Analyze the data by a sign test using $\alpha = 0.05$.

### SOLUTION

We perform a one-sample binomial hypothesis test of signs.

*Population:* All the students at WWT.

*Sample:* The 15 students in the experiment.

$$\text{Let } p = P(\text{a person scores higher on the first test}) = P(X > Y) = P(\text{a sign is } +).$$

$H_0$: The announcement did not affect grades; $p = 0.50$.

$H_a$: The announcement lowered grades on the second test; $p > 0.50$ (a one-tail test).

The data can now be revised as in Table 16-4.

**Table 16-4**

| Student | Sign | Student | Sign |
|---|---|---|---|
| Ellington | + | Mancini | 0 |
| Lombardo | + | Alpert | + |
| Prima | + | Severinsen | + |
| Dorsey | − | Welk | − |
| Basie | + | Floren | + |
| James | + | Cugat | + |
| Kostelanetz | + | Duchin | + |
| Bach | − | | |

Ignoring the pair of scores for Mancini, which were the same, we have $n = 14$.

Since $np = nq = 7 > 5$, we use a normal approximation with

$$\mu_s = 7, \ \sigma_s = \sqrt{7(0.50)} = 1.87.$$

Therefore, $S_c = 7 + 1.65(1.87) = 10.1$. Our decision rule will be to reject the null hypothesis if the outcome is more than 10.1 plus signs.

The outcome is 11 plus signs, and thus we reject the null hypothesis. The evidence supports the idea that anxiety lowered the test grades. ❑

***Helpful Hint***

The sign test for matched pairs of data from two populations reduces to a simple one-sample binomial hypothesis test with $p = 0.50$.

**Exercises**

16-1 Consider the nature of the experiment used in the cheating example (Example 16.2).
(a) Do you think it was ethical on the part of the experimenters to act as they did? Explain why or why not.
(b) Relative to any abuse of rights that occurred, how important do you consider the knowledge obtained by the experiment?
(c) What changes could be made in the procedure to protect the rights of those involved? How would you design an experiment to shed light on the relationship between anxiety and test performance?

16-2 If we are given data from 2 samples, sometimes it is appropriate to perform (a) a sign test, (b) a two-sample test of comparison of means, or (c) a matched-pairs $t$-test. Why did statisticians develop these 3 different tests? How do you decide which one to use?

16-3 A test was done to compare the efficacy of fertilizer #1 with that of fertilizer #2. One season a dozen pear trees in a controlled-environment hothouse were treated with fertilizer #1, and the next season they were given fertilizer #2. The following table shows the number of baskets of marketable pears for each tree.

| Tree Number | 1st Season (#1) | 2d Season (#2) |
|---|---|---|
| 1 | 1.6 | 2.0 |
| 2 | 2.0 | 2.5 |
| 3 | 1.6 | 1.6 |
| 4 | 3.0 | 3.5 |
| 5 | 2.5 | 3.0 |
| 6 | 2.0 | 1.5 |
| 7 | 2.0 | 2.5 |
| 8 | 2.5 | 3.0 |
| 9 | 3.0 | (died) |
| 10 | 1.5 | 2.0 |
| 11 | 1.0 | 1.5 |
| 12 | 2.0 | 1.5 |

(a) Perform a sign test for paired data on this pear data. Use $\alpha = 0.05$.
(b) There is a flaw in the design of this experiment. Even if the data indicate that production is higher with fertilizer #2, the increase may not be due to the fertilizer. What other possible factor was overlooked in the design of this experiment?

16-4 All freshmen are weighed by the medical office when school opens in September. Prof. Haman Deggs teaches a class in nutrition. At the end of the semester in December he asks his students to be weighed again. Testing at the 0.05 significance level, are the variations in weight shown random?

| Student Number | September Weight | December Weight | Student Number | September Weight | December Weight |
|---|---|---|---|---|---|
| 1 | 140 | 137 | 11 | 75 | 81 |
| 2 | 112 | 110 | 12 | 210 | 212 |
| 3 | 176 | 210 | 13 | 193 | 189 |
| 4 | 98 | 98 | 14 | 145 | 140 |
| 5 | 180 | 165 | 15 | 144 | 142 |
| 6 | 140 | 145 | 16 | 139 | 139 |
| 7 | 154 | 150 | 17 | 180 | 164 |
| 8 | 193 | 185 | 18 | 165 | 159 |
| 9 | 128 | 129 | 19 | 98 | 97 |
| 10 | 102 | 101 | 20 | 141 | 136 |

(a) Test by the sign test.
(b) Assume that weights are distributed normally, and test by a matched-pairs $t$-test.

16-5 Do premed students at Leach College score better in physics than in math? The final exam scores for a random sample of these students are given below. Use $\alpha = 0.05$.

| Student | Math | Physics | Student | Math | Physics |
|---|---|---|---|---|---|
| Korn, A. | 90 | 95 | Lash, I. | 78 | 78 |
| Tropey, N. | 78 | 76 | Ective, F. | 90 | 60 |
| Shorr, C. | 80 | 83 | Sera-Sera, K. | 87 | 94 |
| Frost, D. | 81 | 82 | Bow, L. | 78 | 83 |
| Lope, E. | 94 | 90 | Knott, Y. | 72 | 48 |
| Kupp, T. | 30 | 31 | Cleaf, O. | 99 | 98 |
| Nee, G. | 63 | 60 | Kann, P. | 70 | 80 |
| Bohr, R. | 70 | 78 | Kneeaform, Q. | 62 | 78 |

(a) Test using the sign test.
(b) Assume that grades are normally distributed and use the matched-pairs $t$-test.

16-6 Of Mr. Abel Riemann's pupils in geometry, 70 were in his algebra classes last year. To compare their grades in these 2 different subjects, he converted their grades to $z$-scores and then found the differences of their algebra $z$-scores minus their geometry $z$-scores. He got 40 plus signs, 25 minus signs, and 5 zeros. Perform a test at the 0.05 significance level.

16-7 Given that 13 randomly selected sentinels have more friends than foes, 32 have more foes than friends, and 3 have the same number of friends and foes, test at $\alpha = 0.01$ whether the sentinel population splits equally into those with more friends and those with more foes.

*We have previously tested to determine whether population* mean *was a certain value. Since half of the scores fall above the median and half below, we can use a sign test to test the hypothesis that the* median *of a population is some given number. A plus sign corresponds to a score above the median and a minus sign corresponds to a score below the median. Use this approach to solve the next 3 exercises.*

16-8 Eve Wormwood packages apples with a median of 8.5 apples per box. Wormwood samples some packages of her competitor, Adam Upright. She finds the following numbers of apples per box: 10, 12, 8, 7, 15, 9, 8, 7, 12, 8, 8, 9, 9, 9, 9, 10, 12, 7, 8, 15, 14, 18, 12, 6, 8, 9, 9, and 8. Should Wormwood be convinced that Upright's median is greater than hers? Use $\alpha = 0.05$.

16-9 Is the median age of teachers at Gray University 62 years? Test at $\alpha = 0.05$, given a sample of teachers with the following ages: 25, 47, 53, 53, 58, 59, 61, 62, 62, 65, 66, 66, 66, 73, 81, 85, and 94.

16-10 Lloyd is the attendant on a drawbridge from 10 P.M. to 6 A.M. His job is not exciting. He notices that when he raises the bridge it seldom interferes with any traffic, even though he may raise the bridge several times per night. He wonders if the median of the number of times in a week that any traffic stops for the raised bridge is less than 1. In the past 60 weeks, he noticed that in 30 of the weeks no traffic was stopped, in 20 of the weeks traffic was stopped once, and in the remaining 10 weeks traffic was stopped more than once. Test at the 0.01 significance level.

16-11 Given a matched-pairs sign test with 20 nonzero pairs, in a one-tail test on the right, what is the smallest number of plus signs that will cause a rejection of the $H_0$ if $\alpha = 0.05$?

## 16.2 The Runs Test for Randomness

runs test

The **runs test** is used to determine if certain patterns are likely to have occurred just by chance. If the pattern has gone on long enough, then the runs statistic has a normal distribution and can be compared to a table of critical $z$-scores. An illustration is given in Example 16.3.

### ❑ EXAMPLE 16.3 Speeders

Ida Noh, a safety expert, has been monitoring her radar set behind a billboard. Each time a car passes doing the speed limit or less, she writes *S* for slow. Each time a car passes doing more than the speed limit, she writes *F* for fast. The results after 40 cars were

SSFFFFSSSSSSSSSFSSFFFFSSSSSSSFFFFSSSSFSS

Ida wants to know whether or not speeders and nonspeeders occur randomly. That is, do speeders tend to come bunched together? She breaks the series of outcomes into runs of *S*'s and runs of *F*'s, as follows:

SS FFFF SSSSSSSSS F SS FFFF SSSSSSS FFFF SSSS F SS

Does this pattern of runs support the idea that the speeders come bunched together? Test at the 0.05 significance level.

#### SOLUTION

There are 11 runs. If we let $n_1$ equal the number of *S*'s, $n_2$ equal the number of *F*'s, and $R$ equal the number of *r*uns, then $n_1 = 26$, $n_2 = 14$, and $R = 11$.

This is just one possible way that her string of 26 *S*'s and 14 *F*'s might have turned out. Other arrangements would result, perhaps, in a different number of runs. One extreme case is just 2 runs, where all the *S*'s occur in a row and all the *F*'s occur in a row. At the other extreme, alternating *S*'s and *F*'s gives 29 runs. The appearance of a pattern near these extremes implies that the pattern is not random. Thus, we can talk about $R$, the number of runs, as a random variable. If we recorded every possible random arrangement of 26 $S$'s and 14 $F$'s, we would end up with a distribution of $R$'s which has a certain mean $\mu_R$ and a certain standard deviation $\sigma_R$. In a particular experiment, then, if the observed number of runs is near the mean, this supports the hypothesis of randomness. If it is far from the mean, it is evidence against randomness.

The mean number of runs, when you randomly arrange $n_1$ items of one kind and $n_2$ items of another kind, is given by

$$\mu_R = \frac{2n_1n_2}{n_1 + n_2} + 1$$

In this case,

$$\mu_R = \frac{2(26)(14)}{26 + 14} + 1 = 19.2$$

The standard deviation of the number of runs is given by

$$\sigma_R = \sqrt{\frac{2n_1n_2(2n_1n_2 - n_1 - n_2)}{(n_1 + n_2)^2(n_1 + n_2 - 1)}}$$

In this case,

$$\sigma_R = \sqrt{\frac{22(26)(14)[2(26)(14) - 26 - 14]}{(26 + 14)^2(26 + 14 - 1)}} = 2.83$$

Further, the distribution of $R$'s is approximately normal if both $n_1$ and $n_2$ are greater than 10.

We can use this information to set up a hypothesis test as follows:

*Population:* All drivers who pass the billboard.

*Sample:* The 40 cars observed in this experiment.

$H_0$: Fast and slow cars arrive randomly; $\mu_R = 19.2$.

$H_a$: Fast and slow cars do not arrive randomly; $\mu_R \neq 19.2$ (a two-tail test).

This is a two-tail test, since there could be either too many runs or too few runs if they are not random.

From the formulas for mean and standard deviation, and $n_1$ and $n_2 > 10$, we have

$$\text{ND}_R(\mu_R = 19.2, \sigma_R = 2.83)$$

Testing at the 0.05 significance level, we have

$$R_c = \mu_R + z_c\sigma_R = 19.2 \pm 1.96(2.83) = 13.6 \text{ and } 24.8$$

The decision rule will be to reject the null hypothesis if the outcome $R$ is either fewer than 13.6 or greater than 24.8 runs. If Ms. Noh's data contain fewer than 13.6 runs, then she will conclude that fast and slow cars do not arrive randomly *but seem to come in bunches*. If she gets more than 24.8 runs, she will conclude that fast and slow cars do not arrive randomly *but tend to alternate*.

The outcome is $R = 11$ runs, so Ms. Noh can reject the null hypothesis. Fast cars and slow cars arrive in bunches and do not arrive randomly. ❑

---

## ❑ EXAMPLE 16.4 Random Busing and Bussing

Sal, Jean, and Pat leave work together every day and walk to the bus stop. Sal takes the Q43 bus home, Pat rides the Q36, and Jean can take either bus to get home. If the Q36 bus arrives first, Pat and Jean get on and Sal goes home alone. If the Q43 bus arrives first, then Sal and Jean get on and Pat goes home alone. If we let $P$ indicate that Pat went home alone and $S$ that Sal went home alone, then for the past 25 working days we have

PPSSPPPPSSPSPPPSSSSPSSSSSS

Testing with $\alpha = 0.05$, do the buses arrive randomly?

### SOLUTION

We are performing a runs test.

*Population:* The bus arrivals on all working days.

*Sample:* The bus arrivals for the past 25 working days.

$H_0$: The buses arrive randomly.

$H_a$: The buses do not arrive randomly (a two-tail test).

Let $n_1$ = the number of $P$'s and let $n_2$ = the number of $S$'s. We have $n_1 = 11$ and $n_2 = 15$. So, for the mean and standard deviation of $R$, we find

$$\mu_R = \frac{2(11)(15)}{11 + 15} + 1 = 13.7$$

and

$$\sigma_R = \sqrt{\frac{2(11)(15)[2(11)(15) - 11 - 15]}{(11 + 15)^2(11 + 15 - 1)}} = 2.44$$

Since both $n_1$ and $n_2$ are greater than 10, we have

$$\mathrm{ND}_R(\mu_R = 13.7,\ \sigma_R = 2.44)$$

Taking $z_c = \pm 1.96$, because $\alpha = 0.05$, we get

$$R_c = 13.7 \pm 1.96(2.44) = 13.7 \pm 4.8 = 8.9 \text{ and } 18.5$$

Our decision rule is to reject the null hypothesis if the statistic $R$ is less than 8.9 or more than 18.5. The outcome is $R = 10$ runs, and so we fail to reject the null hypothesis. The buses may arrive randomly. (As random luck would have it, Pat and Jean fell in ♥, were wed, and lived happily ever after. They became bus dispatchers and joyfully created statistics on both busing and bussing.) ❑

### Exercises

16-12 To test a coin for fairness, we could do a one-sample binomial hypothesis test to see if $P(\text{heads}) = 1/2$. If the sample data were 10 heads followed by 10 tails, we would not reject the null hypothesis. However, these same data used in a runs test would lead us to reject the null hypothesis. What is the difference between the null hypotheses in the 2 tests?

16-13 Find $n_1$, $n_2$, $R$, $\mu_R$, and $\sigma_R$ based on the following outcomes:
(a) MMMFFMFMMFM
(b) TTTTFFFTTTFFT
(c) SFFSSSFFFFSSSSSFFFFFF
(d) XXYXXYXXYYXXXYX
(e) NONONOONNONNNNOONON

16-14 G. Ringo, a tourist, is gambling in a resort in Acapulco. On the last 25 games, he won and lost in the following order:

WWWLWLWWLLLLWWLLWLLLWWWWL

Are these wins and losses randomly distributed if you test at $\alpha = 0.05$?

16-15 A person, asked to give a "random" string of $A$'s and $B$'s, says: ABAABBAB BABBBAAABAABBA. Does this seem to be random at $\alpha = 0.05$?

16-16 Confined to bed by a broken leg and bored to tears, Xerxes has been watching people go past his window. He has clocked the time a person remains in his view and found the median time to be 6.3 minutes. He wonders if slower persons and faster persons appear randomly. He records the following list of data on the next 25 people: SFFFSSFSSSSSSSFFFFFSSFFSS. Test with $\alpha = 0.05$.

16-17 Carol Louis inspects wabes for the Brillig Brothers Book Group. The median wabe is 15.3 toves. Every $\frac{1}{4}$ hour she takes 1 wabe from the assembly line and determines its toves. Her results are 14.7, 14.9, 15.1, 15.5, 15.6, 15.6, 15.2, 15.7, 15.6, 14.9, 14.7, 14.8, 15.2, 15.4, 15.2, 14.8, 15.0, 16.0, 15.8, 15.5, 15.9, 15.4, and 15.4. Are the variations from the median occurring randomly at the 0.05 significance level?

16-18 In a series of 5 true and false questions, 1 answer is true and 4 are false. There are 5 different arrangements of 1 true and 4 false answers.
(a) List the 5 arrangements and the number of runs in each arrangement. Find the mean of these numbers of runs.
(b) Confirm the formula for the mean number of runs by finding the mean using the formula for $\mu_R$ and comparing your answer with part (a).
(c) Similarly, confirm the formula for the standard deviation of the number of runs by finding it 2 different ways.

# 16.3 The Mann-Whitney *U*-Test for Comparison of Ranks

In some studies a nonparametric approach is called for because the observations are rankings rather than measurements, and in small samples rankings do not have a normal distribution. For instance, in a consumer survey you might give various people a sample of three brands of chewing gum to put in order from their most favorite to their least favorite. Your data then will be a collection of 1s, 2s, and 3s.

Another reason for using a special approach for ranked data is that ranks cannot be interpreted like ordinary measurements in the first place. For example, if brand A is ranked as number 1 (favorite) and brand B as number 2 (next favorite), we cannot tell from this whether someone thought A was a *little* better than B or a *lot* better than B.

### *Frank Wilcoxon*

Frank Wilcoxon (1892–1965) was born in Ireland of American parents. He grew up in Catskill, New York. His invention of the two-sample rank-sum statistic in 1945 marks a major starting point for nonparametric statistics. Though he invented the statistic, others made equally important contributions to a fuller understanding of this approach. For instance, the extension of the method to unequal-sized samples was a result of work by H. Mann and D. Whitney in 1947.

Special tests have been devised for analyzing ranked data. The pioneers in this fairly recent field include Frank Wilcoxon, H. Mann, and D. Whitney. The version of the test we present in this book is usually referred to as the **Mann-Whitney test**. It is used to decide whether the **ranks** in two populations tend to be similar or different.

**Mann-Whitney test**

We treat separately small-sample and large-sample tests, one-tail and two-tail tests, and what to do in case of ties. Small-sample tests are performed if either sample is of size 10 or less; large-sample tests are used when both sample sizes are greater than 10.

## 16.3-1 Small-Sample, One-Tail Tests

Let us examine a typical application, one where a new treatment of some kind is compared to a standard, or "control," treatment.

---

❑ EXAMPLE 16.5 Rat IQs

There are 14 laboratory rats cloned from the same parent. They are split randomly into two groups of size 7. Their diets and living conditions are identical, except for one vitamin which is added to the diet of group 1. The motivated hypothesis is that the vitamin will make the rats in group 1 smarter. Group 1 is considered the treatment group, and group 2 is the control group. During the experiment, two rats in group 1 escape.

After 1 month the rats (on the basis of rat IQ tests) are ranked from smartest (rank = 1) to most stupid (rank = 12). Here are the results.

| | | |
|---|---|---|
| Treatment-group ranks | 1 2 4 5 8 | $n_1 = 5$ |
| Control-group ranks | 3 6 7 9 10 11 12 | $n_2 = 7$ |

**Note** In this book we always assign rank 1 to the "best" outcome, and Table D-15 (see Appendix D) is written from this point of view.

Do these results support the hypothesis that the vitamin makes the rats smarter?

## SOLUTION

You can see that the smarter rats tend to be in the treatment group. This is reflected by the fact that there are *more numerically lower ranks* in the treatment group.

The idea of this statistical test is to summarize this relationship by a single statistic. One way to do this is as follows. Since we are trying to express the fact that the treatment rats tend to have lower ranks than the control ranks, we can count the number of times that a rat in the treatment group does better (that is, gets a lower rank) than a rat in the other group. We will compare in turn each rat in the treatment group to all of the rats in the control group. A rat in the treatment group scores +1 for each rat in the other group that he or she outranks. The total number we get for all rats in the treatment group will be our statistic, called the ***rank-sum statistic***. We show the computation using the data above.

*Computation*

1. The first rat in the treatment group had a rank of 1. This was better than all 7 rats in the other group. So this rat scores 7.
2. The second rat in the treatment group had a rank of 2. This was also better than all 7 rats in the other group. So this rat scores 7.
3. The third rat in the treatment group had a rank of 4. This is better than 6 of the rats in the other group. So this rat scores 6.
4. Similarly, the fourth rat scores 6.
5. Finally, the fifth rat in the treatment group had a rank of 8. This was better than 4 rats in the other group. This rate scores 4.

The total score for the treatment group then is 7 + 7 + 6 + 6 + 4 = 30. This statistic is one way to summarize the amount by which the treatment group had better ranks than the control group. This total is often labeled *U*

**Mann-Whitney *U*-statistic**

and is called the **Mann-Whitney *U*-statistic**. If it is bigger than or equal to* a tabled critical value, $U_c$, we conclude that the population represented by the treatment group is smarter than the population represented by the control group.

This whole approach may be organized as follows:

We will perform a Mann-Whitney U-test of ranks. Since at least one sample is size 10 or less, this is a small-sample test.

*Populations:* All rats *who receive* this vitamin and all those *who do not receive* this vitamin.

*Samples:* The 5 and 7 rats in this experiment.

$H_0$: There is no difference in the IQs of rats who get the vitamin and rats who do not. (Technically, we are testing that there is no difference in the mean rank of rats in the two populations.)

$H_a$: Rats which receive the vitamin have higher IQs than the rats who do not (a one-tail test).

We refer to Table D-15 for one-tail tests, and we find that for samples of size 5 and 7 and $\alpha = 0.05$ the critical value is 29. Hence, our decision rule is to reject the null hypothesis if the outcome is at least 29. Since the sample outcome is 30, we have evidence that the vitamin is effective in raising rat IQs. ❑

*Summary of One-Tail, Small-Sample Mann-Whitney* U*-Test of Ranks*

1. Arrange the ranks in both samples numerically from smallest to largest.
2. Calculate a score for each rank in the group that the motivated hypothesis expects to do better (that is, to have lower ranks).
3. Compute the sum $U$ of these scores.
4. Compare $U$ with the tabled critical value found in Table D-15 (Appendix D). If $U$ is bigger than or equal to the critical value, you can reject the null hypothesis.

Sometimes the experimental data are not ranks to begin with, but the investigator is willing to treat them as such because there is no evidence that the population from which they came is normal. Here is such a case.

## ❑ EXAMPLE 16.6 A Nonnormal Population

Patients with a certain disease are treated with two different medicines. The physicians evaluate the patients in terms of "speed of recovery" in days. The motivated hypothesis is that the new medicine will lead to speedier recovery. Given the results on the next page, evaluate this claim using the Mann-Whitney $U$-test at $\alpha = 0.01$.

---

*Caution: The decision rule for the Mann-Whitney test uses $\geq$ rather than $>$ in order to agree with our table.

| | Number of Days to Recover | | | | | |
|---|---|---|---|---|---|---|
| Old medicine | 13 | 12 | 14 | 16 | 18 | 20 |
| New medicine | 8 | 17 | 9 | 11 | 15 | |

**SOLUTION**

rank

Because we have no reason to assume that the data come from normal distributions, we decide to perform a Mann-Whitney $U$-test of **ranks**. Since at least one sample is 10 or less, this is a small-sample test.

*Populations:* All patients treated with one of these two different medicines.

*Samples:* The 6 and 5 patients in the experiment.

The test statistic is the $U$-score obtained from these samples.

$H_0$: The recovery time is the same for both medicines (there is no difference in the mean ranks of the two groups of patient recovery times).

$H_a$: The new medicine decreases the recovery time (a one-tail test).

***Note*** Remember, we are always seeking $U \geq U_c$ (a one-tail test).

First, put the outcomes in each group in increasing order.

| | Speed of Recovery, Days | | | | | |
|---|---|---|---|---|---|---|
| Old medicine | 12 | 13 | 14 | 16 | 18 | 20 |
| New medicine | 8 | 9 | 11 | 15 | 17 | |

Next, compute $U$ as we did in the previous example. We get a score for each patient on the new medicine by counting the number of patients on the old medicine who had a slower recovery. Notice that we are treating the recovery times as *ranks*—that is, we only consider whether one time is shorter than another and not how much shorter it is. The sum of these scores for the new medicine group is

$$U = 6 + 6 + 6 + 3 + 2 = 23$$

The critical value is found from Table D-15 for $\alpha = 0.01$, $n_1 = 6$, $n_2 = 5$ as $U_c = 28$.

Since the sample outcome is smaller than the critical value, we fail to reject the null hypothesis. Our conclusion is that we have not clearly established the superiority of the new medicine. ❑

## 16.3-2 Small-Sample, Two-Tail Tests

In the case of a *two-tail* test we must calculate $U$ for both groups. If either one is bigger than the appropriate critical value in Table D-16 (see Appendix D), we can reject the hypothesis of equality. Hence we calculate both $U_1$ and $U_2$. This is easier than it may appear, because it is always true that $U_1 + U_2$ must equal $n_1 n_2$. Hence,

$$U_2 = n_1 n_2 - U_1$$

***Note*** A two-tail Mann-Whitney $U$-test reduces to simply performing a one-tail test twice.

---

### ❑ EXAMPLE 16.7 Occult ♥-Potions

A young student of the occult has developed two love-potions. To compare them, she selects 18 subjects at random and gives half ♥-potion 1 and half ♥-potion 2. Then she watches and ranks the subjects from most affected (1) to least affected (18). Analyze the results and test for equality of ♥-potion potency. Use $\alpha = 0.05$.

| Subject | 1 | 2 | 3 | 4 | 5 | 6 | 7 | 8 | 9 | 10 | 11 | 12 | 13 | 14 | 15 | 16 | 17 | 18 |
|---|---|---|---|---|---|---|---|---|---|---|---|---|---|---|---|---|---|---|
| ♥-potion | 1 | 1 | 1 | 1 | 1 | 1 | 1 | 1 | 1 | 2 | 2 | 2 | 2 | 2 | 2 | 2 | 2 | 2 |
| Rank | 4 | 10 | 3 | 12 | 6 | 1 | 11 | 5 | 2 | 16 | 17 | 9 | 15 | 7 | 18 | 14 | 13 | 8 |

SOLUTION

Since the data are ranks, we perform a Mann-Whitney $U$-test of ranks. Since both samples are not greater than 10, this is a small sample test.

*Populations:* All persons given ♥-potion 1 and all persons given ♥-potion 2.

*Samples:* The ranks of the two groups of 9 persons to whom we administered these ♥-potions.

$H_0$: The effects of the two ♥-potions are the same (mean rank for ♥-potion 1 equals mean rank for ♥-potion 2).

$H_a$: The effects of the two ♥-potions are different (a two-tail test).

We organize the data as follows.

| | Ranks | | | | | | | | | |
|---|---|---|---|---|---|---|---|---|---|---|
| ♥-potion 1 | 1 | 2 | 3 | 4 | 5 | 6 | 10 | 11 | 12 | $n_1 = 9$ |
| ♥-potion 2 | 7 | 8 | 9 | 13 | 14 | 15 | 16 | 17 | 18 | $n_2 = 9$ |
| | | | | | | | | | | $N = 18$ |

Since this is a two-tail test, we refer to Table D-16. The critical value for $\alpha = 0.05$ with $n_1 = n_2 = 9$ is $U_c = 64$.

Thus our decision rule is that if *either* $U_1$ or $U_2$ is at least 64, we reject $H_0$.

We compute $U_1$, as usual, by counting:

$$U_1 = 9 + 9 + 9 + 9 + 9 + 9 + 6 + 6 + 6 = 72$$

Then, instead of counting, we compute $U_2$ using $U_2 = n_1 n_2 - U_1$:

$$U_2 = 9(9) - 72 = 81 - 72 = 9$$

The outcomes were $U_1 = 72$ and $U_2 = 9$. Since one of the outcomes, $U_1 = 72$, is greater than 64, we reject $H_0$. We have evidence that ♥-potion 1 is more effective than ♥-potion 2. ❑

## 16.3-3 Large-Sample Tests—Both $n_1$ and $n_2 > 10$

When both sample sizes are greater than 10, it can be shown that the distribution of sample values of $U$ is approximately normal. This means that the Mann-Whitney $U$-test reduces to an ordinary normal distribution hypothesis test, and we can compute the critical values of $U$ using the critical values of normal $z$-test scores. We do not have to use special tables.

The mean and the standard deviation of the normal curve which describes the distribution of $U$ when the null hypothesis is true are

$$\mu_U = \frac{n_1 n_2}{2}$$

$$\sigma_U = \sqrt{\frac{n_1 n_2 (n_1 + n_2 + 1)}{12}}$$

❑ **EXAMPLE 16.8 An Anagram Broad Board**
(A Large-Sample, Two-Tail Experiment)

Arny Ryan, the director of a national anagram contest, is in a rage. He can purchase gear—such as a narrow or broad game board, a stool, or other tools—from two suppliers: Roset's Stores (R) and Tock's Stock (T). Over a period of time he has ranked shipments from dim (poor) to mid (middling), as shown below. Rank 1 equals the better (middling) quality.

| | Ranks | | | | | | | | | | | | |
|---|---|---|---|---|---|---|---|---|---|---|---|---|---|
| Supplier R | 1 | 2 | 6 | 8 | 10 | 11 | 14 | 15 | 17 | 18 | 20 | 23 | $n_1 = 12$ |
| Supplier T | 3 | 4 | 5 | 7 | 9 | 12 | 13 | 16 | 19 | 21 | 22 | | $n_2 = 11$ |
| | | | | | | | | | | | | | $n_1 + n_2 = 23$ |

Do these data indicate a statistically significant difference in quality? Use $\alpha = 0.05$. This is a two-tail test.

SOLUTION

We analyze these rankings by a Mann-Whitney $U$-test of ranks.

*Populations:* All the gear shipments by Roset's Stores and those by Tock's Stock.

*Samples:* The last 23 orders received.

The test statistic is the $U$ score we obtained from this particular ranking of these last 12 and 11 shipments.

$H_0$: The equipment from both suppliers is of the same quality (the mean ranks are equal).

$H_a$: The equipment from both suppliers is not of the same quality (a two-tail test).

Since both samples are large (greater than 10), we use the normal approximation.

$$\mu_U = \frac{n_1 n_2}{2} = \frac{12(11)}{2} = 66$$

$$\sigma_U = \sqrt{\frac{n_1 n_2 (n_1 + n_2 + 1)}{12}} = \sqrt{\frac{12(11)(12 + 11 + 1)}{12}} = 16.248$$

The critical values of $U$ are

$$\begin{aligned} U_c &= \mu_c \pm z_c \sigma_U \\ &= 66 \pm 1.96(16.248) = 66 \pm 31.8 = 34.2 \text{ and } 97.8 \end{aligned}$$

*Decision rule:* It would take an observed value of $U$ either less than 34.2 or more than 97.8 to reject the hypothesis of quality.

$$U_R = 11 + 11 + 8 + 7 + 6 + 6 + 4 + 4 + 3 + 3 + 2 = 65$$

$$U_T = n_1 n_2 - U_R = 12(11) - 65 = 132 - 65 = 67$$

Since neither sample value of $U$ is extreme enough to allow Arny Ryan to reject the hypothesis of equality, he has not established that there is a statistically significant difference in quality between the two suppliers. ❑

## 16.3-4 The Case of Ties

### Small Samples with Ties

Very often, ranked observations result in ties. When this happens, certain adjustments in the calculations must be made.

---

### ❑ EXAMPLE 16.9 The Two-Medicine Example Revisited

Suppose the speed-of-recovery data from Example 16.6 had turned out like this:

| | Number of Days to Recover | | | | | | |
|---|---|---|---|---|---|---|---|
| Old medicine | 10 | 13 | 14 | 16 | 18 | 18 | 20 |
| New medicine | 8 | 8 | 11 | 13 | 18 | | |

SOLUTION

We are interested in $U_{\text{new}}$, so we get a score for each person in the new-medicine group. Ties are present, since it turns out that some in the old-medicine group took the same number of days to recover as people in the new-medicine group. When this happens, the person in the new-medicine group scores 0.5. (This is a compromise between scoring 1 for being better and scoring 0 for being worse.)

The first person in the new-medicine group did better than all 7 of the people on the old medicine, and scores 7. Similarly, the second person also scores a 7. The third person's time of 11 days was better than 6 of the others, meriting a score of 6.

The fourth person's time of 13 days is better than 5 people in the other group and tied with one of them. He scores $5 + 0.5 = 5.5$.

The fifth and last person's time is better than 1 person in the other group and tied with two of them. He scores $1 + 0.5 + 0.5 = 2$. This gives a total sum for the new-medicine group of

$$U = 7 + 7 + 6 + 5.5 + 2 = 27.5$$

The populations, samples, and hypotheses are the same as they were in Example 16.6, so this is still a one-tail, small-sample Mann-Whitney $U$-test of ranks.

From Table D-15, for $\alpha = 0.01$, with $n_1 = 7$ and $n_2 = 5$, we find $U_c = 32$.

Since the sample value is not larger than the critical value, we do not have conclusive evidence that the new medicine is superior. ❑

---

***Note*** Strictly speaking, when ties occur in the ranks, we should not use Table D-15 (Appendix D) because the tabled critical values were computed assuming no ties. But, for most purposes, when there are not very many ties, the tabled results are approximately correct.

### Large Sample with Ties

This is a common situation which occurs in surveys where people are asked to rate something by placing it in one of only a few categories (such as *excellent, fair,* or *poor*). When there are ties, we still have a normal distribution. The mean is still given by

$$\mu_U = \frac{n_1 n_2}{2}$$

However, the formula for the standard deviation of the $U$-distribution must be adjusted. This is because the presence of ties reduces the number of different values $U$ can take on, and so reduces the variability in $U$. Hence, we must modify the previous formula for the standard deviation:

$$\sigma_U = \sqrt{\frac{n_1 n_2 (N+1)}{12} - C}$$

where $N$ equals $n_1 + n_2$, and $C$ is the correction factor for ties, which is calculated by

$$C = \frac{n_1 n_2}{12N(N-1)} \Sigma(T^3 - T)$$

where $T$ counts the number of scores tied at each rank, and $\Sigma$, as usual, indicates a summation. We illustrate the use of these formulas in the next example.

### ❑ EXAMPLE 16.10 Executive Ambition

In a study of executive ambition, a group of 80 middle-level executives in a large corporation were split randomly into two groups. The first group was a control group of 40 who got the usual pep talk and incentives. An experimental group of the other 40 got a newfangled psychodynamic treatment, complete with hypnosis and special diet. At the end of the study, each executive is evaluated as *super, good, fair,* or *poor*. Here are the results.

| | Super | Good | Fair | Poor | |
|---|---|---|---|---|---|
| Special treatment group | 12 | 16 | 7 | 5 | $n_1 = 40$ |
| Control group | 9 | 15 | 9 | 7 | $n_2 = 40$ |
| | | | | | $N = 80$ |

Do these data indicate that the special treatment is producing more-ambitious executives? Use $\alpha = 0.05$. This is a one-tail test.

## SOLUTION

We are performing a Mann-Whitney $U$-test of ranks.

*Populations:* All middle-level executives who are given the usual pep talk and incentives, and all those who receive the newfangled psychodynamic treatment.

*Samples:* The two groups of 40 middle-level executives tested.

Since both samples are greater than 10, this is a large-sample test, and we may use the normal distribution to find the critical values of $U$.

$H_0$: The two training methods are equivalent; the mean ratings for both treatments are the same.

$H_a$: The new method is superior; the mean rating in the special-treatment group is better (a one-tail test).

First, we calculate $U$ as follows:

There were 12 people in the special-treatment group who were ranked *super*. Each of them tied with 9 people in the control group and did better than $15 + 9 + 7 = 31$ people in the control group. Hence, the score for each 1 of the 12 people is

$$9(0.5) + 15 + 9 + 7 = 35.5$$

Since 12 people have this score, we get

$$12[35.5] = 426$$

Similarly, the 16 people in the special-treatment group who were rated *good* score $16[15(0.5) + 9 + 7] = 376$.

The 7 people rated *fair* score $7[9(0.5) + 7] = 80.5$.

The 5 persons rated *poor* score $5[7(0.5)] = 17.5$.

The sum of these numbers is

$$U = 426 + 376 + 80.5 + 17.5 = 900$$

Since we have large samples (both are greater than 10), we can use a normal approximation to find the critical values for $U$.

We now compute the mean and the standard deviation. First, we compute $C$ with the help of the following table.

| | Special-Treatment Group | Control Group | Number of Ties $T$ | $T^3$ | $T^3 - T$ |
|---|---|---|---|---|---|
| Super | 12 | 9 | $12 + 9 = 21$ | 9,261 | 9,240 |
| Good | 16 | 15 | $16 + 15 = 31$ | 29,791 | 29,760 |
| Fair | 7 | 9 | $7 + 9 = 16$ | 4,096 | 4,080 |
| Poor | 5 | 7 | $5 + 7 = 12$ | 1,728 | 1,716 |
| | | | | | $\Sigma(T^3 - T) = 44,796$ |

This gives

$$C = \frac{n_1 n_2}{12N(N-1)} \Sigma(T^3 - T) = \frac{40(40)}{12(80)(79)}(44{,}796) = 945.1$$

Therefore, we have

$$\mu_U = \frac{n_1 n_2}{2} = \frac{40(40)}{2} = 800$$

and

$$\sigma_U = \sqrt{\frac{n_1 n_2 (N+1)}{12} - C} = \sqrt{\frac{40(40)(81)}{12} - 945.1} = 99.27$$

Thus the critical value of $U$ is

$$U_c = \mu + z_c \sigma = 800 + 1.65(99.27) = 800 + 163.8 = 963.8$$

Since the sample value of $U = 900$ is not greater than the critical value, $U_c = 963.8$, we do not have sufficient evidence to reject the null hypothesis. We have not established that the special treatment does a better job than the usual one in producing ambitious corporate executives. ❑

## STUDY AIDS

### Chapter Summary

In this chapter you have learned:

- How to perform the sign test for matched pairs of data, which reduces to an elementary binomial hypothesis test with $p = 0.50$.
- How to perform the runs tests for randomness
- How to perform the Mann-Whitney $U$-test for comparison of ranks in two populations

### Vocabulary

You should be able to explain the meaning of each of these terms:

1. Nonparametric statistic
2. Distribution-free test
3. Robust
4. Sign test
5. Runs test
6. Mann-Whitney test
7. Rank order
8. Mann-Whitney $U$-statistic

**Symbols**

You should understand the meaning of each of these symbols:

1. $R$
2. $\mu_R$
3. $\sigma_R$
4. $U$
5. $C$
6. $T$

**Formulas**

You should know when and how to use each of these formulas:

*For the Runs Test:*

1. $\mu_R = \dfrac{2n_1n_2}{n_1 + n_2} + 1$

2. $\sigma_R = \sqrt{\dfrac{2n_1n_2\,(2n_1n_2 - n_1 - n_2)}{(n_1 + n_2)^2(n_1 + n_2 - 1)}}$

3. $N = n_1 + n_2$

4. $R_c = \mu_R + z_c\sigma_R$

*For the Mann-Whitney* U-*Test:*

5. $U_2 = n_1n_2 - U_1$

6. $\mu_U = \dfrac{n_1n_2}{2}$

7. $\sigma_U = \sqrt{\dfrac{n_1n_2(n_1 + n_2 + 1)}{12}}$ when no ties exist

8. $\sigma_U = \sqrt{\dfrac{n_1n_2(N + 1)}{12} - C}$ when ties exist

9. $C = \dfrac{n_1n_2}{12N(N - 1)}\,\Sigma\,(T^3 - T)$

## EXERCISES

16-19 How many pairs of anagrams can you find in Example 16.8?

16-20 Refer to Example 16.7 and calculate $U_2$ by *counting,* and verify that $U_2 = n_1n_2 - U_1$.

16-21 Given these rankings of Democrats and Republicans in an upcoming race,* calculate $U_{\text{Dem}}$:

| Sample of Democrats | Sample of Republicans |
|---|---|
| 1 | 4 |
| 2 | 6 |
| 3 | 6 |
| 6 | 9 |
| 8 | 10 |

*A 50-yard dash.

16-22 Given these scores of St. Bernards and Newfoundlands in a dog obedience test, calculate $U_{\text{St. Bernards}}$:

| St. Bernards | Newfoundlands |
|---|---|
| 2 | 2 |
| 3 | 2 |
| 4 | 5 |
| 5 | 6 |
| | 6 |

16-23 The following are the scores on the National Manufacturers' annual scholarship exam:

| Student | Score |
|---|---|
| *Debra Draw | 50 |
| Stan Stalemate | 48 |
| *Stu Standoff | 52 |
| Dan Deadlock | 48 |
| *Norma Knott | 50 |
| Ben Bind | 48 |
| Carl Cravat | 46 |

Those marked with an asterisk were tutored by Dr. Noah Vail. Test to see whether this supports the claim that Dr. Vail helps students do well on this exam. Use $\alpha = 0.05$.

16-24 John Smith eats at 2 fast-food places. To find out if one is faster than the other, he times to the nearest minute how long it takes to be served a drink, a burger, and fries. The results of his last 13 visits were as follows.

| McBurgers | 2 | 5 | 13 | 8 | 7 | 12 | 4 | 6 |
|---|---|---|---|---|---|---|---|---|
| King Donalds | 9 | 3 | 11 | 1 | 10 | | | |

Test at $\alpha = 0.05$, using the Mann-Whitney $U$-test.

16-25 Wanda the Witch buys packages of bat wings for her brews. Recently, she has switched suppliers to Winona's Wings because she was getting too many substandard wings from Bertha's Bat Bits. To be fair, she decides to compare the 2 suppliers. Her data are as follows.

**Number of Substandard Wings per Package of 13 Wings**

| Bertha's Bat Bits | Winona's Wings |
|---|---|
| 3 | 0 |
| 2 | 0 |
| 1 | 0 |
| 3 | 2 |
| 0 | 2 |
| 0 | 1 |
| 3 | 0 |
| 1 | 3 |
| 2 | |

Using the Mann-Whitney $U$-test at $\alpha = 0.05$, decide whether there is a difference in quality between the 2 suppliers.

16-26 A waiter at Mount Cupid Honeymoon Lodge kept tabs on the number of meals missed during 1 week by newly married couples. Of the couples that he served during the first week in June, he noted the following. Couples on the first week of their honeymoon missed 7, 5, 11, 8, 7, 11, and 17 meals; while couples on the second week of their honeymoon missed 0, 13, 3, 8, 10, 6, 10, and 6 meals. Test the hypothesis that there is no difference between the number of meals missed by first- and second-week honeymooners. Use $\alpha = 0.05$.

16-27 Two groups of married people were asked to compare how much they were "in love with" their spouse today with how they felt on their wedding day. The data were as follows.

| | Married Less Than 7 Years (110 People) | Married at Least 7 Years (90 People) |
|---|---|---|
| A lot less | 30 | 10 |
| A little less | 20 | 20 |
| The same | 10 | 30 |
| A little more | 20 | 20 |
| A lot more | 30 | 10 |

Calculate $\mu_U$, the correction factor for ties, and the standard deviation.

16-28 In the interest of science, students who studied under 2 professors were asked to rank their jokes. The data were as follow.

| | Funny | Average | Rank | Total |
|---|---|---|---|---|
| Prof. Laurel | 20 | 80 | 100 | 200 |
| Prof. Hardy | 40 | 80 | 80 | 200 |

Use $\alpha = 0.05$ and perform a two-tail $U$-test.

16-29 Espionage agents in 2 different countries are trained to endure pain. Eleven agents from each country enter the 4th Annual International Espionage Agent Pain Endurance Competition. Here are the results.

| Name of Agent | Country | Volts Endured | Name of Agent | Country | Volts Endured |
|---|---|---|---|---|---|
| 001 | A | 100 | 001 | B | 130 |
| 002 | A | 143 | 002 | B | 135 |
| 003 | A | 128 | 003 | B | 142 |
| 004 | A | 118 | 004 | B | 120 |
| 005 | A | 89 | 005 | B | 60 |
| 006 | A | 118 | 006 | B | 140 |
| 007 | A | 132 | 007 | B | 95 |
| 008 | A | 107 | 008 | B | 120 |
| 009 | A | 141 | 009 | B | 60 |
| 010 | A | 93 | 010 | B | 97 |
| 011 | A | 101 | 011 | B | 102 |

Does this indicate at the 0.05 significance level that either country produces a superior agent? Analyze by a $U$-test.

16-30 The vice president of Air Languid wants to measure the results of a recent change in flight scheduling aimed at reducing the number of empty seats per flight. The number of empty seats on the last 14 flights under the old schedule is to be compared with that for the first 14 flights under the new schedule. The data are as follows.

| | Number of Empty Seats per Flight | | | | | | | | | | | | | |
|---|---|---|---|---|---|---|---|---|---|---|---|---|---|---|
| Old schedule | 15 | 17 | 9 | 17 | 7 | 21 | 18 | 25 | 21 | 9 | 27 | 23 | 23 | 27 |
| New schedule | 14 | 14 | 16 | 20 | 24 | 10 | 18 | 12 | 20 | 6 | 16 | 8 | 16 | 14 |

Using $\alpha = 0.05$, decide by a $U$-test if the new schedule reduces the number of empty seats.

16-31 A chemical refinery has developed a new type of gasohol. To see whether it is better than their old gasohol, 50 drivers used the new type in their automobiles for 1000 miles, and then they used the old type for 1000 miles. The drivers reported their average number of miles per gallon. The number of drivers in each category were as follows.

**Number of Drivers**

| | 10–15 mpg | 15–20 mpg | 20–25 mpg | 25–30 mpg | Total |
|---|---|---|---|---|---|
| Old | 10 | 20 | 15 | 5 | 50 |
| New | 0 | 10 | 15 | 25 | 50 |

Test at $\alpha = 0.05$ with a $U$-test.

16-32 (a) A competition between 2 sailors from the Naval Training Center and 2 belly dancers from the Navel Training Center can end in 6 different ways if there are no ties. For example, one outcome could be a belly dancer first, followed by 2 sailors and then a belly dancer in last place. Using B to represent a belly dancer and S to represent a sailor, we represent this outcome as BSSB. Write out the letter strings which represent the other 5 possible outcomes.

(b) If 3 belly dancers and 3 sailors compete, there are 20 possible outcomes:

| | Belly Dancer Ranks | Sailor Ranks | $U$ for Belly Dancers |
|---|---|---|---|
| 1 BBBSSS | 1 2 3 | 4 5 6 | 9 |
| 2 BBSBSS | 1 2 4 | 3 5 6 | 8 |
| 3 BSBBSS | 1 3 4 | 2 5 6 | 7 |
| 4 SBBBSS | 2 3 4 | 1 5 6 | 6 |
| 5 BBSSBS | 1 2 5 | 3 4 6 | 7 |
| 6 BSBSBS | 1 3 5 | 2 4 6 | 6 |
| 7 SBBSBS | 2 3 5 | 1 4 6 | 5 |
| 8 BSSBBS | 1 4 5 | 2 3 6 | 5 |
| 9 SBSBBS | 2 4 5 | 1 3 6 | 4 |
| 10 SSBBBS | 3 4 5 | 1 2 6 | 3 |
| 11 SSSBBB | 4 5 6 | 1 2 3 | 0 |
| 12 SSBSBB | 3 5 6 | 1 2 4 | 1 |
| 13 SBSSBB | 2 5 6 | 1 3 4 | 2 |
| 14 BSSSBB | 1 5 6 | 2 3 4 | 3 |
| 15 SSBBSB | 3 4 6 | 1 2 5 | 2 |
| 16 SBSBSB | 2 4 6 | 1 3 5 | 3 |
| 17 BSSBSB | 1 4 6 | 2 3 5 | 4 |
| 18 SBBSSB | 2 3 6 | 1 4 5 | 4 |
| 19 BSBSSB | ______ | ______ | ______ |
| 20 BBSSSB | ______ | ______ | ______ |

Verify the values of $U$ for the first 2 outcomes above.

(c) Complete the last 2 lines of the table.

(d) Find the critical value of $U$ for $\alpha = 0.05$ for a one-tail test.

(e) Put the 20 values of $U$ in order.

(f) Are 95% of them less than the critical value of $U$?

## CLASS SURVEY QUESTIONS

Compare the heights of the females in the class with the heights of their mothers two ways:

1. Use the sign test for matched pairs. This tests whether or not *a* daughter tends to be taller than *her* mother.
2. Use the Mann-Whitney $U$-test. This tests whether one *generation* tends to be taller than the other.

## FIELD PROJECT

Choose any one of the tables in Appendix D in the back of this book, and outline a test to check if the second digits are randomly odd or even. After your method has been approved by your instructor, proceed to carry out the test.

# Sample Test for Chapters 14, 15, and 16

1. Would you expect the following to have positive, negative, or zero correlation? Explain your choice.
   (a) Height and weight in a population of male college freshmen
   (b) Gas mileage (miles per gallon) and weight of automobiles in a population of American-made automobiles
   (c) Number of cigarettes smoked per day and number of days of work missed because of illness in a population of female factory workers
   (d) Number of polo ponies owned and amount spent on manure removal
   (e) The price of gold and value of the U.S. dollar
2. What would you say if you were told that someone had calculated the coefficient of correlation between height and weight for newborn infants to be 2.38?
3. For the following data find: (a) $r$, (b) $b$, and (c) the formula for the best-fitting line.

| City | Elevation $X$ | Mean High Temperature $Y$ for January 1978 |
|---|---|---|
| Bakersfield, Calif. | 475 | 62.7 |
| Dallas–Ft. Worth, Texas | 551 | 42.3 |
| Denver, Colo. | 5283 | 37.5 |
| Los Angeles, Calif. | 97 | 65.3 |
| San Francisco, Calif. | 8 | 58.3 |
| Seattle, Wash. | 400 | 48.7 |

   (d) Show the scattergram with the best-fitting line superimposed.
4. A production line factory is supposed to fill containers with at least 10 pounds of detergent powder. The standard deviation is supposed to be less than one-tenth of a pound. These are the weights of 10 packages pulled at random from a lot: 10.09, 10.02, 10.02, 10.03, 10.08, 9.95, 10.01, 10.07, 10.03, and 10.00 pounds. Is there evidence at $\alpha = 0.05$ that the line is working erratically?
5. Two different kinds of plastic are being tested for use in an artificial heart valve. Ten samples of each are tried in laboratory animals. Among the questions to be answered is, "Are the two materials equally variable in their working life span?" Answer the above question with $\alpha = 0.05$, based on the following data.

| | Life Span of Valve in Days | | | | | | | | | |
|---|---|---|---|---|---|---|---|---|---|---|
| Material 1 | 1459 | 1458 | 1459 | 1493 | 1415 | 1471 | 1494 | 1465 | 1459 | 1433 |
| Material 2 | 1491 | 1430 | 1426 | 1493 | 1401 | 1469 | 1470 | 1433 | 1401 | 1442 |

6. A refinery produces four grades of gasoline. The gasolines are tested in a sample motor, and certain exhaust pollutants are measured. Is there evidence at $\alpha = 0.05$ that the grades of gasoline differ in average amount of exhaust pollutants?

| | Units of Pollutant | | | |
|---|---|---|---|---|
| Grade A | 100 | 110 | 120 | 113 |
| Grade B | 105 | 115 | 125 | 100 |
| Grade C | 130 | 140 | 145 | 140 |
| Grade D | 135 | 140 | 160 | 150 |

7. People in Disasterville, New Jersey, were called at home and asked if they regularly watched the 6 o'clock news. If they said yes, they were asked whether they watched on channel 0 or on channel 1. The results for 118 calls were as follows:

37 watched channel 0

28 watched channel 1

53 did not usually watch the 6 o'clock news

Analyze these data by the sign test with $\alpha = 0.05$. Among the regular watchers, does the sign test indicate that the 6 o'clock news on channel 0 is more popular in the population? Use the 0.05 significance level.

# Arithmetic Review

It is advisable for you to know what arithmetic skills are needed for the material in this book. The following exercises are a sample of the types of problems you will be asked to do. If you have trouble with these, consult a text on arithmetic or see your instructor for suggestions. The answers to this review are given at the end of the text.

## EXERCISES

A-1 *Decimal Review* Evaluate the following.

(a) $4.1 + 5.03 + 14 + 0.6$ (b) $5 - 0.03$

(c) $0.05 - 0.004$ (d) $6.01(.2)$

(e) $5.12 + 1.96(2.3)$ (f) $\dfrac{0.3(0.04)}{10}$

(g) $\dfrac{17}{0.8}$ (h) $\dfrac{6.1}{12.2}$

(i) $\dfrac{0.18}{9}$ (j) $\dfrac{0}{4.3}$

A-2 *Size of Decimals*

(a) Which is larger, $-2.58$ or $-2.33$? List in order with the smallest first.

(b) 4.7, 0.41, 0.081, 0.6, and 4.51

(c) 0.41, $-0.273$, and 0.273

A-3 *Symbols* Answer true or false.

(a) $5 < 8$ (b) $5 \le 8$

(c) $8 > 9$ (d) $6 \le 6$

If $X$ is any whole number from 0 to 6 inclusive, list which values of $X$ satisfy the following conditions.

(e) $X > 4$ (f) $X \ge 4$

(g) $X < 5$ (h) $X \le 5$

(i) $2 < X < 5$ (j) $2 \le X \le 5$

A-4 *Percent Review*

(a) Change 0.05 to a percentage

(b) Change 0.003 to a percentage

(c) Change 37 percent to a decimal.

(d) Change 3.2 percent to a decimal.

(e) Change 3/8 to a decimal and to a percentage.

(f) Change 5/19 to a decimal and to a percentage.

(g) Find 23 percent of 50.

(h) Find 4 percent of 200.

(i) 15 is what percentage of 50?

(j) 27 is what percentage of 108?

A-5 *Signed Number Review* Evaluate the following.

(a) $4 + (-7) + (13) + (-25)$

(b) $-3.07(-5)$

(c) $-1.65(10)$

(d) $-\dfrac{16}{8}$

(e) $\dfrac{10}{(-2)}$

A-6 *Formulas* Evaluate the following.
Given $x = 7.3$, $y = 1.02$, $\sigma = 0.1$, $\mu = 11.4$, and $z = -2.33$, find:

(a) $xy$ (b) $\dfrac{x - \mu}{\sigma}$ (c) $\mu + z\sigma$

(d) Given $\mu = 3$, $\sigma = 2$, and $z = \pm 1.96$, find $\mu + z\sigma$.

(e) True or false? $\sqrt{pq/n} = \sqrt{\dfrac{pq}{n}}$

A-7 *Exponents* Evaluate the following.

(a) $(-4)^2$ (b) $(0.7)^3$

(c) $(\frac{1}{2})^4$ (d) $10(0.7)^3(0.3)^2$

A-8 *Square Roots* Compute the following.

(a) Given $n = 10$, $p = 0.4$, and $q = 0.6$, find $\sqrt{npq}$.

(b) $\sqrt{\dfrac{50 - \dfrac{(10)^2}{8}}{7}}$ (c) $\sqrt{\dfrac{(3)^2}{5} + \dfrac{(6)^2}{2}}$

# Using a Random Digit Table

How to choose a simple random sample of size $n$.

**simple random sample**

A sample of size $n$ from a population is called a **simple random sample** if it has the same probability of being picked as every other sample of size $n$.

One way to pick such a sample is to use a table of random numbers such as the one shown in Table B-1. This table was generated by a computer, and

**Table B-1 Table of Random Digits**

| *Row* | | | | | | | | | | | | | | | | | | | | | | | | | |
|---|---|---|---|---|---|---|---|---|---|---|---|---|---|---|---|---|---|---|---|---|---|---|---|---|---|
| 1 | 7 | 2 | 4 | 9 | 5 | 7 | 4 | 1 | 6 | 7 | 7 | 1 | 8 | 0 | 3 | 6 | 5 | 5 | 4 | 0 | 5 | 3 | 9 | 6 | 7 |
| 2 | 5 | 9 | 5 | 8 | 8 | 9 | 3 | 1 | 1 | 5 | 5 | 2 | 1 | 0 | 4 | 7 | 6 | 8 | 0 | 0 | 9 | 6 | 3 | 8 | 0 |
| 3 | 9 | 2 | 2 | 1 | 7 | 0 | 1 | 5 | 7 | 4 | 6 | 1 | 5 | 2 | 5 | 3 | 6 | 3 | 6 | 3 | 4 | 8 | 4 | 2 | 8 |
| 4 | 1 | 6 | 5 | 0 | 6 | 3 | 7 | 3 | 1 | 7 | 2 | 4 | 0 | 9 | 2 | 7 | 4 | 1 | 2 | 2 | 2 | 1 | 4 | 3 | 6 |
| 5 | 2 | 4 | 6 | 7 | 6 | 2 | 7 | 2 | 5 | 8 | 9 | 8 | 3 | 1 | 2 | 9 | 8 | 6 | 7 | 6 | 1 | 8 | 5 | 7 | 1 |
| 6 | 0 | 7 | 9 | 8 | 9 | 1 | 4 | 3 | 2 | 0 | 8 | 6 | 1 | 9 | 9 | 5 | 9 | 4 | 7 | 2 | 0 | 2 | 5 | 3 | 9 |
| 7 | 0 | 2 | 4 | 7 | 0 | 5 | 5 | 3 | 5 | 1 | 0 | 6 | 9 | 6 | 2 | 3 | 7 | 4 | 3 | 0 | 9 | 7 | 2 | 0 | 3 |
| 8 | 4 | 2 | 0 | 4 | 8 | 1 | 3 | 3 | 6 | 8 | 1 | 5 | 6 | 8 | 3 | 6 | 1 | 5 | 2 | 8 | 0 | 4 | 6 | 8 | 8 |
| 9 | 9 | 7 | 5 | 0 | 3 | 4 | 1 | 2 | 8 | 0 | 9 | 5 | 1 | 7 | 4 | 9 | 9 | 6 | 2 | 7 | 8 | 7 | 1 | 5 | 6 |
| 10 | 6 | 0 | 2 | 2 | 7 | 0 | 7 | 2 | 2 | 6 | 9 | 6 | 3 | 9 | 9 | 0 | 9 | 0 | 1 | 9 | 3 | 9 | 7 | 6 | 7 |
| 11 | 8 | 5 | 0 | 2 | 3 | 1 | 8 | 8 | 1 | 5 | 6 | 8 | 5 | 5 | 5 | 7 | 4 | 5 | 6 | 5 | 7 | 6 | 2 | 6 | 5 |
| 12 | 8 | 4 | 4 | 0 | 6 | 4 | 2 | 0 | 3 | 2 | 5 | 6 | 3 | 6 | 8 | 8 | 4 | 1 | 7 | 5 | 4 | 4 | 3 | 3 | 6 |
| 13 | 0 | 3 | 1 | 7 | 5 | 4 | 2 | 1 | 3 | 3 | 7 | 9 | 2 | 7 | 8 | 6 | 9 | 5 | 1 | 6 | 5 | 8 | 9 | 8 | 8 |
| 14 | 8 | 7 | 5 | 1 | 4 | 9 | 9 | 2 | 5 | 6 | 0 | 2 | 4 | 0 | 7 | 2 | 4 | 6 | 0 | 5 | 1 | 5 | 4 | 7 | 2 |
| 15 | 4 | 1 | 7 | 0 | 8 | 8 | 9 | 1 | 9 | 7 | 2 | 4 | 2 | 0 | 9 | 7 | 2 | 7 | 1 | 0 | 4 | 8 | 8 | 3 | 1 |
| 16 | 7 | 1 | 3 | 6 | 7 | 1 | 5 | 1 | 9 | 9 | 7 | 8 | 1 | 5 | 1 | 1 | 6 | 9 | 8 | 0 | 2 | 6 | 0 | 7 | 2 |
| 17 | 7 | 4 | 7 | 9 | 7 | 5 | 3 | 4 | 3 | 5 | 7 | 3 | 5 | 7 | 2 | 8 | 0 | 5 | 5 | 5 | 7 | 0 | 4 | 1 | 6 |
| 18 | 1 | 8 | 9 | 7 | 0 | 7 | 3 | 8 | 2 | 7 | 7 | 6 | 9 | 1 | 4 | 9 | 6 | 6 | 9 | 4 | 9 | 6 | 7 | 6 | 8 |
| 19 | 6 | 3 | 7 | 6 | 4 | 4 | 5 | 0 | 0 | 1 | 6 | 1 | 3 | 1 | 6 | 0 | 3 | 6 | 9 | 1 | 1 | 6 | 3 | 0 | 6 |
| 20 | 9 | 3 | 7 | 0 | 0 | 8 | 3 | 3 | 1 | 3 | 2 | 6 | 0 | 1 | 3 | 4 | 6 | 9 | 1 | 7 | 3 | 8 | 4 | 7 | 8 |

the digits in it may be assumed to be distributed randomly, as if picked from a hat containing billions of copies of each digit from 0 to 9, all of which are equally likely to be picked. Many calculators have a key which will produce a sequence of random digits, which can be used for the same purpose.

For ease of reference, each row of the table is given a label. These row numbers are not part of the table.

## How to Use the Table

### ❑ EXAMPLE

Use the random digit table to pick a simple random sample of size 5 from a population which contains 97 members. Suppose, for example, there are 97 students in a lecture hall, and you want to pick 5 of them "at random."

**SOLUTION**

*Step 1.* Number the students from 0 to 96 in any convenient way, perhaps alphabetically, or by seat in the hall. It doesn't matter. Starting with 0 allows the zeros in the table to be fully used.

*Step 2.* Because 96, the largest number assigned to a student, contains **2 digits** we think of the table as a collection of 2-digit numbers, ranging from 00 to 96. You can start anywhere in the table and proceed straight through. For illustration, we start at the beginning of line 6 and proceed to the right. (Note: Because the digits are randomly distributed, we could start anywhere and proceed in any direction.)

We get the following set of 2-digit numbers:

07 98 91 43 20 86 19 95 . . .

*Step 3.* Since we want a sample of size 5, we take the first 5 of these 2-digit numbers that are useful:

07 Student 7

98 No good; there is no such student.

91 Student 91

43 Student 43

20 Student 20

86 Student 86

Thus, the sample consists of students 7, 91, 43, 20, and 86.

*Another Approach* Sometimes there are many numbers in the table that are not useful. For instance, if you have a population of size 20, then the only useful 2-digit numbers are 00 to 19. A more efficient use of the table would then give each member of the population **several** numbers by adding 20 to the original number. In this case,

Student 00 could be assigned 00, 20, 40, 60, and 80

Student 01 could be assigned 01, 21, 41, 61, and 81

.

.

.

Student 19 could be assigned 19, 39, 59, 79, and 99 ❑

## EXERCISES

1. Here is a list of 16 patients from whom you want to select 8 to receive the placebo in an experiment. [The rest will get the real drug.]

| | | | | | | | |
|---|---|---|---|---|---|---|---|
| Bernard | Bess | Betty | Isadore | Jack | Leda | Marilyn | Martin |
| Max | Meyer | Molly | Robert | Rose | Sue | Sylvia | Howard |

   Use the random digit table starting in line 1 to choose the patients for placebo. Describe how you assigned the original numbers to the patients. Then, describe how you used the table.

2. Assume there are 1548 names in the membership list of an organization. Describe how you would use the random number table starting at line 2 to pick 5 of these names at random. Hint: Use 0000 as the lowest number.

APPENDIX

# C

# Probability

This appendix assumes the knowledge of Chapter 4. We are going to consider probability in a little more detail. In Chapter 4 in order to compute a probability we listed the total number of possible outcomes of an event and found the proportion of outcomes that were favorable to that event. Now we will use this method from Chapter 4 and then show another approach to obtain the same results.

## Independent and Dependent Events

independent

Two events are said to be **independent** if the occurrence or nonoccurrence of one has no effect on the occurrence or nonoccurrence of the other.

❑ EXAMPLE C-1

Let us consider 2 machines which operate independently of each other. This means that if 1 machine fails to function, it will have no effect on whether the other machine works or fails. Let us suppose that each machine fails half of the days, so that $P(A) = P(\text{machine } A \text{ fails}) = 1/2$ and $P(B) = P(\text{machine } B \text{ fails}) = 1/2$. Let us find the probability that both machines fail on the same day.

### SOLUTION 1

Using the method of Chapter 4, since the machines operate independently, if machine $A$ works, then machine $B$ may work or fail, and similarly if machine $A$ fails, machine $B$ may work or fail. Since there are 2 possible outcomes for machine $A$ and 2 possible outcomes for machine $B$, there are $2 \times 2 = 4$ possible outcomes. They are listed in the table.

| Outcome | Machine $A$ | Machine $B$ |
|---|---|---|
| 1 | Work | Work |
| 2 | Work | Fail |
| 3 | Fail | Work |
| 4 | Fail | Fail |

Since each of the above possibilities is equally likely, we can see that the probability of both failing is 1/4. We write:

$$P(\text{machine } A \text{ fails and machine } B \text{ fails}) = P(A \text{ and } B) = \frac{1}{4}$$

### SOLUTION 2

We could have reached this result by multiplying the probability that machine $A$ fails times the probability that machine $B$ fails. That is,

$$P(A \text{ and } B) = P(A)P(B) = \frac{1}{2} \cdot \frac{1}{2} = \frac{1}{4}$$

When 2 events are independent, we can always multiply their probabilities together in order to find the probability that they will occur simultaneously. In fact, many authors use this as the definition of independence. They say that 2 events are independent if the probability of their occurring together equals the product of the probabilities of each occurring separately. In symbols we write: events $A$ and $B$ are independent if and only if

$$P(A \text{ and } B) = P(A)P(B)$$ ❑

---

We cannot use this rule for events that are dependent. Let us consider a second example.

---

## ❑ EXAMPLE C-2

Let us suppose that someone decides to play Russian roulette with a six-shooter in which the chambers are numbered 1 through 6. If he places a bullet in 1 chamber, spins the cylinder, and then pulls the trigger twice in a row, find the probability that he lives.

## SOLUTION 1

By the method of Chapter 4, if the cylinder lands so that chamber 3 will be fired first, then chamber 4 will be fired next. Similarly, if chamber 6 is fired first, then chamber 1 is fired next. Thus, there are 6 possible outcomes.

| First Try | Second Try |
|---|---|
| Chamber 1 | Chamber 2 |
| Chamber 2 | Chamber 3 |
| Chamber 3 | Chamber 4 |
| Chamber 4 | Chamber 5 |
| Chamber 5 | Chamber 6 |
| Chamber 6 | Chamber 1 |

Notice that each chamber is listed twice. The bullet is in only 1 chamber. Therefore, there are 4 outcomes of the 6 in which both chambers are empty. If we let $E_1$ stand for the event that the first chamber tried is empty, and $E_2$ stand for the event that the second chamber tried is empty, we have

$$P(\text{both chambers are empty}) = P(E_1 \text{ and } E_2) = \frac{4}{6} = \frac{2}{3}$$

## SOLUTION 2

Let us use $E_1$ and $E_2$ as defined above and calculate the probability that the person will live by computing the probabilities associated with each pull of the trigger separately. The probability that the first chamber is empty is 5/6 since there is only 1 bullet. We write:

$$P(\text{first chamber tried is empty}) = P(E_1) = \frac{5}{6}$$

The probability that the second chamber is empty depends on what happened when we pulled the trigger the first time. If we got the bullet on the first try, we know for sure that the next chamber is empty, but we probably would not care. On the other hand, if the first chamber selected was empty, then of the remaining 5 chambers, 4 are empty. Thus, we write:

$$P(\text{second chamber is empty given first chamber is empty}) = P(E_2 \text{ given } E_1) = \frac{4}{5}$$

**dependent**

Thus, we can say that event $E_2$ is **dependent** on event $E_1$, since the occurrence or nonoccurrence of $E_1$ will affect the occurrence or nonoccurrence of $E_2$.

To find the probability that both chambers are empty, we multiply the probability that $E_1$ occurs times the probability that $E_2$ occurs given that $E_1$ has already occurred:

$$P(E_1 \text{ and } E_2) = P(E_1)P(E_2 \text{ given } E_1) = \frac{5}{6} \cdot \frac{4}{5} = \frac{4}{6} = \frac{2}{3}$$

Note that this is the same result we obtained in Solution 1. ❑

We would also like to mention that we could have solved the problem by considering the second chamber. The probability that the second chamber is empty is 5/6, since there is only 1 bullet. We write:

$$P(\text{second chamber is empty}) = P(E_2) = \frac{5}{6}$$

Given that the second chamber is empty, of the 5 remaining chambers only 1 has a bullet, and so the probability that our first chamber is empty is 4/5. We write:

$$P(\text{first chamber is empty given second chamber is empty}) = P(E_1 \text{ given } E_2) = \frac{2}{3}$$

Therefore,

$$P(E_1 \text{ and } E_2) = P(E_2)P(E_1 \text{ given } E_2) = \frac{5}{6} \cdot \frac{4}{5} = \frac{4}{6} = \frac{2}{3}$$

Here we see that $E_1$ now depends upon the outcome of $E_2$.

In general, if $A$ and $B$ are dependent events, the probability that they occur together is found by

$$P(A \text{ and } B) = P(A)P(B \text{ given } A)$$

or

$$P(A \text{ and } B) = P(B)P(A \text{ given } B)$$

## Replacement and Nonreplacement

Let us consider another type of problem. Suppose we have an urn which contains 3 white balls and 2 red balls. What is the probability that we randomly select 2 red balls in a row? The answer to the question depends upon whether or not we replace the first ball before drawing the second ball.

### ❑ EXAMPLE C-3

Replacement

**Replacement**

Suppose that after we select the first ball we replace it and then select another ball.

SOLUTION

Since the outcome of the first selection has no effect on the outcome of the second selection, the events are independent. Thus, we can use the formula for independent events. Since 2 of the 5 balls are red, the probability of selecting a red is $P(\text{red}) = 2/5$. Thus, the probability of selecting 2 reds is

$$P(\text{red and red}) = P(\text{red})P(\text{red}) = \frac{2}{5} \cdot \frac{2}{5} = \frac{4}{25}$$

❑

❑ EXAMPLE C-4

Nonreplacement

**Nonreplacement**

Let us reconsider the same problem of getting 2 reds from the urn with 3 white and 2 red balls, but this time we will not replace the first ball selected before we randomly select the second ball. The probability of getting a red on the first selection, $P(\text{red})$, is still 2/5, but the probability of getting a red on the second selection depends on what occurred on the first selection.

SOLUTION

If a red ball was selected first, there would be 4 balls remaining of which only 1 was a red. Thus, the probability of a red given a red on the first selection is 1 chance in 4; we write $P(\text{red}_2 \text{ given red}_1) = 1/4$. Thus, according to the rule for dependent events, the probability of selecting 2 reds will be

$$P(\text{red}_1 \text{ and red}_2) = P(\text{red}_1)P(\text{red}_2 \text{ given red}_1)$$
$$= \frac{2}{5} \cdot \frac{1}{4} = \frac{2}{20} = \frac{1}{10}$$

❑

Let us compare our answers in Examples C-3 and C-4. We found that the probability of selecting 2 reds was $4/25 = 16/100 = 0.16$ when we replaced the first ball, and that the probability of 2 reds was $1/10 = 0.10$ when we did not replace the first ball. You can see that there is a difference of $0.06 = 6/100$ between the two methods. As we increase the number of balls in the urn, the difference becomes less pronounced. In fact, for a large number of balls the difference is negligible, as shown in Example C-5:

❑ EXAMPLE C-5

Suppose our urn contains 5000 balls of which 3000 are white and 2000 are red. Find the probability of randomly selecting 2 red balls.

SOLUTION 1

*Replacement*
Since $P(\text{red}) = 2000/5000 = 2/5$ and each selection is independent, we have

$$P(\text{red and red}) = \frac{2}{5} \cdot \frac{2}{5} = \frac{4}{25} = 0.16$$

SOLUTION 2

*Nonreplacement*
On the first selection $P(\text{red}) = 2000/5000 = 2/5$, and the probability of getting a red on the second selection given a red on the first selection is $P(\text{red given red}) = 1999/4999$ since there is 1 less red ball in our urn. Thus

$$\begin{aligned} P(\text{red and red}) &= P(\text{red})P(\text{red given red}) \\ &= \frac{2}{5} \cdot \frac{1999}{4999} \\ &= \frac{3998}{24{,}995} = 0.15995 \end{aligned}$$

You can see that the difference in our answers using the 2 methods is 0.00005 or 1/20,000. ❑

---

The fact that two answers are close to each other becomes important when we take random samples from large populations because in theory we assume replacement (independence), but in practice we often use nonreplacement. For example, if one were asking people whom they planned to vote for, one would not usually ask the same person twice, and so we are not placing this person back into the population.

## Mutually Exclusive Events

Let us consider a regular deck of 52 playing cards. Since there are 4 kings and 4 queens, we can write $P(\text{randomly selecting a king}) = P(K) = 4/52 = 1/13$ and $P(\text{randomly selecting a queen}) = P(Q) = 4/52 = 1/13$. If we wanted to know the probability of selecting a king or a queen, we would count 8 in all so that $P(\text{randomly selecting a king or a queen}) = P(K \text{ or } Q) = 8/52 = 2/13$. (This problem asks the probability of a king *or* a queen on one draw, and the answer is 2/13. The probability of a king *and* a queen on one draw is zero. Do not confuse the word "and" with the word "or.") Note that we could have obtained this last result by adding the individual probabilities of randomly selecting a king and randomly selecting a queen. Thus,

$$P(K \text{ or } Q) = P(K) + P(Q) = \frac{1}{13} + \frac{1}{13} = \frac{2}{13}$$

We must be careful because this procedure will not always work. Suppose we wanted to know the probability of selecting a heart or a jack. Out of the 52 cards there are 13 hearts, 1 of which is a jack, and there are 3 jacks of other suits. Thus there are 16 favorable outcomes. We write $P$(randomly selecting a heart or a jack) = $P(H \text{ or } J) = 16/52 = 4/13$. If we consider hearts and jacks separately, we have $P$(randomly selecting a heart) = $P(H) = 13/52$ since there are 13 hearts and $P$(randomly selecting a jack) = $P(J) = 4/52$ since there are 4 jacks. Note that if we add these results together we do *not* get the right answer of 16/52, since $P(H) + P(J) = 13/52 + 4/52 = 17/52$.

We got the wrong answer because we counted the jack of hearts twice. We counted it once as a heart and once as a jack. In order to count the jack of hearts only once we must subtract the probability of getting the jack of hearts, which is 1/52. We write $P$(jack of hearts) = $P$(randomly selecting a heart and a jack) = $P(H \text{ and } J) = 1/52$. Thus, our formula becomes

$$P(H \text{ or } J) = P(H) + P(J) - P(H \text{ and } J) = \frac{13}{52} + \frac{4}{52} - \frac{1}{52} = \frac{16}{52}$$

Let us compare the two situations. In the first problem there was no card that was both a king and a queen. Thus, we say that selecting kings and selecting queens are **mutually exclusive events.**

**mutually exclusive events**

Formally, we say that two events are mutually exclusive if the probability that they occur simultaneously is zero. We write $P$(selecting a king and a queen in a single draw) = $P(K \text{ and } Q) = 0$. Since the probability of simultaneous occurrence is always zero for mutually exclusive events, many authors use this fact as the definition and say that events $K$ and $Q$ are mutually exclusive if and only if $P(K \text{ and } Q) = 0$.

In the second problem there was a card which is both a heart and a jack, and so we say that the events of selecting a heart and of selecting a jack are not mutually exclusive. Since $P$(jack of hearts) = $P(J \text{ and } H) = 1/52 \neq 0$, we say that the events $J$ and $H$ are not mutually exclusive.

We summarize by considering two events $A$ and $B$. We have seen that

$$P(A \text{ or } B) = P(A) + P(B) - P(A \text{ and } B)$$

If $A$ and $B$ are mutually exclusive, then

$$P(A \text{ and } B) = 0$$

and the formula reduces to

$$P(A \text{ or } B) = P(A) + P(B)$$

## STUDY AIDS

### Vocabulary

1. Independent event
2. Dependent event
3. Replacement
4. Nonreplacement
5. Mutually exclusive events

**Symbols**

1. $P(A \text{ and } B)$
2. $P(A \text{ given } B)$
3. $P(A \text{ or } B)$

**Formulas**

1. $P(A \text{ and } B) = P(A)P(B)$, when $A$ and $B$ are independent
2. $P(A \text{ and } B) = P(A)P(B \text{ given } A)$ or $P(A \text{ and } B) = P(B)P(A \text{ given } B)$
3. $P(A \text{ or } B) = P(A) + P(B) - P(A \text{ and } B)$
4. $P(A \text{ or } B) = P(A) + P(B)$, when $A$ and $B$ are mutually exclusive

## EXERCISES

C-1 Decide intuitively if the following pairs of events are independent or dependent.

(a) Tossing a penny and tossing a dime.

(b) Spinning a spinner twice.

(c) Selecting 2 toadstools from a bag containing 3 mushrooms and 2 toadstools (without replacement).

(d) Part (c) above but with replacement.

C-2 Decide intuitively whether the following pairs of events are independent or dependent.

(a) Playing Russian roulette twice and spinning the cylinder each time.

(b) Playing Russian roulette twice and spinning the cylinder only the first time.

(c) Dealing 2 cards from an ordinary deck.

(d) Rolling 2 dice.

(e) Rolling 1 die twice.

C-3 $P(A) = 1/2$ and $P(B) = 1/3$, while $P(A \text{ and } B)$ is not 1/6. Explain how this is possible.

C-4 Are the following outcomes of the given events mutually exclusive or not?

(a) in picking a card from an ordinary deck, getting a jack and a spade.

(b) in picking a card from an ordinary deck, getting a jack and a queen.

(c) In predicting the sex of your next 2 children, getting two the same sex and at least 1 boy.

(d) In predicting the sex of your next 2 children, getting both girls, and at least 1 boy.

C-5 Are the following outcomes mutually exclusive or not?

(a) Being a surgeon and being a woman.

(b) Being a male and being a mother.

(c) Given that you own 1 pet, owning a cat and a dog.

(d) Given that you own more than 1 pet, owning a boa constrictor and a prairie dog.

C-6 If $P(A \text{ or } B) = 5/7$, $P(A) = 3/7$, and $P(B) = 2/7$

(a) Find $P(A \text{ and } B)$.

(b) Are $A$ and $B$ mutually exclusive?

C-7 If $P(A \text{ or } B) = 0.62$ and $P(A) = 0.41$ and $P(B) = 0.41$:

(a) Find $P(A \text{ and } B)$.

(b) Are $A$ and $B$ mutually exclusive?

C-8 Mauro prefers to date Elena, but if she turns him down, he calls Jill. The probability that Elena says yes is 0.4. If the probability that both Elena and Jill say no is 0.2, find the probability that Jill says no given that Elena said no.

C-9 Two airplane engines operate independently. The plane can fly if either engine works. If engine 1 fails once every 100 flights and engine 2 fails once every 10,000 flights, find the probability that both engines will fail on the same flight.

C-10 In an assortment of 30 chocolates it is impossible to tell the fruits from the nuts. There are 20 fruits and 10 nuts, and the type is identified under each candy. Two chocolates are picked at random.

*Case 1* Replacement:

(a) Find the probability of 2 nuts.

(b) Find the probability of a fruit first and then a nut.

(c) Find the probability of 1 of each in any order.

*Case 2* Without replacement:

(d) Repeat part (a).

(e) Repeat part (b).

(f) Repeat part (c).

C-11 A deck of playing cards is shuffled, and 2 cards are randomly selected.

*Case 1* Replacement (the first card is returned to the deck prior to the second pick and the deck is reshuffled):

(a) Find the probability of 2 hearts.

(b) Find the probability of an ace and then a 7.

(c) Find the probability of an ace and a 7 in any order.

(d) Find the probability of 2 picture cards.

*Case 2* Without replacement:

(e) Repeat part (a).

(f) Repeat part (b).

(g) Repeat part (c).

(h) Repeat part (d).

C-12 Two machines work according to the following rules. The probability that machine $A$ will fail is 1/3. The probability that machine $B$ will fail, given that machine $A$ has failed, is 1/8.

(a) Find the probability that machines $A$ and $B$ both fail.

(b) If the probability that machine $B$ will fail is 1/4, then using the results from part (a), find the probability that machine $A$ will fail, given that machine $B$ has failed.

C-13 A coin-tossing game is played as follows. To win the game you must toss exactly 2 heads. If the first toss produces heads, then the coin is tossed 2 more times. However, if the first toss produces tails, then the coin is tossed 3 more times.

(a) What is the probability of winning, given that the first toss is heads?

(b) What is the probability of winning, given that the first toss is tails?

C-14 What is the probability of picking 3 cards from a shuffled deck such that the first card is an ace, the second card is the king of the same suit, and the third card is the queen of the same suit?

C-15 What is the probability of picking 2 cards at random without replacement from a deck with an outcome of an ace and a king of the same suit, if either card can be picked first?

C-16 From a shuffled deck you draw 1 card.

(a) Find $P$(ace or spade).

(b) Given that it is a picture card, find $P$(10 or jack).

C-17 If you are in an automobile accident, find the probability that you are injured and do not collect insurance if $P$(injured) $= 0.4$ and $P$(not collecting, given that you are injured) $= 0.2$.

APPENDIX D

# Tables

Table D-1 Pascal's Triangle, $\binom{n}{S}$

| n \ S | 0 | 1 | 2 | 3 | 4 | 5 | 6 | 7 | 8 | 9 | 10 | 11 | 12 | 13 | 14 | 15 | 16 | 17 | 18 | 19 | 20 | S / n |
|---|---|---|---|---|---|---|---|---|---|---|---|---|---|---|---|---|---|---|---|---|---|---|
| 0 | 1 | | | | | | | | | | | | | | | | | | | | | 0 |
| 1 | 1 | 1 | | | | | | | | | | | | | | | | | | | | 1 |
| 2 | 1 | 2 | 1 | | | | | | | | | | | | | | | | | | | 2 |
| 3 | 1 | 3 | 3 | 1 | | | | | | | | | | | | | | | | | | 3 |
| 4 | 1 | 4 | 6 | 4 | 1 | | | | | | | | | | | | | | | | | 4 |
| 5 | 1 | 5 | 10 | 10 | 5 | 1 | | | | | | | | | | | | | | | | 5 |
| 6 | 1 | 6 | 15 | 20 | 15 | 6 | 1 | | | | | | | | | | | | | | | 6 |
| 7 | 1 | 7 | 21 | 35 | 35 | 21 | 7 | 1 | | | | | | | | | | | | | | 7 |
| 8 | 1 | 8 | 28 | 56 | 70 | 56 | 28 | 8 | 1 | | | | | | | | | | | | | 8 |
| 9 | 1 | 9 | 36 | 84 | 126 | 126 | 84 | 36 | 9 | 1 | | | | | | | | | | | | 9 |
| 10 | 1 | 10 | 45 | 120 | 210 | 252 | 210 | 120 | 45 | 10 | 1 | | | | | | | | | | | 10 |
| 11 | 1 | 11 | 55 | 165 | 330 | 462 | 462 | 330 | 165 | 55 | 11 | 1 | | | | | | | | | | 11 |
| 12 | 1 | 12 | 66 | 220 | 495 | 792 | 924 | 792 | 495 | 220 | 66 | 12 | 1 | | | | | | | | | 12 |
| 13 | 1 | 13 | 78 | 286 | 715 | 1287 | 1716 | 1716 | 1287 | 715 | 286 | 78 | 13 | 1 | | | | | | | | 13 |
| 14 | 1 | 14 | 91 | 364 | 1001 | 2002 | 3003 | 3432 | 3003 | 2002 | 1001 | 364 | 91 | 14 | 1 | | | | | | | 14 |
| 15 | 1 | 15 | 105 | 455 | 1365 | 3003 | 5005 | 6435 | 6435 | 5005 | 3003 | 1365 | 455 | 105 | 15 | 1 | | | | | | 15 |
| 16 | 1 | 16 | 120 | 560 | 1820 | 4368 | 8008 | 11,440 | 12,870 | 11,440 | 8008 | 4368 | 1820 | 560 | 120 | 16 | 1 | | | | | 16 |
| 17 | 1 | 17 | 136 | 680 | 2380 | 6188 | 12,376 | 19,448 | 24,310 | 24,310 | 19,448 | 12,376 | 6188 | 2380 | 680 | 136 | 17 | 1 | | | | 17 |
| 18 | 1 | 18 | 153 | 816 | 3060 | 8568 | 18,564 | 31,824 | 43,758 | 48,620 | 43,758 | 31,824 | 18,564 | 8568 | 3060 | 816 | 153 | 18 | 1 | | | 18 |
| 19 | 1 | 19 | 171 | 969 | 3876 | 11,628 | 27,132 | 50,388 | 75,582 | 92,378 | 92,378 | 75,582 | 50,388 | 27,132 | 11,628 | 3876 | 969 | 171 | 19 | 1 | | 19 |
| 20 | 1 | 20 | 190 | 1140 | 4845 | 15,504 | 38,760 | 77,520 | 125,970 | 167,960 | 184,756 | 167,960 | 125,970 | 77,520 | 38,760 | 15,504 | 4845 | 1140 | 190 | 20 | 1 | 20 |

## Table D-2 Binomial Probabilities

$n=2$

| S \ p | .05 | .10 | .20 | .30 | .40 | .50 | .60 | .70 | .80 | .90 | .95 |
|---|---|---|---|---|---|---|---|---|---|---|---|
| 0 | .9025 | .8100 | .6400 | .4900 | .3600 | .2500 | .1600 | .0900 | .0400 | .0100 | .0025 |
| 1 | .0950 | .1800 | .3200 | .4200 | .4800 | .5000 | .4800 | .4200 | .3200 | .1800 | .0950 |
| 2 | .0025 | .0100 | .0400 | .0900 | .1600 | .2500 | .3600 | .4900 | .6400 | .8100 | .9025 |

$n=3$

| S \ p | .05 | .10 | .20 | .30 | .40 | .50 | .60 | .70 | .80 | .90 | .95 |
|---|---|---|---|---|---|---|---|---|---|---|---|
| 0 | .8574 | .7290 | .5120 | .3430 | .2160 | .1250 | .0640 | .0270 | .0080 | .0010 | .0001 |
| 1 | .1354 | .2430 | .3840 | .4410 | .4320 | .3750 | .2880 | .1890 | .0960 | .0270 | .0071 |
| 2 | .0071 | .0270 | .0960 | .1890 | .2880 | .3750 | .4320 | .4410 | .3840 | .2430 | .1354 |
| 3 | .0001 | .0010 | .0080 | .0270 | .0640 | .1250 | .2160 | .3430 | .5120 | .7290 | .8574 |

$n=4$

| S \ p | .05 | .10 | .20 | .30 | .40 | .50 | .60 | .70 | .80 | .90 | .95 |
|---|---|---|---|---|---|---|---|---|---|---|---|
| 0 | .8145 | .6561 | .4096 | .2401 | .1296 | .0625 | .0256 | .0081 | .0016 | .0001 | .0000 |
| 1 | .1715 | .2916 | .4096 | .4116 | .3456 | .2500 | .1536 | .0756 | .0256 | .0036 | .0005 |
| 2 | .0135 | .0486 | .1536 | .2646 | .3456 | .3750 | .3456 | .2646 | .1536 | .0486 | .0135 |
| 3 | .0005 | .0036 | .0256 | .0756 | .1536 | .2500 | .3456 | .4116 | .4096 | .2916 | .1715 |
| 4 | .0000 | .0001 | .0016 | .0081 | .0256 | .0625 | .1296 | .2401 | .4096 | .6561 | .8145 |

$n=5$

| S \ p | .05 | .10 | .20 | .30 | .40 | .50 | .60 | .70 | .80 | .90 | .95 |
|---|---|---|---|---|---|---|---|---|---|---|---|
| 0 | .7738 | .5905 | .3277 | .1681 | .0778 | .0313 | .0102 | .0024 | .0003 | .0000 | .0000 |
| 1 | .2036 | .3281 | .4096 | .3602 | .2592 | .1563 | .0768 | .0284 | .0064 | .0005 | .0000 |
| 2 | .0214 | .0729 | .2048 | .3087 | .3456 | .3125 | .2304 | .1323 | .0512 | .0081 | .0011 |
| 3 | .0011 | .0081 | .0512 | .1323 | .2304 | .3125 | .3456 | .3087 | .2048 | .0729 | .0214 |
| 4 | .0000 | .0005 | .0064 | .0284 | .0768 | .1563 | .2592 | .3602 | .4096 | .3281 | .2036 |
| 5 | .0000 | .0000 | .0003 | .0024 | .0102 | .0313 | .0778 | .1681 | .3277 | .5905 | .7738 |

$n=6$

| S \ p | .05 | .10 | .20 | .30 | .40 | .50 | .60 | .70 | .80 | .90 | .95 |
|---|---|---|---|---|---|---|---|---|---|---|---|
| 0 | .7351 | .5314 | .2621 | .1176 | .0467 | .0156 | .0041 | .0007 | .0001 | .0000 | .0000 |
| 1 | .2321 | .3543 | .3932 | .3025 | .1866 | .0938 | .0369 | .0102 | .0015 | .0001 | .0000 |
| 2 | .0305 | .0984 | .2458 | .3241 | .3110 | .2344 | .1382 | .0595 | .0154 | .0012 | .0001 |
| 3 | .0021 | .0146 | .0819 | .1852 | .2765 | .3125 | .2765 | .1852 | .0819 | .0146 | .0021 |
| 4 | .0001 | .0012 | .0154 | .0595 | .1382 | .2344 | .3110 | .3241 | .2458 | .0984 | .0305 |
| 5 | .0000 | .0001 | .0015 | .0102 | .0369 | .0938 | .1866 | .3025 | .3932 | .3543 | .2321 |
| 6 | .0000 | .0000 | .0001 | .0007 | .0041 | .0156 | .0467 | .1176 | .2621 | .5314 | .7351 |

$n=7$

| S \ p | .05 | .10 | .20 | .30 | .40 | .50 | .60 | .70 | .80 | .90 | .95 |
|---|---|---|---|---|---|---|---|---|---|---|---|
| 0 | .6983 | .4783 | .2097 | .0824 | .0280 | .0078 | .0016 | .0002 | .0000 | .0000 | .0000 |
| 1 | .2573 | .3720 | .3670 | .2471 | .1306 | .0547 | .0172 | .0036 | .0004 | .0000 | .0000 |
| 2 | .0406 | .1240 | .2753 | .3177 | .2613 | .1641 | .0774 | .0250 | .0043 | .0002 | .0000 |
| 3 | .0036 | .0230 | .1147 | .2269 | .2903 | .2734 | .1935 | .0972 | .0287 | .0026 | .0002 |
| 4 | .0002 | .0026 | .0287 | .0972 | .1935 | .2734 | .2903 | .2269 | .1147 | .0230 | .0036 |
| 5 | .0000 | .0002 | .0043 | .0250 | .0774 | .1641 | .2613 | .3177 | .2753 | .1240 | .0406 |
| 6 | .0000 | .0000 | .0004 | .0036 | .0172 | .0547 | .1306 | .2471 | .3670 | .3720 | .2573 |
| 7 | .0000 | .0000 | .0000 | .0002 | .0016 | .0078 | .0280 | .0824 | .2097 | .4783 | .6983 |

**Table D-2 (*continued*)**

$n=8$

| S \ p | .05 | .10 | .20 | .30 | .40 | .50 | .60 | .70 | .80 | .90 | .95 |
|---|---|---|---|---|---|---|---|---|---|---|---|
| 0 | .6634 | .4305 | .1678 | .0576 | .0168 | .0039 | .0007 | .0001 | .0000 | .0000 | .0000 |
| 1 | .2793 | .3826 | .3355 | .1977 | .0896 | .0313 | .0079 | .0012 | .0001 | .0000 | .0000 |
| 2 | .0515 | .1488 | .2936 | .2965 | .2090 | .1094 | .0413 | .0100 | .0011 | .0000 | .0000 |
| 3 | .0054 | .0331 | .1468 | .2541 | .2787 | .2188 | .1239 | .0467 | .0092 | .0004 | .0000 |
| 4 | .0004 | .0046 | .0459 | .1361 | .2322 | .2734 | .2322 | .1361 | .0459 | .0046 | .0004 |
| 5 | .0000 | .0004 | .0092 | .0467 | .1239 | .2188 | .2787 | .2541 | .1468 | .0331 | .0054 |
| 6 | .0000 | .0000 | .0011 | .0100 | .0413 | .1094 | .2090 | .2965 | .2936 | .1488 | .0515 |
| 7 | .0000 | .0000 | .0001 | .0012 | .0079 | .0313 | .0896 | .1977 | .3355 | .3826 | .2793 |
| 8 | .0000 | .0000 | .0000 | .0001 | .0007 | .0039 | .0168 | .0576 | .1678 | .4305 | .6634 |

$n=9$

| S \ p | .05 | .10 | .20 | .30 | .40 | .50 | .60 | .70 | .80 | .90 | .95 |
|---|---|---|---|---|---|---|---|---|---|---|---|
| 0 | .6302 | .3874 | .1342 | .0404 | .0101 | .0020 | .0003 | .0000 | .0000 | .0000 | .0000 |
| 1 | .2985 | .3874 | .3020 | .1556 | .0605 | .0176 | .0035 | .0004 | .0000 | .0000 | .0000 |
| 2 | .0629 | .1722 | .3020 | .2668 | .1612 | .0703 | .0212 | .0039 | .0003 | .0000 | .0000 |
| 3 | .0077 | .0446 | .1762 | .2668 | .2508 | .1641 | .0743 | .0210 | .0028 | .0001 | .0000 |
| 4 | .0006 | .0074 | .0661 | .1715 | .2508 | .2461 | .1672 | .0735 | .0165 | .0008 | .0000 |
| 5 | .0000 | .0008 | .0165 | .0735 | .1672 | .2461 | .2508 | .1715 | .0661 | .0074 | .0006 |
| 6 | .0000 | .0001 | .0028 | .0210 | .0743 | .1641 | .2508 | .2668 | .1762 | .0446 | .0077 |
| 7 | .0000 | .0000 | .0003 | .0039 | .0212 | .0703 | .1612 | .2668 | .3020 | .1722 | .0629 |
| 8 | .0000 | .0000 | .0000 | .0004 | .0035 | .0176 | .0605 | .1556 | .3020 | .3874 | .2985 |
| 9 | .0000 | .0000 | .0000 | .0000 | .0003 | .0020 | .0101 | .0404 | .1342 | .3874 | .6302 |

$n=10$

| S \ p | .05 | .10 | .20 | .30 | .40 | .50 | .60 | .70 | .80 | .90 | .95 |
|---|---|---|---|---|---|---|---|---|---|---|---|
| 0 | .5987 | .3487 | .1074 | .0282 | .0060 | .0010 | .0001 | .0000 | .0000 | .0000 | .0000 |
| 1 | .3151 | .3874 | .2684 | .1211 | .0403 | .0098 | .0016 | .0001 | .0000 | .0000 | .0000 |
| 2 | .0746 | .1937 | .3020 | .2335 | .1209 | .0439 | .0106 | .0014 | .0001 | .0000 | .0000 |
| 3 | .0105 | .0574 | .2013 | .2668 | .2150 | .1172 | .0425 | .0090 | .0008 | .0000 | .0000 |
| 4 | .0010 | .0112 | .0881 | .2001 | .2508 | .2051 | .1115 | .0368 | .0055 | .0001 | .0000 |
| 5 | .0001 | .0015 | .0264 | .1029 | .2007 | .2461 | .2007 | .1029 | .0264 | .0015 | .0001 |
| 6 | .0000 | .0001 | .0055 | .0368 | .1115 | .2051 | .2508 | .2001 | .0881 | .0112 | .0010 |
| 7 | .0000 | .0000 | .0008 | .0090 | .0425 | .1172 | .2150 | .2668 | .2013 | .0574 | .0105 |
| 8 | .0000 | .0000 | .0001 | .0014 | .0106 | .0439 | .1209 | .2335 | .3020 | .1937 | .0746 |
| 9 | .0000 | .0000 | .0000 | .0001 | .0016 | .0098 | .0403 | .1211 | .2684 | .3874 | .3151 |
| 10 | .0000 | .0000 | .0000 | .0000 | .0001 | .0010 | .0060 | .0282 | .1074 | .3487 | .5987 |

$n=11$

| S \ p | .05 | .10 | .20 | .30 | .40 | .50 | .60 | .70 | .80 | .90 | .95 |
|---|---|---|---|---|---|---|---|---|---|---|---|
| 0 | .5688 | .3138 | .0859 | .0198 | .0036 | .0005 | .0000 | .0000 | .0000 | .0000 | .0000 |
| 1 | .3293 | .3835 | .2362 | .0932 | .0266 | .0054 | .0007 | .0000 | .0000 | .0000 | .0000 |
| 2 | .0867 | .2131 | .2953 | .1998 | .0887 | .0269 | .0052 | .0005 | .0000 | .0000 | .0000 |
| 3 | .0137 | .0710 | .2215 | .2568 | .1774 | .0806 | .0234 | .0037 | .0002 | .0000 | .0000 |
| 4 | .0014 | .0158 | .1107 | .2201 | .2365 | .1611 | .0701 | .0173 | .0017 | .0000 | .0000 |
| 5 | .0001 | .0025 | .0388 | .1321 | .2207 | .2256 | .1471 | .0566 | .0097 | .0003 | .0000 |
| 6 | .0000 | .0003 | .0097 | .0566 | .1471 | .2256 | .2207 | .1321 | .0388 | .0025 | .0001 |
| 7 | .0000 | .0000 | .0017 | .0173 | .0701 | .1611 | .2365 | .2201 | .1107 | .0158 | .0014 |
| 8 | .0000 | .0000 | .0002 | .0037 | .0234 | .0806 | .1774 | .2568 | .2215 | .0710 | .0137 |
| 9 | .0000 | .0000 | .0000 | .0005 | .0052 | .0269 | .0887 | .1998 | .2953 | .2131 | .0867 |
| 10 | .0000 | .0000 | .0000 | .0000 | .0007 | .0054 | .0266 | .0932 | .2362 | .3835 | .3293 |
| 11 | .0000 | .0000 | .0000 | .0000 | .0000 | .0005 | .0036 | .0198 | .0859 | .3138 | .5688 |

Table D-2 (*continued*)

$n=12$

| S \ p | .05 | .10 | .20 | .30 | .40 | .50 | .60 | .70 | .80 | .90 | .95 |
|---|---|---|---|---|---|---|---|---|---|---|---|
| 0 | .5404 | .2824 | .0687 | .0138 | .0022 | .0002 | .0000 | .0000 | .0000 | .0000 | .0000 |
| 1 | .3413 | .3766 | .2062 | .0712 | .0174 | .0029 | .0003 | .0000 | .0000 | .0000 | .0000 |
| 2 | .0988 | .2301 | .2835 | .1678 | .0639 | .0161 | .0025 | .0002 | .0000 | .0000 | .0000 |
| 3 | .0173 | .0852 | .2362 | .2397 | .1419 | .0537 | .0125 | .0015 | .0001 | .0000 | .0000 |
| 4 | .0021 | .0213 | .1329 | .2311 | .2128 | .1208 | .0420 | .0078 | .0005 | .0000 | .0000 |
| 5 | .0002 | .0038 | .0532 | .1585 | .2270 | .1934 | .1009 | .0291 | .0033 | .0000 | .0000 |
| 6 | .0000 | .0005 | .0155 | .0792 | .1766 | .2256 | .1766 | .0792 | .0155 | .0005 | .0000 |
| 7 | .0000 | .0000 | .0033 | .0291 | .1009 | .1934 | .2270 | .1585 | .0532 | .0038 | .0002 |
| 8 | .0000 | .0000 | .0005 | .0078 | .0420 | .1208 | .2128 | .2311 | .1329 | .0213 | .0021 |
| 9 | .0000 | .0000 | .0001 | .0015 | .0125 | .0537 | .1419 | .2397 | .2362 | .0852 | .0173 |
| 10 | .0000 | .0000 | .0000 | .0002 | .0025 | .0161 | .0639 | .1678 | .2835 | .2301 | .0988 |
| 11 | .0000 | .0000 | .0000 | .0000 | .0003 | .0029 | .0174 | .0712 | .2062 | .3766 | .3413 |
| 12 | .0000 | .0000 | .0000 | .0000 | .0000 | .0002 | .0022 | .0138 | .0687 | .2824 | .5404 |

$n=13$

| S \ p | .05 | .10 | .20 | .30 | .40 | .50 | .60 | .70 | .80 | .90 | .95 |
|---|---|---|---|---|---|---|---|---|---|---|---|
| 0 | .5133 | .2542 | .0550 | .0097 | .0013 | .0001 | .0000 | .0000 | .0000 | .0000 | .0000 |
| 1 | .3512 | .3672 | .1787 | .0540 | .0113 | .0016 | .0001 | .0000 | .0000 | .0000 | .0000 |
| 2 | .1109 | .2448 | .2680 | .1388 | .0453 | .0095 | .0012 | .0001 | .0000 | .0000 | .0000 |
| 3 | .0214 | .0997 | .2457 | .2181 | .1107 | .0349 | .0065 | .0006 | .0000 | .0000 | .0000 |
| 4 | .0028 | .0277 | .1535 | .2337 | .1845 | .0873 | .0243 | .0034 | .0001 | .0000 | .0000 |
| 5 | .0003 | .0055 | .0691 | .1803 | .2214 | .1571 | .0656 | .0142 | .0011 | .0000 | .0000 |
| 6 | .0000 | .0008 | .0230 | .1030 | .1968 | .2095 | .1312 | .0442 | .0058 | .0001 | .0000 |
| 7 | .0000 | .0001 | .0058 | .0442 | .1312 | .2095 | .1968 | .1030 | .0230 | .0008 | .0000 |
| 8 | .0000 | .0000 | .0011 | .0142 | .0656 | .1571 | .2214 | .1803 | .0691 | .0055 | .0003 |
| 9 | .0000 | .0000 | .0001 | .0034 | .0243 | .0873 | .1845 | .2337 | .1535 | .0277 | .0028 |
| 10 | .0000 | .0000 | .0000 | .0006 | .0065 | .0349 | .1107 | .2181 | .2457 | .0997 | .0214 |
| 11 | .0000 | .0000 | .0000 | .0001 | .0012 | .0095 | .0453 | .1388 | .2680 | .2448 | .1109 |
| 12 | .0000 | .0000 | .0000 | .0000 | .0001 | .0016 | .0113 | .0540 | .1787 | .3672 | .3512 |
| 13 | .0000 | .0000 | .0000 | .0000 | .0000 | .0001 | .0013 | .0097 | .0550 | .2542 | .5133 |

$n=14$

| S \ p | .05 | .10 | .20 | .30 | .40 | .50 | .60 | .70 | .80 | .90 | .95 |
|---|---|---|---|---|---|---|---|---|---|---|---|
| 0 | .4877 | .2288 | .0440 | .0068 | .0008 | .0001 | .0000 | .0000 | .0000 | .0000 | .0000 |
| 1 | .3593 | .3559 | .1539 | .0407 | .0073 | .0009 | .0001 | .0000 | .0000 | .0000 | .0000 |
| 2 | .1229 | .2570 | .2501 | .1134 | .0317 | .0056 | .0005 | .0000 | .0000 | .0000 | .0000 |
| 3 | .0259 | .1142 | .2501 | .1943 | .0845 | .0222 | .0033 | .0002 | .0000 | .0000 | .0000 |
| 4 | .0037 | .0349 | .1720 | .2290 | .1549 | .0611 | .0136 | .0014 | .0000 | .0000 | .0000 |
| 5 | .0004 | .0078 | .0860 | .1963 | .2066 | .1222 | .0408 | .0066 | .0003 | .0000 | .0000 |
| 6 | .0000 | .0013 | .0322 | .1262 | .2066 | .1833 | .0918 | .0232 | .0020 | .0000 | .0000 |
| 7 | .0000 | .0002 | .0092 | .0618 | .1574 | .2095 | .1574 | .0618 | .0092 | .0002 | .0000 |
| 8 | .0000 | .0000 | .0020 | .0232 | .0918 | .1833 | .2066 | .1262 | .0322 | .0013 | .0000 |
| 9 | .0000 | .0000 | .0003 | .0066 | .0408 | .1222 | .2066 | .1963 | .0860 | .0078 | .0004 |
| 10 | .0000 | .0000 | .0000 | .0014 | .0136 | .0611 | .1549 | .2290 | .1720 | .0349 | .0037 |
| 11 | .0000 | .0000 | .0000 | .0002 | .0033 | .0222 | .0845 | .1943 | .2501 | .1142 | .0259 |
| 12 | .0000 | .0000 | .0000 | .0000 | .0005 | .0056 | .0317 | .1134 | .2501 | .2570 | .1229 |
| 13 | .0000 | .0000 | .0000 | .0000 | .0001 | .0009 | .0073 | .0407 | .1539 | .3559 | .3593 |
| 14 | .0000 | .0000 | .0000 | .0000 | .0000 | .0001 | .0008 | .0068 | .0440 | .2288 | .4877 |

Table D-2 (*continued*)

$n = 15$

| S \ p | .05 | .10 | .20 | .30 | .40 | .50 | .60 | .70 | .80 | .90 | .95 |
|---|---|---|---|---|---|---|---|---|---|---|---|
| 0 | .4633 | .2059 | .0352 | .0047 | .0005 | .0000 | .0000 | .0000 | .0000 | .0000 | .0000 |
| 1 | .3658 | .3432 | .1319 | .0305 | .0047 | .0005 | .0000 | .0000 | .0000 | .0000 | .0000 |
| 2 | .1348 | .2669 | .2309 | .0916 | .0219 | .0032 | .0003 | .0000 | .0000 | .0000 | .0000 |
| 3 | .0307 | .1285 | .2501 | .1700 | .0634 | .0139 | .0016 | .0001 | .0000 | .0000 | .0000 |
| 4 | .0049 | .0428 | .1876 | .2186 | .1268 | .0417 | .0074 | .0006 | .0000 | .0000 | .0000 |
| 5 | .0006 | .0105 | .1032 | .2061 | .1859 | .0916 | .0245 | .0030 | .0001 | .0000 | .0000 |
| 6 | .0000 | .0019 | .0430 | .1472 | .2066 | .1527 | .0612 | .0116 | .0007 | .0000 | .0000 |
| 7 | .0000 | .0003 | .0138 | .0811 | .1771 | .1964 | .1181 | .0348 | .0035 | .0000 | .0000 |
| 8 | .0000 | .0000 | .0035 | .0348 | .1181 | .1964 | .1771 | .0811 | .0138 | .0003 | .0000 |
| 9 | .0000 | .0000 | .0007 | .0116 | .0612 | .1527 | .2066 | .1472 | .0430 | .0019 | .0000 |
| 10 | .0000 | .0000 | .0001 | .0030 | .0245 | .0916 | .1859 | .2061 | .1032 | .0105 | .0006 |
| 11 | .0000 | .0000 | .0000 | .0006 | .0074 | .0417 | .1268 | .2186 | .1876 | .0428 | .0049 |
| 12 | .0000 | .0000 | .0000 | .0001 | .0016 | .0139 | .0634 | .1700 | .2501 | .1285 | .0307 |
| 13 | .0000 | .0000 | .0000 | .0000 | .0003 | .0032 | .0219 | .0916 | .2309 | .2669 | .1348 |
| 14 | .0000 | .0000 | .0000 | .0000 | .0000 | .0005 | .0047 | .0305 | .1319 | .3432 | .3658 |
| 15 | .0000 | .0000 | .0000 | .0000 | .0000 | .0000 | .0005 | .0047 | .0352 | .2059 | .4633 |

$n = 16$

| S \ p | .05 | .10 | .20 | .30 | .40 | .50 | .60 | .70 | .80 | .90 | .95 |
|---|---|---|---|---|---|---|---|---|---|---|---|
| 0 | .4401 | .1853 | .0281 | .0033 | .0003 | .0000 | .0000 | .0000 | .0000 | .0000 | .0000 |
| 1 | .3706 | .3294 | .1126 | .0228 | .0030 | .0002 | .0000 | .0000 | .0000 | .0000 | .0000 |
| 2 | .1463 | .2745 | .2111 | .0732 | .0150 | .0018 | .0001 | .0000 | .0000 | .0000 | .0000 |
| 3 | .0359 | .1423 | .2463 | .1465 | .0468 | .0085 | .0008 | .0000 | .0000 | .0000 | .0000 |
| 4 | .0061 | .0514 | .2001 | .2040 | .1014 | .0278 | .0040 | .0002 | .0000 | .0000 | .0000 |
| 5 | .0008 | .0137 | .1201 | .2099 | .1623 | .0667 | .0142 | .0013 | .0000 | .0000 | .0000 |
| 6 | .0001 | .0028 | .0550 | .1649 | .1983 | .1222 | .0392 | .0056 | .0002 | .0000 | .0000 |
| 7 | .0000 | .0004 | .0197 | .1010 | .1889 | .1746 | .0840 | .0185 | .0012 | .0000 | .0000 |
| 8 | .0000 | .0001 | .0055 | .0487 | .1417 | .1964 | .1417 | .0487 | .0055 | .0001 | .0000 |
| 9 | .0000 | .0000 | .0012 | .0185 | .0840 | .1746 | .1889 | .1010 | .0197 | .0004 | .0000 |
| 10 | .0000 | .0000 | .0002 | .0056 | .0392 | .1222 | .1983 | .1649 | .0550 | .0028 | .0001 |
| 11 | .0000 | .0000 | .0000 | .0013 | .0142 | .0667 | .1623 | .2099 | .1201 | .0137 | .0008 |
| 12 | .0000 | .0000 | .0000 | .0002 | .0040 | .0278 | .1014 | .2040 | .2001 | .0514 | .0061 |
| 13 | .0000 | .0000 | .0000 | .0000 | .0008 | .0085 | .0468 | .1465 | .2463 | .1423 | .0359 |
| 14 | .0000 | .0000 | .0000 | .0000 | .0001 | .0018 | .0150 | .0732 | .2111 | .2745 | .1463 |
| 15 | .0000 | .0000 | .0000 | .0000 | .0000 | .0002 | .0030 | .0228 | .1126 | .3294 | .3706 |
| 16 | .0000 | .0000 | .0000 | .0000 | .0000 | .0000 | .0003 | .0033 | .0281 | .1853 | .4401 |

**Table D-2 (continued)**

$n = 17$

| S \ p | .05 | .10 | .20 | .30 | .40 | .50 | .60 | .70 | .80 | .90 | .95 |
|---|---|---|---|---|---|---|---|---|---|---|---|
| 0 | .4181 | .1668 | .0225 | .0023 | .0002 | .0000 | .0000 | .0000 | .0000 | .0000 | .0000 |
| 1 | .3741 | .3150 | .0957 | .0169 | .0019 | .0001 | .0000 | .0000 | .0000 | .0000 | .0000 |
| 2 | .1575 | .2800 | .1914 | .0581 | .0102 | .0010 | .0001 | .0000 | .0000 | .0000 | .0000 |
| 3 | .0415 | .1556 | .2393 | .1245 | .0341 | .0052 | .0004 | .0000 | .0000 | .0000 | .0000 |
| 4 | .0076 | .0605 | .2093 | .1868 | .0796 | .0182 | .0021 | .0001 | .0000 | .0000 | .0000 |
| 5 | .0010 | .0175 | .1361 | .2081 | .1379 | .0472 | .0081 | .0006 | .0000 | .0000 | .0000 |
| 6 | .0001 | .0039 | .0680 | .1784 | .1839 | .0944 | .0242 | .0026 | .0001 | .0000 | .0000 |
| 7 | .0000 | .0007 | .0267 | .1201 | .1927 | .1484 | .0571 | .0095 | .0004 | .0000 | .0000 |
| 8 | .0000 | .0001 | .0084 | .0644 | .1606 | .1855 | .1070 | .0276 | .0021 | .0000 | .0000 |
| 9 | .0000 | .0000 | .0021 | .0276 | .1070 | .1855 | .1606 | .0644 | .0084 | .0001 | .0000 |
| 10 | .0000 | .0000 | .0004 | .0095 | .0571 | .1484 | .1927 | .1201 | .0267 | .0007 | .0000 |
| 11 | .0000 | .0000 | .0001 | .0026 | .0242 | .0944 | .1839 | .1784 | .0680 | .0039 | .0001 |
| 12 | .0000 | .0000 | .0000 | .0006 | .0081 | .0472 | .1379 | .2081 | .1361 | .0175 | .0010 |
| 13 | .0000 | .0000 | .0000 | .0001 | .0021 | .0182 | .0796 | .1868 | .2093 | .0605 | .0076 |
| 14 | .0000 | .0000 | .0000 | .0000 | .0004 | .0052 | .0341 | .1245 | .2393 | .1556 | .0415 |
| 15 | .0000 | .0000 | .0000 | .0000 | .0001 | .0010 | .0102 | .0581 | .1914 | .2800 | .1575 |
| 16 | .0000 | .0000 | .0000 | .0000 | .0000 | .0001 | .0019 | .0169 | .0957 | .3150 | .3741 |
| 17 | .0000 | .0000 | .0000 | .0000 | .0000 | .0000 | .0002 | .0023 | .0225 | .1668 | .4181 |

$n = 18$

| S \ p | .05 | .10 | .20 | .30 | .40 | .50 | .60 | .70 | .80 | .90 | .95 |
|---|---|---|---|---|---|---|---|---|---|---|---|
| 0 | .3972 | .1501 | .0180 | .0016 | .0001 | .0000 | .0000 | .0000 | .0000 | .0000 | .0000 |
| 1 | .3763 | .3002 | .0811 | .0126 | .0012 | .0001 | .0000 | .0000 | .0000 | .0000 | .0000 |
| 2 | .1683 | .2835 | .1723 | .0458 | .0069 | .0006 | .0000 | .0000 | .0000 | .0000 | .0000 |
| 3 | .0473 | .1680 | .2297 | .1046 | .0246 | .0031 | .0002 | .0000 | .0000 | .0000 | .0000 |
| 4 | .0093 | .0700 | .2153 | .1681 | .0614 | .0117 | .0011 | .0000 | .0000 | .0000 | .0000 |
| 5 | .0014 | .0218 | .1507 | .2017 | .1146 | .0327 | .0045 | .0002 | .0000 | .0000 | .0000 |
| 6 | .0002 | .0052 | .0816 | .1873 | .1655 | .0708 | .0145 | .0012 | .0000 | .0000 | .0000 |
| 7 | .0000 | .0010 | .0350 | .1376 | .1892 | .1214 | .0374 | .0046 | .0001 | .0000 | .0000 |
| 8 | .0000 | .0002 | .0120 | .0811 | .1734 | .1669 | .0771 | .0149 | .0008 | .0000 | .0000 |
| 9 | .0000 | .0000 | .0033 | .0386 | .1284 | .1855 | .1284 | .0386 | .0033 | .0000 | .0000 |
| 10 | .0000 | .0000 | .0008 | .0149 | .0771 | .1669 | .1734 | .0811 | .0120 | .0002 | .0000 |
| 11 | .0000 | .0000 | .0001 | .0046 | .0374 | .1214 | .1892 | .1376 | .0350 | .0010 | .0000 |
| 12 | .0000 | .0000 | .0000 | .0012 | .0145 | .0708 | .1655 | .1873 | .0816 | .0052 | .0002 |
| 13 | .0000 | .0000 | .0000 | .0002 | .0045 | .0327 | .1146 | .2017 | .1507 | .0218 | .0014 |
| 14 | .0000 | .0000 | .0000 | .0000 | .0011 | .0117 | .0614 | .1681 | .2153 | .0700 | .0093 |
| 15 | .0000 | .0000 | .0000 | .0000 | .0002 | .0031 | .0246 | .1046 | .2297 | .1680 | .0473 |
| 16 | .0000 | .0000 | .0000 | .0000 | .0000 | .0006 | .0069 | .0458 | .1723 | .2835 | .1683 |
| 17 | .0000 | .0000 | .0000 | .0000 | .0000 | .0001 | .0012 | .0126 | .0811 | .3002 | .3763 |
| 18 | .0000 | .0000 | .0000 | .0000 | .0000 | .0000 | .0001 | .0016 | .0180 | .1501 | .3972 |

**Table D-2 (*continued*)**

$n=19$

| S \ p | .05 | .10 | .20 | .30 | .40 | .50 | .60 | .70 | .80 | .90 | .95 |
|---|---|---|---|---|---|---|---|---|---|---|---|
| 0 | .3774 | .1351 | .0144 | .0011 | .0001 | .0000 | .0000 | .0000 | .0000 | .0000 | .0000 |
| 1 | .3774 | .2852 | .0685 | .0093 | .0008 | .0000 | .0000 | .0000 | .0000 | .0000 | .0000 |
| 2 | .1787 | .2852 | .1540 | .0358 | .0046 | .0003 | .0000 | .0000 | .0000 | .0000 | .0000 |
| 3 | .0533 | .1796 | .2182 | .0869 | .0175 | .0018 | .0001 | .0000 | .0000 | .0000 | .0000 |
| 4 | .0112 | .0798 | .2182 | .1491 | .0467 | .0074 | .0005 | .0000 | .0000 | .0000 | .0000 |
| 5 | .0018 | .0266 | .1636 | .1916 | .0933 | .0222 | .0024 | .0001 | .0000 | .0000 | .0000 |
| 6 | .0002 | .0069 | .0955 | .1916 | .1451 | .0518 | .0085 | .0005 | .0000 | .0000 | .0000 |
| 7 | .0000 | .0014 | .0443 | .1525 | .1797 | .0961 | .0237 | .0022 | .0000 | .0000 | .0000 |
| 8 | .0000 | .0002 | .0166 | .0981 | .1797 | .1442 | .0532 | .0077 | .0003 | .0000 | .0000 |
| 9 | .0000 | .0000 | .0051 | .0514 | .1464 | .1762 | .0976 | .0220 | .0013 | .0000 | .0000 |
| 10 | .0000 | .0000 | .0013 | .0220 | .0976 | .1762 | .1464 | .0514 | .0051 | .0000 | .0000 |
| 11 | .0000 | .0000 | .0003 | .0077 | .0532 | .1442 | .1797 | .0981 | .0166 | .0002 | .0000 |
| 12 | .0000 | .0000 | .0000 | .0022 | .0237 | .0961 | .1797 | .1525 | .0443 | .0014 | .0000 |
| 13 | .0000 | .0000 | .0000 | .0005 | .0085 | .0518 | .1451 | .1916 | .0955 | .0069 | .0002 |
| 14 | .0000 | .0000 | .0000 | .0001 | .0024 | .0222 | .0933 | .1916 | .1636 | .0266 | .0018 |
| 15 | .0000 | .0000 | .0000 | .0000 | .0005 | .0074 | .0467 | .1491 | .2182 | .0798 | .0112 |
| 16 | .0000 | .0000 | .0000 | .0000 | .0001 | .0018 | .0175 | .0869 | .2182 | .1796 | .0533 |
| 17 | .0000 | .0000 | .0000 | .0000 | .0000 | .0003 | .0046 | .0358 | .1540 | .2852 | .1787 |
| 18 | .0000 | .0000 | .0000 | .0000 | .0000 | .0000 | .0008 | .0093 | .0685 | .2852 | .3774 |
| 19 | .0000 | .0000 | .0000 | .0000 | .0000 | .0000 | .0001 | .0011 | .0144 | .1351 | .3774 |

$n=20$

| S \ p | .05 | .10 | .20 | .30 | .40 | .50 | .60 | .70 | .80 | .90 | .95 |
|---|---|---|---|---|---|---|---|---|---|---|---|
| 0 | .3585 | .1216 | .0115 | .0008 | .0000 | .0000 | .0000 | .0000 | .0000 | .0000 | .0000 |
| 1 | .3774 | .2702 | .0576 | .0068 | .0005 | .0000 | .0000 | .0000 | .0000 | .0000 | .0000 |
| 2 | .1887 | .2852 | .1369 | .0278 | .0031 | .0002 | .0000 | .0000 | .0000 | .0000 | .0000 |
| 3 | .0596 | .1901 | .2054 | .0716 | .0123 | .0011 | .0000 | .0000 | .0000 | .0000 | .0000 |
| 4 | .0133 | .0898 | .2182 | .1304 | .0350 | .0046 | .0003 | .0000 | .0000 | .0000 | .0000 |
| 5 | .0022 | .0319 | .1746 | .1789 | .0746 | .0148 | .0013 | .0000 | .0000 | .0000 | .0000 |
| 6 | .0003 | .0089 | .1091 | .1916 | .1244 | .0370 | .0049 | .0002 | .0000 | .0000 | .0000 |
| 7 | .0000 | .0020 | .0545 | .1643 | .1659 | .0739 | .0146 | .0010 | .0000 | .0000 | .0000 |
| 8 | .0000 | .0004 | .0222 | .1144 | .1797 | .1201 | .0355 | .0039 | .0001 | .0000 | .0000 |
| 9 | .0000 | .0001 | .0074 | .0654 | .1597 | .1602 | .0710 | .0120 | .0005 | .0000 | .0000 |
| 10 | .0000 | .0000 | .0020 | .0308 | .1171 | .1762 | .1171 | .0308 | .0020 | .0000 | .0000 |
| 11 | .0000 | .0000 | .0005 | .0120 | .0710 | .1602 | .1597 | .0654 | .0074 | .0001 | .0000 |
| 12 | .0000 | .0000 | .0001 | .0039 | .0355 | .1201 | .1797 | .1144 | .0222 | .0004 | .0000 |
| 13 | .0000 | .0000 | .0000 | .0010 | .0146 | .0739 | .1659 | .1643 | .0545 | .0020 | .0000 |
| 14 | .0000 | .0000 | .0000 | .0002 | .0049 | .0370 | .1244 | .1916 | .1091 | .0089 | .0003 |
| 15 | .0000 | .0000 | .0000 | .0000 | .0013 | .0148 | .0746 | .1789 | .1746 | .0319 | .0022 |
| 16 | .0000 | .0000 | .0000 | .0000 | .0003 | .0046 | .0350 | .1304 | .2182 | .0898 | .0133 |
| 17 | .0000 | .0000 | .0000 | .0000 | .0000 | .0011 | .0123 | .0716 | .2054 | .1901 | .0596 |
| 18 | .0000 | .0000 | .0000 | .0000 | .0000 | .0002 | .0031 | .0278 | .1369 | .2852 | .1887 |
| 19 | .0000 | .0000 | .0000 | .0000 | .0000 | .0000 | .0005 | .0068 | .0576 | .2702 | .3774 |
| 20 | .0000 | .0000 | .0000 | .0000 | .0000 | .0000 | .0000 | .0008 | .0115 | .1216 | .3585 |

## Table D-3 Areas to the Left of $z$ under the Normal Curve: Short Form

| $z$-score | proportion of area to the left of $z$ |
|---|---|
| −4 | .00003 |
| −3 | .0013 |
| −2.58 | .0049 |
| −2.33 | .0099 |
| −2 | .0228 |
| −1.96 | .0250 |
| −1.65 | .0495 |
| −1 | .1587 |
| 0 | .5000 |
| 1 | .8413 |
| 1.65 | .9505 |
| 1.96 | .9750 |
| 2 | .9772 |
| 2.33 | .9901 |
| 2.58 | .9951 |
| 3 | .9987 |
| 4 | .99997 |

Area

$z$

Area

$z$

### A QUICK REFERENCE TO SOME IMPORTANT $z$-SCORES

Values of $z$ are given without signs. You must determine whether the critical values are positive, negative, or both from the alternative hypothesis.

| | $\alpha = .05$ | $\alpha = .01$ |
|---|---|---|
| one-tail | 1.65 | 2.33 |
| two-tail | 1.96 | 2.58 |

**Table D-4 Areas to the Left of $z$ under the Normal Curve: Long Form**

| $z$ | area | $z$ | area | $z$ | area |
|---|---|---|---|---|---|
| −4 | .00003 | −2.74 | .0031 | −2.29 | .0110 |
| −3.9 | .00005 | −2.73 | .0032 | −2.28 | .0113 |
| −3.8 | .0001 | −2.72 | .0033 | −2.27 | .0116 |
| −3.7 | .0001 | −2.71 | .0034 | −2.26 | .0119 |
| −3.6 | .0002 | −2.70 | .0035 | −2.25 | .0122 |
| −3.5 | .0002 | −2.69 | .0036 | −2.24 | .0125 |
| −3.4 | .0003 | −2.68 | .0037 | −2.23 | .0129 |
| −3.3 | .0005 | −2.67 | .0038 | −2.22 | .0132 |
| −3.2 | .0007 | −2.66 | .0039 | −2.21 | .0136 |
| −3.1 | .0010 | −2.65 | .0040 | −2.20 | .0139 |
| −3.09 | .0010 | −2.64 | .0041 | −2.19 | .0143 |
| −3.08 | .0010 | −2.63 | .0043 | −2.18 | .0146 |
| −3.07 | .0011 | −2.62 | .0044 | −2.17 | .0150 |
| −3.06 | .0011 | −2.61 | .0045 | −2.16 | .0154 |
| −3.05 | .0011 | −2.60 | .0047 | −2.15 | .0158 |
| −3.04 | .0012 | −2.59 | .0048 | −2.14 | .0162 |
| −3.03 | .0012 | −2.58 | .0049 | −2.13 | .0166 |
| −3.02 | .0013 | −2.57 | .0051 | −2.12 | .0170 |
| −3.01 | .0013 | −2.56 | .0052 | −2.11 | .0174 |
| −3.00 | .0013 | −2.55 | .0054 | −2.10 | .0179 |
| −2.99 | .0014 | −2.54 | .0055 | −2.09 | .0183 |
| −2.98 | .0014 | −2.53 | .0057 | −2.08 | .0188 |
| −2.97 | .0015 | −2.52 | .0059 | −2.07 | .0192 |
| −2.96 | .0015 | −2.51 | .0060 | −2.06 | .0197 |
| −2.95 | .0016 | −2.50 | .0062 | −2.05 | .0202 |
| −2.94 | .0016 | −2.49 | .0064 | −2.04 | .0207 |
| −2.93 | .0017 | −2.48 | .0066 | −2.03 | .0212 |
| −2.92 | .0017 | −2.47 | .0068 | −2.02 | .0217 |
| −2.91 | .0018 | −2.46 | .0069 | −2.01 | .0222 |
| −2.90 | .0019 | −2.45 | .0071 | −2.00 | .0228 |
| −2.89 | .0019 | −2.44 | .0073 | −1.99 | .0233 |
| −2.88 | .0020 | −2.43 | .0075 | −1.98 | .0239 |
| −2.87 | .0021 | −2.42 | .0078 | −1.97 | .0244 |
| −2.86 | .0021 | −2.41 | .0080 | −1.96 | .0250 |
| −2.85 | .0022 | −2.40 | .0082 | −1.95 | .0256 |
| −2.84 | .0023 | −2.39 | .0084 | −1.94 | .0262 |
| −2.83 | .0023 | −2.38 | .0087 | −1.93 | .0268 |
| −2.82 | .0024 | −2.37 | .0089 | −1.92 | .0274 |
| −2.81 | .0025 | −2.36 | .0091 | −1.91 | .0281 |
| −2.80 | .0026 | −2.35 | .0094 | −1.90 | .0287 |
| −2.79 | .0026 | −2.34 | .0096 | −1.89 | .0294 |
| −2.78 | .0027 | −2.33 | .0099 | −1.88 | .0301 |
| −2.77 | .0028 | −2.32 | .0102 | −1.87 | .0307 |
| −2.76 | .0029 | −2.31 | .0104 | −1.86 | .0314 |
| −2.75 | .0030 | −2.30 | .0107 | −1.85 | .0322 |

Table D-4 (*continued*)

| $z$ | area | $z$ | area | $z$ | area |
|---|---|---|---|---|---|
| −1.84 | .0329 | −1.39 | .0823 | − .94 | .1736 |
| −1.83 | .0336 | −1.38 | .0838 | − .93 | .1762 |
| −1.82 | .0344 | −1.37 | .0853 | − .92 | .1788 |
| −1.81 | .0352 | −1.36 | .0869 | − .91 | .1814 |
| −1.80 | .0359 | −1.35 | .0885 | − .90 | .1841 |
| −1.79 | .0367 | −1.34 | .0901 | − .89 | .1867 |
| −1.78 | .0375 | −1.33 | .0918 | − .88 | .1894 |
| −1.77 | .0384 | −1.32 | .0934 | − .87 | .1922 |
| −1.76 | .0392 | −1.31 | .0951 | − .86 | .1949 |
| −1.75 | .0401 | −1.30 | .0968 | − .85 | .1977 |
| −1.74 | .0409 | −1.29 | .0985 | − .84 | .2005 |
| −1.73 | .0418 | −1.28 | .1003 | − .83 | .2033 |
| −1.72 | .0427 | −1.27 | .1020 | − .82 | .2061 |
| −1.71 | .0436 | −1.26 | .1038 | − .81 | .2090 |
| −1.70 | .0446 | −1.25 | .1056 | − .80 | .2119 |
| −1.69 | .0455 | −1.24 | .1075 | − .79 | .2148 |
| −1.68 | .0465 | −1.23 | .1093 | − .78 | .2177 |
| −1.67 | .0475 | −1.22 | .1112 | − .77 | .2206 |
| −1.66 | .0485 | −1.21 | .1131 | − .76 | .2236 |
| −1.65 | .0495 | −1.20 | .1151 | − .75 | .2266 |
| −1.64 | .0505 | −1.19 | .1170 | − .74 | .2296 |
| −1.63 | .0516 | −1.18 | .1190 | − .73 | .2327 |
| −1.62 | .0526 | −1.17 | .1210 | − .72 | .2358 |
| −1.61 | .0537 | −1.16 | .1230 | − .71 | .2389 |
| −1.60 | .0548 | −1.15 | .1251 | − .70 | .2420 |
| −1.59 | .0559 | −1.14 | .1271 | − .69 | .2451 |
| −1.58 | .0571 | −1.13 | .1292 | − .68 | .2483 |
| −1.57 | .0582 | −1.12 | .1314 | − .67 | .2514 |
| −1.56 | .0594 | −1.11 | .1335 | − .66 | .2546 |
| −1.55 | .0606 | −1.10 | .1357 | − .65 | .2578 |
| −1.54 | .0618 | −1.09 | .1379 | − .64 | .2611 |
| −1.53 | .0630 | −1.08 | .1401 | − .63 | .2643 |
| −1.52 | .0643 | −1.07 | .1423 | − .62 | .2676 |
| −1.51 | .0655 | −1.06 | .1446 | − .61 | .2709 |
| −1.50 | .0668 | −1.05 | .1469 | − .60 | .2743 |
| −1.49 | .0681 | −1.04 | .1492 | − .59 | .2776 |
| −1.48 | .0694 | −1.03 | .1515 | − .58 | .2810 |
| −1.47 | .0708 | −1.02 | .1539 | − .57 | .2843 |
| −1.46 | .0722 | −1.01 | .1562 | − .56 | .2877 |
| −1.45 | .0735 | −1.00 | .1587 | − .55 | .2912 |
| −1.44 | .0749 | − .99 | .1611 | − .54 | .2946 |
| −1.43 | .0764 | − .98 | .1635 | − .53 | .2981 |
| −1.42 | .0778 | − .97 | .1660 | − .52 | .3015 |
| −1.41 | .0793 | − .96 | .1685 | − .51 | .3050 |
| −1.40 | .0808 | − .95 | .1711 | − .50 | .3085 |

Table D-4 (*continued*)

| $z$ | area | $z$ | area | $z$ | area |
|---|---|---|---|---|---|
| −.49 | .3121 | −.04 | .4840 | .41 | .6591 |
| −.48 | .3156 | −.03 | .4880 | .42 | .6628 |
| −.47 | .3192 | −.02 | .4920 | .43 | .6664 |
| −.46 | .3228 | −.01 | .4960 | .44 | .6700 |
| −.45 | .3264 | .00 | .5000 | .45 | .6736 |
| −.44 | .3300 | .01 | .5040 | .46 | .6772 |
| −.43 | .3336 | .02 | .5080 | .47 | .6808 |
| −.42 | .3372 | .03 | .5120 | .48 | .6844 |
| −.41 | .3409 | .04 | .5160 | .49 | .6879 |
| −.40 | .3446 | .05 | .5199 | .50 | .6915 |
| −.39 | .3483 | .06 | .5239 | .51 | .6950 |
| −.38 | .3520 | .07 | .5279 | .52 | .6985 |
| −.37 | .3557 | .08 | .5319 | .53 | .7019 |
| −.36 | .3594 | .09 | .5359 | .54 | .7054 |
| −.35 | .3632 | .10 | .5398 | .55 | .7088 |
| −.34 | .3669 | .11 | .5438 | .56 | .7123 |
| −.33 | .3707 | .12 | .5478 | .57 | .7157 |
| −.32 | .3745 | .13 | .5517 | .58 | .7190 |
| −.31 | .3783 | .14 | .5557 | .59 | .7224 |
| −.30 | .3821 | .15 | .5596 | .60 | .7257 |
| −.29 | .3859 | .16 | .5636 | .61 | .7291 |
| −.28 | .3897 | .17 | .5675 | .62 | .7324 |
| −.27 | .3936 | .18 | .5714 | .63 | .7357 |
| −.26 | .3974 | .19 | .5753 | .64 | .7389 |
| −.25 | .4013 | .20 | .5793 | .65 | .7422 |
| −.24 | .4052 | .21 | .5832 | .66 | .7454 |
| −.23 | .4090 | .22 | .5871 | .67 | .7486 |
| −.22 | .4129 | .23 | .5910 | .68 | .7517 |
| −.21 | .4168 | .24 | .5948 | .69 | .7549 |
| −.20 | .4207 | .25 | .5987 | .70 | .7580 |
| −.19 | .4247 | .26 | .6026 | .71 | .7611 |
| −.18 | .4286 | .27 | .6064 | .72 | .7642 |
| −.17 | .4325 | .28 | .6103 | .73 | .7673 |
| −.16 | .4364 | .29 | .6141 | .74 | .7704 |
| −.15 | .4404 | .30 | .6179 | .75 | .7734 |
| −.14 | .4443 | .31 | .6217 | .76 | .7764 |
| −.13 | .4483 | .32 | .6255 | .77 | .7794 |
| −.12 | .4522 | .33 | .6293 | .78 | .7283 |
| −.11 | .4562 | .34 | .6331 | .79 | .7852 |
| −.10 | .4602 | .35 | .6368 | .80 | .7881 |
| −.09 | .4641 | .36 | .6406 | .81 | .7910 |
| −.08 | .4681 | .37 | .6443 | .82 | .7939 |
| −.07 | .4721 | .38 | .6480 | .83 | .7967 |
| −.06 | .4761 | .39 | .6517 | .84 | .7995 |
| −.05 | .4801 | .40 | .6554 | .85 | .8023 |

**Table D-4** (*continued*)

| $z$ | area | $z$ | area | $z$ | area |
|---|---|---|---|---|---|
| .86 | .8051 | 1.31 | .9049 | 1.76 | .9608 |
| .87 | .8078 | 1.32 | .9066 | 1.77 | .9616 |
| .88 | .8106 | 1.33 | .9082 | 1.78 | .9625 |
| .89 | .8133 | 1.34 | .9099 | 1.79 | .9633 |
| .90 | .8159 | 1.35 | .9115 | 1.80 | .9641 |
| .91 | .8186 | 1.36 | .9131 | 1.81 | .9649 |
| .92 | .8212 | 1.37 | .9147 | 1.82 | .9656 |
| .93 | .8238 | 1.38 | .9162 | 1.83 | .9664 |
| .94 | .8264 | 1.39 | .9177 | 1.84 | .9671 |
| .95 | .8289 | 1.40 | .9192 | 1.85 | .9678 |
| .96 | .8315 | 1.41 | .9207 | 1.86 | .9686 |
| .97 | .8340 | 1.42 | .9222 | 1.87 | .9693 |
| .98 | .8365 | 1.43 | .9236 | 1.88 | .9699 |
| .99 | .8389 | 1.44 | .9251 | 1.89 | .9706 |
| 1.00 | .8413 | 1.45 | .9265 | 1.90 | .9713 |
| 1.01 | .8438 | 1.46 | .9278 | 1.91 | .9719 |
| 1.02 | .8461 | 1.47 | .9292 | 1.92 | .9726 |
| 1.03 | .8485 | 1.48 | .9306 | 1.93 | .9732 |
| 1.04 | .8508 | 1.49 | .9319 | 1.94 | .9738 |
| 1.05 | .8531 | 1.50 | .9332 | 1.95 | .9744 |
| 1.06 | .8554 | 1.51 | .9345 | 1.96 | .9750 |
| 1.07 | .8577 | 1.52 | .9357 | 1.97 | .9756 |
| 1.08 | .8599 | 1.53 | .9370 | 1.98 | .9761 |
| 1.09 | .8621 | 1.54 | .9382 | 1.99 | .9767 |
| 1.10 | .8643 | 1.55 | .9394 | 2.00 | .9772 |
| 1.11 | .8665 | 1.56 | .9406 | 2.01 | .9778 |
| 1.12 | .8686 | 1.57 | .9418 | 2.02 | .9783 |
| 1.13 | .8708 | 1.58 | .9429 | 2.03 | .9788 |
| 1.14 | .8729 | 1.59 | .9441 | 2.04 | .9793 |
| 1.15 | .8749 | 1.60 | .9452 | 2.05 | .9798 |
| 1.16 | .8770 | 1.61 | .9463 | 2.06 | .9803 |
| 1.17 | .8790 | 1.62 | .9474 | 2.07 | .9808 |
| 1.18 | .8810 | 1.63 | .9484 | 2.08 | .9812 |
| 1.19 | .8830 | 1.64 | .9495 | 2.09 | .9817 |
| 1.20 | .8849 | 1.65 | .9505 | 2.10 | .9821 |
| 1.21 | .8869 | 1.66 | .9515 | 2.11 | .9826 |
| 1.22 | .8888 | 1.67 | .9525 | 2.12 | .9830 |
| 1.23 | .8907 | 1.68 | .9535 | 2.13 | .9834 |
| 1.24 | .8925 | 1.69 | .9545 | 2.14 | .9838 |
| 1.25 | .8944 | 1.70 | .9554 | 2.15 | .9842 |
| 1.26 | .8962 | 1.71 | .9564 | 2.16 | .9846 |
| 1.27 | .8980 | 1.72 | .9573 | 2.17 | .9850 |
| 1.28 | .8997 | 1.73 | .9582 | 2.18 | .9854 |
| 1.29 | .9015 | 1.74 | .9591 | 2.19 | .9857 |
| 1.30 | .9032 | 1.75 | .9599 | 2.20 | .9861 |

Table D-4 (*continued*)

| $z$ | area | $z$ | area | $z$ | area |
|---|---|---|---|---|---|
| 2.21 | .9864 | 2.66 | .9961 | 3.2 | .9993 |
| 2.22 | .9868 | 2.67 | .9962 | 3.3 | .9995 |
| 2.23 | .9871 | 2.68 | .9963 | 3.4 | .9997 |
| 2.24 | .9875 | 2.69 | .9964 | 3.5 | .9998 |
| 2.25 | .9878 | 2.70 | .9965 | 3.6 | .9998 |
| 2.26 | .9881 | 2.71 | .9966 | 3.7 | .9999 |
| 2.27 | .9884 | 2.72 | .9967 | 3.8 | .9999 |
| 2.28 | .9887 | 2.73 | .9968 | 3.9 | .99995 |
| 2.29 | .9890 | 2.74 | .9969 | 4.0 | .99997 |
| 2.30 | .9893 | 2.75 | .9970 | | |
| 2.31 | .9896 | 2.76 | .9971 | | |
| 2.32 | .9898 | 2.77 | .9972 | | |
| 2.33 | .9901 | 2.78 | .9973 | | |
| 2.34 | .9904 | 2.79 | .9974 | | |
| 2.35 | .9906 | 2.80 | .9974 | | |
| 2.36 | .9909 | 2.81 | .9975 | | |
| 2.37 | .9911 | 2.82 | .9976 | | |
| 2.38 | .9913 | 2.83 | .9977 | | |
| 2.39 | .9916 | 2.84 | .9977 | | |
| 2.40 | .9918 | 2.85 | .9978 | | |
| 2.41 | .9920 | 2.86 | .9979 | | |
| 2.42 | .9922 | 2.87 | .9979 | | |
| 2.43 | .9925 | 2.88 | .9980 | | |
| 2.44 | .9927 | 2.89 | .9981 | | |
| 2.45 | .9929 | 2.90 | .9981 | | |
| 2.46 | .9931 | 2.91 | .9982 | | |
| 2.47 | .9932 | 2.92 | .9982 | | |
| 2.48 | .9934 | 2.93 | .9983 | | |
| 2.49 | .9936 | 2.94 | .9984 | | |
| 2.50 | .9938 | 2.95 | .9984 | | |
| 2.51 | .9940 | 2.96 | .9985 | | |
| 2.52 | .9941 | 2.97 | .9985 | | |
| 2.53 | .9943 | 2.98 | .9986 | | |
| 2.54 | .9945 | 2.99 | .9986 | | |
| 2.55 | .9946 | 3.00 | .9987 | | |
| 2.56 | .9948 | 3.01 | .9987 | | |
| 2.57 | .9949 | 3.02 | .9987 | | |
| 2.58 | .9951 | 3.03 | .9988 | | |
| 2.59 | .9952 | 3.04 | .9988 | | |
| 2.60 | .9953 | 3.05 | .9989 | | |
| 2.61 | .9955 | 3.06 | .9989 | | |
| 2.62 | .9956 | 3.07 | .9989 | | |
| 2.63 | .9957 | 3.08 | .9990 | | |
| 2.64 | .9959 | 3.09 | .9990 | | |
| 2.65 | .9960 | 3.10 | .9990 | | |

**Table D-5 Critical Values of $t$ for a Two-Tail Test (Values of $t_c$ in this table are given without signs. All values are both positive and negative, that is $t_c = \pm 12.71$.)**

| degrees of freedom $\sigma$ | $t_c$ for $\alpha = .05$ | $t_c$ for $\alpha = .01$ |
|---|---|---|
| 1 | 12.71 | 63.66 |
| 2 | 4.30 | 9.92 |
| 3 | 3.18 | 5.84 |
| 4 | 2.78 | 4.60 |
| 5 | 2.57 | 4.03 |
| 6 | 2.45 | 3.71 |
| 7 | 2.36 | 3.50 |
| 8 | 2.31 | 3.36 |
| 9 | 2.26 | 3.25 |
| 10 | 2.23 | 3.17 |
| 11 | 2.20 | 3.11 |
| 12 | 2.18 | 3.06 |
| 13 | 2.16 | 3.01 |
| 14 | 2.14 | 2.98 |
| 15 | 2.13 | 2.95 |
| 16 | 2.12 | 2.92 |
| 17 | 2.11 | 2.90 |
| 18 | 2.10 | 2.88 |
| 19 | 2.09 | 2.86 |
| 20 | 2.09 | 2.84 |
| 21 | 2.08 | 2.83 |
| 22 | 2.07 | 2.82 |
| 23 | 2.07 | 2.81 |
| 24 | 2.06 | 2.80 |
| 25 | 2.06 | 2.79 |
| 26 | 2.06 | 2.78 |
| 27 | 2.05 | 2.77 |
| 28 | 2.05 | 2.76 |
| 29 | 2.04 | 2.76 |
| 30 | 2.04 | 2.75 |
| 40 | 2.02 | 2.70 |
| 50 | 2.01 | 2.68 |
| 60 | 2.00 | 2.66 |
| 80 | 1.99 | 2.64 |
| 100 | 1.98 | 2.63 |
| 120 | 1.98 | 2.62 |
| 200 | 1.97 | 2.60 |
| 500 | 1.96 | 2.59 |
| infinity | 1.96 | 2.58 |

**Table D-6 Critical Values of $t$ for a One-Tail Test (Values of $t_c$ are given without signs. You must determine whether $t_c$ is positive or negative from the alternative hypothesis.)**

| degrees of freedom $\sigma$ | $t_c$ for $\alpha = .05$ | $t_c$ for $\alpha = .01$ |
|---|---|---|
| 1 | 6.31 | 31.82 |
| 2 | 2.92 | 6.96 |
| 3 | 2.35 | 4.54 |
| 4 | 2.13 | 3.75 |
| 5 | 2.02 | 3.36 |
| 6 | 1.94 | 3.14 |
| 7 | 1.90 | 3.00 |
| 8 | 1.86 | 2.90 |
| 9 | 1.83 | 2.82 |
| 10 | 1.81 | 2.76 |
| 11 | 1.80 | 2.72 |
| 12 | 1.78 | 2.68 |
| 13 | 1.77 | 2.65 |
| 14 | 1.76 | 2.62 |
| 15 | 1.75 | 2.60 |
| 16 | 1.75 | 2.58 |
| 17 | 1.74 | 2.57 |
| 18 | 1.73 | 2.55 |
| 19 | 1.73 | 2.54 |
| 20 | 1.72 | 2.53 |
| 21 | 1.72 | 2.52 |
| 22 | 1.72 | 2.51 |
| 23 | 1.71 | 2.50 |
| 24 | 1.71 | 2.49 |
| 25 | 1.71 | 2.48 |
| 26 | 1.71 | 2.48 |
| 27 | 1.70 | 2.47 |
| 28 | 1.70 | 2.47 |
| 29 | 1.70 | 2.46 |
| 30 | 1.70 | 2.46 |
| 40 | 1.68 | 2.42 |
| 50 | 1.68 | 2.40 |
| 60 | 1.67 | 2.40 |
| 80 | 1.66 | 2.37 |
| 100 | 1.66 | 2.36 |
| 120 | 1.66 | 2.36 |
| 200 | 1.65 | 2.34 |
| 500 | 1.65 | 2.33 |
| infinity | 1.65 | 2.33 |

## Table D-7 Critical Values of $X^2$ for a One-Tail Test

NOTE: 2 TAIL TEST ON 484

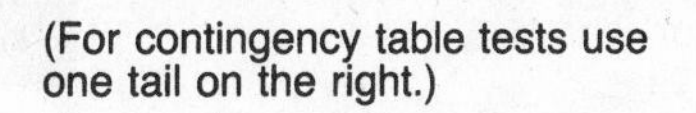


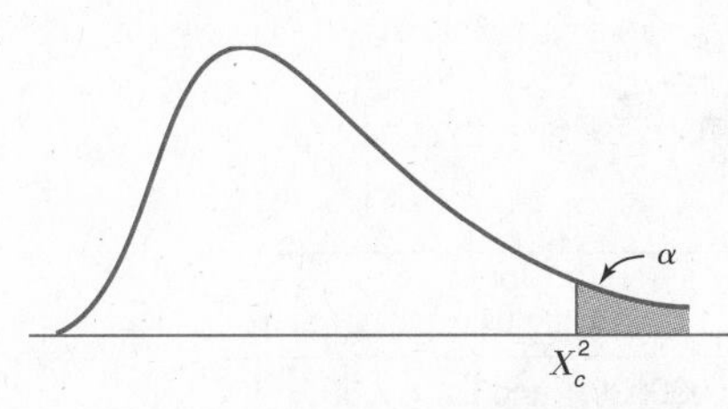


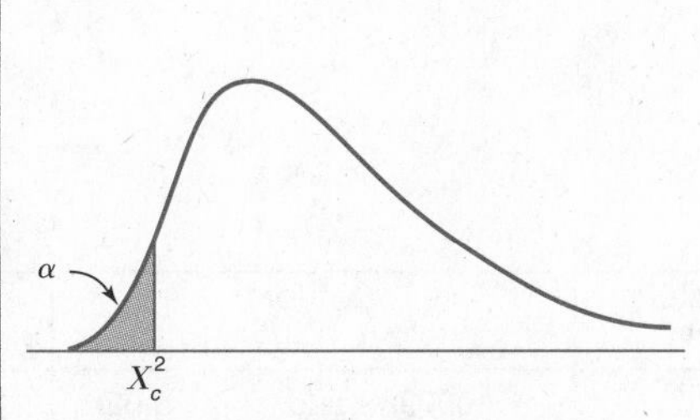


| degrees of freedom | one tail on the right: $X_c^2$ for $\alpha = .05$ | one tail on the right: $X_c^2$ for $\alpha = .01$ | one tail on the left: $X_c^2$ for $\alpha = .05$ | one tail on the left: $X_c^2$ for $\alpha = .01$ |
|---|---|---|---|---|
| 1 | 3.84 | 6.63 | .0039 | .00016 |
| 2 | 5.99 | 9.21 | .1026 | .0201 |
| 3 | 7.81 | 11.34 | .352 | .115 |
| 4 | 9.49 | 13.28 | .711 | .297 |
| 5 | 11.07 | 15.09 | 1.15 | .554 |
| 6 | 12.59 | 16.81 | 1.64 | .872 |
| 7 | 14.07 | 18.48 | 2.17 | 1.24 |
| 8 | 15.51 | 20.09 | 2.73 | 1.65 |
| 9 | 16.92 | 21.67 | 3.33 | 2.09 |
| 10 | 18.31 | 23.21 | 3.94 | 2.56 |
| 11 | 19.68 | 24.73 | 4.57 | 3.05 |
| 12 | 21.03 | 26.22 | 5.23 | 3.57 |
| 13 | 22.36 | 27.69 | 5.89 | 4.11 |
| 14 | 23.68 | 29.14 | 6.57 | 4.66 |
| 15 | 25.00 | 30.58 | 7.26 | 5.23 |
| 16 | 26.30 | 32.00 | 7.96 | 5.81 |
| 18 | 28.87 | 34.81 | 9.39 | 7.01 |
| 20 | 31.41 | 37.57 | 10.85 | 8.26 |
| 24 | 36.42 | 42.98 | 13.85 | 10.86 |
| 30 | 43.77 | 50.89 | 18.49 | 14.95 |
| 40 | 55.76 | 63.69 | 26.51 | 22.16 |
| 60 | 79.08 | 88.38 | 43.19 | 37.48 |
| 120 | 146.57 | 158.95 | 95.70 | 86.92 |

For critical values of $X^2$ whose degrees of freedom exceed 30 we can approximate $X_c^2$ by the formula

$$X_c^2 = \frac{1}{2}\left(z_c + \sqrt{2d - 1}\right)^2$$

where $d$ is the degrees of freedom and $z_c$ is either $\pm 1.65$ or $\pm 2.33$.

## Table D-8 Critical Values of $X^2$ for a Two-Tail Test

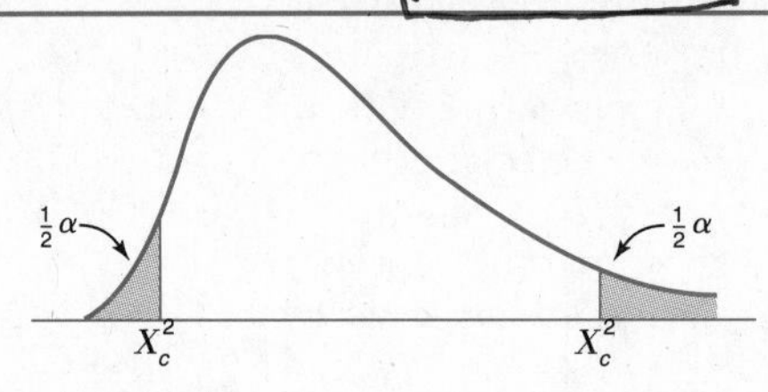


| degrees of freedom | $X_c^2$ for $\alpha = .05$ | | | $X_c^2$ for $\alpha = .01$ | | |
|---|---|---|---|---|---|---|
| 1 | .00098 | and | 5.02 | .000039 | and | 7.88 |
| 2 | .0506 | | 7.38 | .0100 | | 10.60 |
| 3 | .216 | | 9.35 | .0717 | | 12.84 |
| 4 | .484 | | 11.14 | .207 | | 14.86 |
| 5 | .831 | | 12.83 | .412 | | 16.75 |
| 6 | 1.24 | | 14.45 | .676 | | 18.55 |
| 7 | 1.69 | | 16.01 | .989 | | 20.28 |
| 8 | 2.18 | | 17.53 | 1.34 | | 21.96 |
| 9 | 2.70 | | 19.02 | 1.73 | | 23.59 |
| 10 | 3.25 | | 20.48 | 2.16 | | 25.19 |
| 11 | 3.82 | | 21.92 | 2.60 | | 26.76 |
| 12 | 4.40 | | 23.34 | 3.07 | | 28.30 |
| 13 | 5.01 | | 24.74 | 3.57 | | 29.82 |
| 14 | 5.63 | | 26.12 | 4.07 | | 31.32 |
| 15 | 6.26 | | 27.49 | 4.60 | | 32.80 |
| 16 | 6.91 | | 28.85 | 5.14 | | 34.27 |
| 18 | 8.23 | | 31.53 | 6.26 | | 37.16 |
| 20 | 9.59 | | 34.17 | 7.43 | | 40.00 |
| 24 | 12.40 | | 39.36 | 9.89 | | 45.56 |
| 30 | 16.79 | | 46.98 | 13.79 | | 53.67 |
| 40 | 24.43 | | 59.34 | 20.71 | | 66.77 |
| 60 | 40.48 | | 83.30 | 35.53 | | 91.95 |
| 120 | 91.58 | | 152.21 | 83.85 | | 163.64 |

For critical values of $X^2$ whose degrees of freedom exceed 30 we can approximate $X_c^2$ by the formula

$$X_c^2 = \frac{1}{2}\left(z_c + \sqrt{2d - 1}\right)^2$$

where $d$ is the degrees of freedom and $z_c$ is $\pm 1.96$ or $\pm 2.58$.

2

**Table D-9 Critical Values of $r$ for a Two-Tail Test (Values of $r$ are given without signs. All values are both positive and negative, that is, $r_c = \pm 1.00$.)**

| $n$ | $r_c$ for $\alpha = .05$ | $r_c$ for $\alpha = .01$ |
|---|---|---|
| 3 | 1.00 | 1.00 |
| 4 | .95 | .99 |
| 5 | .88 | .96 |
| 6 | .81 | .92 |
| 7 | .75 | .87 |
| 8 | .71 | .83 |
| 9 | .67 | .80 |
| 10 | .63 | .76 |
| 11 | .60 | .73 |
| 12 | .58 | .71 |
| 13 | .53 | .68 |
| 14 | .53 | .66 |
| 15 | .51 | .64 |
| 16 | .50 | .61 |
| 17 | .48 | .61 |
| 18 | .47 | .59 |
| 19 | .46 | .58 |
| 20 | .44 | .56 |
| 21 | .43 | .55 |
| 22 | .42 | .54 |
| 23 | .41 | .53 |
| 24 | .40 | .52 |
| 25 | .40 | .51 |
| 26 | .39 | .50 |
| 27 | .38 | .49 |
| 28 | .37 | .48 |
| 29 | .37 | .47 |
| 30 | .36 | .46 |

For values of $r_c$, when $n$ is greater than 30, use

$$r_c = \frac{t_c}{\sqrt{t_c^2 + (n - 2)}}$$

where $t_c$ is the corresponding critical value of $t$ for $(n - 2)$ degrees of freedom in Table D-5.

1

**Table D-10 Critical Values of $r$ for a One-Tail Test (Values of $r$ are given without signs. You must determine whether $r_c$ is positive or negative from the alternative hypothesis.)**

| $n$ | $r_c$ for $\alpha = .05$ | $r_c$ for $\alpha = .01$ |
|---|---|---|
| 3 | .99 | 1.00 |
| 4 | .90 | .98 |
| 5 | .81 | .93 |
| 6 | .73 | .88 |
| 7 | .67 | .83 |
| 8 | .62 | .79 |
| 9 | .58 | .75 |
| 10 | .54 | .72 |
| 11 | .52 | .69 |
| 12 | .50 | .66 |
| 13 | .48 | .63 |
| 14 | .46 | .61 |
| 15 | .44 | .59 |
| 16 | .42 | .57 |
| 17 | .41 | .56 |
| 18 | .40 | .54 |
| 19 | .39 | .53 |
| 20 | .38 | .52 |
| 21 | .37 | .50 |
| 22 | .36 | .49 |
| 23 | .35 | .48 |
| 24 | .34 | .47 |
| 25 | .34 | .46 |
| 26 | .33 | .45 |
| 27 | .32 | .45 |
| 28 | .32 | .44 |
| 29 | .31 | .43 |
| 30 | .31 | .42 |

For values of $r_c$, when $n$ is greater than 30, use

$$r_c = \frac{t_c}{\sqrt{t_c^2 + (n - 2)}}$$

where $t_c$ is the corresponding critical value of $t$ for $(n - 2)$ degrees of freedom in Table D-6.

F

1 TAIL .05

**Table D-11**
**Critical Values of $F$**
**for $\alpha = .05$**
**(for a One-Tail Test)**



Degrees of freedom for numerator

| Degrees of freedom for denominator | 1 | 2 | 3 | 4 | 5 | 6 | 7 | 8 | 9 | 10 | 12 | 15 | 20 | 24 | 30 | 40 | 50 | ∞ |
|---|---|---|---|---|---|---|---|---|---|---|---|---|---|---|---|---|---|---|
| 1 | 161 | 200 | 216 | 225 | 230 | 234 | 237 | 239 | 241 | 242 | 244 | 246 | 248 | 249 | 250 | 251 | 252 | 254 |
| 2 | 18.5 | 19.0 | 19.2 | 19.2 | 19.3 | 19.3 | 19.4 | 19.4 | 19.4 | 19.4 | 19.4 | 19.4 | 19.4 | 19.5 | 19.5 | 19.5 | 19.5 | 19.5 |
| 3 | 10.1 | 9.55 | 9.28 | 9.12 | 9.01 | 8.94 | 8.89 | 8.85 | 8.81 | 8.79 | 8.74 | 8.70 | 8.66 | 8.64 | 8.62 | 8.59 | 8.58 | 8.53 |
| 4 | 7.71 | 6.94 | 6.59 | 6.39 | 6.26 | 6.16 | 6.09 | 6.04 | 6.00 | 5.96 | 5.91 | 5.86 | 5.80 | 5.77 | 5.75 | 5.72 | 5.70 | 5.63 |
| 5 | 6.61 | 5.79 | 5.41 | 5.19 | 5.05 | 4.95 | 4.88 | 4.82 | 4.77 | 4.74 | 4.68 | 4.62 | 4.56 | 4.53 | 4.50 | 4.46 | 4.44 | 4.37 |
| 6 | 5.99 | 5.14 | 4.76 | 4.53 | 4.39 | 4.28 | 4.21 | 4.15 | 4.10 | 4.06 | 4.00 | 3.94 | 3.87 | 3.84 | 3.81 | 3.77 | 3.75 | 3.67 |
| 7 | 5.59 | 4.74 | 4.35 | 4.12 | 3.97 | 3.87 | 3.79 | 3.73 | 3.68 | 3.64 | 3.57 | 3.51 | 3.44 | 3.41 | 3.38 | 3.34 | 3.32 | 3.23 |
| 8 | 5.32 | 4.46 | 4.07 | 3.84 | 3.69 | 3.58 | 3.50 | 3.44 | 3.39 | 3.35 | 3.28 | 3.22 | 3.15 | 3.12 | 3.08 | 3.04 | 3.03 | 2.93 |
| 9 | 5.12 | 4.26 | 3.86 | 3.63 | 3.48 | 3.37 | 3.29 | 3.23 | 3.18 | 3.14 | 3.07 | 3.01 | 2.94 | 2.90 | 2.86 | 2.83 | 2.80 | 2.71 |
| 10 | 4.96 | 4.10 | 3.71 | 3.48 | 3.33 | 3.22 | 3.14 | 3.07 | 3.02 | 2.98 | 2.91 | 2.85 | 2.77 | 2.74 | 2.70 | 2.66 | 2.64 | 2.54 |
| 11 | 4.84 | 3.98 | 3.59 | 3.36 | 3.20 | 3.09 | 3.01 | 2.95 | 2.90 | 2.85 | 2.79 | 2.72 | 2.65 | 2.61 | 2.57 | 2.53 | 2.50 | 2.40 |
| 12 | 4.75 | 3.89 | 3.49 | 3.26 | 3.11 | 3.00 | 2.91 | 2.85 | 2.80 | 2.75 | 2.69 | 2.62 | 2.54 | 2.51 | 2.47 | 2.43 | 2.40 | 2.30 |
| 13 | 4.67 | 3.81 | 3.41 | 3.18 | 3.03 | 2.92 | 2.83 | 2.77 | 2.71 | 2.67 | 2.60 | 2.53 | 2.46 | 2.42 | 2.38 | 2.34 | 2.32 | 2.21 |
| 14 | 4.60 | 3.74 | 3.34 | 3.11 | 2.96 | 2.85 | 2.76 | 2.70 | 2.65 | 2.60 | 2.53 | 2.46 | 2.39 | 2.35 | 2.31 | 2.27 | 2.24 | 2.13 |
| 15 | 4.54 | 3.68 | 3.29 | 3.06 | 2.90 | 2.79 | 2.71 | 2.64 | 2.59 | 2.54 | 2.48 | 2.40 | 2.33 | 2.29 | 2.25 | 2.20 | 2.18 | 2.07 |
| 16 | 4.49 | 3.63 | 3.24 | 3.01 | 2.85 | 2.74 | 2.66 | 2.59 | 2.54 | 2.49 | 2.42 | 2.35 | 2.28 | 2.24 | 2.19 | 2.15 | 2.13 | 2.01 |
| 17 | 4.45 | 3.59 | 3.20 | 2.96 | 2.81 | 2.70 | 2.61 | 2.55 | 2.49 | 2.45 | 2.38 | 2.31 | 2.23 | 2.19 | 2.15 | 2.10 | 2.08 | 1.96 |
| 18 | 4.41 | 3.55 | 3.16 | 2.93 | 2.77 | 2.66 | 2.58 | 2.51 | 2.46 | 2.41 | 2.34 | 2.27 | 2.19 | 2.15 | 2.11 | 2.06 | 2.04 | 1.92 |
| 19 | 4.38 | 3.52 | 3.13 | 2.90 | 2.74 | 2.63 | 2.54 | 2.48 | 2.42 | 2.38 | 2.31 | 2.23 | 2.16 | 2.11 | 2.07 | 2.03 | 2.00 | 1.88 |
| 20 | 4.35 | 3.49 | 3.10 | 2.87 | 2.71 | 2.60 | 2.51 | 2.45 | 2.39 | 2.35 | 2.28 | 2.20 | 2.12 | 2.08 | 2.04 | 1.99 | 1.96 | 1.84 |
| 25 | 4.24 | 3.39 | 2.99 | 2.76 | 2.60 | 2.49 | 2.40 | 2.34 | 2.28 | 2.24 | 2.16 | 2.09 | 2.01 | 1.96 | 1.92 | 1.87 | 1.84 | 1.71 |
| 30 | 4.17 | 3.32 | 2.92 | 2.69 | 2.53 | 2.42 | 2.33 | 2.27 | 2.21 | 2.16 | 2.09 | 2.01 | 1.93 | 1.89 | 1.84 | 1.79 | 1.76 | 1.62 |
| 40 | 4.08 | 3.23 | 2.84 | 2.61 | 2.45 | 2.34 | 2.25 | 2.18 | 2.12 | 2.08 | 2.00 | 1.92 | 1.84 | 1.79 | 1.74 | 1.69 | 1.66 | 1.51 |
| 50 | 4.03 | 3.18 | 2.79 | 2.56 | 2.40 | 2.29 | 2.20 | 2.13 | 2.07 | 2.02 | 1.95 | 1.87 | 1.78 | 1.74 | 1.69 | 1.63 | 1.60 | 1.44 |
| ∞ | 3.84 | 3.00 | 2.60 | 2.37 | 2.21 | 2.10 | 2.01 | 1.94 | 1.88 | 1.83 | 1.75 | 1.67 | 1.57 | 1.52 | 1.46 | 1.39 | 1.35 | 1.00 |

F

2 Tail .05

Table D-12
Critical Values of $F$
for $\alpha = .025$
(for a Two-Tail Test
with $\alpha = .05$)

.025

$F$

Degrees of freedom for numerator (columns); Degrees of freedom for denominator (rows)

| | 1 | 2 | 3 | 4 | 5 | 6 | 7 | 8 | 9 | 10 | 11 | 12 | 15 | 20 | 24 | 30 | 40 | 50 | ∞ |
|---|---|---|---|---|---|---|---|---|---|---|---|---|---|---|---|---|---|---|---|
| 1 | 648 | 800 | 864 | 900 | 922 | 937 | 948 | 957 | 963 | 969 | 973 | 977 | 985 | 993 | 997 | 1000 | 1010 | 1010 | 1020 |
| 2 | 38.5 | 39.0 | 39.2 | 39.2 | 39.3 | 39.3 | 39.4 | 39.4 | 39.4 | 39.4 | 39.4 | 39.4 | 39.4 | 39.4 | 39.5 | 39.5 | 39.5 | 39.5 | 39.5 |
| 3 | 17.4 | 16.0 | 15.4 | 15.1 | 14.9 | 14.7 | 14.6 | 14.5 | 14.5 | 14.4 | 14.3 | 14.3 | 14.3 | 14.2 | 14.1 | 14.1 | 14.0 | 14.0 | 13.9 |
| 4 | 12.2 | 10.6 | 9.98 | 9.60 | 9.36 | 9.20 | 9.07 | 8.98 | 8.90 | 8.84 | 8.79 | 8.75 | 8.66 | 8.56 | 8.51 | 8.46 | 8.41 | 8.38 | 8.26 |
| 5 | 10.0 | 8.43 | 7.76 | 7.39 | 7.15 | 6.98 | 6.85 | 6.76 | 6.68 | 6.62 | 6.57 | 6.52 | 6.43 | 6.33 | 6.28 | 6.23 | 6.18 | 6.14 | 6.02 |
| 6 | 8.81 | 7.26 | 6.60 | 6.23 | 5.99 | 5.82 | 5.70 | 5.60 | 5.52 | 5.46 | 5.41 | 5.37 | 5.27 | 5.17 | 5.12 | 5.07 | 5.01 | 4.98 | 4.85 |
| 7 | 8.07 | 6.54 | 5.89 | 5.52 | 5.29 | 5.12 | 4.99 | 4.90 | 4.82 | 4.76 | 4.71 | 4.67 | 4.57 | 4.47 | 4.42 | 4.36 | 4.31 | 4.27 | 4.14 |
| 8 | 7.57 | 6.06 | 5.42 | 5.05 | 4.82 | 4.65 | 4.53 | 4.43 | 4.36 | 4.30 | 4.25 | 4.20 | 4.10 | 4.00 | 3.95 | 3.89 | 3.84 | 3.80 | 3.67 |
| 9 | 7.21 | 5.71 | 5.08 | 4.72 | 4.48 | 4.32 | 4.20 | 4.10 | 4.03 | 3.96 | 3.91 | 3.87 | 3.77 | 3.67 | 3.61 | 3.56 | 3.51 | 3.47 | 3.33 |
| 10 | 6.94 | 5.46 | 4.83 | 4.47 | 4.24 | 4.07 | 3.95 | 3.85 | 3.78 | 3.72 | 3.67 | 3.62 | 3.52 | 3.42 | 3.37 | 3.31 | 3.26 | 3.22 | 3.08 |
| 11 | 6.72 | 5.26 | 4.63 | 4.28 | 4.04 | 3.88 | 3.76 | 3.66 | 3.59 | 3.53 | 3.48 | 3.43 | 3.33 | 3.23 | 3.17 | 3.12 | 3.06 | 3.02 | 2.88 |
| 12 | 6.55 | 5.10 | 4.47 | 4.12 | 3.89 | 3.73 | 3.61 | 3.51 | 3.44 | 3.37 | 3.32 | 3.28 | 3.18 | 3.07 | 3.02 | 2.96 | 2.91 | 2.87 | 2.72 |
| 13 | 6.41 | 4.97 | 4.35 | 4.10 | 3.77 | 3.60 | 3.48 | 3.39 | 3.31 | 3.25 | 3.20 | 3.15 | 3.05 | 2.95 | 2.89 | 2.84 | 2.78 | 2.74 | 2.60 |
| 14 | 6.30 | 4.86 | 4.24 | 3.89 | 3.66 | 3.50 | 3.38 | 3.29 | 3.20 | 3.15 | 3.10 | 3.05 | 2.95 | 2.84 | 2.79 | 2.73 | 2.67 | 2.64 | 2.49 |
| 15 | 6.20 | 4.77 | 4.15 | 3.80 | 3.58 | 3.41 | 3.29 | 3.20 | 3.12 | 3.06 | 3.01 | 2.96 | 2.86 | 2.76 | 2.70 | 2.64 | 2.59 | 2.55 | 2.40 |
| 16 | 6.12 | 4.69 | 4.08 | 3.73 | 3.50 | 3.34 | 3.22 | 3.12 | 3.05 | 2.99 | 2.93 | 2.89 | 2.79 | 2.68 | 2.63 | 2.57 | 2.51 | 2.47 | 2.32 |
| 17 | 6.04 | 4.62 | 4.01 | 3.66 | 3.44 | 3.28 | 3.16 | 3.06 | 2.98 | 2.92 | 2.87 | 2.82 | 2.72 | 2.62 | 2.56 | 2.50 | 2.44 | 2.41 | 2.25 |
| 18 | 5.98 | 4.56 | 3.95 | 3.61 | 3.38 | 3.22 | 3.10 | 3.01 | 2.93 | 2.87 | 2.81 | 2.77 | 2.67 | 2.56 | 2.50 | 2.44 | 2.38 | 2.35 | 2.19 |
| 19 | 5.92 | 4.51 | 3.90 | 3.56 | 3.33 | 3.17 | 3.05 | 2.96 | 2.88 | 2.82 | 2.76 | 2.72 | 2.62 | 2.51 | 2.45 | 2.39 | 2.33 | 2.30 | 2.13 |
| 20 | 5.87 | 4.46 | 3.86 | 3.51 | 3.29 | 3.13 | 3.01 | 2.91 | 2.84 | 2.77 | 2.72 | 2.68 | 2.57 | 2.46 | 2.41 | 2.35 | 2.29 | 2.25 | 2.09 |
| 25 | 5.69 | 4.29 | 3.69 | 3.35 | 3.13 | 2.97 | 2.85 | 2.75 | 2.68 | 2.61 | 2.56 | 2.51 | 2.41 | 2.30 | 2.24 | 2.18 | 2.45 | 2.08 | 1.91 |
| 30 | 5.57 | 4.18 | 3.59 | 3.25 | 3.03 | 2.87 | 2.75 | 2.65 | 2.57 | 2.51 | 2.46 | 2.41 | 2.31 | 2.20 | 2.14 | 2.07 | 2.01 | 1.97 | 1.79 |
| 40 | 5.42 | 4.05 | 3.46 | 3.13 | 2.90 | 2.74 | 2.62 | 2.53 | 2.45 | 2.39 | 2.33 | 2.29 | 2.18 | 2.07 | 2.01 | 1.94 | 1.88 | 1.83 | 1.64 |

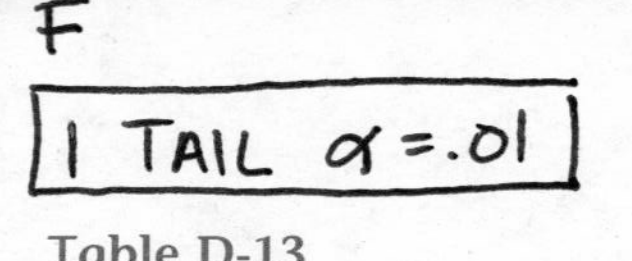


**Table D-13**
**Critical Values of $F$**
**for $\alpha = .01$**
**(for a One-Tail Test)**

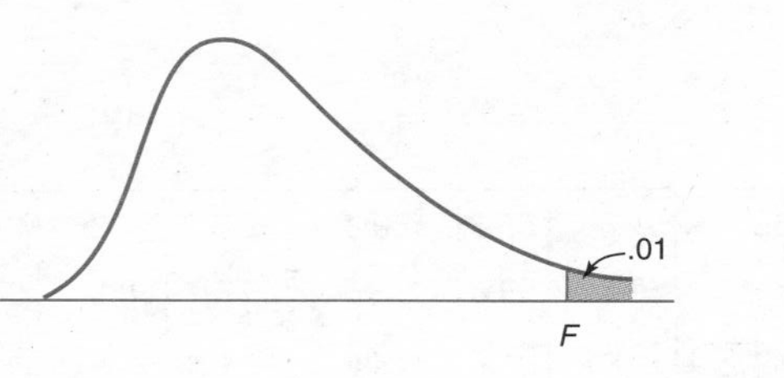


Degrees of freedom for numerator

| Degrees of freedom for denominator | 1 | 2 | 3 | 4 | 5 | 6 | 7 | 8 | 9 | 10 | 12 | 15 | 20 | 24 | 30 | 40 | 50 | ∞ |
|---|---|---|---|---|---|---|---|---|---|---|---|---|---|---|---|---|---|---|
| 1 | 4052 | 5000 | 5403 | 5625 | 5764 | 5859 | 5928 | 5982 | 6023 | 6056 | 6106 | 6157 | 6209 | 6235 | 6261 | 6287 | 6302 | 6366 |
| 2 | 98.5 | 99.0 | 99.2 | 99.2 | 99.3 | 99.3 | 99.4 | 99.4 | 99.4 | 99.4 | 99.4 | 99.4 | 99.4 | 99.5 | 99.5 | 99.5 | 99.5 | 99.5 |
| 3 | 34.1 | 30.8 | 29.5 | 28.7 | 28.2 | 27.9 | 27.7 | 27.5 | 27.3 | 27.2 | 27.1 | 26.9 | 26.7 | 26.6 | 26.5 | 26.4 | 26.4 | 26.1 |
| 4 | 21.2 | 18.0 | 16.7 | 16.0 | 15.5 | 15.2 | 15.0 | 14.8 | 14.7 | 14.5 | 14.4 | 14.2 | 14.0 | 13.9 | 13.8 | 13.7 | 13.7 | 13.5 |
| 5 | 16.3 | 13.3 | 12.1 | 11.4 | 11.0 | 10.7 | 10.5 | 10.3 | 10.2 | 10.1 | 9.89 | 9.72 | 9.55 | 9.47 | 9.38 | 9.29 | 9.24 | 9.02 |
| 6 | 13.7 | 10.9 | 9.78 | 9.15 | 8.75 | 8.47 | 8.26 | 8.10 | 7.98 | 7.87 | 7.72 | 7.56 | 7.40 | 7.31 | 7.23 | 7.14 | 7.09 | 6.88 |
| 7 | 12.2 | 9.55 | 8.45 | 7.85 | 7.46 | 7.19 | 6.99 | 6.84 | 6.72 | 6.62 | 6.47 | 6.31 | 6.16 | 6.07 | 5.99 | 5.91 | 5.85 | 5.65 |
| 8 | 11.3 | 8.65 | 7.59 | 7.01 | 6.63 | 6.37 | 6.18 | 6.03 | 5.91 | 5.81 | 5.67 | 5.52 | 5.36 | 5.28 | 5.20 | 5.12 | 5.06 | 4.86 |
| 9 | 10.6 | 8.02 | 6.99 | 6.42 | 6.06 | 5.80 | 5.61 | 5.47 | 5.35 | 5.26 | 5.11 | 4.96 | 4.81 | 4.73 | 4.65 | 4.57 | 4.51 | 4.31 |
| 10 | 10.0 | 7.56 | 6.55 | 5.99 | 5.64 | 5.39 | 5.20 | 5.06 | 4.94 | 4.85 | 4.71 | 4.56 | 4.41 | 4.33 | 4.25 | 4.17 | 4.12 | 3.91 |
| 11 | 9.65 | 7.21 | 6.22 | 5.67 | 5.32 | 5.07 | 4.89 | 4.74 | 4.63 | 4.54 | 4.40 | 4.25 | 4.10 | 4.02 | 3.94 | 3.86 | 3.80 | 3.60 |
| 12 | 9.33 | 6.93 | 5.95 | 5.41 | 5.06 | 4.82 | 4.64 | 4.50 | 4.39 | 4.30 | 4.16 | 4.01 | 3.86 | 3.78 | 3.70 | 3.62 | 3.56 | 3.36 |
| 13 | 9.07 | 6.70 | 5.74 | 5.21 | 4.86 | 4.62 | 4.44 | 4.30 | 4.19 | 4.10 | 3.96 | 3.82 | 3.66 | 3.59 | 3.51 | 3.43 | 3.37 | 3.17 |
| 14 | 8.86 | 6.51 | 5.56 | 5.04 | 4.70 | 4.46 | 4.28 | 4.14 | 4.03 | 3.94 | 3.80 | 3.66 | 3.51 | 3.43 | 3.35 | 3.27 | 3.21 | 3.00 |
| 15 | 8.68 | 6.36 | 5.42 | 4.89 | 4.56 | 4.32 | 4.14 | 4.00 | 3.89 | 3.80 | 3.67 | 3.52 | 3.37 | 3.29 | 3.21 | 3.13 | 3.07 | 2.87 |
| 16 | 8.53 | 6.23 | 5.29 | 4.77 | 4.44 | 4.20 | 4.03 | 3.89 | 3.78 | 3.69 | 3.55 | 3.41 | 3.26 | 3.18 | 3.10 | 3.02 | 2.96 | 2.75 |
| 17 | 8.40 | 6.11 | 5.19 | 4.67 | 4.34 | 4.10 | 3.93 | 3.79 | 3.68 | 3.59 | 3.46 | 3.31 | 3.16 | 3.08 | 3.00 | 2.92 | 2.86 | 2.65 |
| 18 | 8.29 | 6.01 | 5.09 | 4.58 | 4.25 | 4.01 | 3.84 | 3.71 | 3.60 | 3.51 | 3.37 | 3.23 | 3.08 | 3.00 | 2.92 | 2.84 | 2.78 | 2.57 |
| 19 | 8.19 | 5.93 | 5.01 | 4.50 | 4.17 | 3.94 | 3.77 | 3.63 | 3.52 | 3.43 | 3.30 | 3.15 | 3.00 | 2.92 | 2.84 | 2.76 | 2.70 | 2.49 |
| 20 | 8.10 | 5.85 | 4.94 | 4.43 | 4.10 | 3.87 | 3.70 | 3.56 | 3.46 | 3.37 | 3.23 | 3.09 | 2.94 | 2.86 | 2.78 | 2.69 | 2.63 | 2.42 |
| 25 | 7.77 | 5.57 | 4.68 | 4.18 | 3.86 | 3.63 | 3.46 | 3.32 | 3.22 | 3.13 | 2.99 | 2.85 | 2.70 | 2.62 | 2.53 | 2.45 | 2.40 | 2.17 |
| 30 | 7.56 | 5.39 | 4.51 | 4.02 | 3.70 | 3.47 | 3.30 | 3.17 | 3.07 | 2.98 | 2.84 | 2.70 | 2.55 | 2.47 | 2.39 | 2.30 | 2.24 | 2.01 |
| 40 | 7.31 | 5.18 | 4.31 | 3.83 | 3.51 | 3.29 | 3.12 | 2.99 | 2.89 | 2.80 | 2.66 | 2.52 | 2.37 | 2.29 | 2.20 | 2.11 | 2.05 | 1.80 |
| 50 | 7.17 | 5.06 | 4.20 | 3.72 | 3.41 | 3.18 | 3.02 | 2.88 | 2.78 | 2.70 | 2.56 | 2.42 | 2.26 | 2.18 | 2.10 | 2.00 | 1.94 | 1.69 |
| ∞ | 6.63 | 4.61 | 3.78 | 3.32 | 3.02 | 2.80 | 2.64 | 2.51 | 2.41 | 2.32 | 2.18 | 2.04 | 1.88 | 1.79 | 1.70 | 1.59 | 1.52 | 1.00 |

F

2 TAIL $\alpha$ = .01

**Table D-14**
**Critical Values of $F$ for $\alpha = .005$ (for a Two-Tail Test with $\alpha = .01$)**

.005

$F$

Degrees of freedom for numerator (columns); Degrees of freedom for denominator (rows)

| | 1 | 2 | 3 | 4 | 5 | 6 | 7 | 8 | 9 | 10 | 11 | 12 | 15 | 20 | 24 | 30 | 40 | 50 | ∞ |
|---|---|---|---|---|---|---|---|---|---|---|---|---|---|---|---|---|---|---|---|
| 1 | 16,200 | 20,000 | 21,600 | 22,500 | 23,100 | 23,400 | 23,700 | 23,900 | 24,100 | 24,200 | 24,300 | 24,400 | 24,630 | 24,836 | 24,940 | 25,040 | 25,148 | 25,211 | 25,465 |
| 2 | 198 | 199 | 199 | 199 | 199 | 199 | 199 | 199 | 199 | 199 | 199 | 199 | 199 | 199 | 199 | 199 | 199 | 199 | 199 |
| 3 | 55.6 | 49.8 | 47.5 | 46.2 | 45.4 | 44.8 | 44.4 | 44.1 | 43.9 | 43.7 | 43.5 | 43.4 | 43.1 | 42.8 | 42.6 | 42.5 | 42.3 | 42.2 | 41.8 |
| 4 | 31.3 | 26.3 | 24.3 | 23.2 | 22.4 | 22.0 | 21.6 | 21.4 | 21.1 | 21.0 | 20.8 | 20.7 | 20.4 | 20.2 | 20.0 | 19.9 | 19.8 | 19.7 | 19.3 |
| 5 | 22.8 | 18.3 | 16.5 | 15.6 | 14.9 | 14.5 | 14.2 | 14.0 | 13.8 | 13.6 | 13.5 | 13.4 | 13.1 | 12.9 | 12.8 | 12.7 | 12.5 | 12.4 | 12.1 |
| 6 | 18.6 | 14.5 | 12.9 | 12.0 | 11.5 | 11.1 | 10.8 | 10.6 | 10.4 | 10.3 | 10.1 | 10.0 | 9.81 | 9.59 | 9.47 | 9.36 | 9.24 | 9.17 | 8.88 |
| 7 | 16.2 | 12.4 | 10.9 | 10.1 | 9.52 | 9.16 | 8.89 | 8.68 | 8.52 | 8.38 | 8.27 | 8.18 | 7.97 | 7.75 | 7.65 | 7.53 | 7.42 | 7.35 | 7.08 |
| 8 | 14.7 | 11.0 | 9.60 | 8.81 | 8.30 | 7.95 | 7.69 | 7.50 | 7.34 | 7.21 | 7.10 | 7.01 | 6.81 | 6.61 | 6.50 | 6.40 | 6.29 | 6.22 | 5.95 |
| 9 | 13.6 | 10.1 | 8.72 | 7.96 | 7.47 | 7.13 | 6.88 | 6.69 | 6.54 | 6.42 | 6.32 | 6.23 | 6.03 | 5.83 | 5.73 | 5.62 | 5.52 | 5.45 | 5.19 |
| 10 | 12.8 | 9.43 | 8.08 | 7.34 | 6.87 | 6.54 | 6.30 | 6.12 | 5.97 | 5.85 | 5.75 | 5.66 | 5.47 | 5.27 | 5.17 | 5.07 | 4.97 | 4.90 | 4.64 |
| 11 | 12.2 | 8.91 | 7.60 | 6.88 | 6.42 | 6.10 | 5.86 | 5.68 | 5.54 | 5.42 | 5.32 | 5.24 | 5.05 | 4.86 | 4.76 | 4.65 | 4.55 | 4.49 | 4.23 |
| 12 | 11.8 | 8.51 | 7.23 | 6.52 | 6.07 | 5.76 | 5.52 | 5.35 | 5.20 | 5.09 | 4.99 | 4.91 | 4.72 | 4.53 | 4.43 | 4.33 | 4.23 | 4.16 | 3.90 |
| 13 | 11.4 | 8.19 | 6.93 | 6.23 | 5.79 | 5.48 | 5.25 | 5.08 | 4.94 | 4.82 | 4.73 | 4.64 | 4.46 | 4.27 | 4.17 | 4.07 | 3.97 | 3.91 | 3.65 |
| 14 | 11.1 | 7.92 | 6.68 | 6.00 | 5.53 | 5.26 | 5.03 | 4.86 | 4.72 | 4.60 | 4.51 | 4.43 | 4.25 | 4.06 | 3.96 | 3.86 | 3.76 | 3.70 | 3.44 |
| 15 | 10.8 | 7.70 | 6.48 | 5.80 | 5.37 | 5.07 | 4.85 | 4.67 | 4.54 | 4.42 | 4.33 | 4.25 | 4.07 | 3.88 | 3.79 | 3.69 | 3.59 | 3.52 | 3.26 |
| 16 | 10.6 | 7.51 | 6.30 | 5.64 | 5.21 | 4.91 | 4.69 | 4.52 | 4.38 | 4.27 | 4.18 | 4.10 | 3.92 | 3.73 | 3.64 | 3.54 | 3.44 | 3.37 | 3.11 |
| 17 | 10.4 | 7.35 | 6.16 | 5.50 | 5.07 | 4.78 | 4.56 | 4.39 | 4.25 | 4.14 | 4.05 | 3.97 | 3.79 | 3.61 | 3.51 | 3.41 | 3.31 | 3.25 | 2.98 |
| 18 | 10.2 | 7.21 | 6.03 | 5.37 | 4.96 | 4.66 | 4.44 | 4.28 | 4.14 | 4.03 | 3.94 | 3.86 | 3.68 | 3.50 | 3.40 | 3.30 | 3.20 | 3.14 | 2.87 |
| 19 | 10.1 | 7.09 | 5.92 | 5.27 | 4.85 | 4.56 | 4.34 | 4.18 | 4.04 | 3.93 | 3.84 | 3.76 | 3.59 | 3.40 | 3.31 | 3.21 | 3.11 | 3.04 | 2.78 |
| 20 | 9.94 | 6.99 | 5.82 | 5.17 | 4.76 | 4.47 | 4.26 | 4.09 | 3.96 | 3.85 | 3.76 | 3.68 | 3.50 | 3.32 | 3.22 | 3.12 | 3.02 | 2.96 | 2.69 |
| 25 | 9.48 | 6.60 | 5.46 | 4.84 | 4.43 | 4.15 | 3.94 | 3.78 | 3.64 | 3.54 | 3.44 | 3.37 | 3.20 | 3.01 | 2.92 | 2.82 | 2.72 | 2.65 | 2.38 |
| 30 | 9.18 | 6.35 | 5.24 | 4.62 | 4.23 | 3.95 | 3.74 | 3.58 | 3.45 | 3.34 | 3.25 | 3.18 | 3.01 | 2.82 | 2.73 | 2.63 | 2.52 | 2.46 | 2.18 |
| 40 | 8.83 | 6.07 | 4.98 | 4.37 | 3.99 | 3.71 | 3.51 | 3.35 | 3.22 | 3.12 | 3.03 | 2.95 | 2.78 | 2.60 | 2.50 | 2.40 | 2.30 | 2.23 | 1.93 |

## Table D-15 Critical Values of $U$ for a One-Tail Test

| $n_1$ | $n_2$ | $\alpha = .05$ | $\alpha = .01$ |
|---|---|---|---|
| 3 | 2 | | |
| | 3 | 9 | |
| 4 | 2 | | |
| | 3 | 12 | |
| | 4 | 15 | |
| 5 | 2 | 10 | |
| | 3 | 14 | |
| | 4 | 18 | 20 |
| | 5 | 21 | 24 |
| 6 | 2 | 12 | |
| | 3 | 16 | |
| | 4 | 21 | 23 |
| | 5 | 25 | 28 |
| | 6 | 29 | 33 |
| 7 | 2 | 14 | |
| | 3 | 19 | 21 |
| | 4 | 24 | 27 |
| | 5 | 29 | 32 |
| | 6 | 34 | 38 |
| | 7 | 38 | 43 |
| 8 | 2 | 15 | |
| | 3 | 21 | 24 |
| | 4 | 27 | 30 |
| | 5 | 32 | 36 |
| | 6 | 38 | 42 |
| | 7 | 43 | 49 |
| | 8 | 49 | 55 |
| 9 | 1 | | |
| | 2 | 17 | |
| | 3 | 23 | 26 |
| | 4 | 30 | 33 |
| | 5 | 36 | 40 |
| | 6 | 42 | 47 |
| | 7 | 48 | 54 |
| | 8 | 54 | 61 |
| | 9 | 60 | 67 |
| 10 | 1 | | |
| | 2 | 19 | |
| | 3 | 26 | 29 |
| | 4 | 33 | 37 |
| | 5 | 39 | 44 |
| | 6 | 46 | 52 |
| | 7 | 53 | 59 |
| | 8 | 60 | 67 |
| | 9 | 66 | 74 |
| | 10 | 73 | 81 |

Table adapted from *Handbook of Statistical Tables*, D. B. Owen, Addison-Wesley Publishing Co., with permission.

**Table D-16 Critical Values of $U$ for a Two-Tail Test**

| $n_1$ | $n_2$ | $\alpha$ = .05 | $\alpha$ = .01 |
|---|---|---|---|
| 3 | 2 | | |
| | 3 | | |
| 4 | 2 | | |
| | 3 | | |
| | 4 | 16 | |
| 5 | 2 | | |
| | 3 | 15 | |
| | 4 | 19 | |
| | 5 | 23 | 25 |
| 6 | 2 | | |
| | 3 | 17 | |
| | 4 | 22 | 24 |
| | 5 | 27 | 29 |
| | 6 | 31 | 34 |
| 7 | 2 | | |
| | 3 | 20 | |
| | 4 | 25 | 28 |
| | 5 | 30 | 34 |
| | 6 | 36 | 39 |
| | 7 | 41 | 45 |
| 8 | 2 | 16 | |
| | 3 | 22 | |
| | 4 | 28 | 31 |
| | 5 | 34 | 38 |
| | 6 | 40 | 44 |
| | 7 | 46 | 50 |
| | 8 | 51 | 57 |
| 9 | 1 | | |
| | 2 | 18 | |
| | 3 | 25 | 27 |
| | 4 | 32 | 35 |
| | 5 | 38 | 42 |
| | 6 | 44 | 49 |
| | 7 | 51 | 56 |
| | 8 | 57 | 63 |
| | 9 | 64 | 70 |
| 10 | 1 | | |
| | 2 | 20 | |
| | 3 | 27 | 30 |
| | 4 | 35 | 38 |
| | 5 | 42 | 46 |
| | 6 | 49 | 54 |
| | 7 | 56 | 61 |
| | 8 | 63 | 69 |
| | 9 | 70 | 77 |
| | 10 | 77 | 84 |

Table adapted from *Handbook of Statistical Tables,* D. B. Owen, Addison-Wesley Publishing Co., with permission.

# Answers to Odd-Numbered Exercises*

## Chapter 1: Introduction

1-1 Many answers are possible; here are some possibilities.

(a) An experiment to see if a new cancer treatment is better than an old one.

(b) Surveys to see how voters are planning to vote.

(c) Statistics to follow trends in SAT scores.

(d) Baseball: Statistics to see if it is really better to bunt with no outs and a player on first.

(e) Quality control procedures to see if a manufacturing process is working up to specifications.

(f) Charts for tracking trends in sales over several years.

(g) Unemployment statistics help describe the state of the economy.

(h) Statistics for summarizing and combining laboratory data.

1-3 Many answers are possible. People admire statistics when it allows a clear pattern to emerge from a mass of data. People fear statistics when they think it fails to take their individual needs into account.

---

*Do not be concerned if your answer is off by one digit in the last place. This is often due to rounding.

1-5 Possibly not, because neighborhoods tend to differ from one another if the city is large enough. So the feelings of the people in one neighborhood may not represent those in other neighborhoods. As a sample, the neighborhood may not be representative of the whole city, which is the population.

1-7 Yes, if he is not cheating in some way, because the coin had an equal chance of falling heads or tails each time. Runs of heads sometimes happen just by random luck.

1-9 (a) This might mean that one report was just reusing the results of the other one. Check that these are independent surveys.

(b) The accounting firm probably did the most accurate job. The restaurant association studies may be biased in the direction of lost sales, and the city may be biased toward results that imply high tax returns.

1-11 No. The letter writers are a volunteer sample and probably are not representative of the population, which is ALL the senator's constituents.

1-15 (a) Statistical inference; not all districts were used.

(b) Statistical inference; not all fish were examined.

(c) Descriptive statistics; all the SAT scores were recorded.

(d) Statistical inference; based on past events which are assumed to be representative.

(e) Descriptive statistics; uses all the data.

1-19 (a) 16.4 (b) 50,600 (c) 40,540 (d) 18.1 (e) $4\bar{0},000$ (f) 19.90 (g) $106\bar{0}$ (h) 1060 (i) 1100 (j) 1000

1-21 (a) 0.4 (b) 0.3

(c) Rounding at different stages of the calculation may produce different final answers.

1-23 The machine cannot accurately combine very large and very small numbers because of the limit on the number of digits in the display. The machine essentially ignores the small number.

1-25 Just because event $B$ happens *after* event $A$ does not mean that the first event caused the second one. A more complete study would ask people *why* they bought the product. Two other possible reasons why the second survey found a higher percentage of users: Random variation in the surveys, but no real difference in sales; the toothpaste went on sale, more people bought it because of price, not because of advertising.

1-27 Just because event $B$ happens *after* event $A$ does not mean that the first event caused the second one. It would be helpful to know the percentage of milk drinkers who become marijuana users; this is not the percentage given in the problem.

## Chapter 2: Common Statistical Measures

2-1 Mean ≈ $27,000 (It is doubtful that the salaries were exact to the nearest penny as given, hence the mean is rounded.)
Median = $26,000
Mode = $26,000

The median or mode salaries seem more "typical."

2-3 Mean 3.3
Median 3.25 ≈ 3.3
No mode

2-5 Possible answers include:

| | | | |
|---|---|---|---|
| 70 | 50 | 1 | −1 |
| 70 | 60 | 2 | 0 |
| 70 | 70 | 3 | 0 |
| 70 | 80 | 4 | 0 |
| 70 | 90 | 345 | 351 |
| 350 | 350 | 350 | 350 |

We can create an unlimited number of such distributions starting with any 4 numbers, as long as the sum of the 5 numbers is 350.

2-7 $85.6 \approx 86$

2-9 If the higher wages increase a lot, while the lower wages decrease. One possible answer is:

| ID | Hourly Wage, $ | |
|---|---|---|
| 1 | 6 | 6 |
| 2 | 8 | 8.50 |
| 3 | 9 | 8 |
| 4 | 12 | 18 |
| 5 | 15 | 20 |
| Mean | 10 | 12.10 → up $2.10 |
| Median | 9 | 8.50 → down $0.50 |

2-11 $n = 1000 + 500 + 500 = 2000$

$$\text{Mean} = \frac{5000 + 3000 + 4000}{2000} = 6$$

Median = mean of 1000th and 1001st values

$$= \frac{5 + 6}{2} = 5.5$$

Mode = 5

2-13 (a) The year the data represent, the median age at first marriage of brides, and the median age at first marriage of grooms.

(b) Both medians appear to be increasing over the years reported.

(c) Higher. The ages most likely included more older people, say aged 30 or 40 at their first marriage, than it did 15- or 12-year-olds.

2-15 Although a median is usually used for average incomes, an advertising agency may prefer to use the mean if it is higher. The average was most likely calculated from a sample of some of the magazine's subscribers or from subscription information.

2-17 Probably not. $75 + 78 + 82 = 235$, and $4(89.5) = 358$, therefore he needs $358 - 235 = 123$.

2-19 If 10 families had 23 children, then we should correctly refer to the "average number of children" rather than the "average family."

2-21 (a) $\Sigma Y = 35$ (b) $\Sigma Y^2 = 203$ (c) $(\Sigma Y)^2 = 35^2 = 1225$ (d) and (e) 5 (f) $\Sigma(Y - 2) = 21$ (g) $\Sigma(Y - \bar{Y}) = 0$ (h) $\Sigma Y - \bar{Y} = 30$ (i) and

(j) $\frac{14}{3}$ or $4.666\cdots$ (k) $(\Sigma Y)^2 = 35^2 = 1225 > \Sigma Y^2 = 203$

2-23 (a) $\Sigma X = 12$ (b) and (c) 2 (d) $\Sigma(X - m) = 0$ (e) $(\Sigma X)^2 = 144$ (f) $\Sigma X^2 = 46$ (g) 4.4

2-25 (a) 5.2 (b) and (c) 18.7

2-27 Some possible answers $(0, 0, x)$ for any number $x$; $(4, 12, -3)$, or $(a, b, -ab/(a + b))$ for any numbers $a$ and $b$ where $a \neq -b$.

2-29

| | Northern Africa | Northern Europe |
|---|---|---|
| Median | 64 | 75 |
| Range | 15 | 8 |

In 1993 the northern Europeans had a median life expectancy of about 11 more years than the north Africans had. In addition, there was only about half the variability in the northern European countries. (Note that the data for western Sahara were not available.) Thus the people in northern Europe have a greater life expectancy and their life expectancy is more consistent.

2-31 Although the two groups had the same average score, the score for the psychotic children varied more than the score for the other group.

2-33 Pythagorean brand with the smaller standard deviation is better.

2-35 The small standard deviation in Experiment I indicates that most likely almost all Valentine brand artificial hearts are better than almost all Cor brand hearts. However, the larger standard deviation in the second experiment indicates that although it is likely that most $X$ hearts are better than $Y$ hearts, in many cases this might not be so. The results are not as conclusive.

2-37 (a) $s = 2$; both are easy.

(b) $s = 6.9$; the computational formula is easier.

2-39 (a) $\bar{X} \pm s = 85.53 \pm 6.7 = 78.8$ and 92.2; and 13 scores are between these 2 values.

(b) $13/17 = 76\%$

(c) $\bar{X} \pm 2s = 72.1$ and 98.9; 16 scores are in this interval.

(d) $16/17 = 94\%$

2-41 (a) $m = 5$, $s^2 = 14/3$, $s = 2.16$

(b) $m = 5 + 10 = 15$, $s^2 = 14/3$, $s = 2.16$

2-43 (a) $11{,}000 + 500 = 11{,}500$ marks and 800 marks

(b) $11{,}500(1.1) = 12{,}650$ marks and $800(1.1) = 880$ marks

2-45 (a) $m_F = 25(1.8) + 32 = 77°F$, $s = 10(1.8) = 18°F$

(b) $18(1.8) = 32.4°F$

2-47 (a) 81.44 cents (b) 3.986 cents

2-49

| | |
|---|---|
| Eastern Europe | $m = 15.4$, $s = 4.1$ |
| East Asia | $m = 21.5$, $s = 20.3$ |

East Asian countries have a higher and a more variable infant mortality rate than eastern European countries.

2-51 (a) $z_{99.1°} = 1$ (b) $z_{97.6°} = -2$ (c) $z_{98.6°} = 0$ (d) $z_{100°} = 2.8$ (e) $z_{98°} = -1.2$

2-53 (a) Bastion, Carmulon, and Elfremde

(b) Adelbert and Joe

(c) Bastion, Elfremde, Carmulon, Adelbert, and Joe

2-55 (a) Xgol, 1; Zib, −5; Mni, −5; Rfd, 2

(b) 3.5 Martian inches

(c) Ms. Zar: 3.92 > 3.9 Martian inches, or 1.6 > 1.5

2-57 (a) Math, 3.44; verbal, 0.77; geography, −0.21

(b) Math was highest, geography was lowest.

2-59 Assuming a 5-day workweek, we have

$$\frac{5(5) - 3.5(5)}{4} = 1.875$$

2-61 (a) $\mu = 0$ (b) $\sigma = \sqrt{na^2/n} = a$ (c) $= \pm 1$

2-63 $m = 5.9$, $s = 3.33$, $z_{13.3} = 2.22$

2-65 Yes, if there were a very few exceptionally high scores and more lower scores (for example, 100, 100, 5, 5, 1 where $z_5 < 0$ and $5 = P_{60}$).

2-67 (a) $P_{84}$ (b) $P_{15}$

2-69 (a) 25% (b) 50% (c) 75% (d) 90% (e) 10% (f) 50%

2-71 (a) Not from the table (but $n = 500$)

(b) $m = 100$ when $z = 0$

(c) $P_{50} = 116$

(d) 2% scored below $P_2$ or 68; 50% − 16% = 34% scored between 84 and 116

(e) $s = 16$ since $z_1 - z_0 = 116 - 100 = 16$

(f) $m + 1.5s = 100 + 1.5(16) = 124$

(g) $m - 1.20s = 100 - 1.2(16) = 80.8 \approx 81$

2-73 b

2-75 (a) College graduates $550, high school graduates $275

(b) College graduates $500, high school graduates $300

(c) No, the top 25% of the high school graduates' salaries have the same range as the bottom 25% of the college graduates' salaries.

(d) There is less variation in the salaries of the 25% of the people just below the mean than in the 25% just above the mean.

(e) The middle half of the college graduates' salaries is 4 times more spread out than the middle half of the high school graduates' salaries.

2-77 (a) (1) The year 1994 (2) People in the United States (3) Number of births

(b) (1) The year 1994 (2) People in the United States who are aged 65 and over (3) Number of deaths

(c) (1) The month of July 1995 (2) People in the United States (3) Number of divorces

(d) (1) The year 1 (2) Number of people in the garden (3) Number of apples consumed

2-79 (a) $\dfrac{400}{400{,}000} = 1$ per 1000 (b) 12 per 1000

2-81 19.4 accidents per person

2-83 The first rate includes women over 44 years of age, the second does not. The rate of marriages among older women is lower.

2-85

| Food | Attack Rate for | Attack Rate against |
|---|---|---|
| Baked ham | 0.63 | 0.59 |
| Spinach | 0.60 | 0.63 |
| Mashed potatoes | 0.62 | 0.61 |
| Cabbage salad | 0.64 | 0.60 |
| Jello | 0.70 | 0.58 |
| Rolls | 0.57 | 0.66 |
| Bread | 0.67 | 0.58 |
| Milk | 0.50 | 0.62 |
| Coffee | 0.61 | 0.61 |
| Water | 0.54 | 0.65 |
| Cakes | 0.68 | 0.54 |
| Vanilla ice cream | 0.80 | 0.14 |
| Chocolate ice cream | ~~0.51~~ 0.53 | 0.75 |
| Fruit salad | 0.67 | 0.61 |

2-87 −17.9%

## Chapter 3: Frequency Tables and Graphs

3-1 (a)

| Interval | Frequency |
|---|---|
| 40–44 | 2 |
| 45–49 | 6 |
| 50–54 | 12 |
| 55–59 | 12 |
| 60–64 | 7 |
| 65–69 | 3 |

(b)

|  |  |
|---|---|
| 4 | 23 |
| 4 | 667899 |
| 5 | 001111224444 |
| 5 | 555566677778 |
| 6 | 0111244 |
| 6 | 589 |

(d) Mean = 54.8, median = 55

(e) 57.1%

3-3 Note that because the first boundary is not given, you may have a different set of intervals from this answer.

(a)

| Interval | Midpoint | Frequency |
|---|---|---|
| 93–95 | 94 | 3 |
| 95–97 | 96 | 14 |
| 97–99 | 98 | 25 |
| 99–101 | 100 | 25 |
| 101–103 | 102 | 18 |
| 103–105 | 104 | 11 |
| 105–107 | 106 | 4 |

(b)

| Interval | Midpoint | Frequency |
|---|---|---|
| 93.5–94.5 | 94 | 2 |
| 94.5–95.5 | 95 | 4 |
| 95.5–96.5 | 96 | 6 |
| 96.5–97.5 | 97 | 10 |
| 97.5–98.5 | 98 | 11 |
| 98.5–99.5 | 99 | 18 |
| 99.5–100.5 | 100 | 13 |
| 100.5–101.5 | 101 | 9 |
| 101.5–102.5 | 102 | 8 |
| 102.5–103.5 | 103 | 6 |
| 103.5–104.5 | 104 | 5 |
| 104.5–105.5 | 105 | 8 |

3-5 (a) The population is aging. The categories of older people are increasing in size, and those of younger people are shrinking.

(b)

| Year | Percentage of Population Less than 20 Years Old | Year | Percentage of Population Less than 20 Years Old |
|---|---|---|---|
| 1860 | 51.2 | 1930 | 38.8 |
| 1870 | 49.7 | 1940 | 34.4 |
| 1880 | 48.1 | 1950 | 33.9 |
| 1890 | 46.1 | 1960 | 38.4 |
| 1900 | 44.4 | 1970 | 37.8 |
| 1910 | 42.0 | 1980 | 32.0 |
| 1920 | 40.8 | 1990 | 28.9 |

3-7 (a)

| Interval | Percentage of Households |
|---|---|
| 0–14,999 | 24.3 |
| 15,000–24,999 | 17.4 |
| 25,000–49,999 | 32.5 |
| 50,000 and over | 25.8 |

(c) 24.3%; 49.9%

(d) Median is between $25,000 and $35,000.

3-9 (a) 39 (b) 9; 23.1% (c) 53.8%

3-11 (a) 62% (b) 5%

3-13 (b) Graph 2 makes the companies look most alike. Graph 3 makes Monimo look much worse than the other two. Monimo would probably prefer graph 2.

3-15 (a) 40% (b) 50% (c) 20% (d) 20% (e) 90% (f) 60% (g) 40% (h) 30% (i) 30% (j) 70%

## Answers to Sample Test for Chapters 1, 2, and 3

1. (a) Median since the distribution is probably not symmetric.
   (b) Mean. This should be a reasonably symmetric distribution.
   (c) Mode. The most frequent time is probably the correct time.
   (d) Mean or median. Depends on the shape of the distribution.
2. $z$-Score, percentile rank, raw scores
3. Statistic
4. (a) 2.1; 2; 1 and 2 (b) 5; 2.77; 1.66 (c) 2.34; −1.26 (d) 3
5. (b) $\frac{5 + 0.5}{14} \times 100 = 39$
6. Households with no children will be left out of the study, so his estimate will be too large.
7. (a) $\mu$ (b) $\bar{x}$ or $m$
8. (a) Probably, unless there is some connection between wrist size and enrollment in this class, which is unlikely.
   (b) It would probably be a little small, since it is unlikely that this class has both the person with the largest wrist on campus and the person with the smallest wrist.
9. Many answers are possible. An opinion poll on voters for U.S. President for an election 3 days from now. Estimation of the mean mercury level in all swordfish in the ocean.
10. The 400 patients are the sample; they represent all patients who suffer from disease $X$.
11. The listeners who called constitute a volunteer sample; they are especially interested in the topic and feel strongly about it. They are not likely to represent a cross section of the population. It is very rare for 85% of people in the United States to agree about anything. A less biased method would be based on taking a random sample of citizens.
12. Statistical inference is the use of numerical data to draw conclusions about a larger group than just those from whom the data were drawn.
13. 14.62
14. Many answers are possible. $X$: 1, 2, 3, 4, 5 and $Y$: 3, 3, 3, 3, 3.
15. Variables: age, sex, and systolic blood pressure. The parameter is the mean systolic blood pressure of all patients who use the clinic (not just those who use it that week).
16. There are many more female than male teachers.
17. (a)

|  |  |
|---|---|
| 6 | 5 |
| 6 | |
| 5 | |
| 5 | 234 |
| 4 | 56 |
| 4 | 2133 |
| 3 | 576789 |
| 3 | 4440444 |
| 2 | 866 |
| 2 | 3145333 |
| 1 | 7855 |
| 1 | 3 |
| 0 | |
| 0 | 3 |

(b)

|  |  |
|---|---|
| 6 | 5 |
| 5 | 234 |
| 4 | 123356 |
| 3 | 0444444567789 |
| 2 | 1333345668 |
| 1 | 35578 |
| 0 | 3 |

(c) The second display gives a better image of a symmetric mound-shaped distribution.

## Chapter 4: Probability

4-1 (a) 1/52 (b) 1/4 (c) 1/13

4-3 (a) 31/365 (b) 1/365 (c) 61/365 (d) 334/365

4-5 (a) 8/15 (b) 7/15 (c) 4/15 (d) 2/3 (e) 2/15 (f) 9/15 or 3/5

4-7 (a) 4 or 5 over 28, 29, 30, or 31 (In the long run, the answer is 1/7.)
(b) 1 minus the answer to part (a)

4-9 (a) 0.20 (b) 0.80 (c) 0.52

4-11 (a) (i) 0.50 (ii) 0.45 (iii) 0.05
(b) $1/9 \approx 0.11$
(c) $1/10 = 0.10$

4-13 (a) HH, HT, TH, TT (b) 0.50 (c) 0.25 (d) 0.50 (e) 0.75

4-15 (a) 5/36 (b) 1/6

4-17 (a) 1/4 (b) 1/2

(c)

| | | | |
|---|---|---|---|
| Jakob, Jakob | Johann, Jakob | Daniel, Jakob | Nicholaus, Jakob |
| Jakob, Johann | Johann, Johann | Daniel, Johann | Nicholaus, Johann |
| Jakob, Daniel | Johann, Daniel | Daniel, Daniel | Nicholaus, Daniel |
| Jakob, Nicholaus | Johann, Nicholaus | Daniel, Nicholaus | Nicholaus, Nicholaus |

(d) 1/16 (e) 1/16 (f) 3/8 (g) 7/16

4-19

| | | | | |
|---|---|---|---|---|
| G1, G2 | G2, G1 | G3, G1 | B1, G1 | B2, G1 |
| G1, G3 | G2, G3 | G3, G2 | B1, G2 | B2, G2 |
| G1, B1 | G2, B1 | G3, B1 | B1, G3 | B2, G3 |
| G1, B2 | G2, B2 | G3, B2 | B1, B2 | B2, B1 |

3/10 = probability

4-21 (a)

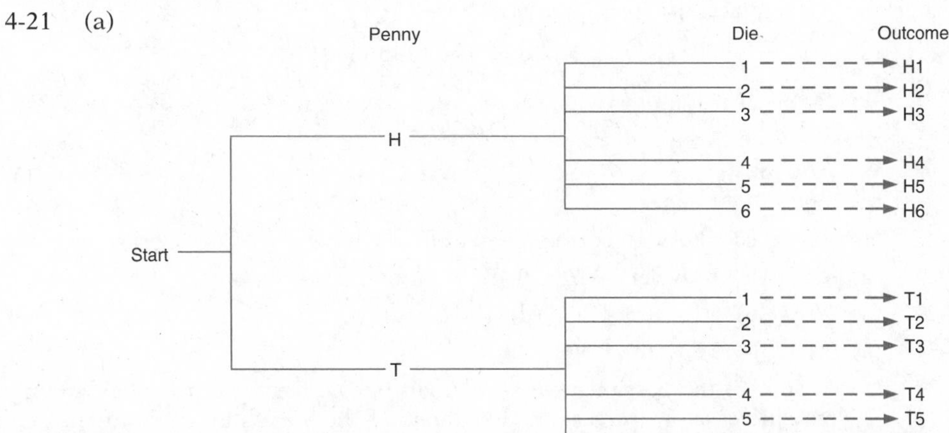

(b)

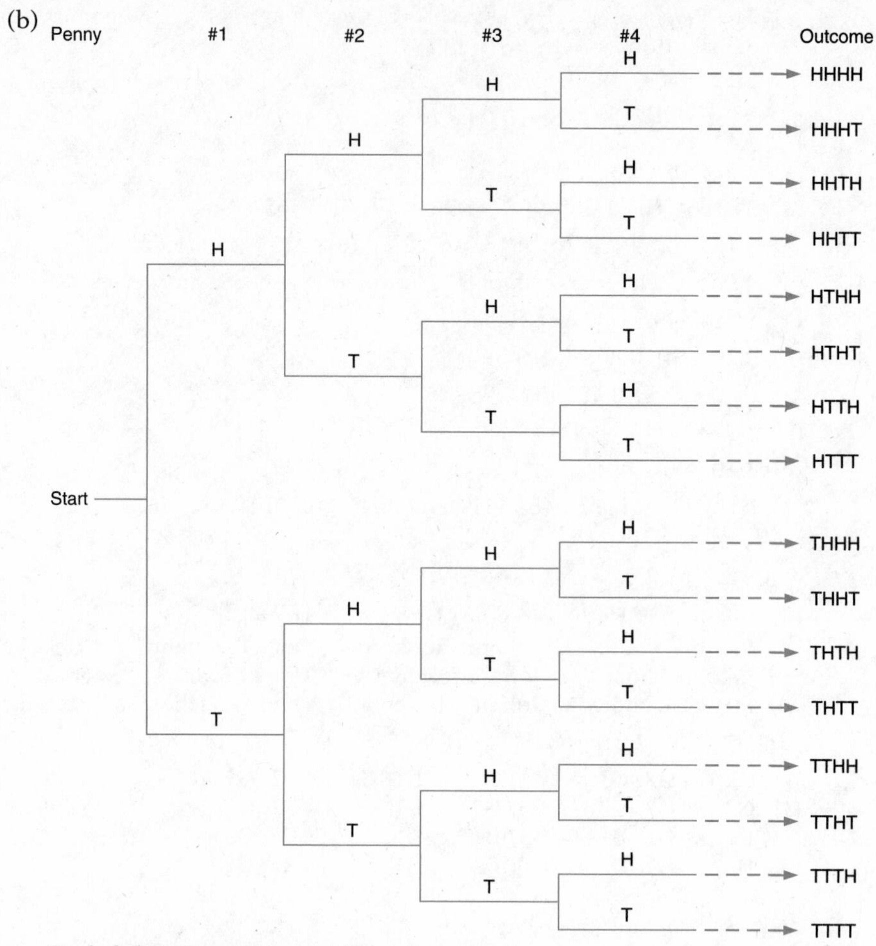


Probabilities are 3/8 and 5/16.

(c) No, it would need 6 × 6 × 6 × 6 = 1296 lines!

4-23 (a) 1 − 4/5(1/4) = 0.80

(b) 1 − [(4/5)(2/5) + (1/5)(1/5)] = 16/25 = 0.64

(c) 3/4 = 0.75

(d) 1.00

4-25 No, we expect about 10 correct guesses *if* random.

4-27 (a) 4 per 1000 travelers (b) 0.004

4-29 (a) 1/16 (b) 6/16 (c) 1/16 (d) 11/16 (e) 1/2

4-31 (a) 0.16 (b) 0.68 (c) 0.50

4-33 (a) In spite of the pessimists, it well might be 1/2. On the other hand, adding butter may increase or decrease the likelihood of the bread's falling buttered side down.

(b) The number of professors of each sex may not be the same.

(c) The probability of obtaining a 2 will depend upon the specific method of altering the die.

4-35 (a) 1/6 (b) 1/6 (c) 25/36

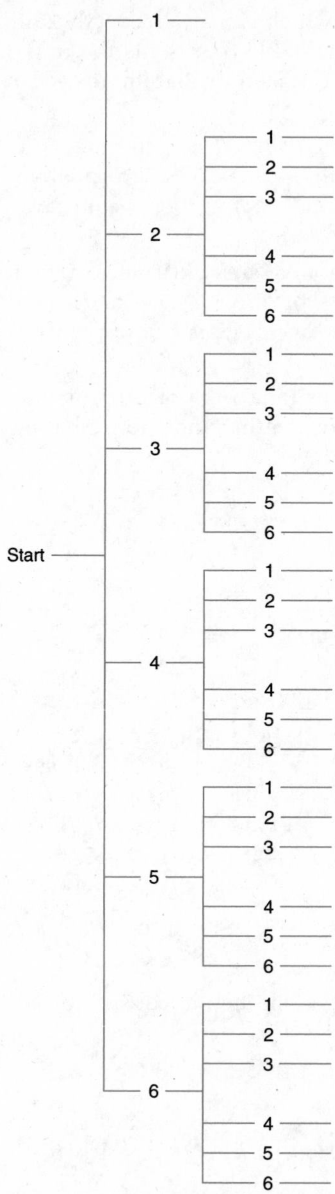


Because those who "die" on the first roll do not get a second chance.

4-37 (a) 1 (b) 6 (c) 12 (d) 8 (e) 0 (f) $\dfrac{6 \times 9}{6 \times 27} = \dfrac{1}{3}$

## Chapter 5: The Binomial Distribution

5-1 A dichotomous random variable is one which can attain exactly 2 different values (success and failure). A binomial random variable represents the number of successes in $n$ trials of a dichotomous variable with a fixed probability of success, so it may take on $n + 1$ values from 0 to $n$.

5-3 Success is rolling a 3. $S$ is binomial since $P(3) = 1/6$ each time, and the trials are independent. The number of rolls of the die is $n = 25$. The probability of getting a 3 is $p = 1/6$. The probability of not getting a 3 is $q = 5/6$. $S$ can have the values 0, 1, 2, . . . , 25.

5-5 Success is drawing a face card. $S$ is binomial since $P(\text{face card}) = 4/52$ each time, and the draws are independent. The number of draws is $n = 5$. The probability of getting a face card is $p = 4/52$. The probability of not getting a face card is $q = 48/52$. $S$ can have the values 0, 1, 2, . . . , 5.

5-7 Success is selecting a rabbit with an ear tag. $S$ is not binomial since the number of rabbits will change because of births or deaths, and the probability of selecting one with a tag will change.

5-9 1, 21, 210, 1330

5-11 rrrr bbbb gggg

rrrg rrrb bbbg rggg rbbb bggg
rrgr rrbr bbgb grgg brbb gbgg
rgrr rbrr bgbb ggrg bbrb ggbg
grrr brrr gbbb gggr bbbr gggb

rrgg rrbb bbgg rrgb rrbg ggbr ggrb bbgr bbrb
rgrg rbrb bgbg rgrb rbrg gbgr grgb bbgr brbg
grrg brrb gbbg rgbr rbgr gbrg grbg bgrb brgb
rggr rbbr bggb grbr brgr bgrg rgbg gbbr rbbg
grgr brbr gbgb gbrr bgrr bggr rggb gbrb rbgb
ggrr bbrr ggbb grrb brrg brgg rbgg grbb rgbb

5-13 (a) 6/32; 19%

(b) 6/32; 19%

(c) 19%; these 3 problems have the same mathematical structure.

5-15 (a) 0.317 (b) 0.913

5-17 (a) 0.04 (b) $1 - 0.64 = 0.36$

5-19 $1 - (0.8)^4 = 0.59$

5-21 (a) 1/16 (b) 9/16 (c) $1 - 9/16 = 7/16$ (d) $1 - 1/16 = 15/16$

5-23 (a) 0.1536 (b) 0.3456 (c) 0.1552

5-25 (a) 0.96 (b) 0.00000001 (c) 0.00000397

5-27 (a) $0.3125 \approx 0.31$ (b) $0.3125(0.5) \approx 0.16$

5-29 0.89

5-31 (c) 6545 (d) 19,600

## Chapter 6: The Normal Distribution

6-1 Heights, weights of adults, plants, etc.

6-3 (a) 97.5% (f) $\approx$ 0%

(b) 59.1% (g) $\approx$ 0

(c) 0.0735 − 0.0166 = 56.9% (h) 0.0668

(d) 0.13% (i) 0.1401

(e) 95.44% (j) $\approx$ 0

(k) They would be the same IF the lengths of Martian middle arms were normally distributed.

6-5 (a) 100 (b) Nothing changes.

6-7 (a) 1.28 (b) −0.84 (c) ±0.39 (d) −0.13 (e) 0.39 (f) ±0.52

6-9 (a) 1.65 (b) 1.96 (c) 2.33 (d) 2.58

6-11

| $z$ | −4 | −2 | 0 | 2 | 4 |
|---|---|---|---|---|---|
| $X$ | 0.2 | 0.4 | 0.6 | 0.8 | 1.0 |

6-13 (a) $0 + 1.65(10) = 16.5$ (b) $0 - 1.65(10) = -16.5$ (c) $0 - 1.96(10) = -19.6$
(d) $0 + 1.96(10) = 19.6$

6-15 (a) $z = \dfrac{80 - 74}{6} = 1 \quad \rightarrow \quad 0.8413$

$z = \dfrac{70 - 74}{6} = -0.67 \quad \rightarrow \quad 0.2514$

0.5899 or about 59%

(b) $\dfrac{86 - 74}{6} = 2 \rightarrow 0.0228$ or about 0.02

(c) $z = \dfrac{79 - 74}{6} = 0.83 \quad \rightarrow \quad 0.7967$

$z = \dfrac{69 - 74}{6} = -0.83 \quad \rightarrow \quad 0.2033$

0.5934

$1 - 0.5934 = 0.4066$, or about 0.41

6-17 (a) $z = \dfrac{4 - 5.7}{0.6} = -2.83 \rightarrow 0.9977$, almost 100%

(b) $z = \dfrac{6.5 - 5.7}{0.6} = 1.33 \rightarrow 0.0918$, about 9%

(c) $z = \dfrac{6 - 5.7}{0.6} = 0.5 \quad \rightarrow \quad 0.6915$

$z = \dfrac{5 - 5.7}{0.6} = -1.17 \quad \rightarrow \quad 0.1210$

0.5705, about 57%

6-19 (a) $z = \dfrac{300 - 210}{56} = 1.61 \rightarrow 0.0537$

(b) $z = \dfrac{100 - 210}{56} = -1.96 \rightarrow 0.0250$

(c) $z = \ \ 1.61 \rightarrow 0.9463$

$z = -1.96 \rightarrow \underline{0.0250}$

$0.9213$

(d) $z = \dfrac{120 - 210}{56} = -1.61 \rightarrow 0.0537$, about 5% or less, since many customers will not return a defective bulb

6-21 (a)

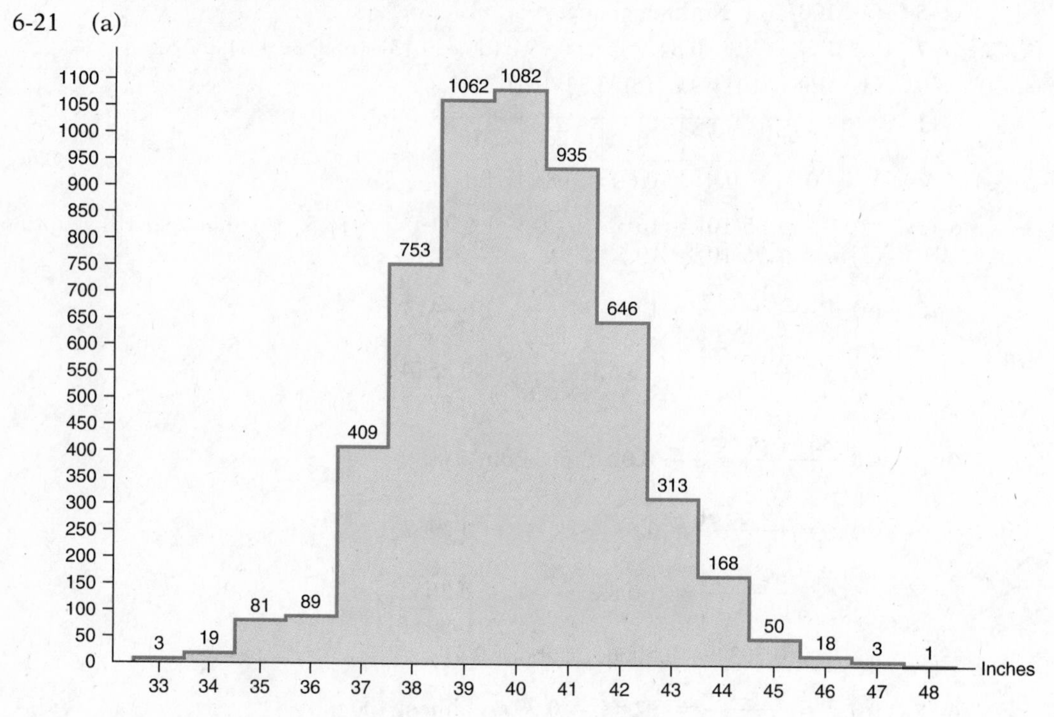


(c) $z = \dfrac{48 - 39.85}{2.07} = 3.94$

$z = \dfrac{33 - 39.85}{2.07} = -3.31$

(d) $39.85 \pm 1(2.07) = 37.78$ and $41.92$

$\dfrac{3832}{5732} \approx 67\%$

6-23 (a) True (b) false (c) false

6-25 The freshman grades are bell-shaped with few very high or very low grades. Most of the grades are in the middle. The mean will be near the median.

The advanced course grades are bimodal. There are many high grades, followed by many low grades. Middle grades are more scarce.

6-27 We do not know the shape of the graph of the distribution. It may not be normal.

6-29 (a) $z = \dfrac{6 - 5}{2.3} = 0.43$

$$\frac{69 + 0.5(16)}{110}(100) = 0.70, \text{ or } P_{70}$$

(b) $z = 0.43 \rightarrow 0.6664$, $P_{67}$ which is close to $P_{70}$

(c)

| Number of Hits | Koufax Percentiles | Normal Curve Percentiles |
|---|---|---|
| 0 | $P_2$ | $P_2$ |
| 1 | $P_5$ | $P_4$ |
| 2 | $P_9$ | $P_{10}$ |
| 3 | $P_{20}$ | $P_{19}$ |
| 4 | $P_{34}$ | $P_{33}$ |
| 5 | $P_{52}$ | $P_{50}$ |
| 6 | $P_{65}$ | $P_{67}$ |
| 7 | $P_{83}$ | $P_{81}$ |
| 8 | $P_{91}$ | $P_{90}$ |
| 9 | $P_{95}$ | $P_{96}$ |
| 10 | $P_{97}$ | $P_{99}$ |
| 11 | $P_{98}$ | $P_{100}$ |
| 12 | $P_{99}$ | $P_{100}$ |
| 13 | $P_{100}$ | $P_{100}$ |

## Chapter 7: Approximation of the Binomial Distribution by Use of the Normal Distribution

7-1 (a) 12.5 (b) 11.5 (c) 11.5 (d) 12.5

7-3 Binomial: $79(0.5)^{12} = 0.0193 \approx 0.02$

Normal: $P\left(z > \dfrac{9.5 - 6}{\sqrt{12(0.5)(0.5)}}\right) = P(z > 2.02) = 0.0217 \approx 0.02$

These answers agree to the nearest hundredth. They are so close mainly because $p = 0.5$, and the binomial graph is symmetric.

7-5 (a) $np = 6$, $nq = 14$ (b) 6 (c) 2.05

(d) $z = (3.5 - 6)/2.05 = -1.22$; $P(z > -1.22) = 0.89$

(e) $z = (9.5 - 6)/2.05 = 1.71$; $P(z > 1.71) = 0.04$

(f) $X = 6 - 2.33(2.05) = 1.22$; $S = 1$

7-7 (a) $\mu = 270$; $\sigma = 13.75$; $z = \dfrac{360.5 - 270}{13.75} = 6.58$; $P(z > 6.58) \approx 0$

(b) $0.33(900) = 297$; $z = \dfrac{297.5 - 270}{13.75}$; $P(z > 2) \approx 0.02$

(c) $0.28(900) = 252$; $z = \dfrac{251.5 - 270}{13.75} = -1.35$; $P(z < -1.35) \approx 0.09$

7-9 (a) $\mu = 10$; $\sigma = 3$; $z = \dfrac{14.5 - 10}{3} = 1.5$; $P(z > 1.5) \approx 0.07$

(b) Exact-answer binomial approach: $0.9^{100} = 0.0000266 \approx 0.00003$.

By normal approximation: $z = \dfrac{0.5 - 10}{3} = -3.17$; $P(z < -3.17) \approx 0.0007$

(c) $z_c = -1.65$; $S \approx -1.65(3) + 10 = 5.05$; $S_? = 5$. there is a 95% probability that at least 5 tadpoles will survive.

7-11 $\mu = 10$; $\sigma = 2.24$; $z = (11.5 - 10)/2.24 = 0.67$; $P(z > 0.67) = 0.2514 \approx 0.25$

7-13 It is possible, but the probability is only $(0.5)^{10} = 0.00098 \approx 0.001$, or about 1 in 1000.

7-15 (a) $\mu = 6$; $\sigma = 2.12$; $z = (17.5 - 6)/2.12 = 5.42$; $P(z > 5.42) \approx 0.00$. It is extremely unlikely that Bob won 18 times just by luck.

(b) This is the same situation as in part (a); the probability is still close to zero. $\mu = 18$; $\sigma = 2.12$; $z = (6.5 - 18)/2.12 = -5.42$; $P(z < -5.42) \approx 0.00$

7-17 $n = 20$; $p = 0.8$; $\mu = 16$; $\sigma = 1.79$; $z = (15.5 - 16)/1.79 = -0.28$; $P(z > -0.28) = 0.6103$

7-19 (a) $z = (18 - 14)/3 = 1.33$; $P(z > 1.33) = 0.0918$

(b) $z = -0.25$; $A = 14 - 0.25(3) = 13.25$

(c) $z = (18.5 - 14)/3 = 1.5$; $P(z > 1.5) = 0.0668$; $A = 13$ 14

7-21 (a) 0.90

(b) $n = 500$; $p = 0.10$; $\mu = 50$; $\sigma = 6.708$; $S = 0.12(500) + 0.5 = 60.5$; $z = (60.5 - 50)/6.708 = 1.57$; $P(S > 1.57) = 0.0582 \approx 0.06$

(c) Find smallest $n$ so that $0.9^n < 0.50$; $n = 7$.

7-23 (a) $n = 300$; $p = 1/30$; $\mu = 10$; $\sigma = 3.11$; $z = (10.5 - 10)/3.11 = 0.16$; $P(S > 10.5) = 0.4364$

(b) $(1/30)^2 = 1/900$

(c) $(1/900)(1/4) = 1/3600$

## Answers to Sample Test for Chapters 4, 5, 6, and 7

1. The probability that heads appears on one toss.
2. (a)
3. Random
4. $3 \times 2 \times 5 = 30$
5. under 21, male, Christian
   under 21, male, Jewish
   under 21, male, Muslim
   under 21, male, Buddhist
   under 21, male, Other

5. (Continued)
under 21, female, Christian
under 21, female, Jewish
under 21, female, Muslim
under 21, female, Buddhist
under 21, female, Other

21 to 39, male, Christian
21 to 39, male, Jewish
21 to 39, male, Muslim
21 to 39, male, Buddhist
21 to 39, male, Other

21 to 39, female, Christian
21 to 39, female, Jewish
21 to 39, female, Muslim
21 to 39, female, Buddhist
21 to 39, female, Other

40 & up, male, Christian
40 & up, male, Jewish
40 & up, male, Muslim
40 & up, male, Buddhist
40 & up, male, Other

40 & up, female, Christian
40 & up, female, Jewish
40 & up, female, Muslim
40 & up, female, Buddhist
40 & up, female, Other

6. The results of 12 trials are quite unpredictable. Outcomes of experiments match theory better as $n$ increases.

7. (a) $(1/2)^3 = 1/8$ (b) 1/8

8. (a) 43% (b) 57%

9. (a) 62% (b) 38%

10. $\mu = 50(0.70) = 35$. We expect about 35 heads.

11. (a) $P(z > 1.5) = 0.0668$; about 7%

 (b) $z = 1.28$; $W = 6.7 + 1.28(2.30) = 9.6$ pounds. The heaviest 10% are those that weigh more than 9.6 pounds.

 (c) $z = -1.04$; $W = 6.7 - 1.04(2.3) = 4.3$ pounds. The lightest 15% are those that weigh less than 4.3 pounds.

 (d) $z_8 = (8 - 6.7)/2.3 = 0.56$; $z_7 = (7 - 6.7)/2.3 = 0.13$;
 $P(7 < W < 8) = 0.7123 - 0.5517 = 0.1606 \approx 0.16$

12. $z = 1.56$; $X = 5.10 + 1.56(1.29) = 7.11$; no, the people in the top 6% are each paid at least \$7.11 per smile.

13. There are 4 different ways to have exactly 1 success in 4 trials; there are 4 different ways to have exactly 3 successes in 4 trials.

14. (a) 120 (b) 2 or 3 (c) 1 (d) 1 (e) 117 (f) 231

15. $0.1669 + 0.1855 + 0.1669 = 0.5193$

16. $1(0.58^{12})(0.42)^0 = 0.0014$

17. (a) $n = 500$; $p = 0.7$; $\mu = 350$; $\sigma = 10.25$; $z = (314.5 - 350)/10.25 = -3.460$; $P(S < 314.5) = 0.0003$

(b) Yes

(c) No, his claim is very unlikely.

(d) One possibility is to take another sample and compare the two.

## Chapter 8: Hypothesis Testing: One-Sample Tests of Percentages in Binomial Experiments

*Note:* One-tail tests can be done in two ways depending upon your choice of variables. For example, testing for too many heads (one tail on the right) is equivalent to testing for too few tails (one tail on the left).

In addition, there are often good arguments as to whether a test should be a one-tail test or a two-tail test. Some researchers perform only two-tail tests. Use your common sense.

8-1 (b) A one-sample test

(c) $\mu$

(d) $\mu_{pop}$ is the mean workweek of all workers in Centerville.

$H_0$: The average workweek in Centerville is 40 hours; $\mu_{pop} = 40$ hours.
$H_a$: The average workweek in Centerville is less than 40 hours; $\mu_{pop} < 40$ hours (one-tail).

8-3 (a) Asks for a number, is not a hypothesis test.

8-5 (b) A two-sample test

(c) $\mu$

(d) $\mu_1$ is the average height of all 6-year-old male horses.
$\mu_2$ is the average height of all 6-year-old female horses.

$H_0$: The average height of 6-year-old male horses is the same as the average height of 6-year-old female horses; $\mu_1 = \mu_2$.
$H_a$: The average height of 6-year-old male horses is not the same as the average height of 6-year-old female horses; $\mu_1 \neq \mu_2$ (two-tail).

8-7 (b) A two-sample test

(c) $p$

(d) $p_1 = P(\text{a 20-year-old female diets})$
$p_2 = P(\text{a 20-year-old male diets})$

$H_0$: The percentage of 20-year-old females who diet is the same as the percentage of 20-year-old males who diet; $p_1 = p_2$.
$H_a$: The percentage of 20-year-old females who diet is greater than the percentage of 20-year-old males who diet; $p_1 > p_2$ (one-tail).

8-9 (a) Type I error: case ii. A true $H_0$ is rejected.

(b) Type II error: case iii. A false $H_0$ appears to be true.

8-11 Assuming that the null hypothesis is that the player should be picked, $\alpha$ will increase as more good players will be rejected.

8-13 Type II error

8-15 Type I errors. The drug is OK, and the company doesn't sell it.

8-17 Mart is correct, Type II errors are involved with *accepting* a false $H_0$. Marv is wrong. Hopefully neither error was made.

8-19 (a) Population: Children in the age bracket 1–4 who die
$p = P$(a child who dies does so as a result of a motor accident)
$H_0$: The newspaper is correct; $p = 0.60$.

(b) $H_a$: The newspaper is wrong; $p \neq 0.60$ (a two-tail test).

(c) $\mu = 30(0.60) = 18$
$\sigma = \sqrt{30(0.60)(0.40)} = 2.68$
$np = 18$, $nq = 12$, both $> 5$
$S_c = 18 \pm 2.58(2.68) = 18 \pm 6.9 = 11.1$ and $24.9$
Reject the $H_0$ if the outcome is either less than 12 deaths or greater than 24 deaths.

8-21 (a) Population: All jail inmates in her state
$p = P$(an inmate had not completed high school)
$H_0$: 61% of the inmates had not completed high school; $p = 0.61$.
$H_a$: The percentage of inmates who had not completed high school was less than 61%; $p < 0.61$ (one-tail test).

(b) $\mu = 400(0.61) = 244$
$\sigma = \sqrt{400(0.61)(0.39)} = 9.75$
$np = 244$, $nq = 156$, both $> 5$
$S_c = 244 - 1.65(9.75) = 244 - 16.0957 = 227.9$
Reject $H_0$ if the outcome is less than 228 inmates.

(c) $S_c = 244 - 2.33(9.75) = 244 - 22.7 = 221.3$
Reject $H_0$ if the outcome is less than 222 inmates.

8-23 $\mu = 100(0.42) = 42$
$\sigma = \sqrt{100(0.42)(0.58)} = 4.94$
$np = 42$, $nq = 58$, both $> 5$
$S_c = 42 + 2.33(4.94) = 42 + 11.5 = 53.5$
Reject $H_0$ if outcome exceeds 53 women.

8-25 (a) $p = P$(a train is late)
$H_a$: More than 4% of trains are late; $p > 0.04$. (one-tail test)

(b) $H_0$: 4% of trains are late; $p = 0.04$

(c) $\mu = 500(0.04) = 20$
$\sigma = \sqrt{500(0.04)(0.96)} = 4.38$
$np = 20$, $nq = 480$, both $> 5$
$$z_c = \frac{40.5 - 20}{4.38} = 4.68$$

(d) $\alpha \approx 0$ ($< 0.00003$)

(e) $$z_c = \frac{30.5 - 20}{4.38} = 2.40$$
$\alpha = 0.0082$

(f) Decrease

(g) $S_c = 20 + 1.65(4.38)$
$= 20 + 7.2 = 27.2$

8-27 $p = P(\text{a nut is a peanut})$
$\mu = 100(0.30) = 30$
$\sigma = \sqrt{100(0.30)(0.70)} = 4.58$
$np = 30, nq = 70$, both $> 5$
$$z = \frac{35.5 - 30}{4.58} = 1.20 \rightarrow 0.1151$$
Only if you are willing to risk more than an 11½% chance of being wrong.

8-29 $p = P(\text{a student cuts at least once a week})$
$\mu = 200(0.30) = 60$
$\sigma = \sqrt{200(0.30)(0.70)} = 6.48$
$np = 60, nq = 140$, both $> 5$
$$z_c = \frac{49.5 - 60}{6.48} = -1.62$$ which is to the right of $-1.65$, therefore Danny's $\alpha > 0.05$ (about 5¼%)

8-31

| $z_c$ | One-Tail | Two-Tail |
|---|---|---|
| $\alpha = 0.10$ | $\pm 1.28$ | $\pm 1.65$ |

8-33 (a) $\mu = 40(0.60) = 24$
$\sigma = \sqrt{24(0.40)} = 3.10$
$np = 24, nq = 16$, both $> 5$
$$z_{25.5} = \frac{25.5 - 24}{3.098} = 0.48 \rightarrow \beta = 0.6844 \approx 0.68\checkmark$$
(b) For $\alpha = 0.01$, $S_c = 20 + 2.33(3.16) = 27.4$
$\mu = 40(0.80) = 32$
$\sigma = \sqrt{32(0.20)} = 2.53$
$np = 32, nq = 8$, both $> 5$
$$z_{27.5} = \frac{27.5 - 32}{2.53} = -1.78 \rightarrow \beta = 0.0375 \approx 0.04\checkmark$$

8-35 (a) $p = P(\text{a spark plug is defective})$
$\mu = 100(0.30) = 30$
$\sigma = \sqrt{30(0.70)} = 4.58$
$np = 30, nq = 70$, both $> 5$
$S_c = 30 - 1.65(4.58) = 30 - 7.6 = 22.4$
Reject $H_0$ if the outcome is less than 23.
(b) $\mu = 100(0.20) = 20$
$\sigma = \sqrt{20(0.80)} = 4$
$np = 20, nq = 80$, both $> 5$
$$z_{22.5} = \frac{22.5 - 20}{4} = 0.625 \rightarrow \beta = 0.2643 \approx 0.26$$

8-37 $\mu = 60(0.35) = 21$ $\beta = 0.0681$
$\sigma = \sqrt{21(0.65)} = 3.69$
$np = 21, nq = 39$, both $> 5$
$$z_{15.5} = \frac{15.5 - 21}{3.69} = -1.49 \rightarrow 0.0681$$
$$z_{4.5} = \frac{4.5 - 21}{3.69} = -4.5 \rightarrow \quad 0.0000$$
$0.0681$ $\beta \approx 0.087$

8-39 (a) Since Dr. Senserd had an estimate of the power of the test, he was safer in using the stronger phrase, "accepted the null hypothesis." Dr. Bulschmidt, lacking this knowledge, satisfied himself with the weaker "failed to reject."

(b) Dr. Senserd has learned more than Dr. Bulschmidt. Of course, Dr. Senserd had to have some idea of the possible values of $p$, and using this extra knowledge, he was able to calculate the power of the test performed. Dr. Bulschmidt knows that the null hypothesis *may* be true. Dr. Senserd is 96% sure that it *is* true.

8-41 $\mu = 50(0.60) = 30$
$\sigma = \sqrt{30(0.40)} = 3.46$
$np = 30,\ nq = 20$, both $> 5$
$S_c = 30 \pm 1.96(3.46) = 30 \pm 6.8 = 23.2$ and $36.8$

If $p = 0.40$,

$$\mu = 50(0.40) = 20$$
$$\sigma = \sqrt{20(0.60)} = 3.46$$
$$np = 20,\ nq = 30,\ \text{both} > 5$$
$$z_{36.5} = \frac{36.5 - 20}{3.46} = 4.8 \rightarrow 1.0000$$
$$z_{23.5} = \frac{23.5 - 20}{3.46} = 1.01 \rightarrow \underline{0.8438}$$
$$0.1562 \qquad \text{thus } \beta \approx 0.16 \qquad \text{power} \approx 0.84$$

If $p = 0.50$,

$$\mu = 50(0.50) = 25$$
$$\sigma = \sqrt{25(0.50)} = 3.54$$
$$np = 25,\ nq = 25,\ \text{both} > 5$$
$$z_{36.5} = \frac{36.5 - 25}{3.54} = 3.25 \rightarrow 0.9995$$
$$z_{23.5} = \frac{23.5 - 25}{3.54} = -0.42 \rightarrow \underline{0.3372}$$
$$0.6623 \qquad \beta \approx 0.66 \qquad \text{power} \approx 0.34$$

If $p = 0.70$,

$$\mu = 50(0.70) = 35$$
$$\sigma = \sqrt{35(0.30)} = 3.24$$
$$np = 35,\ nq = 15,\ \text{both} > 5$$
$$z_{36.5} = \frac{36.5 - 35}{3.24} = 0.46 \rightarrow 0.6772$$
$$z_{23.5} = \frac{23.5 - 35}{3.24} = -3.5 \rightarrow \underline{0.0002}$$
$$0.6770 \qquad \beta \approx 0.68 \qquad \text{power} \approx 0.32$$

If $p = 0.80$,

$$\mu = 50(0.80) = 40$$
$$\sigma = \sqrt{40(0.20)} = 2.83$$
$$np = 40,\ nq = 10,\ \text{both} > 5$$
$$z_{36.5} = \frac{36.5 - 40}{2.83} = -1.24 \to 0.1075$$
$$z_{23.5} = \frac{23.5 - 40}{2.83} = -5.8 \to \underline{0.0000}$$
$$0.1075 = \beta \qquad \text{power} \approx 0.89$$

| $p$ | $\beta$ | Power |
|---|---|---|
| 0.40 | 0.16 | 0.84 |
| 0.50 | 0.66 | 0.34 |
| 0.70 | 0.68 | 0.32 |
| 0.80 | 0.11 | 0.89 |

8-43 (a) $\mu = 50(0.50) = 25$

$\sigma = \sqrt{25(0.50)} = 3.54$

$np = nq = 25$, both $> 5$

$S_c = 25 \pm 1.96(3.54) = 18.1$ and $31.9$

(b) $\mu = 50(0.80) = 40$

$\sigma = \sqrt{40(0.20)} = 2.83$

$np = 40$, $nq = 10$, both $> 5$

$$z_{31.5} = \frac{31.5 - 40}{2.83} = -3.00 \to 0.0013$$
$$z_{18.5} = \frac{18.5 - 40}{2.83} = -7.6 \to \underline{0}$$
$$0.0013 = \beta \qquad \text{power} \approx 99.9\%$$

(c) $\mu = 100(0.50) = 50$

$\sigma = \sqrt{50(0.50)} = 5$

$np = nq = 50$, both $> 5$

$S_c = 50 \pm 1.96(5) = 40.2$ and $59.8$

$\mu = 100(0.80) = 80$

$\sigma = \sqrt{80(0.20)} = 4$

$np = 80$, $nq = 20$, both $> 5$

$$z_{59.5} = \frac{59.5 - 80}{4} = -5.1 \to 0$$
$$z_{40.5} = \frac{40.5 - 80}{4} = -9.9 \to \underline{0}$$
$$0 = \beta \qquad \text{power} \approx 100\%$$

(d) $\beta$ decreases and the power increases.

8-45 $\mu = 10 \qquad \sigma = 2.83$

$$z_{18.5} = \frac{18.5 - 10}{2.83} = 3.01 \rightarrow 0.9987$$

$$z_{6.5} = \frac{6.5 - 10}{2.83} = -1.24 \rightarrow \underline{0.1075}$$

$$0.8912 = \beta \qquad \text{power} = 0.11$$

8-47 (a) $\mu = 6.25 \qquad \sigma = 2.17$

$S_c = 6.25 + 1.65(2.17) = 9.8$ hence more than 9 correct

(b) The closer value to the assumed 25% gives the larger $\beta$, hence we use 0.40.

$\mu = 10 \qquad \sigma = 2.45$

$$z_c = \frac{9.5 - 10}{2.45} = -0.20 \rightarrow 0.4207 \qquad \beta \approx 0.42$$

8-49 (a) $\mu_{\text{fair}} = 10 \qquad \sigma_{\text{fair}} = 2.24$

$$z_{11.5} = \frac{11.5 - 10}{2.24} = 0.67 \rightarrow 0.7486$$

$$z_{8.5} = \frac{8.5 - 10}{2.24} = -0.67 \rightarrow \underline{0.2514}$$

$$0.4972 \qquad \text{hence } \alpha \approx 0.50$$

Therefore he will discard about 50% of the good coins.

(b) Since we have no approximate value for $p$, we cannot estimate $\beta$.

(c) The wife

8-51 (a) The $H_0$ is false, therefore Type II.

(b) The $H_0$ is true, therefore Type I.

8-53 The sample was not random (the 24th book or the 53d book could not be picked). It is possible, for example, that due to a faulty gear every assembly line produced a series such as this, where G stands for a good book and D stands for a defective one:

DDGDGDDGDGDDGDGDDGDG · · ·

8-55 (a) $p = P$(a rattlesnake has broken fangs)

$H_a$: Less than 3% of the rattlesnakes found on Park Avenue have broken fangs; $p < 0.03$ (a one-tail test on the left).

$H_0$: 3% of the rattlesnakes found on Park Avenue have broken fangs; $p = 0.03$.

(b) $p = P$(a teacher is tardy)

$H_a$: More than 18% of the teachers at the On-Time Railroad Dispatchers School are tardy; $p > 0.18$ (a one-tail test on the right).

$H_0$: 18% of the teachers at the On-Time Railroad Dispatchers School are tardy; $p = 0.18$.

(c) $H_a$: Joining the Sandworm Pickers Union will change the average take-home pay of a sandworm picker (a two-tail test).

$H_0$: Joining the Sandworm Pickers Union will not change the average take-home pay of a sandworm picker.

(d) $H_a$: Dr. Meany's practice of surprise tests in his course "Do-It-Yourself Open-Heart Surgery" increases his students' grades (a one-tail test on the right).

$H_0$: Dr. Meany's practice of surprise tests in his course "Do-It-Yourself Open-Heart Surgery" effects no change in his students' grades.

(e) $H_a$: Coach Aquanut's policy of having his basketball players wear flippers during practice sessions improves their average number of points per game (a one-tail test on the right).
$H_0$: Coach Aquanut's policy of having his basketball players wear flippers during practice sessions leaves their average number of points per game unchanged.

*We do not present every step in the solutions to the hypothesis tests, but rather the important results so that you can verify your calculations. For more complete solutions refer to the* Student Solutions Manual. *A sample solution is given for Problem 8-57. Your solutions should be in a similar format accompanied by an appropriate graph.*

8-57 A binomial one-sample hypothesis test.
Population: All students at the university
Sample: 100 students
Let $p = P$(a student is left-handed)
$H_a$: The percentage of lefties is more than 10%; $p > 0.10$ (a one-tail test).
$H_0$: The dean is correct, the percentage is 10%; $p = 0.10$.
$\alpha = 0.05$
$n = 100$
$np = 100(0.10) = 10$, $nq = 90$, both $> 5$, therefore a normal distribution
$\mu = 10$
$\sigma = \sqrt{10(0.90)} = 3$
$S_c = 10 > 1.65(3) = 14.95$
Conclusion:
a: Decision rule: Reject $H_0$ if $S > 14$.
b: $S = 16$
c: Reject $H_0$, lefties make up more than 10% of the student body.

8-59 $S_c = 16 - 1.65(3.58) = 10.1$
Since $S = 3 < 11$, reject $H_0$; the new technique does reduce the number of unfit cars.

8-61 $S_c = 300 - 1.65(12.2) = 279.8$
Since $S = 260 < 280$, reject $H_0$; the new policies lowered the dropout rate.

8-63 $S_c = 12 + 2.33(2.68) = 18.3$
Since $S = 14$ is not greater than 18, we fail to reject $H_0$; we do not have sufficient evidence to show that more than 40% favor the bill.

8-65 $S_c = 12.5 + 1.65(3.06) = 17.6$
Since $S = 9$, reject $H_0$; the percentage is less than 25%.

8-67 $S_c = 60 - 1.65(4.90) = 51.9$
Since $S = 10 < 52$, reject $H_0$; 60% is too high. Note, this is what the freshmen *said*.

8-69 $S_c = 9.8 + 1.65(2.97) = 14.7$
If $S = 12$ is not greater than 14, we fail to reject $H_0$; even though more than 10 survived, there would not be enough evidence at the 0.05 significance level.
If $S = 27 > 14$, reject $H_0$; vitamins would affect survival.

8-71 (a) $\mu = 10$ $\quad \sigma = \sqrt{10(5/6)} = 2.87$

$$z_{15} = z_{14.5} = \frac{14.5 - 10}{2.87} = 1.56 \rightarrow 0.0594$$

$P$(fair dice will produce 15 or more doubles in 60 rolls) $= 0.594$

(b) This is $\alpha$.

## Chapter 9: Hypothesis Testing: Two-Sample Tests of Percentages in Binomial Experiments

9-1 No. He doesn't know the sample size for the New York City sample.

9-3 (a) $\hat{p} = 250/400 = 0.625$; $d\hat{p}_c = 0 \pm 2.58(0.048) = \pm 0.124$. Decision rule: Reject $H_0$ if the magnitude of the observed difference is more than 0.124 (or 12.4%).

(b) $d\hat{p} = 0.65 - 0.60 = 0.05$, which is smaller in magnitude than 0.124. We fail to reject $H_0$. The proportions of business passengers on the 2 flights do not differ enough to be statistically significant.

9-5 $\hat{p} = 0.545$; $d\hat{p}_c = 0 + 2.33(0.152) = 0.354$; $d\hat{p} = 0.404$. Since 0.404 is greater than 0.354, we reject $H_0$; at $\alpha = 0.01$, the evidence supports the claim that aspirin reduces clotting.

9-7 $\hat{p} = 0.592$; $d\hat{p}_c = 0 \pm 1.96(0.09) = \pm 0.176$; $d\hat{p} = 0.217$. Since 0.217 is greater than 0.176, we reject $H_0$; at $\alpha = 0.05$ we can conclude that a higher proportion of Roman Catholics than Orthodox Jews opposes birth control.

9-9 $\hat{p} = 0.833$; $d\hat{p}_c = 0 + 2.33(0.024) = 0.056$; $d\hat{p} = 0.062$. Since 0.062 is greater than 0.056, we reject $H_0$; we have evidence at the 0.01 significance level that a higher proportion of male students eats 3 meals a day.

9-11 $\hat{p} = 0.191$; $d\hat{p}_c = 0 + 1.65(0.075) = 0.124$; $d\hat{p} = 0.13$. Since 0.13 is greater than 0.124, we reject $H_0$; at $\alpha = 0.05$ we have evidence that the proportion of lung cancer cases is higher among the smokers.

9-13 (a) $\hat{p} = 0.292$; $d\hat{p}_c = 0 \pm 1.96(0.083) = \pm 0.163$; $d\hat{p} = 0.083$. Since 0.083 is less than 0.163, we fail to reject $H_0$; at $\alpha = 0.05$ the difference between the proportions of correct tries for Wally and Berta is not statistically significant.

(b) $H_0$: $p = 1/6$. ND($\mu = 10$; $\sigma = 2.89$). $z_c = 2.33$; $S_c = 16.7$; $S = 20$. Since 20 is greater than 16.7, we reject $H_0$; at $\alpha = 0.01$ we have evidence that Wally has ESP.

(c) $S = 15$. Even though 15 is above average (10), it is less than 16.7, so we cannot say we have demonstrated conclusively that Berta has ESP.

9-15 $\hat{p} = 0.322$; $d\hat{p}_c = 0 + 2.33(0.015) = 0.035$; $d\hat{p} = 0.27$. Since 0.27 is greater than 0.035, we reject $H_0$; at $\alpha = 0.01$ we have evidence that a greater proportion of younger people reports infirmities.

9-17 For last year and for this year evaluated separately, we find that a higher percentage of girls prefer peanut butter and jelly, but the difference between boys and girls is not statistically significant. When the data are combined, however, we see that a higher percentage of boys prefer peanut butter and jelly and that the difference is statistically significant. This is an illustration of Simpson's paradox, which occurs when the results for a large combined group are opposite from the results for each of the smaller subgroups. Because this can happen, it is important to decide whether the results for the subgroups are important on their own. Very often the subgroup results are more meaningful than the combined results.

9-19 (a) $\hat{p} = 0.379$; $d\hat{p}_c = 0 + 1.65(0.037) = 0.06$; $d\hat{p} = 0.09$. Since 0.09 is greater than 0.06, we reject $H_0$; at $\alpha = 0.05$ we have established that the 20- to 24-year-old age group is responsible for a higher proportion of bicycle-automobile accidents.

(b) $\hat{p} = 0.875$; $d\hat{p}_c = 0 + 1.65(0.024) = 0.04$; $d\hat{p} = 0.12$. Since 0.12 is greater than 0.04, we reject $H_0$; at $\alpha = 0.05$ we have evidence that the safety program reduces the proportion of accidents caused by the young cyclers.

9-21 (a) $\hat{p} = 0.183$; $d\hat{p}_c = 0 + 2.33(0.016) = 0.037$; $d\hat{p} = 0.19$. Since 0.19 is greater than 0.037, we reject $H_0$; at $\alpha = 0.01$ we have evidence that the proportion of deaths is lowered by the use of anticoagulants.

(b) $\hat{p} = 0.30$; $d\hat{p}_c = 0 \pm 2.58(0.094) = \pm 0.243$; $d\hat{p} = 0.083$. Since 0.083 is less than 0.243, we fail to reject $H_0$; at $\alpha = 0.01$ we cannot conclude that there is a difference in the percentages of grandparents and nongrandparents who are visited regularly.

9-23 $\hat{p} = 0.22$; $d\hat{p}_c = 0 \pm 2.58(0.046) = \pm 0.119$; $d\hat{p} = 0.15$. Since 0.15 is greater than 0.119, we reject $H_0$; at $\alpha = 0.01$, evidently a higher proportion of students than teachers approves of school bussing.

9-25 $H_0$; $p_1 - p_2 = 0.20$; so $\mu = 0.20$ instead of 0.

$$\hat{\sigma}_{d\hat{p}} = \sqrt{\frac{0.60(0.40)}{200} + \frac{0.375(0.625)}{400}} = 0.042 \qquad z_c = \pm 1.96$$

$$d\hat{p}_c = 0.20 \pm 1.96(0.042) = 0.118,\ 0.282 \qquad d\hat{p} = 0.225$$

Since 0.225 is between 0.118 and 0.282, we fail to reject $H_0$. The observed difference between the percentages of young male and young female smokers is not statistically significant.

## Chapter 10: Hypothesis Testing with Sample Means: Large Samples

10-1 The means should be approximately the same. The standard deviation and the range parameters of the population are probably larger than your sample statistics.

10-3 (a) All purchases from this machine

(b) $m = 438/40 = 10.95$ (c) $\mu \approx m = 10.95$ (d) $s = 6.23$

(e) $n = 40 > 30$, hence approximately a normal distribution

(f) $\mu_m \approx 10.95$ (g) $s_m = 6.23/\sqrt{40} \approx 0.99$

10-5 $s_m = \dfrac{0.5}{\sqrt{100}} = 0.05$

$m_c = 2.6 \pm 1.96(0.05) = 2.5$ and $2.7$

Since $m = 2.8 > 2.7$, reject $H_0$. It took longer for Americans to develop eye strain. The actual difference on average is about $2.8 - 2.6 = 0.2$ hour, or about 12 minutes. This may be of no practical importance.

10-7 (a) $z_{39.5} = \dfrac{39.5 - 50}{7} = -1.5 \rightarrow 0.9332$ (b) $z_{45} = \dfrac{45 - 50}{1.17} = 4.3 \rightarrow$ about 100%

10-9 (a) $m_c = 56 \pm 1.96(54/\sqrt{144}) = \$47.18,\ \$64.82$

(b) No, since $m = \$70 > \$64.82$. The average is more than \$56.

10-11 $s_m = \dfrac{4.80}{\sqrt{36}} = \$0.80 \qquad m_c = 195 - 1.65(0.80) = \$193.68$

Since $m = \$193.20 < \$193.68$, reject $H_0$. The average spent is less than \$195 (but maybe only \$1.80 less).

10-13 $z_{0.55} = \dfrac{0.55 - 0.52}{0.04/\sqrt{50}} = 5.3 \rightarrow 1.0000$

$z_{0.53} = \dfrac{0.53 - 0.52}{0.04/\sqrt{50}} = 1.77 \rightarrow \underline{0.9616}$

$\qquad 0.0384$

10-15 $s_m = \dfrac{2.89}{\sqrt{36}} = 0.048 \qquad m_c = 4.1 - 2.33(0.048) = 3.99$

Since $m = 3.8 < 3.99$, reject $H_0$. The mean is less than 4.1 ounces.

10-17 $s_{\text{dm}} = \sqrt{\dfrac{75^2}{50} + \dfrac{75^2}{50}} = 15 \qquad \text{dm}_c = 0 + 2.33(15) = 34.95$

Since dm = $50 > 34.95$, reject $H_0$. Model 1 withstands greater stress.

10-19 $S_{\text{dm}} = \sqrt{\dfrac{10^2}{36} + \dfrac{7^2}{36}} = 2.03 \qquad \text{dm}_c = 0 \pm 2.58(2.03) = \pm 5.2$

Since dm = $-8 < -5.2$, reject $H_0$. Lying on the floor is better than hanging upside down during a test.

10-21 $s_{\text{dm}} = \sqrt{\dfrac{20^2}{40} + \dfrac{20^2}{40}} = 4.47 \qquad \text{dm}_c = 0 + 1.65(4.47) = 7.4$

Since dm = $30 > 7.4$, reject $H_0$. The mean of 50-year-old men is lower.

10-23 $s_{\text{dm}} = \sqrt{\dfrac{25^2}{75} + \dfrac{40^2}{75}} = 5.45 \qquad \text{dm}_c = 0 + 2.33(5.45) = 12.7$

Since dm = $50 > 12.7$, reject $H_0$. Those who were educated in Viet Nam score higher.

10-25 $s_{\text{dm}} = \sqrt{\dfrac{1^2}{35} + \dfrac{1^2}{35}} = 0.24 \qquad \text{dm}_c = 0 + 1.65(0.24) = 0.4$

Since dm = 0.3 in., which is not greater than 0.4, we fail to reject $H_0$. Although the first mean is higher, there still may be no difference in the growth due to the poetry.

10-27 $s_1 = 2.56 \qquad s_2 = 2.27 \qquad s_{\text{dm}} = \sqrt{\dfrac{2.56^2}{50} + \dfrac{2.27^2}{50}} = 0.484$

$\text{dm}_c = 0 \pm 2.58(0.484) = \pm 1.25$

Since dm = 0.50, which is between $-1.25$ and 1.25, we fail to reject $H_0$; the word lengths may be the same.

## Chapter 11: Hypothesis Testing with Sample Means: Small Samples

*Note:* In these answers "df" denotes degrees of freedom.

11-1 Statement (a) is true.

11-3 df = 19 $\qquad m_c = 3 + \dfrac{1.2}{\sqrt{20}}(1.73) = 3.46$

Since $m = 3.8$ is greater than 3.46, we reject $H_0$. At $\alpha = 0.05$ these data do support the claim of discrimination.

11-5 df = 7 $\qquad m_c = 260 - \dfrac{8}{\sqrt{8}}(3.00) = 251.5$

Since $m = 250$ is less than 251.5, we reject $H_0$. At $\alpha = 0.01$ these data do support the claim that the mean Martian IQ is less than 260.

11-7 df = 24 $m_c = 71 \pm \frac{3}{\sqrt{25}}(2.06) = 69.76, 72.24$

Since $m = 70$ is between 69.76 and 72.24, we fail to reject $H_0$. At $\alpha = 0.05$, we have insufficient evidence to dispute the claim that the mean height of officers is 5 feet 10 inches.

11-9 This is a one-tail test on the right, but the observed mean is smaller than that given by $H_0$, so we cannot reject $H_0$. The evidence does not support the claim that the mean score for psychologists is greater than 50.

11-11 $m = 30.83$ $s = 1.47$ df = 5 $m_c = 30 + \frac{1.47}{\sqrt{6}}(3.36) = 32.01$

Since $m = 30.83$ is not greater than 32.01, we fail to reject $H_0$. At $\alpha = 0.01$ these data do not support the claim that the mean brightness is greater than 30 units.

11-13 $p = 0.95$ $\mu = 475$ $\sigma = 4.87$ $S_c = 475 \pm 1.96(4.87) = 465.45, 484.54$

Since $S = 468$ is between 465.45 and 484.54, we fail to reject $H_0$. At $\alpha = 0.05$ these data do not support the claim that the teacher was wrong.

11-15 $s_p^2 = 0.034$ df = 53 $m_c = 0 \pm 2.68(0.051) = \pm 0.14$

Since dm = 0.10 is between −0.14 and 0.14, we fail to reject $H_0$. At $\alpha = 0.01$ these data indicate that the difference between mean grade point averages for male and female math majors is not statistically significant.

11-17 $s_p^2 = 105.79$ df = 38 $m_c = 0 + 1.68(3.25) = 5.46$

Since dm = 2.60 is less than 5.46, we fail to reject $H_0$. At $\alpha = 0.05$ these data do not provide conclusive evidence that the promise of exemptions from the final increases the mean grade.

11-19 $s_p^2 = 52.625$ df = 48 $m_c = 0 \pm 2.01(2.24) = \pm 4.5$

Since dm = 4.4 is between −4.5 and 4.5, we fail to reject $H_0$. At $\alpha = 0.05$ these data indicate that the difference between the mean ages for sedan and convertible owners is not statistically significant.

11-21 $s_p^2 = 0.003$ df = 73 $m_c = 0 + 2.37(0.014) = 0.03$

Since dm = 1.00 is greater than 0.03, we reject $H_0$. At $\alpha = 0.01$ these data indicate that the mean weight of toads in August is greater than the mean weight in June.

11-23 $s_p^2 = 100$ df = 23 $m_c = 0 \pm 2.07(4.08) = \pm 8.45$

Since dm = 10 is greater than 8.45, we reject $H_0$. At $\alpha = 0.05$ these data indicate that the mean score for male nurses is higher than the mean score for businessmen.

11-25 $m_{\text{No-Burn}} = 66$ $s_{\text{No-Burn}} = 9.62$ $m_{\text{Krispy}} = 64$ $s_{\text{Krispy}} = 4.18$
$s_p^2 = 55.00$ df = 8 $m_c = 0 + 2.90(4.69) = 13.60$

Since dm = 2 is not greater than 13.60, we fail to reject $H_0$. At $\alpha = 0.01$ these data show no significant difference between the mean toasting times for the two brands. We cannot conclude that the Krispy brand is quicker.

11-27 The matched-pairs analysis is more sensitive to small but consistent differences. When it is an appropriate design, it is more powerful than the independent-samples design.

11-29 (a) $s_p^2 = 16.82$ df = 18 $m_c = 0 \pm 2.88(1.83) = \pm 5.28$

Since dm = −0.60 is between −5.28 and 5.28, we fail to reject $H_0$. At $\alpha = 0.01$ these data indicate that the difference in mean gas mileage is not statistically significant.

(b) df = 9 $\quad m_c = 0 \pm 3.25\dfrac{0.39}{\sqrt{10}} = \pm 0.40$

Since $m = -0.60$ is less than $-0.40$, we reject $H_0$. At $\alpha = 0.01$ these data indicate that the cabs get higher mileage with Flug gas.

11-31 df = 9 $\quad m_c = 0 + 1.83\dfrac{5}{\sqrt{10}} = \pm 2.89$

Since $m = 5.8$ is greater than 2.89, we reject $H_0$. At $\alpha = 0.05$ these data indicate that the program works.

11-33 df = 9 $\quad m_c = 0 + 1.83\dfrac{6}{\sqrt{10}} = 3.47$

Since $m = 10$ is greater than 3.47, we reject $H_0$. At $\alpha = 0.05$ these data indicate that the mean number of errors is greater on the backhand.

11-35 $d = \{3, 4, 2, -1, -2, 2, 14, -1, -2, -1\} \quad m = 1.8 \quad s = 4.80 \quad$ df = 9

$m_c = 0 + 1.83\dfrac{4.80}{\sqrt{10}} = 2.78$

Since $m = 1.8$ is not greater than 2.78, we fail to reject $H_0$. At $\alpha = 0.05$ these data do not support the claim that Dr. Fleece's course increases the mean grade.

11-37 Assume that a higher score indicates a better mood.

$d = \{4, 3, 3, 4, 0, 1, 3, -2, 5\} \quad m = 2.3 \quad s = 2.50 \quad$ df = 8

$m_c = 0 + 1.86\dfrac{2.24}{\sqrt{9}} = 1.39$

Since $m = 2.3$ is greater than 1.39, we reject $H_0$. At $\alpha = 0.05$ these data support the claim that this combination of music improves the mood of teenagers.

## Answers to Sample Test for Chapters 8, 9, 10, and 11

1. Parameter
2. The probability of rejecting the null hypothesis even though it is true. The numerical value of the Type I error is the same as the significance level of the test.
3. The only way to have $\alpha = 0$ is to sample the entire population.
4. Each population is normal. The 2 population variances are approximately equal.
5. (a) $H_0$: $p = 0.25$; the probability the wheel stops on red is 1/4.
   $H_a$: $p \neq 0.25$; the probability the wheel stops on red is not 1/4 (a two-tail test).

   (b) Reject $H_0$ if $S$ is less than 9 or more than 21.

   (c) $\mu = 15 \quad \sigma = 3.35 \quad S_c = 15 \pm 1.96(3.35) = 8.43, 21.57$

   Since $S = 12$ is between 8.43 and 21.57, we fail to reject $H_0$. There is insufficient evidence to establish conclusively that the wheel is not honest.
6. (a) $H_0$: $\mu = 135$; the mean parking time is 135 minutes.
   $H_a$: $\mu > 135$; the mean parking time is more than 135 minutes (a one-tail test).

   (b) $m_c = 135 + 1.65(45/\sqrt{100}) = 142.43$
   Reject $H_0$ if $m$ is more than 142.43 minutes.

   (c) Since $m = 156$ minutes, we reject $H_0$. There is sufficient evidence to establish at $\alpha = 0.05$ that the mean parking time is more than 2 hours 15 minutes.

7. (a) $H_0$: $\mu_1 - \mu_2 = 0$; the mean time to boil water is the same with both brands of cookers.

   $H_a$: $\mu_1 - \mu_2 \neq 0$; the mean time to boil water is not the same with both brands of cookers.

   (b) $m_1 = 5$ $m_2 = 4.4$ $s_1 = 0.678$ $s_2 = 0.212$ $s_p^2 = 0.252$ df = 8
   $\text{dm}_c = 0 \pm 2.31(0.318) = \pm 0.734$
   Reject $H_0$ if dm is not between −0.73 and 0.73 minute.

   (c) Since $m = 0.60$ minute, we fail to reject $H_0$. There is not sufficient evidence at $\alpha = 0.05$ to establish the superiority of either brand.

8. (a) $H_0$: $\mu_1 - \mu_2 = 0$; the mean time to endure the noise is the same after both films.
   $H_a$: $\mu_1 - \mu_2 \neq 0$; the mean time to endure the noise is not the same after both films.

   (b) $m_1 = 4.4$ $m_2 = 3.6$ $s_1 = 1$ $s_2 = 1$ $s_p^2 = 1$ df = 98
   $\text{dm}_c = 0 \pm 1.98(0.20) = \pm 0.40$
   Reject $H_0$ if dm is not between −0.40 and 0.40 minute.

   (c) Since $m = 0.80$ minute, we reject $H_0$. There is sufficient evidence at $\alpha = 0.05$ to claim that after watching the movie where the soldiers win, students can endure the noise longer.

9. (a) Twins are used so each member of the pair would be equal except for the teaching method. This is a good way to control extraneous variables that might confuse the interpretation of the data.

   (b) Scores are normally distributed, and variances are approximately equal.

   (c) Paired $t$-test:
   $m = 6.7$ $s = 43.13$ df = 9 $m_c = 0 \pm 2.26(43.13/\sqrt{10}) = \pm 30.82$
   Since $m = 6.7$ is between −30.82 and 30.82, we fail to reject $H_0$. The evidence does not conclusively establish the superiority of either teaching method.

   Unpaired $t$-test:
   $m_1 = 153.5$ $m_2 = 146.8$ $s_1 = 34.32$ $s_2 = 24.79$ $s_p^2 = 896.25$
   df = 18 $\text{dm}_c = 0 \pm 2.10(13.39) = \pm 28.12$

   Since $m = 6.7$ is between −28.12 and 28.12, we fail to reject $H_0$. The evidence does not conclusively establish the superiority of either teaching method.

10. Your answer should include the idea that to prove a statement is true, one assumes the opposite to be true, and then shows that this leads to a result which is so unbelievable that the assumption is probably false and therefore the original statement is true. For example:

    Dear Aunt Tilly,

    In response to your question about hypothesis tests, to show that smoking adversely affects breathing, one could obtain a large *random* sample of people, both smokers and non-smokers. Assuming the opposite, that smoking and lung disorders are *NOT* related, we would expect approximately similar percentages of both groups to be healthy. If the difference in the percentages of healthy people in the two groups is *extraordinarily* large, then we must either:

    1) accept the truth of our assumption and the occurrence of a "miraculous" run of luck in our random samples, or

2) refuse to believe that such a rare event occurred, but rather reject our assumption and accept the original thesis.

Your loving nephew,
Stan DeVeashun

## Chapter 12: Confidence Intervals

12-1 Because $\alpha$ refers to an error about a null hypothesis. We don't have any hypothesis.

12-3 $\hat{p} - 1.65\sigma < p < \hat{p} + 1.65\sigma$

12-5 $n\hat{p} = 53$, $n\hat{q} = 47$, both $> 5$

$$\hat{p} - 2.58\hat{\sigma} < p < \hat{p} + 2.58\hat{\sigma}$$

$$0.53 - 2.58\sqrt{\frac{0.53(0.47)}{100}} < p < 0.53 + 2.58\sqrt{\frac{0.53(0.47)}{100}}$$

$$0.53 - 0.13 < p < 0.53 + 0.13$$

$$0.40 < p < 0.66$$

12-7 $n\hat{p} = 30$, $n\hat{q} = 20$, both $> 5$

$$\hat{p} - 1.96\hat{\sigma} < p < \hat{p} + 1.96\hat{\sigma}$$

$$0.60 - 1.96\sqrt{\frac{0.60(0.40)}{50}} < p < 0.60 + 1.96\sqrt{\frac{0.60(0.40)}{50}}$$

$$0.60 - 0.14 < p < 0.60 + 0.14$$

$$0.46 < p < 0.74$$

12-9 $n\hat{p} = 80$, $n\hat{q} = 20$, both $> 5$

$$\hat{p} - 1.65\hat{\sigma} < p < \hat{p} + 1.65\hat{\sigma}$$

$$0.80 - 1.65\sqrt{\frac{0.80(0.20)}{100}} < p < 0.80 + 1.65\sqrt{\frac{0.80(0.20)}{100}}$$

$$0.80 - 0.066 < p < 0.80 + 0.066$$

$$0.73 < p < 0.87$$

12-11 (a) $n\hat{p} = 12$, $n\hat{q} = 68$, both $> 5$

$$\hat{p} - 1.96\hat{\sigma} < p < \hat{p} + 1.96\hat{\sigma}$$

$$0.15 - 1.96\sqrt{\frac{0.15(0.85)}{80}} < p < 0.15 + 1.96\sqrt{\frac{0.15(0.85)}{80}}$$

$$0.15 - 0.8 < p < 0.15 + 0.8$$

$$0.07 < p < 0.23$$

(b) About 0.15(700) ± 0.08(700), that is, about 105 ± 56, or between 49 and 161 (about 50 to 160).

12-13 (a) $n\hat{p} = 32$, $n\hat{q} = 38$, both $> 5$

$$\hat{p} - 1.96\hat{\sigma} < p < \hat{p} + 1.96\hat{\sigma}$$

$$0.32 - 1.96\sqrt{\frac{0.32(0.68)}{100}} < p < 0.32 + 1.96\sqrt{\frac{0.32(0.68)}{100}}$$

$$0.32 - 0.09 < p < 0.32 + 0.09$$

$$0.23 < p < 0.41$$

(b) $n\hat{p} = 139$, $n\hat{q} = 361$, both $> 5$

$$\hat{p} - 1.96\hat{\sigma} < p < \hat{p} + 1.96\hat{\sigma}$$

$$0.28 - 1.96\sqrt{\frac{0.28(0.72)}{500}} < p < 0.28 + 1.96\sqrt{\frac{0.28(0.72)}{500}}$$

$$0.28 - 0.04 < p < 0.28 + 0.04$$

$$0.24 < p < 0.32$$

(c) $n\hat{p} = 297$, $n\hat{q} = 703$, both $> 5$

$$\hat{p} - 1.96\hat{\sigma} < p < \hat{p} + 1.96\hat{\sigma}$$

$$0.30 - 1.96\sqrt{\frac{0.30(0.70)}{1000}} < p < 0.30 + 1.96\sqrt{\frac{0.30(0.70)}{1000}}$$

$$0.30 - 0.03 < p < 0.30 + 0.03$$

$$0.27 < p < 0.33$$

(d) Probably not

(e) The error term has a maximum error when $p = q = 0.5$. Thus we have $1.96\sqrt{\frac{0.5(0.5)}{38{,}000}} = 0.005$ or $\pm\frac{1}{2}\%$.

12-15 $n > \left(\frac{2.58}{a}\right)^2(0.25) > \left(\frac{2.58}{0.05}\right)^2(0.25) > 666$

Thus a sample size of at least 666 times will suffice.

12-17 $m = 23.03$ glymphs $\quad n = 36 > 30$

$$m - 1.96s_m < \mu < m + 1.96s_m$$

$$23.03 - \frac{1.96(5.28)}{\sqrt{36}} < \mu < 23.03 + \frac{1.96(5.28)}{\sqrt{36}}$$

$$23.03 - 1.72 < \mu < 23.03 + 1.72$$

$$21.3 < \mu < 24.8$$

12-19 $n = 52 > 30 \quad m - 1.96s_m < \mu < m + 1.96s_m$

$$3523 - \frac{1.96(162)}{\sqrt{52}} < \mu < 3523 + \frac{1.96(162)}{\sqrt{52}}$$

$$3523 - 44 < \mu < 3523 + 44$$

$$3479 < \mu < 3567$$

12-21 $n = 49 > 30 \quad m - 2.58s_m < \mu < m + 2.58s_m$

$$4.6 - \frac{2.58(2.9)}{\sqrt{49}} < \mu < 4.6 + \frac{2.58(2.9)}{\sqrt{49}}$$

$$4.6 - 1.1 < \mu < 4.6 + 1.1$$

$$3.5 < \mu < 5.7$$

12-23 $n = 48 > 30 \quad m - 2.33s_m < \mu < m + 2.33s_m$

$$15.2 - \frac{2.33(1.5)}{\sqrt{48}} < \mu < 15.2 + \frac{2.33(1.5)}{\sqrt{48}}$$

$$15.2 - 0.5 < \mu < 15.2 + 0.5$$

$$14.7 < \mu < 15.7$$

12-25 $n = 400 > 30 \quad m - 2.58s_m < \mu < m + 2.58s_m$

$$110.3 - \frac{2.58(12.2)}{\sqrt{400}} < \mu < 110.3 + \frac{2.58(12.2)}{\sqrt{400}}$$

$$110.3 - 1.6 < \mu < 110.3 + 1.6$$

$$108.7 < \mu < 111.9$$

12-27 Too narrow

*Note*: In the following answers df means degrees of freedom.

12-29 $n = 12$, not $n > 30$

$\text{df} = 11$

$$m - 2.20s_m < \mu < m + 2.20s_m$$

$$814 - \frac{2.20(148)}{\sqrt{12}} < \mu < 814 + \frac{2.20(148)}{\sqrt{12}}$$

$$814 - 93.99 < \mu < 814 + 93.99$$

$$\$720 < \mu < \$908$$

12-31 $n = 19$, not $n > 30$

$\text{df} = 18$

$$m - 2.86s_m < \mu < m + 2.86s_m$$

$$3.97 - \frac{2.86(3.2)}{\sqrt{19}} < \mu < 39.7 + \frac{2.86(3.2)}{\sqrt{19}}$$

$$39.7 - 2.1 < \mu < 39.7 + 2.1$$

$$37.6 \text{ years} < \mu < 41.8 \text{ years}$$

12-33 $n = 16$, not $n > 30$

$\text{df} = 15$

$$m - 1.75s_m < \mu < m + 1.75s_m$$

$$123.1 - \frac{1.75(8.0)}{\sqrt{16}} < \mu < 123.1 + \frac{1.75(8.0)}{\sqrt{16}}$$

$$123.1 - 3.5 < \mu < 123.1 + 3.5$$

$$119.6 < \mu < 126.6$$

12-35 (a) $n_1 = 160 + 340 = 500$ aggies

$n_2 = 220 + 360 = 580$ miners

$n_1\hat{p}_1 = 160,\ n_1\hat{q}_1 = 340,\ n_2\hat{p}_2 = 220,\ n_2\hat{q}_2 = 360$, all $> 5$

$$\hat{\sigma} = \sqrt{\frac{(160/500)(340/500)}{500} + \frac{(220/580)(360/580)}{580}} = 0.03$$

$$d\hat{p} - zs_{d\hat{p}} < p_1 - p_2 < d\hat{p} + zs_{d\hat{p}}$$

$$\frac{160}{500} - \frac{220}{580} - 1.96(0.03) < p_1 - p_2 < \frac{160}{500} - \frac{220}{580} + 1.96(0.03)$$

$$-0.06 - 0.06 < p_1 - p_2 < -0.06 + 0.06$$

$$-0.12 < p_1 - p_2 < 0.00$$

(b) $$\frac{340}{500} - \frac{360}{580} - 1.96(0.03) < p_1 - p_2 < \frac{340}{500} - \frac{360}{580} + 1.96(0.03)$$

$$0.06 - 0.06 < p_1 - p_2 < 0.06 + 0.06$$

$$0.00 < p_1 - p_2 < 0.12$$

(c) Obviously!

12-37 $$d\hat{p} = \frac{350}{500} - \frac{250}{500} = 0.20$$

$n_1 = 500 \quad n_2 = 500$

$n_1\hat{p}_1 = 350 \quad n_1\hat{q}_1 = 150 \quad n_2\hat{p}_2 = 250 \quad n_2\hat{q}_2 = 250$ all $> 5$

$$\hat{\sigma} = \sqrt{\frac{(350/500)(150/500)}{500} + \frac{(250/500)(250/500)}{500}} = 0.03$$

$$d\hat{p} - zs_{d\hat{p}} < p_1 - p_2 < d\hat{p} + zs_{d\hat{p}}$$

$$0.20 - 1.96(0.03) < p_1 - p_2 < 1.96(0.03)$$

$$0.20 - 0.06 < p_1 - p_2 < 0.20 + 0.06$$

$$0.14 < p_1 - p_2 < 0.26$$

12-39 $m_1 = \$58 \quad s_1 = \$9.84 \quad m_2 = \$88 \quad s_2 = \$9.84$
$n_1 = 10 \quad n_2 = 10$ not both > 30
$\text{df} = 10 + 10 - 2 = 18$

$$s_p^2 = \frac{9(9.84)^2 + 9(9.84)^2}{18} = 9.84^2$$

$$s_{\text{dm}} = \sqrt{9.84^2\left(\frac{1}{10} + \frac{1}{10}\right)} \approx \$4.40$$

$$\text{dm} - 2.10s_{\text{dm}} < \mu_1 - \mu_2 < \text{dm} + 2.10s_{\text{dm}}$$
$$(58 - 88) - 2.10(4.40) < \mu_1 - \mu_2 < (58 - 88) + 2.10(4.40)$$
$$-30 - 9.24 < \mu_1 - \mu_2 < -30 + 9.24$$
$$-\$39.24 < \mu_1 - \mu_2 < -\$20.76$$

12-41 $n_1 = 19 \quad n_2 = 13$ not both > 30
$\text{df} = 19 + 13 - 2 = 30$

$$s_p^2 = \frac{18(1.4)^2 + 12(2.1)^2}{30} = 2.94$$

$$s_{\text{dm}} = \sqrt{2.94\left(\frac{1}{19} + \frac{1}{13}\right)} \approx 0.617$$

$$\text{dm} - 2.75s_{\text{dm}} < \mu_1 - \mu_2 < \text{dm} + 2.75s_{\text{dm}}$$
$$(5.1 - 3.8) - 2.75(0.617) < \mu_1 - \mu_2 < (5.1 - 3.8) + 2.75(0.617)$$
$$1.3 - 1.70 < \mu_1 - \mu_2 < 1.3 + 1.70$$
$$-0.4 < \mu_1 - \mu_2 < 3.0$$

12-43 $n_1 = 50 \quad n_2 = 50$ both > 30

$$s_{\text{dm}} = \sqrt{\frac{1.1^2}{50} + \frac{1.3^2}{50}} = 0.24$$

$$\text{dm} - zs_{\text{dm}} < \mu_1 - \mu_2 < \text{dm} + zs_{\text{dm}}$$
$$(17.5 - 14.7) - 1.96(0.24) < \mu_1 - \mu_2 < (17.5 - 14.7) + 1.96(0.24)$$
$$2.8 - 0.47 < \mu_1 - \mu_2 < 2.8 + 0.47$$
$$2.3 < \mu_1 - \mu_2 < 3.3$$

## Chapter 13: Chi Square

13-1 (a) You cannot determine both row and column totals before the survey is conducted.

(b) $H_0$: Age and opinion are independent.

$$\text{df} = 2 \quad \alpha = 0.05 \quad X_c^2 = 5.99$$

Since $X^2 = 38.67$ is greater than 5.99, reject $H_0$. The evidence is sufficient to establish a relation between age and opinion. Evidently, younger people are more likely to favor the new highway.

(c) Three possible options: Continue the survey to increase $n$; combine the last two columns into a "not for" category; drop the last column from the survey.

13-3 (a) Chi-square test of independence

(b) All students who take English composition at Uptudate College

(c) $H_0$: Grade and use of word processor are independent.
$H_a$: Grade is related to use of word processor.

(d) df = 4

(e) $X_c^2 = 9.49$

(f)

| | | |
|---|---|---|
| 20.33 | 19.67 | 20 |
| 20.33 | 19.67 | 20 |
| 20.33 | 19.67 | 20 |

(g) $$X^2 = \frac{38^2 + 20^2 + 3^2}{20.33} + \frac{20^2 + 18^2 + 21^2}{19.67} + \frac{2^2 + 22^2 + 36^2}{20} - 180 = 59.57$$

(h) Since $X^2 = 59.57$ is greater than 9.49, we reject $H_0$. There is evidence that grade and use of the word processor are related.

(i) Not exactly. Perhaps the use of the word processor does improve grades. But because of the design of the experiment we cannot rule out other possibilities. For instance, suppose students who are already better are the ones who choose to use the word processor.

13-5 df = 3 $X_c^2 = 7.81$ $X^2 = 22.63$

Since 22.63 is greater than 7.81, we reject $H_0$. At $\alpha = 0.05$, we have sufficient evidence to establish a relation between grades in Calculus 1 and passing Calculus 2. Evidently, people with A's in Calculus 1 are likely to pass Calculus 2, and people with D's in Calculus 1 are more likely to fail Calculus 2.

13-7 df = 2 $X_c^2 = 5.99$ $X^2 = 0.40$

Since 0.40 is not greater than 5.99, we fail to reject $H_0$. At $\alpha = 0.05$, we do not have sufficient evidence to establish a relation between a cutoff admission point of 600 and eventual likelihood of graduation.

13-9 df = 6 $X_c^2 = 12.59$ $X^2 = 116.38$

Since 116.38 is greater than 12.59, we reject $H_0$. At $\alpha = 0.05$, we have sufficient evidence to establish a relation between the type of manuscript used and the symbol used for footnotes.

13-11 df = 14 $X_c^2 = 29.14$ $X^2 = 26.27$

Since 26.27 is not greater than 29.14, we fail to reject $H_0$. At $\alpha = 0.01$, we do not have sufficient evidence to establish a relation between pole position and outcome of the race.

13-13 df = 1 $X_c^2 = 3.84$ $X^2 = 0.23$

Since 0.23 is not greater than 3.84, we fail to reject $H_0$. At $\alpha = 0.05$, we do not have sufficient evidence to establish a relation between sex and degree of competitiveness. These results support Pat's argument.

13-15 df = 1 $X_c^2 = 3.84$ $X^2 = 3.92$

Since 3.92 is greater than 3.84, we reject $H_0$. At $\alpha = 0.05$, we have sufficient evidence to establish a relation between sex and the likely direction the person turns. It appears that males are more likely to turn left.

13-17 $X^2 = 62.22 - 60 = 2.22$

13-19 df = 3 $X_c^2 = 7.81$ $X^2 = 62.20$

Since 62.20 is greater than 7.81, we reject $H_0$. At $\alpha = 0.05$, we have sufficient evidence to reject his theory.

13-21 df = 2 $X_c^2 = 5.99$ $X^2 = 120$

Since 120 is greater than 5.999, we reject $H_0$. At $\alpha = 0.05$, we have sufficient evidence to conclude that the sex distribution in Irtusk is different.

13-23 Combine the last 3 categories so all expected values are greater than 5. df = 3; $X_c^2 = 7.81$ if $\alpha = 0.05$ or $X_c^2 = 11.34$ if $\alpha = 0.01$. $X^2 = 182.25$. Since 182.25 is greater than both critical values, we have sufficient evidence to conclude that the distribution of bugs is not random.

13-25 (a) $dp$ = present percent − former percent = 0.10 $\hat{\sigma}_{d\hat{p}} = 0.054$

95% confidence interval: $0.10 \pm 1.96(0.054) = 0.10 \pm 0.11$
$= -0.01, 0.21$

Because this interval contains 0, the observed difference is not statistically significant.

(b) $H_0$: the percentages of present and former students who said they would pose nude are the same.

$\hat{p} = 0.23$ $\hat{\sigma}_{d\hat{p}} = 0.052$ $d\hat{p}_c = 0 \pm 1.96(0.052) = \pm 0.102$

Since 0.10 is not greater than 0.102, we fail to reject $H_0$. At $\alpha = 0.05$ we do not have conclusive evidence that the percentage has changed.

(c) $H_0$: Willingness to pose nude and date of attendance at Lax University are independent.

df = 1 $X_c^2 = 3.84$ $X^2 = 3.73$

Since 3.73 is not greater than 3.84, we fail to reject $H_0$. At $\alpha = 0.05$, we do not have sufficient evidence to conclude that there is a relation between willingness to pose nude and date of attendance at Lax University.

13-27 (a) $n = 10{,}654$ all expected values are $\dfrac{10{,}654}{12} = 887.833$

$X^2 = 113.77$ df = 11 $X_c^2 = 24.73$

Since 113.77 is greater than 24.73, we reject $H_0$. At $\alpha = 0.01$, we have sufficient evidence to conclude that homicides are not equally likely in every month.

(b) This question is equivalent to asking if more than half the murders that occur in these 2 months occur in December.
$H_0$: $p = 0.5$; the probability that a murder in these 2 months occurs in December is 0.5. $\mu = 1033$; $\sigma = 22.73$; $S_c = 1033 + 2.33(22.73) = 1085.96$. since 1042 is not greater than 1085.96, we fail to reject $H_0$. At $\alpha = 0.01$, we cannot establish that a greater percentage of murders occurs in December than in July.

(c) There were 744 murders in 283 February days for a rate of 2.63 murders per day. There were 789 murders in 310 March days for a rate of 2.55 murders per day. Thus the statement is true.

13-29 (a)

| Cell | (1) | (2) | (3) | (4) | (5) | (6) |
|---|---|---|---|---|---|---|
| E | 5.93 | 35.33 | 88.74 | 88.74 | 35.33 | 5.93 |
| O | 15 | 30 | 85 | 80 | 40 | 10 |

(b) df = 5 $\quad X_c^2 = 11.07 \quad X^2 = 19.11$

Since 19.11 is greater than 11.07, we reject $H_0$. At $\alpha = 0.05$, we have sufficient evidence to conclude that the speeds are not distributed normally.

13-31 df = 2 $\quad X_c^2 = 5.99 \quad X^2 = 6.87$

Since 6.87 is greater than 5.99, we reject $H_0$. At $\alpha = 0.05$, we have sufficient evidence to conclude that there is a relation between type of medication and achievement of relief from chronic pain. The greatest percentage of relieved patients occurred in the experimental medicine group.

13-35 (a) df = 1 $\quad X_c^2 = 3.84 \quad$ with correction factor, $X^2 = 3.38$

Since 3.38 is not greater than 3.84, Tim does not reject $H_0$. At $\alpha = 0.05$, he does not have sufficient evidence to establish a relation between sex and the likely direction the person turns.

(b) $\hat{p} = \dfrac{57 + 43}{200} = 0.50 \quad d\hat{p} = 0.14 \quad \hat{\sigma}_{d\hat{p}} = 0.0707$

$d\hat{p}_c = 0 \pm 1.96(0.071) = \pm 0.139$

Since 0.14 is greater than 0.139, Chet can reject $H_0$. At $\alpha = 0.05$, he has sufficient evidence to establish a relation between sex and the likely direction the person turns. He concludes that males are more likely to turn left.

(c) Using the correction factor yields a smaller value of $X^2$, which makes it less likely that the null hypothesis will be rejected. Statisticians call this a more "conservative" approach, because it needs stronger evidence to reject $H_0$. The test without the correction factor is equivalent to the binomial two-sample test.

## Answers to Sample Test for Chapters 12 and 13

1. A 100% confidence interval based on a sample will include every possible value of the parameter. No possible values will be eliminated.

2. (a) $\dfrac{30}{200} \pm 1.96\sqrt{\dfrac{0.15(0.85)}{200}} = 0.101, 0.199$

   The confidence interval is from 10.1% to 19.9%.

   (b) It would be a good estimate if the membership in the association were representative of the profession, and not biased by sex.

3. $n \geq \left(\dfrac{19.6}{0.03}\right)^2 (0.25) = 1067.1$

   The sample should contain at least 1068 people.

4. (a) The response time of all emergency calls received between 8 A.M. and 4 P.M. by the Emergency Medical Service in New York City. This is a reasonable estimate if this is a typical day and all calls are reported. Otherwise it might be a poor estimate.

   (b) $328.8 \pm 1.96(15/\sqrt{88})$. The 95% confidence interval is from 25.7 to 31.9 minutes.

5. $\dfrac{36}{50} - \dfrac{18}{50} \pm 1.96\sqrt{\dfrac{0.72(0.28)}{50} + \dfrac{0.36(0.64)}{50}}$

The 95% confidence interval is from 18% to 54%. Some possible explanations: bias against women in promotion; women are new to managerial positions, and many have not been there long enough to be promoted.

6. $X^2 = 77.99$ df $= 4$ $X_c^2 = 9.49$

Since 77.99 is greater than 9.49, we reject $H_0$. There is some relation between the heights of the fathers and the heights of the sons. It appears that there are more short-father/short-son pairs and more tall-father/tall-son pairs than expected from random pairings.

7. $X^2 = 0$. This supports the hypothesis of independence.

8. (a)

| Number of Heads: | 0 | 1 | 2 | 3 |
|---|---|---|---|---|
| Expected number of times: | 5 | 15 | 15 | 5 |
| Observed number of times | 4 | 9 | 16 | 11 |

(b) $X^2 = 4^2/5 + 9^2/15 + 16^2/15 + 11^2/5 = 49.8666$ df $= 3$

$X_c^2 = 7.81$, reject fairness. The coin appears biased; a higher than expected proportion of the experiments yielded 3 heads.

## Chapter 14: Correlation and Prediction

14-1 A value of $r$ near $-1$ would mean that those with high grades in Spanish II got low grades in Spanish III and vice versa.

A value of $r$ near 0 would mean that there is little or no connection between grades in Spanish II and grades in Spanish III.

A value of $r$ near $+1$ would mean that those with high grades in Spanish II got high grades in Spanish III and vice versa.

14-3 No. Bradley has changed "no correlation" to a negative correlation.

14-5 (b) It appears to be positive, but not very strong, thus about +0.03.

(c)

| $X$ | $Y$ | $X^2$ | $Y^2$ | $XY$ |
|---|---|---|---|---|
| 10 | 1 | 100 | 1 | 10 |
| 5 | 9 | 25 | 81 | 45 |
| 3 | 3 | 9 | 9 | 9 |
| 8 | 9 | 64 | 81 | 72 |
| 0 | 0 | 0 | 0 | 0 |
| 26 | 22 | 198 | 172 | 136 |

(d) $n = 5$ (e) $\Sigma X = 26$ (f) $\Sigma Y = 22$ (g) $\Sigma XY = 136$ (h) $\Sigma X^2 = 198$

(i) $\Sigma Y^2 = 172$ (j) $(\Sigma X)^2 = 676$ (k) $(\Sigma Y)^2 = 484$

(l) $n\Sigma XY - (\Sigma X)(\Sigma Y) = 5(136) - 26(22) = 108$

(m) $\sqrt{n\Sigma X^2 - (\Sigma X)^2}\sqrt{n\Sigma Y^2 - (\Sigma Y)^2} = \sqrt{5(198) - 676}\sqrt{5(172) - 484} = 343.6$

(n) $r = 0.31$

14-7 (b) +1

(c) $n = 6$ $\Sigma X = 252$ $\Sigma Y = 21$

$\Sigma XY = 1092$ $\Sigma X^2 = 13{,}104$ $\Sigma Y^2 = 91$

$$n\Sigma XY - (\Sigma X)(\Sigma Y) = 6(1092) - 252(21) = 1260$$

$$\sqrt{n\Sigma X^2 - (\Sigma X)^2}\,\sqrt{n\Sigma Y^2 - (\Sigma Y)^2} = \sqrt{6(13{,}104) - 252^2}\,\sqrt{6(91) - 21^2} = 1260$$

$r = 1$

14-9 High positive; the older mothers would tend to have older children.

14-11 $$r = \frac{6(30{,}201) - 419(414)}{\sqrt{6(30{,}319) - 419^2}\,\sqrt{6(30{,}204) - 414^2}} = \frac{7740}{\sqrt{6353}\,\sqrt{9828}} = 0.98$$

14-13 (a) $$r = \frac{7(441) - 21(91)}{\sqrt{7(91) - 21^2}\,\sqrt{7(2275) - 91^2}} = \frac{1176}{\sqrt{196}\,\sqrt{7644}} = 0.96$$

(b) $$r = \frac{4(432) - 18(86)}{\sqrt{4(86) - 18^2}\,\sqrt{4(2258) - 86^2}} = \frac{180}{\sqrt{20}\,\sqrt{1636}} = 0.995$$

14-15 $$r = \frac{5(17{,}274) - 348(248)}{\sqrt{5(24{,}258 - 348^2}\,\sqrt{5(12{,}306) - 248^2}} = \frac{66}{\sqrt{186}\,\sqrt{26}} = 0.95$$

14-17 Both sexes:

$$r = \frac{20(3690) - 110(635)}{\sqrt{20(770) - 110^2}\,\sqrt{20(21{,}137 - 635^2}} = \frac{3950}{\sqrt{3300}\,\sqrt{19{,}515}} = 0.49$$

$r^2 = 0.24$

Females:

$$r = \frac{10(2172) - 56(367)}{\sqrt{10(394) - 56^2}\,\sqrt{10(13{,}715) - 367^2}} = \frac{1168}{\sqrt{804}\,\sqrt{2461}} = 0.83$$

$r^2 = 0.69$

Males:

$$r = \frac{10(1518) - 54(268)}{\sqrt{10(376) - 54^2}\,\sqrt{10(7422) - 268^2}} = \frac{708}{\sqrt{844}\,\sqrt{2396}} = 0.50$$

$r^2 = 0.25$

14-19 (a) Weight of small packages and the number that you can carry at one time. As they get heavier, you will be able to carry fewer.

(b) Age and height of adults. Older adults may be either taller or shorter than younger adults.

(c) Age and height of children. As children grow older, they usually grow taller.

14-21 See Exercise 14-10. We will perform a hypothesis test on the significance of $r$.

Population: All male customers of this shopping center.

Sample: 10 such men

$H_0$: There is no correlation between shoe size and the number of ties owned; $\rho = 0$.

$H_a$: There is some nonzero correlation between shoe size and the number of ties owned; $\rho \neq 0$ (a two-tail test).

$r_c = \pm 0.76$

Decision rule: We will reject the null hypothesis if $r$ is either less than $-0.76$ or greater than $+0.76$.

Since $r = -0.22$ (from Exercise 14-10), we fail to reject $H_0$. There may be no correlation between shoe size and the number of ties owned.

14-23 (See Exercise 14-12) $r_c = \pm 0.96$

Decision rule: We will reject the null hypothesis if $r$ is either less than $-0.96$ or greater than $+0.96$.

Since $r = 0.99$ (from Exercise 14-12), we reject $H_0$. There is some positive correlation between the volume of water and the length of the goldfish.

14-25 (See Exercise 14-14a.) $r_c = 0.98$

Decision rule: We will reject the null hypothesis if $r$ is greater than $+0.98$.

Since $r = 0.995$ (from Exercise 14-13b), we reject $H_0$. There is some positive correlation between the time elapsed and the distance traveled for 3 seconds or more.

14-27 $r_c = 0.34$

Decision rule: We will reject the null hypothesis if $r$ is greater than $+0.34$.

Outcome:

$$r = \frac{25(8380) - 67(2800)}{\sqrt{25(212) - 67^2}\,\sqrt{25(342{,}900) - 2800^2}} = \frac{21{,}900}{\sqrt{811}\,\sqrt{732{,}500}} = 0.90$$

Conclusion: Reject $H_0$; there is some positive correlation between the length of a male college student's hair and his political beliefs.

14-29 (a) $r_c = 0.67$

Decision rule: We will reject the null hypothesis if $r$ is greater than $+0.67$.

Outcome:

$$r = \frac{7(1915) - 59(225)}{\sqrt{7(507) - 59^2}\,\sqrt{7(7875) - 225^2}} = \frac{130}{\sqrt{68}\,\sqrt{4500}} = 0.24$$

Conclusion: Fail to reject $H_0$. There may be no correlation between the reading and the spelling scores.

(b) $r_c = -0.73$

Decision rule: We will reject the null hypothesis if $r$ is less than $-0.73$.

Outcome:

$$r = \frac{6(3950) - 50(480)}{\sqrt{6(426) - 50^2}\,\sqrt{6(39{,}400) - 480^2}} = \frac{-300}{\sqrt{56}\,\sqrt{6000}} = -0.52$$

Conclusion: Fail to reject $H_0$. There may be no correlation between the spelling and the math scores.

(c) $r_c = \pm 0.81$

Decision rule: We will reject the null hypothesis if $r$ is either greater than $+0.81$ or less than $-0.81$.

Outcome:

$$r = \frac{6(522) - 52(62)}{\sqrt{6(458) - 52^2}\,\sqrt{6(840) - 62^2}} = \frac{-92}{\sqrt{44}\,\sqrt{1196}} = -0.40$$

Conclusion: Fail to reject $H_0$. There may be no correlation between the spelling and the music scores.

14-31 (a) $Y_{\text{predicted}} = 1.5(82) - 40 = 83$

(b) The predicted value of 83 does not refer to one individual grade, but to the *average* of all persons who scored an 82.

14-33 (a) $r = \dfrac{15(9416) - 133(1116)}{\sqrt{15(1361) - 133^2}\sqrt{15(86{,}424) - 1116^2}} = \dfrac{-7188}{\sqrt{2726}\sqrt{50{,}904}} = -0.61$

(b) $b = -\dfrac{7188}{2726} = -2.64$

(c) $Y_{\text{predicted}} = -2.64\left(X - \dfrac{133}{15}\right) + \dfrac{1116}{15} = -2.64(X - 8.87) + 74.4$

(d) $Y_{\text{predicted}} = -2.64(11 - 8.87) + 74.4 = 69$

14-35 $Y_{\text{predicted}} = -1.4(48 - 36) + 50 = 33.2$ barrels

14-37 (b) $r = \dfrac{8(-21{,}281) - (-556)372.5}{\sqrt{8(41{,}812) - (-556)^2}\sqrt{8(24{,}311.25) - 372.5^2}} = 0.98$

(c) $Y_{\text{predicted}} = 1.45(X + 69.5) + 46.6$ (d) $Y_{\text{predicted}} = 60.4 \approx 60$

14-39 $1.00 - 0.8^2 = 0.36$ or 36%

14-41 $r_1 = 0.4$

Variation due to speed $= 0.4^2 = 0.16$
Variation due to inattention $= 4(0.16) = 0.64 = r_2^2$

Hence $r_2 = \sqrt{0.64} = \pm 0.8$. Assuming inattention and errors have a positive correlation, $r_2 = +0.8$.

## Chapter 15: Tests Involving Variance

15-1 The 2 population variances are equal.

15-3 (a) $X^2 = \dfrac{30(8)^2}{25} = 76.8$ df = 30 $X_c^2 = 43.77$

Since 76.8 is greater than 43.77, we reject $H_0$. At $\alpha = 0.05$ the data support the conclusion that the standard deviation of the distances is more than 5 feet.

(b) df = 30 $m_c = 125 \pm \dfrac{2.04(8)}{31} = 122.1, 127.9$ $m = 150$

Since 150 is greater than 127.9, we reject $H_0$. At $\alpha = 0.05$ the data support the conclusion that the mean distance is more than 125 feet.

15-5 (a) $m = 152.5$ $s = 6.40$ $X^2 = \dfrac{9(6.4)^2}{10^2} = 3.7$ df = 9 $X_c^2 = 2.70, 19.02$

Since 3.7 is between 2.70 and 19.02, we fail to reject $H_0$. At $\alpha = 0.05$ the data are consistent with the hypothesis that the standard deviation of the bone lengths is 10 millimeters.

(b) $z = \dfrac{(162 - 152.5)}{10} = 0.95$; $P(z > 0.95) = 0.171$

15-7 $X^2 = \dfrac{7(8)^2}{10^2} = 4.48$ df = 7 $X_c^2 = 2.17$

Since 4.48 is not less than 2.17, we fail to reject $H_0$. At $\alpha = 0.05$ the data do not support the conclusion that $\sigma$ is less than 10.

15-9 $X^2 = \dfrac{24(3)^2}{1^2} = 216 \quad \text{df} = 24 \quad X_c^2 = 36.42$

Since 216 is greater than 36.42, we reject $H_0$. At $\alpha = 0.05$ the data support the conclusion that $\sigma$ is more than 1 inch.

15-11 (a) $F = 1.78 \quad \text{df} = 34, 19 \quad F_c = 2.39$

Since 1.78 is not greater than 2.39, we fail to reject $H_0$. At $\alpha = 0.05$, these data are consistent with the hypothesis of equal variances.

(b) $F = 1.65 \quad \text{df} = 14, 14 \quad F_c = 2.95$

Since 1.65 is not greater than 2.95, we fail to reject $H_0$. At $\alpha = 0.05$, these data are consistent with the hypothesis of equal variances.

(c) $F = 1.39 \quad \text{df} = 19, 19 \quad F_c = 2.51$

Since 1.39 is not greater than 2.51, we fail to reject $H_0$. At $\alpha = 0.05$, these data are consistent with the hypothesis of equal variances.

(d) $F = 1.14 \quad \text{df} = 9, 9 \quad F_c = 4.03$

Since 1.14 is not greater than 4.03, we fail to reject $H_0$. At $\alpha = 0.05$, these data are consistent with the hypothesis of equal variances.

15-13 $F = 6.25 \quad \text{df} = 19, 19 \quad F_c = 3.00$

Since 6.25 is greater than 3.00, we reject $H_0$. At $\alpha = 0.01$, these data are consistent with the hypothesis that the chemical acts more erratically in oat plants than in corn plants.

15-15 $F = 3.41 \quad \text{df} = 9, 9 \quad F_c = 6.54$

Since 3.41 is not greater than 6.54, we fail to reject $H_0$. At $\alpha = 0.05$, these data are consistent with the hypothesis of equal variances.

15-17 $s_W^2 = 3.95 \quad s_m^2 = 1.45 \quad s_A^2 = 37.7 \quad F = 9.54 \quad \text{df} = 2, 75 \quad F_c = 5.06$

Since 9.54 is greater than 5.06, we reject $H_0$. At $\alpha = 0.01$, these data support the hypothesis that running affects frequency of menstruation. Increased running appears to be associated with decreased menstrual frequency.

15-19 (a) $dm = 38 \quad d_c = 0 \pm 1.96\sqrt{\dfrac{65^2}{51} + \dfrac{80^2}{51}} = 28.3$

Since 38 is greater than 28.36, we reject $H_0$. At $\alpha = 0.05$, these data support the conclusion that the birds are sensitive to the magnetic field.

(b) $s_W^2 = 5312.5 \quad s_m^2 = 722.0 \quad s_A^2 = 36{,}822 \quad F = 6.93 \quad \text{df} = 1, 100$
$F_c = 3.84$

Since 6.93 is greater than 3.84, we reject $H_0$. At $\alpha = 0.05$, these data support the conclusion that the birds are sensitive to the magnetic field.

15-21 $s_W^2 = 0.00025 \quad s_m^2 = 0.000666 \quad s_A^2 = 0.00467 \quad F = 18.67 \quad \text{df} = 3, 24$
$F_c = 2.99$

Since 18.67 is greater than 2.99, we reject $H_0$. At $\alpha = 0.05$, these data support the conclusion that the mouse strains do differ in their mean consumptions of chocolate. If the only differences among strains are due to genetics, then this experiment supports the hypothesis that desire for chocolate is genetically determined.

15-23 (a) You could do several two-sample tests to compare the means two at a time. This approach is called *multiple comparisons.* If you do use this appraoch, you should also adjust the individual significance levels to control the overall chance of a Type I error. This topic is covered in more advanced texts.

(b) $dm = 0.66 \quad s_p^2 = 1.37 \quad \text{df} = 10 \quad m_c = 0 \pm 2.23\sqrt{1.37\left(\frac{1}{6} + \frac{1}{6}\right)} = \pm 1.51$

Since 0.66 is between −1.51 and 1.51, we fail to reject $H_0$. At $\alpha = 0.05$, these data do not support the conclusion that methods 2 and 3 produce different amounts of rainfall. *Note:* to make better use of all the data, it is standard to use $s_W^2$ in place of $s_p^2$ in this analysis and to use the corresponding critical $t$-value for 15 degrees of freedom. These kinds of adjustments are discussed in more advanced texts.

## Chapter 16: Nonparametric Tests

16-1 The answer depends upon your values and can lead to an informative class discussion.

(a) The authors believe that this would not be ethical because the situation is not serious enough to warrant such a deception and the concomitant anxiety placed upon the uninformed students.

(b) This experiment seems trivial and therefore abusive.

(c) One suggestion is this: Before the start of the semester, students could be asked if they were willing to be part of several experiments during the term. Those who agree could be assigned to the same section. They would not know exactly when they were part of an experiment.

16-3 (a) This is a one-sample binomial hypothesis test of signs.

The population is all pear trees.

The sample is 12 pear trees.

$p = P(\text{a pear tree produces more fruit with fertilizer 1})$
$= P(X > Y) = P(\text{a sign is } +)$

$H_0$: There is no difference in the efficacy of the two fertilizers. A plus sign is just as likely as a minus sign; $p = 0.50$.

$H_a$: There is a difference in the efficacy of the two fertilizers; $p \neq 0.50$ (a two-tailed test).

| Tree Number | $X$ First Season | $Y$ Second Season | Sign |
|---|---|---|---|
| 1 | 1.6 | 2.0 | − |
| 2 | 2.0 | 2.5 | − |
| 3 | 1.6 | 1.6 | 0 |
| 4 | 3.0 | 3.5 | − |
| 5 | 2.5 | 3.0 | − |
| 6 | 2.0 | 1.5 | + |
| 7 | 2.0 | 2.5 | − |
| 8 | 2.5 | 3.0 | − |
| 9 | 3.0 | (died) | omit |
| 10 | 1.5 | 2.0 | − |
| 11 | 1.0 | 1.5 | − |
| 12 | 2.0 | 1.5 | + |

Since we have ignored the results for 2 trees in the sample, we have $n = 12 - 2 = 10$.

$\mu = np = 10(0.50) = 5$ which is *not* greater than 5

since we have a two-tail test with $\alpha = 0.05$, we want approximately 2½% on the top and 2½% on the bottom.

Referring to the table below, we see that we obtain $P(S > 8) = P(S = 9) + P(S = 10) = 0.011$ and $P(S > 7) = 0.055$. By the symmetry of the table, we conclude that more than 95% of the random outcomes would occur for $2 \leq S \leq 8$.

| $S$ | $\binom{6}{S}$ Number of Ways $S$ Can Occur | | $P(S)$ |
|---|---|---|---|
| 10 | 1 | $1(0.7)^{10}$ | $= 1(0.000976562)$ |
| 9 | 10 | $10(0.5)^{9}(0.5)^{1}$ | $= 10(0.000976562)$ |
| 8 | 45 | $45(0.5)^{8}(0.5)^{2}$ | $= 45(0.000976562)$ |
| 7 | 120 | $120(0.5)^{7}(0.5)^{3}$ | $=$ |
| 6 | 210 | $210(0.5)^{6}(0.5)^{4}$ | $=$ |
| 5 | 252 | $252(0.5)^{5}(0.5)^{5}$ | $=$ |
| 4 | 210 | $210(0.5)^{4}(0.5)^{6}$ | $=$ |
| 3 | 120 | $120(0.5)^{3}(0.5)^{7}$ | $=$ |
| 2 | 45 | $45(0.5)^{2}(0.5)^{8}$ | $=$ |
| 1 | 10 | $10(0.5)^{1}(0.5)^{9}$ | $=$ |
| 0 | 1 | $1\ (0.5)^{10}$ | $=$ |

Therefore, our decision rule is: we will reject $H_0$ if we get more than 8 plus signs or less than 2.

Since the outcome is 2 plus signs, we fail to reject $H_0$. We have failed to show that one fertilizer is more effective than the other.

(b) One possibility is that young (old) trees may produce more (less) fruit each year. We could do 2 experiments, one using fertilizer 1 first and the other using fertilizer 2 first.

16-5 (a)

| Student | $Y$ Math | $X$ Physics | Sign | Student | $Y$ Math | $X$ Physics | Sign |
|---|---|---|---|---|---|---|---|
| Korn, A. | 90 | 95 | + | Lash, I. | 78 | 78 | 0 |
| Tropey, N. | 78 | 76 | − | Ective, F. | 90 | 60 | − |
| Shorr, C. | 80 | 83 | + | Sera-Sera, K. | 87 | 94 | + |
| Frost, D. | 81 | 82 | + | Bow, L. | 78 | 83 | + |
| Lope, E. | 94 | 90 | − | Knott, Y. | 72 | 48 | − |
| Kupp, T. | 30 | 31 | + | Cleaf, O. | 99 | 98 | − |
| Nee, G. | 63 | 60 | − | Kann, P. | 70 | 80 | + |
| Bohr, R. | 70 | 78 | + | Kneeaform, Q. | 62 | 78 | + |

Since we have ignored the results for 1 person in the sample, we have $n = 16 - 1 = 15$.

$\mu = 7.5 \qquad \sigma = 1.94$
$S_c = \mu_S + z_c\sigma_S = 7.5 + 1.65(1.94) = 10.7$

Since the outcome is 9 plus signs, we fail to reject $H_0$. We have failed to show that the premed students at Leach College score better in physics than in math.

(b) A test of paired differences.

| Student | $Y$ Math | $X$ Physics | $d$ | $d^2$ |
|---|---|---|---|---|
| Korn, A. | 90 | 95 | −5 | 25 |
| Tropey, N. | 78 | 76 | 2 | 4 |
| Shorr, C. | 80 | 83 | −3 | 9 |
| Frost, D. | 81 | 82 | −1 | 1 |
| Lope, E. | 94 | 90 | 4 | 16 |
| Kupp, T. | 30 | 31 | −1 | 1 |
| Nee, G. | 63 | 60 | 3 | 9 |
| Bohr, R. | 70 | 78 | −8 | 64 |
| Lash, I. | 78 | 78 | 0 | 0 |
| Ective, F. | 90 | 60 | 30 | 900 |
| Sera-Sera, K. | 87 | 94 | −7 | 49 |
| Bow, L. | 78 | 83 | −5 | 25 |
| Knott, Y. | 72 | 48 | 24 | 576 |
| Cleaf, O. | 99 | 98 | 1 | 1 |
| Kann, P. | 70 | 80 | −10 | 100 |
| Kneeaform, Q. | 62 | 78 | −16 | 256 |
| | | | 8 | 2036 |

Since the mean is positive, 8/16, we obviously fail to prove that the physics scores are higher.

16-7 Since we have ignored the results for 3 people in the sample, we have $n = 48 - 3 = 45$.

$\mu = 22.5 \qquad \sigma = 3.35$
$S_c = \mu_S + z_c\sigma_S = 22.5 \pm 2.58(3.35) = 13.8, 31.2$

Since the outcome is 32 plus signs, we reject $H_0$. We have shown that sentinels have more foes than friends.

16-9 Since we have ignored the results for 2 scores in the sample, we have $n = 17 - 2 = 15$.

$\mu = 7.5 \qquad \sigma = 1.94$
$S_c = \mu_S + z_c\sigma_S = 7.5 \pm 1.96(1.94) = 3.7, 11.3$

Since the outcome is 8 plus signs, we fail to reject $H_0$. The median may indeed be 62.

16-11 $\mu = 10 \qquad \sigma = 2.24$
$S_c = 10 + 1.65(2.24) = 13.7$, that is, at least 14

16-13 (a) $n_1 = 7 \quad n_2 = 4 \quad R = 7$

$$\mu_R = \frac{2(7)(4)}{7 + 4} + 1 = 6.06$$

$$\sigma_R = \sqrt{\frac{2(7)(4)[2(7)(4) - 7 - 4]}{(7 + 4)^2(7 + 4 - 1)}} = 1.44$$

(b) $n_1 = 8 \quad n_2 = 5 \quad R = 5$

$$\mu_R = \frac{2(8)(5)}{8 + 5} + 1 = 7.15$$

$$\sigma_R = \sqrt{\frac{2(8)(5)[2(8)(5) - 8 - 5]}{(8 + 5)^2(8 + 5 - 1)}} = 1.63$$

(c) $n_1 = 9 \quad n_2 = 12 \quad R = 6$

$$\mu_R = \frac{2(9)(12)}{9 + 12} + 1 = 11.29$$

$$\sigma_R = \sqrt{\frac{2(9)(12)[2(9)(12) - 9 - 12]}{(9 + 12)^2(9 + 12 - 1)}} = 2.19$$

(d) $n_1 = 10 \quad n_2 = 5 \quad R = 9$

$$\mu_R = \frac{2(10)(5)}{10 + 5} + 1 = 7.67$$

$$\sigma_R = \sqrt{\frac{2(10)(5)[2(10)(5) - 10 - 5]}{(10 + 5)^2(10 + 5 - 1)}} = 1.64$$

(e) $n_1 = 11 \quad n_2 = 8 \quad R = 13$

$$\mu_R = \frac{2(11)(8)}{11 + 8} + 1 = 10.26$$

$$\sigma_R = \sqrt{\frac{2(11)(8)[2(11)(8) - 11 - 8]}{(11 + 8)^2(11 + 8 - 1)}} = 2.13$$

16-15 Let $n_1$ = the number of A's and let $n_2$ = the number of B's. We have $n_1 = 11$ and $n_2 = 11$. So for the mean and standard deviation of $R$ we find

$$\mu_R = \frac{2(11)(11)}{11 + 11} + 1 = 12$$

$$\sigma_R = \sqrt{\frac{2(11)(11)[2(11)(11) - 11 - 11]}{(11 + 11)^2(11 + 11 - 1)}} = 2.29$$

$R_c = 12 \pm 1.96(2.29) = 7.5, 16.5$

The outcome is $R = 13$ runs, and so we fail to reject the null hypothesis. The runs of A's and B may be random.

16-17 Let $n_1$ = the number of wabes with toves below (denoted B) the median of 15.3, and let $n_2$ = the number of wabes with toves above (denoted A) the median. Her outcomes become BBBAAABAABBBBABBBAAAAAA.

$$\mu_R = \frac{2(12)(11)}{12 + 11} + 1 = 12.5$$

$$\sigma_R = \sqrt{\frac{2(12)(11)[2(12)(11) - 12 - 11]}{(12 + 11)^2(12 + 11 - 1)}} = 2.34$$

$$R_c = 12.5 \pm 1.96(2.34) = 7.9, 17.1$$

The outcome is $R = 8$ runs, and so we fail to reject the null hypothesis. The variation from the median may appear randomly.

16-19 Seven: Arny and Ryan, rage and gear, broad and board, stool and tools, Roset's and Stores, Tock's and Stock, dim and mid

16-21 $U_{\text{Dem}} = 5 + 5 + 5 + 2 + 2(\frac{1}{2}) + 2 = 20$

16-23 The critical value is found from Table D-15 for $\alpha = 0.05$, $n_1 = 3$, $n_2 = 4$ as $U_C = 12$. Outcome: The students who were tutored ranked 1, 2, 2. The others ranked 3, 3, 3, 4.

$$U = 4 + 4 + 4 = 12$$

Since the sample outcome is bigger than or equal to the critical value, we reject the null hypothesis. Our conclusion is that Dr. Vail's tutoring helped.

16-25 The critical value is found from Table D-15 for $\alpha = 0.05$, $n_1 = 9$, $n_2 = 8$ as $U_C = 57$.
Outcomes: $U_{\text{Winona}} = 8 + 8 + 8 + 4 + 4 + 6 + 8 + 1.5 = 47.5$
$U_{\text{Bertha}} = 8(5) - 21 = 19$

Since both sample outcomes are smaller than the critical value, we fail to reject the null hypothesis. Our conclusion is that we have not clearly established the superiority of either merchant.

16-27 $\mu_U = \dfrac{n_1 n_2}{2} = \dfrac{110(90)}{2} = 4950$

| | Less than 7 Years | More than 7 Years | $T$ | $T^3$ | $T^3 - T$ |
|---|---|---|---|---|---|
| A lot less | 30 | 10 | 40 | 64,000 | 63,960 |
| A little less | 20 | 20 | 40 | 64,000 | 63,960 |
| The same | 10 | 30 | 40 | 64,000 | 63,960 |
| A little more | 20 | 20 | 40 | 64,000 | 63,960 |
| A lot more | 30 | 10 | 40 | 64,000 | 63,960 |
| | | | | | $\Sigma(T^3 - T) = 319{,}800$ |

This gives $$C = \frac{n_1 n_2}{12N(N-1)}\Sigma(T^3 - T) = \frac{110(90)}{12(200)(199)}(319{,}800) = 6629.02$$

and thus $$\sigma_U = \sqrt{\frac{n_1 n_2 (N+1)}{12} - C} = \sqrt{\frac{110(90)(201)}{12} - 6629.02} = 398.99$$

16-29

$$\mu_U = \frac{n_1 n_2}{2} = \frac{11(11)}{2} = 60.5$$

$$\sigma_U = \sqrt{\frac{n_1 n_2 (n_1 + n_2 + 1)}{12}} = \sqrt{\frac{11(11)(11 + 11 + 1)}{12}} = 15.23$$

$$U_c = \mu + z_c\sigma = 60.5 \pm 1.96(15.23) = 60.5 \pm 29.8 = 30.7,\ 90.3$$

Putting the data in order, we obtain the following:

$A$: 089, 093, 100, 101, 107, 118, 118, 128, 132, 141, 143

$B$: 060, 060, 095, 097, 102, 120, 120, 130, 135, 140, 142

and $U_1 = 11 + 10 + 8 + 7 + 5 + 5 + 5 + 4 + 4 + 2 + 2 = 63$

$U_2 = 11(11) - 63 = 58$

Conclusion: Since both sample values of $U$ are between 30.7 and 90.3, we do not have sufficient evidence to reject the null hypothesis. We do not have enough evidence to show that the agents of one country endure more pain than those from the second country.

16-31

| | Old | New | $T$ | $T^3$ | $T^3 - T$ |
|---|---|---|---|---|---|
| 10–15 miles per gallon | 10 | 0 | 10 | 1,000 | 990 |
| 15–20 miles per gallon | 20 | 10 | 30 | 27,000 | 26,970 |
| 20–25 miles per gallon | 15 | 15 | 30 | 27,000 | 26,970 |
| 25–30 miles per gallon | 5 | 25 | 30 | 27,000 | 26,970 |
| | | | | | $\Sigma(T^3 - T) = 81{,}900$ |

This gives

$$C = \frac{n_1 n_2}{12N(N-1)}\Sigma(T^3 - T) = \frac{50(50)}{12(100)(99)}(81{,}900) = 1732.5$$

Therefore we have

$$\mu_U = \frac{n_1 n_2}{2} = \frac{50(50)}{2} = 1250$$

$$\sigma_U = \sqrt{\frac{n_1 n_2 (N+1)}{12} - C} = \sqrt{\frac{50(50)(101)}{12} - 1732.5} = 138.99$$

$$U_c = \mu + z_c \sigma = 1250 + 1.65(138.99) = 1250 + 229.3 = 1479.3$$

Putting the data in order, we obtain the following:

$$U = 25(45 + 2.5) + 15(30 + 7.5) + 10(10 + 10) = 1950$$

Since the sample value of $U = 1950$ is greater than the critical value $U_c = 1340.6$, we reject the null hypothesis. The new gasohol gets better mileage.

## Answers to the Sample Test for Chapters 14, 15, and 16

1. (a) Positive (b) negative (c) positive (d) positive (e) negative
2. They goofed! The coefficient of correlation $\leq 1$.
3. (a) $b = \dfrac{6(277{,}482.8) - 6814(314.8)}{6(28{,}608{,}788) - 6814^2} = \dfrac{-480{,}150.4}{125{,}222{,}132} = -0.004$

   (b) $r = \dfrac{-480{,}150.4}{\sqrt{125{,}222{,}132}\sqrt{6(17{,}161.5) - 314.8^2}} = -0.69$

   (c) $Y_{\text{predicted}} = -0.004(X - 1135.7) + 52.5$
4. Procedure: This is a one-sample test of variance via chi-square.

   Population: All the weights of 10-pound packages of detergents produced by this factory. We are assuming that the distribution of $Y$ is near normal.

The sample is the 10 packages pulled at random.

$H_0$: The variance is equal to (or less than) $(0.1)^2$, $\sigma^2 = 0.01$.
$H_a$: The variance is more than 0.01; $\sigma^2 > 0.01$ (a one-tail test on the right).

$$s^2 = \frac{\Sigma Y^2 - (\Sigma Y)^2/n}{n-1} = \frac{1006.0246 - 100.3^2/10}{9} = 0.0017333 \cdots$$

$$X^2 = \frac{(n-1)s^2}{\sigma^2} = \frac{9(0.00173 \cdots)}{0.01} = 1.56$$

Degrees of freedom: We have $n = 10$, and therefore $n - 1 = 9$ degrees of freedom.

$\alpha = 0.05$

Decision rule: The critical value of $X^2$ for 9 degrees of freedom with $\alpha = 0.05$ is $X_c^2 = 16.92$.

We will reject $H_0$ if $X^2$ is greater than 16.92.

Outcome: $X^2 = 1.56$

Conclusion: Since 1.56 is not greater than 16.92, we fail to reject the null hypothesis. The standard deviation may well be less than 0.1 pound ($\sqrt{0.00173} \approx 0.04$).

5. Procedure: This is a two-sample comparison of variance hypothesis test (via the $F$-test).

   Populations: All heart valves made from these two types of plastic.

   Because $s = 23.917$ for material 1 and $s = 33.711$ for material 2, we choose the larger one for $s_1$, so that material 2 becomes population 1.

   The samples are the numbers of days each that the 20 heart valves lasted.

   $H_0$: There is no difference in the variability of the two samples' life spans; $\sigma_1^2 = \sigma_2^2$.

   $H_a$: There is a difference in the variability of the life spans; $\sigma_1^2 \neq \sigma_2^2$ (a two-tail hypothesis which leads to a one-tail test on the right).

   Decision rule: The degrees of freedom for $n_1 = 10$ and $n_2 = 10$ will be

$$\text{Degrees of freedom (numerator)} = n_1 - 1 = 9$$
$$\text{Degrees of freedom (denominator)} = n_2 - 1 = 9$$

   At $\alpha = 0.05$ on a two-tail test we use $\alpha = 0.025$ in Table D-12 and obtain $F_c = 4.03$. Hence our decision rule is: We will reject the null hypothesis if our value of $F$ is greater than 4.03.

   Outcome: The test statistic is

$$F = \frac{s_1^2}{s_2^2} = \frac{(33.711)^2}{(23.917)^2} = 1.99$$

   Conclusion: Since the $F$ statistic is smaller than the critical value of $F$, we fail to reject the null hypothesis of equal variances. The variability of the two types of heart valves may be the same.

6. Procedure: Assuming that all four populations are normal with the same variance, we use ANOVA to perform an $F$-test for equal means.

   Population: All the gasoline produced by the 4 methods.
   Samples: the results of the 16 trials, 4 for each method.

$H_0$: The average pollutants produced by each grade are the same; $\mu_1 = \mu_2 = \mu_3 = \mu_4$.
$H_a$: Not all the $\mu$'s are equal; at least 2 of them differ (always a one-tail test).

| | | | | | Mean | Variance |
|---|---|---|---|---|---|---|
| Grade A | 100 | 110 | 120 | 113 | 110.75 | 68.9167 |
| Grade B | 105 | 115 | 125 | 100 | 111.25 | 122.9167 |
| Grade C | 130 | 140 | 145 | 140 | 138.75 | 39.5833 |
| Grade D | 135 | 140 | 160 | 150 | 146.25 | 122.9167 |

Part 1: Mean squared within groups (method 1). Since we are considering 4 populations, $N = 4$.

$$s_W^2 = \frac{\Sigma s^2}{N} = \frac{68.917 + 122.917 + 39.583 + 122.917}{4} = 88.6$$

Part 2: Mean squared among groups (method 2).

$$s_m^2 = \frac{\Sigma m^2 - (\Sigma m)^2/N}{N - 1} = \frac{65{,}282.75 - (507)^2/4}{3} = 340.17$$

Since each sample consists of 4 trials, $n = 4$, and

$$s_A^2 = ns_m^2 = 4(340.17) = 1360.68$$

Part 3: Comparison of estimates.

$$F = \frac{1360.68}{88.6} = 15.36$$

$$\begin{aligned} \text{Degrees of freedom (numerator)} &= N - 1 = 4 - 1 = 3 \\ \text{Degrees of freedom (denominator)} &= n_1 - 1 + n_2 - 1 + n_3 - 1 + n_4 - 1 \\ &= 3 + 3 + 3 + 3 = 12 \end{aligned}$$

$$F_c = 3.49$$

The decision rule: Reject the null hypothesis if the outcome exceeds 3.49.

Outcome: $F = 15.36$

Conclusion: Reject the null hypothesis. At least 2 of the average amounts of exhaust pollutants are different.

7. We arbitrarily assign plus signs to the 37 who watch channel 0, minus signs to the 28 who watch channel 1, and 0's to the other 53.

   This is a one-sample *binomial* hypothesis test.

   The population is the people in Disasterville, NJ, who regularly watched the six o'clock news.

   The sample is the 65 plus and minus signs.

   Let $p = P(\text{a person watches channel 0}) = P(\text{a sign is } +)$.

   $H_0$: The 2 channels are equally popular; $p = 0.50$.
   $H_a$: Channel 0 is more popular; $p > 0.50$ (a one-tail test).

Since $np = nq = 32.5 > 5$, we have a normal distribution of $p$'s, with

$\mu_s = 32.5 \qquad \sigma_s = \sqrt{32.5(0.50)} = 4.03$

Therefore, $S_c = 32.5 + 1.65(4.03) = 39.2$. Our decision rule will be to reject the null hypothesis if the outcome is more than 39.2 plus signs.

The outcome is 37 plus signs, and thus we fail to reject the null hypothesis. The evidence does not support the idea that channel 0 news is more popular.

## Answers to Exercises in Appendix A

1. (a) 23.73 (b) 4.97 (c) 0.046 (d) 1.202 (e) 9.628 (f) 0.0012 (g) 21.25 (h) 0.5 (i) 0.02 (j) 0
2. (a) −2.33 (b) 0.081, 0.41, 0.6, 4.51, 4.7 (c) −0.273, 0.273, 0.41
3. (a) true (b) true (c) false (d) true (e) 5, 6 (f) 4, 5, 6 (g) 0, 1, 2, 3, 4 (h) 0, 1, 2, 3, 4, 5 (i) 3, 4 (j) 2, 3, 4, 5
4. (a) 5% (b) .3% (c) 0.37 (d) 0.032 (e) 0.375, 37.5% (f) 0.263, 26.3% (g) 11.5 (h) 8 (i) 30 (j) 25
5. (a) −15 (b) +15.35 (c) −16.5 (d) −2 (e) −5
6. (a) 7.446 (b) −41 (c) 11.167 (d) 6.92 and −0.92 (e) true
7. (a) +16 (b) 343 (c) $\frac{1}{18}$ (d) .3087
8. (a) 1.550 (b) 2.315 (c) 4.45

## Answers to Exercises in Appendix B

1. *Step 1.* Assign numbers to the patients. You can do this any way you like. Here is one example.

| | | | | |
|---|---|---|---|---|
| 00. Bernard | 01. Bess | 02. Betty | 03. Isadore | 04. Jack |
| 05. Leda | 06. Marilyn | 07. Martin | 08. Max | 09. Meyer |
| 10. Molly | 11. Robert | 12. Rose | 13. Sue | 14. Sylvia |
| 15. Howard | | | | |

   *Step 2.* Since the highest assigned number is a 2-digit number, we go through the table 2 digits at a time starting at the first number in line 1. Any numbers from 00 to 15 are useful. The first useful 2-digit number is 11, which occurs in line 2. We continue until we have selected 8 useful numbers.

   Result: 11, 04, 09, 15, 06, 12, 14, 10.

   Thus Robert, Jack, Meyer, Howard, Marilyn, Rose, Sylvia, and Molly are chosen to get the placebo. The other 8 get the real drug.

2. Because the highest assigned number is 1547, which has 4 digits, all names are assigned a 4-digit number. So the list of useful numbers goes from 0000 to 1547. We therefore go through the table 4 digits at a time. Starting at line 2 the first useful number is 1047. Continuing this way until we have 5 useful numbers yields: 1047, 0922, 1525, 1298, 0798. Thus the 5 people with these numbers are the ones chosen at random.

## Answers to Exercises in Appendix C

1. (a) independent (b) independent (c) dependent (d) independent
3. *A* and *B* are dependent events.
5. (a) no (b) yes (c) yes (d) no
7. (a) 0.20 (b) no
9. 1/1,000,000
11. (a) 1/16 (b) 1/169 (c) 2/169 (d) 9/169 (e) 1/17 (f) 4/663 (g) 8/663 (h) 11/221
13. (a) 1/2 (b) 3/8
15. 2/663
17. 0.08

# Index

## Chapter 11

*One-Sample Tests*

1. Degrees of freedom $= n - 1$
2. $\mu_m = \mu_{\text{pop}}$
3. $s_m = \dfrac{s}{\sqrt{n}}$
4. $m_c = \mu_m + t_c s_m$
5. Experimental outcome, $m = \dfrac{\Sigma X}{n}$

*Two-Sample Tests*

6. Degrees of freedom $= n_1 + n_2 - 2$
7. $\mu_{\text{dm}} = \mu_1 - \mu_2$ (if $H_0$ states that $\mu_1 = \mu_2$, then $\mu_1 - \mu_2 = 0$)
8. $s_{\text{dm}} = \sqrt{s_p^2 \left(\dfrac{1}{n_1} + \dfrac{1}{n^2}\right)}$
9. $s_p^2 = \dfrac{(n_1 - 1)(s_1^2) + (n_2 - 1)(s_2^2)}{n_1 + n_2 - 2}$
10. $\text{dm}_c = \mu_{\text{dm}} + t_c s_{\text{dm}}$
11. Experimental outcome, $\text{dm} = m_1 - m_2$

## Chapter 12

*Distribution of Sample Proportions*

1. $\mu = p$
2. $\sigma = \sqrt{\dfrac{pq}{n}}$
3. $\hat{\sigma} = \sqrt{\dfrac{\hat{p}\hat{q}}{n}}$
4. $p$ lies between $\hat{p} - z_c\hat{\sigma}$ and $\hat{p} + z_c\hat{\sigma}$
5. $n > \left(\dfrac{z_c}{a}\right)^2 pq$

*Distribution of Sample Means*

6. $s_m = \dfrac{s}{\sqrt{n}}$
7. $\mu$ lies between $m - z_c s_m$ and $m + z_c s_m$ (large samples)
8. $\mu$ lies between $m - t_c s_m$ and $m + t_c s_m$ (small samples)

*Differences between Two Proportions*

9. $\hat{\sigma} = \sqrt{\dfrac{\hat{p}_1\hat{q}_1}{n_1} + \dfrac{\hat{p}_2\hat{q}_2}{n_2}}$
10. $p_1 - p_2$ lies between $d\hat{p} - z_c\hat{\sigma}$ and $d\hat{p} + z_c\hat{\sigma}$

*Differences between Two Means*

11. $s_{\text{dm}} = \sqrt{\dfrac{s_1^2}{n_1} + \dfrac{s_2^2}{n_2}}$ (large samples)
12. $s_p^2 = \dfrac{(n_1 - 1)s_1^2 + (n_2 - 1)s_2^2}{n_1 + n_2 - 2}$
13. $s_{\text{dm}} = \sqrt{s_p^2\left(\dfrac{1}{n_1} + \dfrac{1}{n_2}\right)}$ (small samples)
14. $\mu_1 - \mu_2$ lies between $\text{dm} - z_c s_{\text{dm}}$ and $\text{dm} + z_c s_{\text{dm}}$ (large samples)
15. $\mu_1 - \mu_2$ lies between $\text{dm} - t_c s_{\text{dm}}$ and $\text{dm} + t_c s_{\text{dm}}$ (small samples)